Algebra 1

Explorations and Applications

1 knot = 1.15 mi/h

$y = 0.013x^2$

Authors

Senior Authors
Miriam A. Leiva   Richard G. Brown

Loring Coes III
Shirley Frazier Cooper
Joan Ferrini-Mundy
Amy T. Herman
Patrick W. Hopfensperger
Celia Lazarski
Stuart J. Murphy
Anthony Scott
Marvin S. Weingarden

McDougal Littell
A HOUGHTON MIFFLIN COMPANY
ston, Illinois ◆ **Boston** ◆ **Dallas**

Authors

Senior Authors

Richard G. Brown
Mathematics Teacher, Phillips Exeter Academy, Exeter, New Hampshire

Miriam A. Leiva
Cone Distinguished Professor for Teaching and Professor of Mathematics, University of North Carolina at Charlotte

Loring Coes III
Chair of the Mathematics Department, Rocky Hill School, E. Greenwich, Rhode Island

Shirley Frazier Cooper
Curriculum Specialist, Secondary Mathematics, Dayton Public Schools, Dayton, Ohio

Joan Ferrini-Mundy
Professor of Mathematics and Mathematics Education, University of New Hampshire, Durham, New Hampshire

Amy T. Herman
Mathematics Teacher, Atherton High School, Louisville, Kentucky

Patrick W. Hopfensperger
Mathematics Teacher, Homestead High School, Mequon, Wisconsin

Celia Lazarski
Mathematics Teacher, Glenbard North High School, Carol Stream, Illinois

Stuart J. Murphy
Visual Learning Specialist, Evanston, Illinois

Anthony Scott
Assistant Principal, Orr Community Academy, Chicago, Illinois

Marvin S. Weingarden
Supervisor of Secondary Mathematics, Detroit Public Schools, Detroit, Michigan

Editorial Advisors

Martha A. Brown
Mathematics Supervisor, Prince George's County Public Schools, Capitol Heights, Maryland

Diana Garcia
Mathematics Supervisor, Laredo Independent School District, Laredo, Texas

Sue Ann McGraw
Mathematics Teacher, Lake Oswego High School, Lake Oswego, Oregon

Editorial Advisors helped plan the concept, teaching approach, and format of the book. They also reviewed draft manuscripts.

ISBN: 0-395-86296-5   23456789—VH—01 00 99 98

Review Panel

Judy B. Basara — Curriculum Chair, St. Hubert's High School, Philadelphia, Pennsylvania

Dane Camp — Mathematics Teacher, New Trier High School, Winnetka, Illinois

Pamela W. Coffield — Mathematics Teacher, Brookstone School, Columbus, Georgia

Kathleen Curran — Mathematics Teacher, Ball High School, Galveston, Texas

Randy Harter — Mathematics Specialist, Buncombe County Schools, Asheville, North Carolina

William Leonard — Assistant Director of Mathematics, Fresno Unified School District, Fresno, California

Betty McDaniel — Coordinator of Mathematics, Florence School District, Florence, South Carolina

Roger O'Brien — Mathematics Supervisor, Polk County Schools, Bartow, Florida

Leo Ramirez — Mathematics Teacher, McAllen High School, McAllen, Texas

Michelle Rohr — Director of Mathematics, Houston Independent School District, Houston, Texas

May Samuels — Mathematics Chairperson, Weequahic High School, Newark, New Jersey

Betty Takesuye — Mathematics Teacher, Chaparral High School, Scottsdale, Arizona

Members of the review panel read and commented upon outlines, sample lessons, and research versions of the chapters.

Manuscript Reviewers

Elda López — Jefferson Davis High School, Houston, Texas

Anita G. Morris — Coordinator of Mathematics, Anne Arundel County Schools, Annapolis, Maryland

Michelle Rohr — Director of Mathematics, Houston Independent School District, Houston, Texas

May Samuels — Mathematics Chairperson, Weequahic High School, Newark Public Schools, Newark, New Jersey

Straightline Editorial Development, Inc. — Editorial Consultants, San Francisco, California

Manuscript Reviewers read and reacted to draft manuscript, focusing on its effectiveness from a teaching/learning viewpoint.

Student Advisors

Iruma Bello, Henry H. Filer Middle School, Hialeah, FL; Leslie Blaha, Walnut Springs Middle School, Westerville, OH; Tai-Ling Bloomfield, Alyeska Central School, Anchorage, AK; Gabriel Bonilla, G.W. Carver Middle School, Miami, FL; Anne Burke, Nichols School, Buffalo, NY; Eric de Armas, G.W. Carver Middle School, Miami, FL; Beth Donaldson, Bettendorf Middle School, Bettendorf, IA; Megan Foreman, Allen High School, Allen, TX; Clifton Gray, Riverside University High School, Milwaukee, WI; Colleen Kelly, Chaparral High School, Scottsdale, AZ; Tony Liberati, Hampton Middle School, Gibsonia, PA; Christian Maiden, Hathaway Brown School, Shaker Heights, OH; J.P. Marshall, Lincoln High School, Tallahassee, FL; Jeff Phi, Irving Junior High School, Kansas City, MO; Jaret Radford, Fondren Middle School, Houston, TX; Marisa A. Sharer, Chaparral High School, Scottsdale, AZ; Scott Terrill, Swartz Creek School, Swartz Creek, MI; Gabriela Zúñiga, San Marcos High School, San Marcos, TX.

Acknowledgment

The authors wish to thank Pam Coffield and her students at Brookstone School in Columbus, Georgia, for using and providing comments on the Portfolio Projects in this book. Ms. Coffield's students are Kate Baker, Bo Bickerstaff, Lizzie Bowles, Bradford Carmack, Lucy Cartledge, Henry Dunn, Patrick Graffagnino, Charles Haines, Daniel McFall, Sravanthi Meka, Blake Melton, Jason Pease, Ann Phillips, Dorsey Staples, and Jeffrey Usman.

Contents

Using Algebra to Work with Data

1.4 *Climbing Annapurna* 19

Applications

Additional applications include: statistics, history, child care, advertising, biology, sports, personal finance, recreation, nutrition, fundraising, geometry

2

Equations and Functions

2.1   Solving One-Step Equations   55

2.2   Solving Two-Step Equations   61
Exploration:
Solving Equations with Algebra Tiles   61

2.3   Applying Functions   66

2.4   Coordinate Graphs   72

2.5   Representing Functions   78
Exploration: *Graphing Equations*   78

2.6   Using Graphs to Solve Problems   84

Review   Chapter Review   92

Assessment   Ongoing Assessment   60, 65, 71,
                                                        77, 83, 89

                 Assess Your Progress   71, 89

                 Journal writing   71, 89

                 Portfolio Project: *Functions of Time*   90

                 Chapter Assessment   94

2.2   *Modeling with algebra tiles*   61

Applications

Interview: *Bob Hall*
*Athlete and Racing
Wheelchair Designer*
57, 60, 82

—————— Connection ——————

History	59	Geography	75
Cars	64	History	76
Economics	70	Biology	88

Additional applications include: zoology, sports,
language arts, medicine, history, geometry, personal
finance, aviation, biology, consumer economics,
energy use, business, recreation, science

Graphing Linear Equations

Animal Speeds

d

Distance (m)

Cheetah

Race horse

0 1 2 3 4 *t*
Time (s)

3.3 *Comparing rates* 113

Applications

Interview:
Mae Jemison
Astronaut
99, 100, 104, 110

—————— Connection ——————

Additional applications include: aviation, sports, astronomy, weather, employment, consumer economics, physics, business, geometry, energy use, civil engineering, recreation, geology

Solving Equations and Inequalities

4.1   *Travel in Quebec*   152

Applications

Interview:

Barbara Romanowicz
Seismologist
165, 169

———— Connection ————

Energy Use	151	**Meteorology**	174
Music	156	**Transportation**	179
History	163		

Additional applications include: consumer economics, personal finance, recreation, geometry, fuel economy, transportation, sports, architecture, communications, business

Connecting Algebra and Geometry

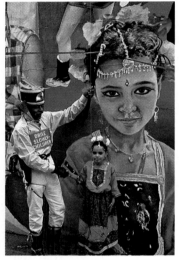

5.1   *Ratios in art*   195

Applications

Interview: *Daniel Galvez*
                Artist   195, 203

——————— Connection ———————

Government	196	**Social Studies**	222
History	210	**Sports**	227
Statistics	216		

Additional applications include: history, ecology, forestry, language arts, sports, languages, zoology, business, geometry, architecture, construction, photography

Working with Radicals

6.1 *Parasailing*   240

250 ft

200 ft

Applications

Interview:
Nathan Shigemura
*Accident
Reconstructionist*
241, 245, 249, 255

——— Connection ———

Additional applications include: recreation, urban planning, aviation, language arts, biology, physics, geometry, medicine, construction

Systems of Equations and Inequalities

7.5 *Cooperative learning*   306

Applications

Interview:
Nancy Clark
Sports Nutritionist
281, 285, 309, 314

——— Connection ———

Business	292	**History**	303
Cooking	297	**Literature**	304

Additional applications include: sports, fitness,
aviation, finance, space science, geometry,
personal finance, nutrition

Quadratic Functions

line of symmetry

8.2 *Symmetry in art* 333

Applications

Interview:
Bill Pinkney
Sailor
341, 347

Additional applications include: geometry, physics, cooking, sports, firefighting, science

9

Exponential Functions

9.3 *Tennis*   386

Applications

Interview: *Lori Arviso Alvord*
          Surgeon   380, 387, 399

———— Connection ————

Computers	374	**Literature**	404
Cooking	388	**History**	410
Consumer Economics	394		

Additional applications include: architecture, geometry, carpentry, number theory, ecology, biology, personal finance, traffic, social studies, physics, meteorology, sports, occupational safety, fuel economy, population, history, astronomy

Polynomial Functions

10.2 *Modeling* 431

Applications

Interview:
Linda Pei
Financial Adviser
427, 446

───── Connection ─────

Crafts	433	**Sports**	452
Business	440		

Additional applications include: geometry, crafts, physics, sports

Rational Functions

Average Points Per Bean

LISA $\dfrac{18(10)+6(5)+3(0)}{18+6+3} \approx 7.78$

11.2 *Comparing game scores* 469

Applications

Interview: *Alex MacLean*
Aerial Photographer
467, 486

——— Connection ———

Recreation	468	**Probability**	491
Sports	473	**Engineering**	493
Recreation	481		

Additional applications include: transportation, physics, test scores, business, geometry, recreation, sports, optics, cars, cooking

Discrete Mathematics

12

12.1 *Algorithmic art* 514

Applications

Interview:
Phil Roman
Animator
509, 513

——————— Connection ———————

Art	514	**Braille**	533
Architecture	520	**Social Studies**	534
Business	522	**Social Studies**	539
Volunteering	527	**Literature**	546
Politics	528		

Additional applications include: fundraising, fine arts, cars, business, fashion design, sports, government, manufacturing, geometry, photography

Student Resources

About the *Interviews*

Each chapter of this book starts with a personal interview with someone who uses mathematics in his or her daily life. Does this mean that all the interviews are with mathematicians or scientists? Not at all. You may be surprised by the wide range of careers that are included. These are the people you will be reading about:

- Journalist *Carrie Teegardin*
- Racing Wheelchair Designer *Bob Hall*
- Astronaut *Mae Jemison*
- Seismologist *Barbara Romanowicz*
- Artist *Daniel Galvez*
- Accident Reconstructionist *Nathan Shigemura*
- Sports Nutritionist *Nancy Clark*
- Sailor *Bill Pinkney*
- Surgeon *Lori Arviso Alvord*
- Financial Adviser *Linda Pei*
- Aerial Photographer *Alex MacLean*
- Animator *Phil Roman*

Phil Roman

At the end of each interview, there are *Explore and Connect* questions that guide you in learning more about the career being discussed. Some of these questions involve research that is done outside of class. In addition, in each chapter there are *Related Examples and Exercises* that show how the mathematics you are learning is directly related to the career highlighted in the interview.

Lori Arviso Alvord

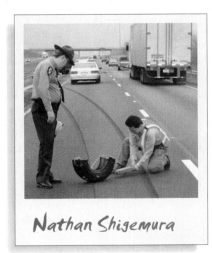

Nathan Shigemura

Welcome *to* Algebra 1
Explorations and Applications

GOALS OF THE COURSE

This book will help you use mathematics in your daily life and prepare you for success in future courses and careers.

In this course you will:
- Study the algebra concepts that are most important for today's students
- Apply these concepts to solve many different types of problems
- Learn how calculators and computers can help you find solutions

You will have a chance to develop your skills in:
- Reasoning and problem solving
- Communicating orally and in writing
- Studying and learning independently and as a team member

MATHEMATICAL CONTENT

This contemporary algebra course gives you a strong background in the types of mathematical reasoning and problem solving that will be important in your future.

The book emphasizes:
- Using functions, equations, and graphs to model problem situations
- Investigating connections of algebra to geometry, statistics, probability, and discrete mathematics

ACTIVE LEARNING

To learn algebra successfully, you need to get involved!

There will be many opportunities in this course for you to participate in:
- Explorations of mathematical concepts
- Cooperative learning activities
- Small-group and whole-class discussions

So don't sit back and be a spectator. If you join in and share your ideas, everyone will learn more.

Course Overview

To get an overview of your course, turn to pages xx–xxix to see some of the types of problems you will solve and topics you will explore.

"What does algebra have to do with me?"

Applications and Connections

Algebra is about you and the world around you.

In this course you'll learn how algebra can help answer many different types of questions in daily life and in careers.

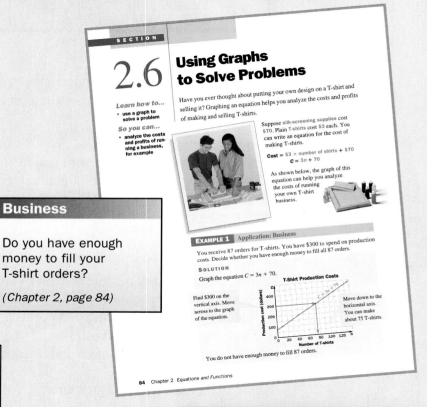

SECTION
2.6
Using Graphs to Solve Problems

Learn how to...
• use a graph to solve a problem

So you can...
• analyze the costs and profits of running a business, for example

Have you ever thought about putting your own design on a T-shirt and selling it? Graphing an equation helps you analyze the costs and profits of making and selling T-shirts.

Suppose silk-screening supplies cost $70. Plain T-shirts cost $3 each. You can write an equation for the cost of making T-shirts.

Cost = $3 × number of shirts + $70
C = 3n + 70

As shown below, the graph of this equation can help you analyze the costs of running your own T-shirt business.

EXAMPLE 1 Application: Business

You receive 87 orders for T-shirts. You have $300 to spend on production costs. Decide whether you have enough money to fill all 87 orders.

SOLUTION

Graph the equation C = 3n + 70.

Find $300 on the vertical axis. Move across to the graph of the equation.

Move down to the horizontal axis. You can make about 75 T-shirts.

You do not have enough money to fill 87 orders.

84   Chapter 2 Equations and Functions

Business

Do you have enough money to fill your T-shirt orders?

(Chapter 2, page 84)

Sports Nutrition

Can a teenage athlete get enough calcium from milk and ice cream without eating too much fat?

(Chapter 7, page 309)

Fuel Economy

At what speeds should a car be driven to get more than 25 mi/gal?

(Chapter 4, page 173)

Speed and Fuel Economy

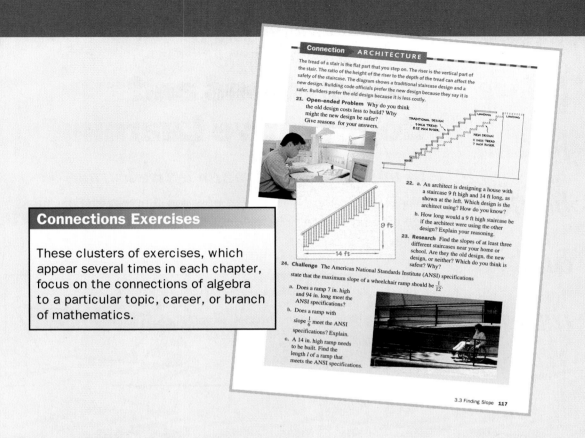

Connection ARCHITECTURE

The tread of a stair is the flat part that you step on. The riser is the vertical part of the stair. The ratio of the height of the riser to the depth of the tread can affect the safety of the staircase. The diagram shows a traditional staircase design and a new design. Building code officials prefer the new design because they say it is safer. Builders prefer the old design because it is less costly.

21. Open-ended Problem Why do you think the old design costs less to build? Why might the new design be safer? Give reasons for your answers.

TRADITIONAL DESIGN
9 INCH TREAD
8.25 INCH RISER.

NEW DESIGN
11 INCH TREAD
7 INCH RISER.

LANDING   LANDING

22. a. An architect is designing a house with a staircase 9 ft high and 14 ft long, as shown at the left. Which design is the architect using? How do you know?

b. How long would a 9 ft high staircase be if the architect were using the other design? Explain your reasoning.

9 ft

14 ft

23. Research Find the slopes of at least three different staircases near your home or school. Are they the old design, the new design, or neither? Which do you think is safest? Why?

24. Challenge The American National Standards Institute (ANSI) specifications state that the maximum slope of a wheelchair ramp should be $\frac{1}{12}$.

a. Does a ramp 7 in. high and 94 in. long meet the ANSI specifications?

b. Does a ramp with slope $\frac{1}{6}$ meet the ANSI specifications? Explain.

c. A 14 in. high ramp needs to be built. Find the length l of a ramp that meets the ANSI specifications.

3.3 Finding Slope **117**

Connections Exercises

These clusters of exercises, which appear several times in each chapter, focus on the connections of algebra to a particular topic, career, or branch of mathematics.

Earth Science

How can you predict when Old Faithful geyser will erupt next?

(Chapter 3, pages 130–131)

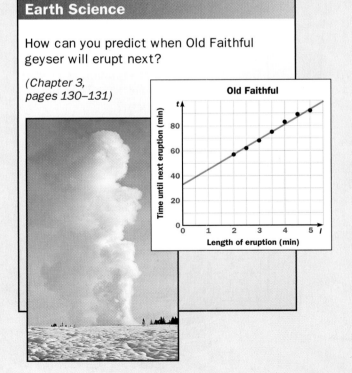

Old Faithful

Time until next eruption (min) vs. Length of eruption (min)

Social Studies

What is New Jersey's quota of seats in the United States House of Representatives?

(Chapter 5, page 196)

"Do we just sit back and listen?"

Explorations and Cooperative Learning

In this course you'll be an active learner.

Working individually and in groups, you'll investigate questions and then present and discuss your results. Here are some of the topics you'll explore.

Writing Variable Expressions

How many toothpicks do you need to make *n* triangles?

(Chapter 1, page 30)

SECTION

2.5 Representing Functions

Take any number and subtract 1 from it. You can think of this command as a function. The number is the input. Your answer is the output. One way to represent this function is to use a graph.

Learn how to...
- graph equations

So you can...
- visually represent functions, such as costs at a swimming pool

EXPLORATION
COOPERATIVE LEARNING

Graphing Equations

Work with your class.

1 Turn your class into a coordinate grid, as shown. What are your coordinates?

2 Stand up if your coordinates make the equation $y = x - 1$ true. What do you notice about the people standing up?

3 Sit down again. Your teacher will give you other equations. For each equation given, stand up if your coordinates make the equation true. In each case, describe what you observe.

This student is standing because her coordinates, (1, 0), make the equation true.
$$y = x - 1$$
$$0 = 1 - 1$$

Graphing Equations

How can your class model the graph of an equation?

(Chapter 2, page 78)

Creating a Linear Model

Can you find a linear equation to predict how many marbles are needed to break the pieces of spaghetti?

(Chapter 3, page 132)

Number of pieces of spaghetti	Number of marbles needed to break spaghetti
1	?
2	?
3	?
4	?
5	?

PORTFOLIO PROJECT

Investigating Pendulum Length

The pendulum of a clock swings back and forth, causing the clock to tick off the seconds. A pendulum clock runs down over time. With each swing, the pendulum travels a slightly smaller distance.

Even as the clock starts to run down, it can still keep good time. This is because the **period** of a pendulum, or the amount of time it takes the pendulum to swing from point *A* to point *B* and back to point *A*, is the same for large swings and small swings.

Small swing

Large swing

PROJECT GOAL Investigate how the length of a pendulum affects the period.

Conducting an Experiment

Work with a partner to design and carry out an experiment in which you measure the period of a pendulum for at least 10 different lengths of string. Here are some hints for organizing your work:

1. MAKE your pendulum by taping several coins to a piece of string. Use a pencil to hold the pendulum away from the table.

2. DETERMINE how many seconds it takes your pendulum to make 10 full swings. Divide by 10 to find the period.

3. ORGANIZE your data in a table. Then make a scatter plot. Put string length on the horizontal axis and period on the vertical axis.

Use a range of string lengths between 10 cm and 30 cm.

Based on your observations, find a string length that produces a period of one second. Can you keep time with this pendulum?

270   Chapter 6 Working with Radicals

Portfolio Projects

These open-ended projects give you a chance to explore applications of the topics you have studied.

Exploring Exponents

How many grains of rice will be on the 64th square of the chessboard?

(Chapter 9, page 369)

Using Models to Factor Trinomials

How can algebra tiles be used to model the process of factoring?

(Chapter 10, page 442)

$x - 3$

$x - 1$

"How can I *visualize* the problem?"

Using Technology

Calculators and computers can help you see mathematical relationships.

In this course there are many opportunities to use technology to model problem situations, identify patterns, and find solutions.

Graphing Calculator

What information do the numbers m and b give you about the graph of $y = mx + b$?

(Chapter 3, page 119)

Calculator

Why did Robin and Jeff's calculators give different answers for the amount they owe for the pizza?

(Chapter 1, page 8)

Technology Exercises

In these exercises you will be using graphing technology or spreadsheets to practice, apply, and extend what you have learned.

Technology Use a graphing calculator to solve each equation. If an equation has no solution, write *no solution.*

9. $x^2 + 2x - 8 = 0$ 10. $2x^2 + 7x - 4 = 0$ 11. $-x^2 + 9 = 0$

12. $x^2 - 9x + 18 = 0$ 13. $x^2 - 8x + 16 = 0$ 14. $2x^2 - 3x + 5 = 0$

15. $-4x^2 - 12x = 9$ 16. $x^2 - 4x + 9 = 4$ 17. $2x^2 + 6x - 4 = 1$

18. $-3x^2 - 6x + 8 = 3$ 19. $3x^2 - 15x = 24$ 20. $-x^2 + 8 = 12x$

21. **SAT/ACT Preview** If S represents the greater of the two solutions of the equation $x^2 - 4x - 3 = 5$, then:

   A. $S > 3$ **B.** $S < 3$ **C.** $S = 3$

   D. relationship cannot be determined

Connection   HIGHWAY SAFETY

Speed limits vary from country to country. The speed limit on highways in Mexico is 100 km/h. In France the limits vary from 90 to 130 km/h. In Germany, some sections of the highway have no speed limit. Other sections are limited to either 180 km/h or 130 km/h.

The equation $d = 0.0056s^2 + 0.14s$ models the relationship between a vehicle's stopping distance d in meters and the vehicle's speed s in kilometers per hour.

Technology For Exercises 22–24, use a graphing calculator.

22. How many solutions does the equation $0.0056s^2 + 0.14s = 0$ have? How many of the solutions make sense in this situation?

23. About how fast should you be driving in order to be able to stop within 100 m? In which of the above countries is this speed within the speed limit?

24. Suppose a car is traveling at the speed of 130 km/h on a German highway. What distance should the driver of the car leave between the driver's car and the car in front of it?

25. **Research** Find out what is a reasonable length in meters for a car. Then estimate the number of car lengths that corresponds to the stopping distance you found in Exercise 24.

348 Chapter 8 *Quadratic Functions*

Graphing Technology

How has the population of Alaskan peregrine falcons grown since 1980?

(Chapter 9, page 378)

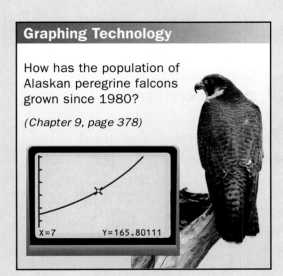

Spreadsheet

Which region of the United States had the greatest increase in the number of in-line skaters?

(Chapter 1, page 46)

" Can I solve this problem with algebra?"

Integrating Math Topics

Sometimes you need to combine algebra with other math topics in order to find a solution.

In this course you'll see how you can solve problems by integrating algebra with geometry, statistics, probability, and discrete mathematics.

CHAPTER

5 Connecting Algebra and Geometry

Painting a neighborhood

INTERVIEW Daniel Galvez

Daniel Galvez always wanted to be an artist, but he didn't know what kind of artist until he learned in college about the great Mexican muralists of the 1900s. Galvez admired the way these artists used colorful paintings on public walls to reach all types of people. "I liked the idea of making a painting that you don't have to go see in a gallery," he says, "the idea that it was part of people's lives whether or not they understood art. It would be something they would see on their way to school or work."

66 I liked the idea of making a painting that you don't have to go see in a gallery. 99

21. a. The size of a television screen is given by the length of the diagonal of the screen. What size is a television screen that is 21.6 in. wide and 16.2 in. high?

16.2 in.

21.6 in.

b. Open-ended Problem Why do you think that manufacturers might choose to measure the size of television screens this way? Use algebra to support your answer.

Toolbox p. 595
Angles

Connection / GEOMETRY

The Pythagorean theorem is important in both algebra and geometry. Exercises 22 and 23 will help you see why the theorem applies only to right triangles.

Acute triangle
All angles < 90°

Right triangle
1 angle = 90°

Obtuse triangle
1 angle > 90°

The area of a triangle is $\frac{1}{2}$ × base × height.

22. Cooperative Learning Work in a group of 3–4 students. You will need at least six pieces of graph paper and a ruler.

Work together to repeat the steps below at least six times. Use two right triangles, two acute triangles, and two obtuse triangles.

Step 1 Draw a triangle on each piece of graph paper so that each corner is on a grid point. Label the sides of each triangle a, b, and c, with c being the longest side (if there is a longest side). Label each triangle *right, acute,* or *obtuse*.

Step 2 Create a square on each side of a triangle.

Step 3 Find the area of each square. If a square is not aligned with the grid, divide it into right triangles and a small square, as shown. Add the areas of the pieces to find the area of the square.

Step 4 Compare the sum of the areas of the two smaller squares to the area of the largest square.

right triangle

Area of this triangle
$\frac{1}{2} \cdot 5 \cdot 3 = \frac{15}{2}$

Area of large blue square =
$4 \cdot \frac{15}{2} + 4 = 34$

Area = 25

right triangle

Area = 9

$25 + 9 = 34$

...complete each

...angle, $a^2 + b^2 \underline{?} c^2$.

...ythagorean Theorem **243**

Geometry

Why doesn't the Pythagorean theorem work for obtuse triangles?

(Chapter 6, page 243)

68 67 70 66 69 71 67 70 62 69

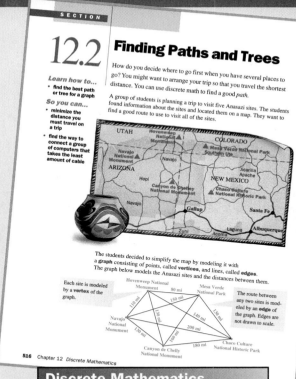

SECTION

12.2

Finding Paths and Trees

Learn how to...
- find the best path or tree for a graph

So you can...
- minimize the distance you must travel on a trip
- find the way to connect a group of computers that takes the least amount of cable

How do you decide where to go first when you have several places to go? You might want to arrange your trip so that you travel the shortest distance. You can use discrete math to find a good *path*.

A group of students is planning a trip to visit five Anasazi sites. The students found information about the sites and located them on a map. They want to find a good route to use to visit all of the sites.

The students decided to simplify the map by modeling it with a **graph** consisting of points, called **vertices**, and lines, called **edges**. The graph below models the Anasazi sites and the distances between them.

Each site is modeled by a **vertex** of the graph.

The route between any two sites is modeled by an **edge** of the graph. Edges are not drawn to scale.

516 Chapter 12 *Discrete Mathematics*

Probability

How many times will each number on a die occur in 24 rolls?

(Chapter 5, page 218)

Discrete Mathematics

How can you find a short path for visiting all five of the Anasazi sites on the map?

(Chapter 12, pages 516-517)

Statistics

What is the life expectancy of a person born in the United States in the year 2000?

(Chapter 3, page 133)

Life Expectancy	
Year of birth	Life expectancy (years)
1900	47.3
1910	50.0
1920	54.1
1930	59.7
1940	62.9
1950	68.2
1960	69.7
1970	70.8
1980	73.7
1990	75.4

Data Analysis

What two inequalities would you use to describe the heights of these students?

(Chapter 1, page 3)

67   68   62   66   63   63   69   66   62   66

"When will I ever use this ?"

Building for the Future

The skills you'll learn in this course will form a strong foundation for the future.

They'll prepare you for more advanced courses and increase your career opportunities.

Problem Solving and Communication

There are many exercises to help you develop your problem solving skills and your ability to communicate orally and in writing.

Open-ended Problems give you a chance to share your thinking.

SAT/ACT Preview exercises help you prepare for these college entrance tests.

Writing You can be creative in answering these questions.

WEATHER For Exercises 23–25, use the table showing rainfall amounts.

• • • • World Record Rainfall Amounts • • • •

Location	Amount	Duration
Unionville, Maryland (U.S.)	3.1 cm	1 min
Curtea-de-Arges (Romania)	20.5 cm	20 min
Holt, Missouri (U.S.)	30.5 cm	42 min
Cilaos, La Réunion	188 cm	24 h
Cherrapunji (India)	930 cm	1 month

23. Write a rate and a unit rate for each location. Use the same units for all the unit rates.

24. In Exercise 23, what information does each rate give you that the unit rate does not? What information does the unit rate give you that the rate does not?

25. Suppose rain continued to fall at a rate of 3.1 cm/min in Unionville. How long would it take to break Cherrapunji's record? Explain.

26. **Open-ended Problem** A child's height is an example of a variable showing a positive rate of change over time. Give two examples of a variable showing a negative rate of change over time. For one of these variables, estimate a rate of change. Explain your answer.

27. **SAT/ACT Preview** Vera has $20. The cost of a pair of socks ranges from $1.75 to $2.60. What is the greatest number of pairs she can buy?

A. 7   B. 8   C. 11   D. 9   E. 10

ONGOING ASSESSMENT

28. **Writing** Write a story that relates the graph and the picture shown at the right. Be sure to include rates in your story.

SPIRAL REVIEW

Write each fraction as a decimal. *(Toolbox, page 587)*

29. $\frac{3}{8}$ 30. $\frac{1}{2}$ 31. $\frac{5}{4}$ 32. $\frac{7}{11}$

Graph each point in a coordinate plane. Label each point with its letter. Name the quadrant (if any) in which the point is located. *(Section 2.4)*

33. $A(4, 7)$ 34. $B(3, 23)$ 35. $C(28, 25)$ 36. $D(-12, 6)$

3.1 Applying Rates   **105**

---

12. Use the graph at the right to estimate the average rate of change in the number of job openings for each time period. Round your answers to the nearest ten thousand.
    a. 1987–1988
    b. 1988–1989
    c. 1989–1990
    d. 1990–1991

Job Openings per Year

SPORTS For Exercises 13–16, use the news clipping.

13. Give two rates that indicate the team's scoring ability. Be sure to include units in your answers.

14. Express home game attendance as a unit rate.

15. Suppose the fullback carries the ball an average of 20 times per game. About how many yards does he gain per game?

16. A quarterback's pass completion rate is the ratio of passes completed to passes attempted. Find Fred Sorenson's pass completion rate. Give your answer as a percent.

Giants off to a Great Start

By Lisa Hodgden
Almanac Staff Writer

Coach George Winn says the Giants have never had a better season. The football team has won 4 of its first 5 games, scoring at least one touchdown in each half of every game. Fullback Johnny Palombo has gained an average of 4.5 yards per carry. Quarterback Fred Sorenson has completed 17 out of the 31 passes he has attempted. Total attendance for the team's first 3 home games has been about 4200 people.
"It's been a ...

Connection MUSIC

The compact disk (CD) was first sold to the public in 1982. During the years that followed, music buyers could choose to buy long-playing records (LPs) or CDs. Use the graph for Exercises 17 and 18.

Sales of LPs and CDs in the United States

● CD Sales
■ LP Sales

17. For each time period, find the average rate of change for LP sales and the average rate of change for CD sales. Organize your results in a table. Round your answers to the nearest ten million.
    a. 1984–1985   b. 1985–1986
    c. 1986–1987   d. 1987–1988

18. **Writing** Take yourself back in time to the year 1988. A music producer wants to know how many CDs and LPs to produce in 1989. Write a letter describing a good business strategy. Use rates of change to support your suggestions.

3.1 Applying Rates   **103**

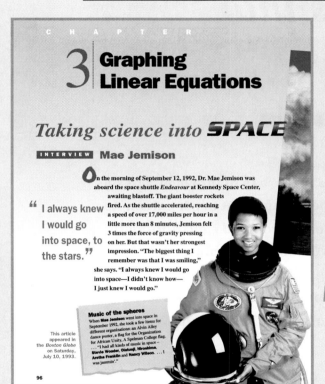

C H A P T E R

3 Graphing Linear Equations

Taking science into SPACE

INTERVIEW Mae Jemison

On the morning of September 12, 1992, Dr. Mae Jemison was aboard the space shuttle *Endeavour* at Kennedy Space Center, awaiting blastoff. The giant booster rockets fired. As the shuttle accelerated, reaching a speed of over 17,000 miles per hour in a little more than 8 minutes, Jemison felt 3 times the force of gravity pressing on her. But that wasn't her strongest impression. "The biggest thing I remember was that I was smiling," she says. "I always knew I would go into space—I didn't know how—I just knew I would go."

" I always knew I would go into space, to the stars. "

Music of the spheres

When **Mae Jemison** went into space in September 1992, she took a few items for different organizations: an Alvin Ailey dance poster, a flag for the Organization for African Unity, A Spelman College flag. ... "I had all kinds of music in space – **Stevie Wonder, Olatunji, Hiroshima, Aretha Franklin** and **Nancy Wilson**. . . . I was jammin'."

This article appeared in the *Boston Globe* on Saturday, July 10, 1993.

96

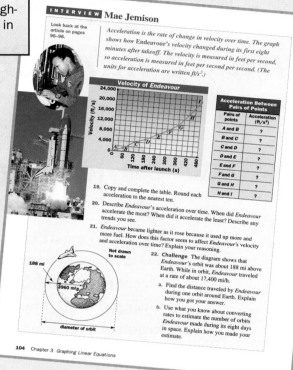

INTERVIEW Mae Jemison

Look back at the article on pages 96–98.

Acceleration is the rate of change in velocity over time. The graph shows how Endeavour's velocity changed during its first eight minutes after takeoff. The velocity is measured in feet per second, so acceleration is measured in feet per second per second. (The units for acceleration are written ft/s².)

Velocity of Endeavour

Acceleration Between Pairs of Points	
Pairs of points	Acceleration (ft/s²)
A and B	?
B and C	?
C and D	?
D and E	?
E and F	?
F and G	?
G and H	?
H and I	?

19. Copy and complete the table. Round each acceleration to the nearest ten.

20. Describe *Endeavour*'s acceleration over time. When did *Endeavour* accelerate the most? When did it accelerate the least? Describe any trends you see.

21. *Endeavour* became lighter as it rose because it used up more and more fuel. How does this factor seem to affect *Endeavour*'s velocity and acceleration over time? Explain your reasoning.

22. **Challenge** The diagram shows that *Endeavour*'s orbit was about 188 mi above Earth. While in orbit, *Endeavour* traveled at a rate of about 17,400 mi/h.

a. Find the distance traveled by *Endeavour* during one orbit around Earth. Explain how you got your answer.

b. Use what you know about converting rates to estimate the number of orbits *Endeavour* made during its eight days in space. Explain how you made your estimate.

Not drawn to scale

188 mi

3960 mi

diameter of orbit

104 Chapter 3 Graphing Linear Equations

Daniel Galvez
Mural Artist

Carrie Teegardin
Journalist

Bob Hall
Athlete and Racing Wheelchair Designer

1 Using Algebra to Work with Data

Turning numbers into NEWS

Carrie Teegardin

*C*arrie Teegardin liked algebra in high school, but she never dreamed it would be so important to her career. As a reporter at the *Atlanta Journal and Constitution* in Georgia, Teegardin does database reporting, a job that requires finding meaning hidden in large sets of numbers. In 1993, Teegardin produced an eight-part series called "Growing Up Southern." She used data and interviews to investigate the lives of American children. On a computer, she analyzed data from 3248 counties.

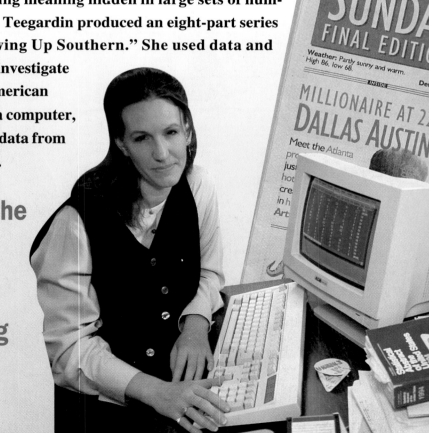

" Math helps you see the world clearly ... you need math to figure out what's happening around you. "

Turning Numbers into Percents

For each county, Teegardin turned numbers into percents. The percents helped her compare aspects of life across the country, from income level to the percent of working mothers. "It's like organizing batting averages," says Teegardin. "You may know somebody's hitting well, but you don't know exactly how well until you quantify their hitting with a number and compare it to others."

Practical Problem Solvers

Teegardin also interviewed people about the real life behind the statistics. She talked to a cross section of teens about their families, their sense of the future, and current problems such as drugs and poverty. "Teens hear people saying they're going to be the first generation that won't be better off than their parents," says Teegardin. "They worry about being able to afford college. They're the first generation not to know life without AIDS. Even so, I got the impression that a lot of kids want to be practical problem solvers for themselves and society." One student she interviewed joined the Marines so he could afford to go to college and become a pediatrician. He wants to help kids like himself who don't get the attention he thinks they deserve from their doctors. "I like that," says Teegardin. "I like the fact that teens are determined to make their lives work and make the world a better place in a very practical way."

> **"...teens are determined to make their lives work..."**

Organizing Data

As Teegardin weaves together personal interviews and data, she has to organize the data to make sense of them. "The power of numbers is that you can combine them, and manipulate them, and boil thousands of them down to four or five," says Teegardin. "Then you can convert them to forms that readers can understand, like maps, charts, and graphs." For example, during the 1980s, families with children tended to move from rural areas to urban areas. The map below gives you a picture of this shift in population.

Where children live
The number of children in the United States remained almost unchanged over the 1980s, while the elderly population grew rapidly. Rural areas saw an exodus of nearly 1 million children, while metro areas -- particularly Atlanta, Boston, Dallas, Los Angeles and Washington -- saw their under-18 population grow substantially.

Percent change in child population, 1980 to 90
-10% or more   -10% to 0   0 to 10%   + 10% or more   Missing

---

EXPLORE AND CONNECT

Carrie Teegardin took six months to gather and analyze the data to create a statistical portrait of Southern children.

1. Writing According to the map above, in what parts of the United States did the largest percent change in child population take place? In what parts did the smallest percent change take place? In what parts did the number of children stay about the same? How does using percents help you compare population changes in different parts of the country?

2. Research What percent of the students in your class have lived in at least one other state? How did you calculate the percent?

3. Project Find a newspaper or magazine article about a topic that interests you. The article should include at least one table, graph, or map. Share the article with your class.

Mathematics
& Carrie Teegardin

In this chapter, you will learn more about how mathematics is related to database reporting.

Related Exercises

Section 1.1
• **Exercises 12–16**

Section 1.8
• **Exercises 8–14**

1.1

Working with Variables and Data

Numbers can describe your height, your grades, even where you live. Numbers that give you information are called **data**. For example, the heights of the students in the photo are data.

Notice that the heights vary from student to student. You can use a letter called a **variable** to stand for a quantity that varies.

Let the variable h stand for the heights of the students in inches. Each student's height is called a **value** of h.

The value of h for this student is **68**.

EXAMPLE 1   Application: Statistics

a. What is the smallest value of h for the students above? What is the largest?

b. Write two inequalities that describe the heights of the students.

SOLUTION

a. The shortest students are 62 in. Therefore, the smallest value of h is 62. The tallest student is 71 in. Therefore, the largest value of h is 71.

b. All students are 62 in. or taller. $h \geq 62$
All students are 71 in. or shorter. $h \leq 71$

THINK AND COMMUNICATE

1. Explain why you can write $h > 61$ and $h < 72$ to describe the students' heights. How are these two inequalities different from the ones in part (b) of Example 1?

2. Let d = the distance in miles traveled from home to school.
Let n = the number of telephone calls someone made yesterday.
Give some reasonable values of each of these variables.

Reading Data from Graphs

One way to organize data is in a graph. If you group the students by height, you can make a graph called a *histogram*.

The bars show the number of students for each height. The bar at $h = 69$ is three units high. Therefore, **3** students are **69 in.** tall.

Toolbox p. 596
Statistical Graphs

The **values of h** are shown along the bottom of the histogram.

EXAMPLE 2   **Application: Statistics**

Use the histogram.

a. How many students are shorter than 66 in.?

b. How many students are at least 68 in. tall?

SOLUTION

a. Count the number of students represented by the bars to the left of $h = 66$. Five students are shorter than 66 in.

b. Two students are 68 in. tall. Six students are more than 68 in. tall. A total of eight students are at least 68 in. tall.

☑ CHECKING KEY CONCEPTS

For Questions 1 and 2, let t = outdoor temperature.

1. Give several reasonable values of t for your part of the world.

2. Use the variable t to write two inequalities describing temperatures where you live.

3. Sandra says that you can use the inequalities $h < 100$ and $h > 0$ to describe the heights of the students shown on page 3. Explain why you agree or disagree.

For Questions 4–6, use the histogram on page 4.

4. How many students are 63 in. tall? How do you know?

5. How many students are at least 63 in. tall?

6. For how many students is $h < 70$? For how many students is $h \geq 68$? How can you tell from the histogram?

7. Let h = the number of hours you spend visiting a friend. What does $h \leq 6$ mean?

8. Compare the variables in Questions 6 and 7 above. How are they different?

1.1 | Exercises and Applications

Extra Practice exercises on page 557

HISTORY For Exercises 1–3, use the histogram at the right.

1. How many states were admitted between 1925 and 1949?

2. How many states were there at the end of 1849?

3. How many states were admitted after 1874?

CHILD CARE For Exercises 4–6, use the data in the chart at the left.

Age at which Infants First Walk

Name	Age (months)	Name	Age (months)
John	14	Cora	12
Lynne	12	Theo	11
Parker	12	Bryan	14
Andrea	13	Yi Hsa	13
Brad	11	Alison	12

4. Write two inequalities that describe the ages at which the infants began walking.

5. What percent of the infants began walking before 12 months? after 12 months?

6. **Writing** Who might be interested in these data and your answers to Exercises 4 and 5? Explain.

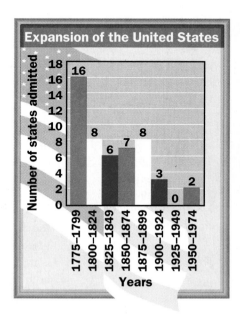

Expansion of the United States

Number of states admitted (vertical axis: 0, 2, 4, 6, 8, 10, 12, 14, 16, 18)

Years (horizontal axis): 1775–1799 (16), 1800–1824 (8), 1825–1849 (6), 1850–1874 (7), 1875–1899 (8), 1900–1924 (3), 1925–1949 (0), 1950–1974 (2)

ADVERTISING Choose a variable and write an inequality for each ad. Explain what each variable stands for.

7.
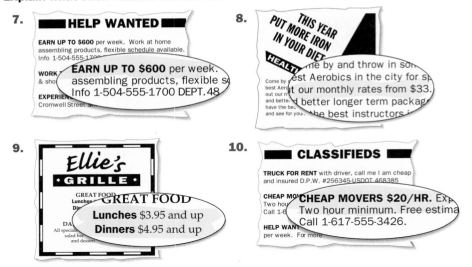

HELP WANTED

EARN UP TO $600 per week. Work at home assembling products, flexible schedule available. Info 1-504-555-1700

WORK...
& sho...

EXPERIEN...
Cromwell Street...

EARN UP TO $600 per week. assembling products, flexible s... Info 1-504-555-1700 DEPT. 48

8.

THIS YEAR PUT MORE IRON IN YOUR DIET

HEALT...
...me by and throw in so...
Come by ...est Aerobics in the city for s...
out our m...t our monthly rates from $33.
and bette...d better longer term packag...
have the be...
and see for your...he best instructors i...

9.

Ellie's
•GRILLE•

GREAT FOOD
Lunches ...
Di...

DA...
All specials...
salad bar...
and dessert...

GREAT FOOD
Lunches $3.95 and up
Dinners $4.95 and up

10.

CLASSIFIEDS

TRUCK FOR RENT with driver, call me I am cheap and insured D.P.W. #256345-USDOT 468385

CHEAP MO...
Two hou...
Call 1-6...

HELP WAN...
per week. For more

CHEAP MOVERS $20/HR. Ex... Two hour minimum. Free estima... Call 1-617-555-3426.

11. a. **Writing** Look back at the ads in Exercises 7–10. Why do you think advertisers use phrases that suggest inequalities rather than providing specific prices?

b. **Open-ended Problem** Write your own ad for an imaginary product. Explain how your ad suggests an inequality.

Look back at the article on pages xxx–2.

INTERVIEW Carrie Teegardin

In "Growing Up Southern," Carrie Teegardin reported on the amount of TV eighth-grade students watch. The graph shows some data she used. Let h = the number of hours spent watching TV each day.

Carrie Teegardin at work in the *Atlanta Journal and Constitution* office.

12. For about what percent of eighth graders is $h = 3$?

13. For about what percent of eighth graders is $h \geq 6$?

14. **Challenge** Tom says that, according to the graph, most eighth graders in the South watch four or five hours of TV daily. Do you agree? Use the graph to support your answer.

15. What is the greatest amount of TV you watch in a day? the least? Write two inequalities that describe your daily viewing habits.

16. **Writing** The graph shown below is a *bar graph*. How is it different from a histogram?

Daily TV Viewing in the South

Percent of students: 30, 25, 20, 15, 10, 5, 0

Hours spent watching TV: 1 or less, 2, 3, 4–5, 6 or more

17. a. Research Find out how many pets each student in your class has. Make a histogram of the data.

   b. Let p = the number of pets each student has. Write two inequalities to describe the values of p.

BIOLOGY **For each statement, choose a variable and write at least two inequalities. Explain what each variable stands for and what units you are using. Remember that 1 kg = 1000 g.**

18. Throughout the history of Earth, birds have ranged in size from the bee hummingbird (2 g) to the now extinct elephant bird (about 500 kg).

19. The smallest land mammal is Savi's pygmy shrew (2 g). The largest land mammal ever was the Baluchitherium, a prehistoric giant rhinoceros (about 30,000 kg).

For Exercises 20–22, give several reasonable values of each variable.

20. the number of words w a person says in an hour

21. the cost c in dollars of a pair of sneakers

22. the number of people n at a concert

23. Open-ended Problem Many factors can cause the values of a variable to vary. For example, the amount a of your paycheck may depend on your pay rate and how many hours you work. Choose one of the variables in Exercises 20–22. Describe some factors that could cause the values of the variable to vary.

24. SAT/ACT Preview If 2 identical packages contain a total of 12 bagels, how many bagels are there in 5 packages?
   A. 12   **B.** 24   **C.** 30   **D.** 36   **E.** 60

ONGOING ASSESSMENT

25. Writing Describe three ways to represent data. Compare the advantages and disadvantages of the three representations. When might each be used?

SPIRAL REVIEW

For Exercises 26–28, use the bar graph at the right. *(Toolbox, page 596)*

26. Which country shown had the highest gasoline price in 1992?

27. Which had the lowest price?

28. Writing How else can the bars on this graph be arranged? Why doesn't it make sense to rearrange the bars of the histogram on page 5?

Write each number as a decimal rounded to the nearest hundredth.
(Toolbox, page 587)

29. $\frac{8}{3}$ **30.** $\frac{3}{7}$ **31.** $\frac{100}{18}$ **32.** $5\frac{7}{9}$ **33.** $\frac{1}{8}$ **34.** $3\frac{1}{3}$

1992 Gasoline Prices

(Bar graph, Price per gallon (U.S. dollars) vs Country)
U.S. 1.13, Turkey 2.82, Japan 3.70, France 3.76, Norway 4.86

1.2 Using the Order of Operations

Learn how to...

- simplify an expression using the order of operations

So you can...

- solve real-world problems, such as estimating the height of a waterfall or a tall bridge

While studying for their math exam, Robin and Jeff had a pizza delivered. The pizza cost $12.50, and they gave the driver a $2.00 tip. To figure out how much each of them should pay, they both used the keystrokes 12.5 [+] 2 [÷] 2 [=] on their calculators. Why did they get two different answers, as shown below?

EXPLORATION
COOPERATIVE LEARNING

Calculator Order of Operations

Work with a partner.
You will need:

- a calculator

1 Reread the paragraph above about Robin and Jeff. Whose answer makes more sense? Why?

2 Try using Robin's and Jeff's keystrokes on your calculator. Did your calculator add first, or did it divide first? How can you tell?

3 Which operation did Robin's calculator do first? Jeff's? How do you know?

4 Press the following keys on your calculator:

24 [÷] 7 [−] 3 [=]

Which operation did the calculator do first?

Now try: 24 [÷] [(] 7 [−] 3 [)] [=]

Which operation did the calculator do first this time?

5 Explain how Jeff can use parentheses keys to figure out how much to contribute for pizza. Test your answer on your calculator.

Jeff needs to use parentheses because his calculator follows the *order of operations*, a set of rules people agree to use so that an expression has only one value. You should use the order of operations when you simplify expressions.

1. First do all work inside parentheses.

2. Then do all multiplications and divisions in order from left to right. Multiplication and division are usually written like this:

   Multiplication: $5 \cdot 3$, $5(3)$, or $(5)(3)$ **Division:** $10 \div 5$ or $\frac{10}{5}$

3. Then do all additions and subtractions in order from left to right.

EXAMPLE 1

Simplify each expression.

a. $7 + 3 \cdot 5 \cdot 4$ **b.** $\frac{9}{4} + 3(6 - 3)$

SOLUTION

a. First do the multiplications in order from left to right.

$$7 + 3 \cdot 5 \cdot 4 = 7 + 15 \cdot 4$$
$$= 7 + 60 \qquad \text{Then do the addition.}$$
$$= 67$$

b. First do the subtraction inside the parentheses.

$$\frac{9}{4} + 3(6 - 3) = \frac{9}{4} + 3(3) \qquad \text{Then divide and multiply.}$$
$$= 2.25 + 9$$
$$\qquad\qquad\qquad \text{Do the addition last.}$$
$$= 11.25$$

Variable Expressions

Suppose two students buy a pizza for p dollars and give a tip of t dollars. If they split the total cost, they each pay $\frac{p + t}{2}$ dollars. The fraction $\frac{p + t}{2}$ is an example of a *variable expression*.

A fraction bar groups operations. For example, $\frac{p + t}{2}$ means $(p + t) \div 2$.

A **variable expression** is made up of **variables**, **numbers**, and **operations**. The value of a variable expression depends on what values you choose for the variables. For example, you can **evaluate** the variable expression $s + r$ when $s = 20$ and $r = 3$:

$$s + r = 20 + 3$$
$$= 23$$

Examples of Variable Expressions	
$a + b + c$	$10x - 3$
$\frac{t}{60}$	$s + r$
$2(l + w)$	x^2
$2lw$	$-3v + 1.5$

EXAMPLE 2

a. Let h = a person's height in inches. Write a variable expression for the person's height after the person grows 5 in.

b. James is 63 in. tall. Evaluate the expression from part (a) when $h = 63$.

SOLUTION

a. $h + 5$

b. $h + 5 = 63 + 5$ — Substitute **63** for h.

$= 68$

James will be 68 in. tall after he grows 5 in.

EXAMPLE 3

Evaluate each variable expression when $a = 4$ and $b = 3$.

a. $2(a + b)$

b. $\dfrac{b^2 + ab}{7}$

SOLUTION

a. Substitute **4** for a and **3** for b.

$$2(a + b) = 2(4 + 3)$$

Simplify inside the parentheses first.

$$= 2(7)$$

Remember, 2(7) means $2 \cdot 7$.

$$= 14$$

b. The "2" in this expression is an **exponent**. It tells you to use b as a factor two times: $b^2 = b \cdot b$. You read b^2 as "b squared."

Substitute 4 for a and 3 for b. Notice that ab means $a \cdot b$.

$$\frac{b^2 + ab}{7} = \frac{3^2 + 4 \cdot 3}{7}$$

Simplify exponents before other multiplications and divisions.

$$= \frac{9 + 4 \cdot 3}{7}$$

Simplify the numerator completely before dividing by 7.

$$= \frac{9 + 12}{7}$$

$$= \frac{21}{7}$$

$$= 3$$

☑ CHECKING KEY CONCEPTS

Simplify each expression.

1. $8 + 2 \div 2$

2. $2 \cdot 4 - 6 \div 2$

3. $15 - 5 \cdot 4 \div 2$

4. $2(1 + 3)$

5. $3.4(5 \div 2 - 1)$

6. $1 + 3(5 - 2)$

7. $\dfrac{1 + 2}{6}$

8. $\dfrac{1 + 3}{2(4 - 3)}$

9. $2 - \dfrac{6}{2(1.5)}$

Evaluate each variable expression when $m = 4$, $n = 7$, and $p = 10$.

10. $m + n + p$

11. $m^2 - 3m$

12. $p - (5 + m)$

13. $\dfrac{10p + n}{p}$

14. $\dfrac{mn}{p}$

15. $\dfrac{p - n + m}{n + p}$

1.2 ⋮ **Exercises and Applications**

Simplify each expression.

Extra Practice exercises on page 557

1. a. $8 \div 4 - 2$

   b. $8 \div (4 - 2)$

2. a. $6 - 2 \cdot 3$

   b. $(6 - 2) \cdot 3$

3. a. $5 \cdot 8 + 2$

   b. $2 + 5 \cdot 8$

4. a. $6 \cdot 3 - 2 \cdot 2$

   b. $6 \cdot (3 - 2) \cdot 2$

5. a. $\dfrac{7 + 9}{2}$

   b. $7 + 9 \div 2$

6. a. $12 - 2(9 - 7)$

   b. $(12 - 2)(9 - 7)$

7. Writing Use your calculator to simplify each expression in Exercise 4. Describe your keystrokes and explain what the calculator does.

8. a. Simplify the expression $6 + 9 \div 3 - 2$.

   b. Add parentheses to the expression so that it equals 3 when simplified.

   c. Add parentheses to the expression so that it equals 1 when simplified.

Simplify each expression.

9. $3(7 - 2.5) - 10$

10. $24 \div 6 + 3.2 \div 4$

11. $0.80(7 + 7 \cdot 0.05)$

12. $15 + \dfrac{95}{100} \cdot 3$

13. $\dfrac{3}{15} + 2(3)$

14. $\dfrac{5(7)}{4(3)}$

15. $\dfrac{5 - 3}{3 + 8}$

16. $\dfrac{4(4) + 5}{5}$

17. $\dfrac{12 - 2}{4} + 3(10)$

18. $\dfrac{12.25 - 2 \cdot 5}{3 + 2}$

19. $2\left(3 \cdot 8 - \dfrac{15}{16}\right)$

20. $\dfrac{(5 + 2)(6 - 3)}{3 + 2}$

21. Open-ended Problem Describe how to test a calculator to see if it follows the order of operations. Explain why your test works.

22. Let $g =$ the number of gallons of gasoline in a car's tank. Write a variable expression for the amount in the tank if 3 gal are added.

23. Let g = the number of gallons of gasoline in a car's tank. Write a variable expression for the number of miles the car can travel if it gets 25 miles to the gallon.

"If you want to plant kudzu, drop it and run," is a joke you might hear in the southeastern United States. The kudzu vine grows very quickly. A vine can grow up to 12 in. per day. Use this information for Exercises 24–26.

24. Write a variable expression for how much a kudzu vine can grow in d days.

25. How much can kudzu grow in 3 days?

26. How much can kudzu grow in 10 days?

Evaluate each variable expression when $a = 2$, $b = 7$, and $c = 6$.

27. $a^2 + b^2$

28. $6c - b$

29. $\dfrac{5b}{3c - 5a}$

30. $\dfrac{7a + b}{c}$

31. $\dfrac{a^2 + b}{c}$

32. $\dfrac{6b - 2a}{c}$

33. $b^2 + c^2 - 2bc$

34. $(b + a)(b - a) + a^2$

35. $c^2 - ab + c - 4b$

Connection PHYSICS

You can tell how high a bridge is, how deep a hole is, or how high a cliff is by dropping a pebble and counting the time it takes to hit the bottom. In t seconds, an object falls about $16t^2$ feet.

36. A stone dropped into Mexico's **Sótano de las Golondrinas**, possibly the deepest natural pit in the world, takes about nine seconds to reach the bottom. About how deep is the pit?

37. A stone dropped from the Golden Gate Bridge in San Francisco takes about four seconds to reach the water. About how high above the water is the Golden Gate Bridge?

38. A stick going over **Victoria Falls** on the Zambia-Zimbabwe border in Africa takes about five seconds to reach the bottom. About how high is the waterfall?

39. **Writing** A log going over Niagara Falls plummets about 170 ft. Use a calculator to estimate how long it takes the log to reach the bottom of the falls. Explain how you found your answer.

40. Challenge Add parentheses to this expression so that it equals 1 when simplified: $5 - 4 \cdot 3 \div 2 + 1$.

41. 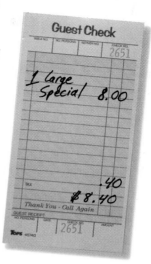 **Technology** Write the keystrokes you would use to simplify this expression with a calculator: $\dfrac{2(5) - 7}{15}$.

42. **Technology** Ben and Diana are splitting the cost of an $8.40 pizza. They want to leave a 15% tip. Using her calculator, Diana finds that they should each pay $9.03.

   a. Does Diana's answer make sense? Explain.

   b. Diana's calculator has parentheses keys. What keystrokes should she use to find her share of the cost?

   c. What keystrokes do you think Diana used?

43. SAT/ACT Preview $(8 - 6 \div 2) \cdot (1 + 2 \cdot 2) = \underline{\ ?\ }$.

   A. 5   **B.** 18   **C.** 25   **D.** 6   **E.** 30

"YOUR SPAGHETTI DINNER IS READY, SIR."

44. Open-ended Problem Write a short story or draw a cartoon like the one shown to show what can go wrong if you do something nonmathematical in the wrong order.

ONGOING ASSESSMENT

45. Writing Explain the difference between simplifying a numerical expression and evaluating a variable expression. Give an example of each.

SPIRAL REVIEW

Exercises 46–48 give you information about tropical rain forests. For each statement, choose a variable and write an inequality. Explain what each variable stands for. *(Section 1.1)*

46. At least 60 in. of rain falls in a single year.

47. Scientists have counted up to 80 species of trees per acre in some forests.

48. Trees may be as tall as 150 ft.

1.3 Mean, Median, and Mode

According to a national report, people under the age of 18 typically watch between 3 and 4 hours of TV daily. Todd made a histogram showing how much TV his friends watch daily.

THINK AND COMMUNICATE

1. Each X represents one student. On average, how many hours of TV do Todd's friends watch? Explain your answer.

2. A report says that in 1993, the average family in the United States had 3.16 members. What do you think this statement means?

Number of hours watched each day

Three Kinds of Averages

When people talk about averages, they often mean different things.

Mean To find the *mean* of the data, first find the sum of all the data. Then divide by the number of data items.

Median The *median* is the middle number when you put the data in order from smallest to largest.

Mode The *mode* is the data value that appears most often.

EXAMPLE 1

Use the histogram of Todd's data to find each kind of average.

a. the mean   **b. the median**   **c. the mode**

SOLUTION

a. Use this formula to find the mean of Todd's data:

Two students watch **4 h** of TV.

$$\text{Mean} = \frac{\text{Total number of hours watched}}{\text{Number of students}}$$

$$= \frac{2(0) + 4(1) + 2(2) + 3(3) + 2(4) + 1(6) + 1(9)}{2 + 4 + 2 + 3 + 2 + 1 + 1}$$

$$= \frac{40}{15}$$

The symbol ≈ means *approximately equal to.*

$$\approx 2.7 \quad \text{The mean number of hours watched is about 2.7.}$$

b. To find the median, put the data values in order from smallest to largest. For Todd's data, the median is 2.

0 0 1 1 1 1 2 2 3 3 3 4 4 6 9

7 data items
below the median

median

7 data items
above the median

c. In a histogram, the mode is represented by the tallest bar. For Todd's data, the mode is 1.

THINK AND COMMUNICATE

3. Explain why the tallest bar of a histogram represents the mode.

4. Do you think the mean, the median, or the mode best describes the average hours of TV watched by Todd's class? Explain your choice.

EXAMPLE 2 | **Application: Sports**

Find the median and the mode(s) of the ages shown in the histogram at the right.

Ages of U.S. Women Basketball Players, 1992 Summer Olympics

Number of players / Age (years)

SOLUTION

To find the median, put the ages in order from smallest to largest.

There is no middle number.

22 23 24 25 25 **25 26** 26 28 29 29 29

$$\frac{25 + 26}{2}$$

25.5

When the number of data items is even, the median is the **mean of the two middle numbers**.

The median age of the players is 25.5.

To find the mode, look for the data values that appear most often.

22 23 24 **25 25 25** 26 26 28 **29 29 29**

The modes are ages 25 and 29.

25 and **29** each appear three times. There can be more than one mode.

5. Look back at Example 1. Sharon told Todd that she was joking when she said she watches nine hours of TV each day. Find the mean, the median, and the mode of the data if you ignore Sharon's data. Which average changes? Why?

6. Is the mean always greater than the median and the mode(s)? Give examples to support your answer.

✓ CHECKING KEY CONCEPTS

Find the mean, the median, and the mode(s) of each set of data.

1. Days absent from school

   2, 6, 1, 5, 2, 2, 4, 3, 0, 7, 3

2. Ages of students in a cooking class

   23, 17, 18, 19, 23, 25, 67, 19, 22, 18

3. Ages of U.S. Male Gymnasts, 1992 Summer Olympics

1.3 | Exercises and Applications

Extra Practice exercises on page 557

For Exercises 1 and 2, find the mean, the median, and the mode(s) of the data.

1. Number of sitcoms watched in one weekend
   3, 0, 3, 2, 1, 0, 0, 4, 4, 8, 2, 2, 2, 1, 0, 0

2. Number of books read in one year
   10, 8, 25, 5, 14, 10, 5, 5, 14, 14, 5, 8, 8, 14

3. Find the mean, the median, and the mode(s) of the data shown on the map. Which kind of average do you think best describes the population density of the islands of the Lesser Antilles? Explain your reasoning.

Estimated Population Density of the Lesser Antilles, 1991 (people/km²)

British Virgin Islands 85
Anguilla 73
Atlantic Ocean
U.S. Virgin Islands 345
Antigua and Barbuda 173
St. Christopher and Nevis 169
Guadeloupe 202
Montserrat 118
Dominica 111
Martinique 311
Caribbean Sea
St. Lucia 246
Aruba 306
St. Vincent and the Grenadines 302
Barbados 593
Netherlands Antilles 236
Grenada 244
Trinidad and Tobago 244

4. a. Find the mean, the median, and the mode(s) of the data in the histogram.

b. Writing Paul Pelletier, the restaurant manager, wonders how long a typical customer must wait in line. What answer would you give him? Explain.

5. Challenge Find a set of five data values with modes 0 and 2, median 2, and mean 2. Explain how you found your answer.

Time Spent in a Fast Food Restaurant Line

Connection ECOLOGY

The Gulf of California harbor porpoise, or vaquita (pronounced bah-KEE-tah), is an endangered species. About 30 to 35 vaquitas die each year after being captured in fishing nets. The histogram shows the ages of a group of captured vaquitas. The age groupings of the captured vaquitas may represent the ages of the entire population of vaquitas.

Gulf of California

Ages of Captured Vaquitas

Report 1

The average vaquita is less than one year old. Fishing with nets should stop for at least twenty years until these baby vaquitas have a chance to reproduce.

Report 2

The average vaquita is six years old. This is healthy, given that the common porpoise has a normal life span of only 15 years. Fishing with nets can continue.

6. Cooperative Learning Work with a partner. Each of you should choose one of the reports at the left. Explain how the data in the histogram supports the report. Then discuss together which report you think is more accurate. Explain your conclusions.

7. a. Open-ended Problem How would you describe the typical age of the captured vaquitas? Give reasons for your answer.

b. Write your own report like the ones at the left, or write some questions you would want answered before writing a report.

8. One scientist decides not to include the 25-year-old vaquita in an analysis of the data. How does this affect the mean of the data?

9. **Writing** Which histogram represents data for which the mean, the median, and the mode are equal? Explain your reasoning.

A. B. C.

For Exercises 10–15, write each decimal as a fraction in lowest terms. Then write each decimal as a percent. *(Toolbox, page 587)*

10. 0.765   **11.** 0.05   **12.** 0.2

13. 0.25   **14.** 0.125   **15.** 1.75

ASSESS YOUR PROGRESS

VOCABULARY

data (p. 3)   **exponent** (p. 10)
variable (p. 3)   **mean** (p. 14)
value of a variable (p. 3)   **median** (p. 14)
variable expression (p. 9)   **mode** (p. 14)
evaluate (p. 9)

Simplify each expression. *(Section 1.2)*

1. $12 \div 3 \cdot 2 + 8$ **2.** $26 - 3(6 + 4 \div 2)$ **3.** $\dfrac{2(5) - 7}{15}$

Evaluate each variable expression when $x = 3$. *(Section 1.2)*

4. $6x + \dfrac{2}{x}$ **5.** $x^2 - 2x + 7$ **6.** $\dfrac{x - 2}{8 - x}$

Use the histogram for Exercises 7–10. *(Sections 1.1 and 1.3)*

7. How many students are represented by the histogram?

8. How many students saw three or more movies during the month?

9. Write two inequalities that describe the data in the histogram. What does the variable represent?

10. Find the mean, the median, and the mode(s) of the data.

11. **Journal** Review the meanings of the words *mean*, *median*, and *mode*. What are some everyday uses of these words that can help you remember their meanings?

1.4

Working with Integers

In 1978, a group of American women climbed Annapurna, one of the highest mountains in the world. The illustration below shows some of the camps the women made. The scale is labeled with *integers*.

Learn how to...
- **find the absolute value of a number**
- **add and subtract integers**

So you can...
- **use negative numbers to find differences on number scales, for example**

Integers are numbers like these:
... −4, −3, −2, −1, 0, 1, 2, 3, 4 ...

Compared to base camp, Arlene Blum is at 400 ft and Piro Kramar is at −400 ft. Although they are in different places, they are both 400 ft from base camp because distance is never negative.

The **absolute value** of a number is its **distance from zero** on a number line. The symbol $|x|$ means *the absolute value of x.*

Elev. 26,504 ft   Area of Detail

The absolute value of a number is never negative.

$$|-400| = |400| = 400$$

$$-800 \qquad -400 \qquad 0 \qquad 400 \qquad 800$$

The absolute value of 0 is 0.

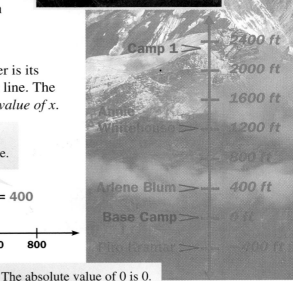

Camp 1 — 2400 ft
2000 ft
1600 ft
Annie Whitehouse — 1200 ft
800 ft
Arlene Blum — 400 ft
Base Camp — 0 ft
Piro Kramar — −400 ft

EXAMPLE 1

Evaluate each expression when $x = -2$.

a. $|x|$

b. $2|1 + x| + |4|$

SOLUTION

a. $\qquad |x| = |-2|$

Substitute −2 for x.

$$-4 \quad -3 \quad -2 \quad -1 \quad 0 \quad 1 \quad 2 \quad 3 \quad 4$$

−2 is a **distance of 2** away from zero.

$$|x| = 2$$

Solution continued on next page.

Toolbox p. 588
Integers

S O L U T I O N *continued*

b. Substitute -2 for x.

$$2|1 + x| + |4| = 2|1 + (-2)| + |4|$$
$$= 2|-1| + |4|$$
$$= 2 \cdot 1 + 4$$
$$= 6$$

> The absolute value bars act like parentheses. Do the operations inside them first.

> $2|-1|$ means $2 \cdot |-1|$.

Adding and Subtracting Negatives

The integers 400 and -400 are **opposites** because their sum is zero. Every number has an opposite.

Property of Opposites

For every number a, there is a number $-a$ such that $a + (-a) = 0$.

The expression $-a$ means "the opposite of a."

For example: $3 + (-3) = 0$ ◄────── The numbers 3 and -3 are opposites.

$0 + 0 = 0$ ◄────── The opposite of 0 is 0.

Subtracting a **number** is the same as **adding** its **opposite**.

$$5 - 3 = 5 + (-3)$$

EXAMPLE 2

Simplify each expression.

a. $6 - 10$ **b.** $-700 - (-400)$

S O L U T I O N

> Subtracting 10 is the same as adding -10.

a. $\quad 6 - 10 = 6 + (-10)$

> **1.** Start at 0. Move $+6$ on a number line.

To add a positive number, move to the right. To add a negative number, move to the left.

> **2.** Then move -10 on the number line.

$$6 + (-10) = -4$$

> Subtracting -400 is the same as adding 400.

b. $-700 - (-400) = -700 + 400$
$$= -300$$

THINK AND COMMUNICATE

1. Decide whether the statement below is *always*, *sometimes*, or *never* true. Give examples to support your answer.

$$|a + b| = |a| + |b|$$

2. Look back at the illustration on page 19. Write and simplify an expression to find each distance.

   a. Annie Whitehouse's distance above Arlene Blum

   b. Arlene Blum's distance above Piro Kramar

3. Is $-x$ *always*, *sometimes*, or *never* a negative number? Explain.

☑ CHECKING KEY CONCEPTS

State the opposite of each number or variable.

1. 24   2. -24 3. $\frac{1}{4}$

4. -12.5 5. n 6. 0

Simplify each expression. Which of your answers are *not* integers?

7. a. $10 - 4$ 8. a. $12 - 6$ 9. a. $-4 - 2$

   b. $10 + (-4)$ b. $6 - 12$ b. $-4 - (-2)$

10. a. $|-2|$ 11. a. $|3 + (-5)|$ 12. a. $\frac{2|5 - 7|}{3}$

    b. $|2|$ b. $|3| + |-5|$ b. $\frac{2}{3}|5 - 7|$

1.4 **Exercises and Applications**

Simplify each expression.

1. $5 - 15$ 2. $-2 + 5$ 3. $11 + (-24)$

4. $10 + (-2)$ 5. $-6 - 2$ 6. $-3 + 1$

7. $4 + (-8)$ 8. $-6 + 4$ 9. $-2 - (-5)$

Extra Practice exercises on page 558

10. What is the sum of any number and its opposite?

11. **Open-ended Problem** Draw a number line with your home at zero. Label neighboring buildings or landmarks on the number line according to their distance and direction from your home.

12. What two numbers have an absolute value of 18? How many numbers have an absolute value of zero?

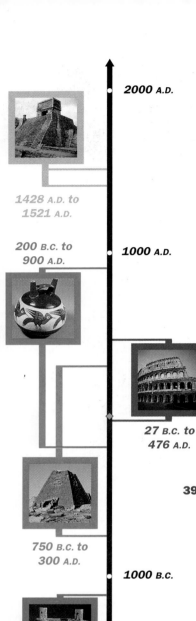

2000 A.D.

1428 A.D. to
1521 A.D.

200 B.C. to
900 A.D.

1000 A.D.

1000 B.C.

750 B.C. to
300 A.D.

1766 B.C. to
1122 B.C.

2000 B.C.

27 B.C. to
476 A.D.

HISTORY The time line shows the approximate beginning and ending dates of some civilizations from the past. Each civilization is matched by color to the time line. About how long did each civilization last?

13. Kingdom of Kush   14. Shang Dynasty   15. Roman Empire

16. Aztec Empire   17. Nacza Civilization

18. **Research** Write a report about one of the civilizations in Exercises 13–17.

Simplify each expression.

19. $8 - 20$ 20. $4 + (-5)$ 21. $72 - (-6)$

22. $-5 - (-14)$ 23. $1.9 - (-2.4)$ 24. $|7 - 3 \cdot 3|$

25. $|-14| - |17|$ 26. $|7 - 2| + 3|-4|$ 27. $\dfrac{3|2 - 5|}{4}$

28. Which of the answers to Exercises 19–27 are *not* integers?

29. Find values of x and y for which the value of $x - y$ is:

    a. positive   **b.** negative   **c.** zero

Evaluate each variable expression when $t = 4$, $u = 13$, and $v = -5$.

30. $t + (-u)$ 31. $12 - u$ 32. $u - v$

33. $u + v$ 34. $t - (-v) + 1$ 35. $5 - |v|$

36. $|t - u|$ 37. $|t + u| + v$ 38. $|t - v| - u$

39. **a. Challenge** Simplify each expression. Describe any patterns you see.

    $|7 - 4|$ $|4 - 7|$ $|8 - 2|$ $|2 - 8|$

    b. Suppose a and b are positive numbers. What can you say about $|a - b|$ and $|b - a|$?

    c. Suppose a is a positive number and b is a negative number. Do you think there is a relationship between $|a - b|$ and $|b - a|$? Design and carry out an investigation to find out.

40. Decide whether the statement $a \le |a|$ is *always*, *sometimes*, or *never* true. Give examples to support your answer.

41. **SAT/ACT Preview** $-3 - (-4) + 5 - 2 = \underline{\ ?\ }$.

    A. 0   **B.** -2 **C.** -4 **D.** 4   **E.** -14

42. **Technology** Many calculators have a key that changes the sign of a number. On some calculators, this key looks like this: (−) . On others, it looks like this: +/− .

    a. Which key does your calculator use to change the sign of a number? How can you use the key to enter -6? How do you enter $-(-6)$?

    b. Use the change sign key to simplify each expression. Check your work to make sure you are using the key correctly.

    $12 + (-6)$ $8 - (-6)$ $3 - (-2)$

    c. Write an expression that means "the opposite of the opposite of x."

The United States women's basketball team won the gold medal at the 1984 Olympics. The table gives the number of points scored in six games by each player.

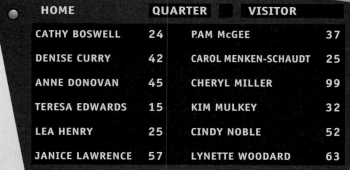

HOME	QUARTER	VISITOR	
CATHY BOSWELL	24	P.AM McGEE	37
DENISE CURRY	42	CAROL MENKEN-SCHAUDT	25
ANNE DONOVAN	45	CHERYL MILLER	99
TERESA EDWARDS	15	KIM MULKEY	32
LEA HENRY	25	CINDY NOBLE	52
JANICE LAWRENCE	57	LYNETTE WOODARD	63

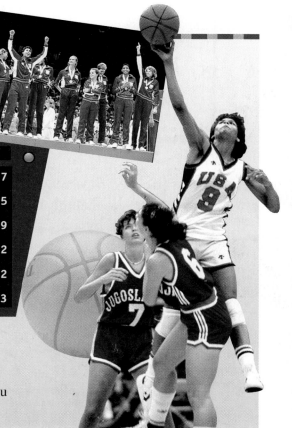

43. Find the mean of the scores.

44. Find each player's *deviation from the mean*.
Note: Deviation from the mean = Score − Mean

45. Writing What does deviation from the mean tell you about a player?

46. Writing Yael says that $|-x| - x = 0$ for all values of x. Explain why you agree or disagree.

Find the perimeter and area of each figure. *(Toolbox, page 593)*

47.

2

square

48.

2

0.8

rectangle

49.

2

circle

50.

2
1
1
3
2
3

51.

1

$\frac{5}{2}$

rectangle

52.

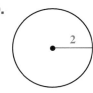

4

5

3

triangle

1.5 Exploring Negative Numbers

Learn how to...

- **multiply and divide integers**
- **use properties of addition and multiplication**

So you can...

- **solve problems involving negative numbers, such as finding very cold temperatures**

If you multiply a number by 1, nothing happens to it. But what if you multiply the number by -1? You can use the multiplicative property of -1 to explore multiplication by any negative number.

Identity Property of Multiplication

For every number *a*:

$$1 \cdot a = a$$

For example:

$$1 \cdot 23 = 23$$
$$(1)(-14) = -14$$

Multiplicative Property of -1

For every number *a*:

$$-1 \cdot a = -a$$

For example:

$$(-1)(100) = -100$$
$$(-1)(-6) = 6$$

THINK AND COMMUNICATE

For Questions 1–3, use the multiplicative property of -1 to complete each simplification. Then answer Question 4.

1. a. $(-1)(2) = \underline{\ ?\ }$

   b. $(-1)(-5) = \underline{\ ?\ }$

   c. $(-1)(-3) = \underline{\ ?\ }$

2. a. $-3 = (-1)(\underline{\ ?\ })$

   b. $-6 = (-1)(\underline{\ ?\ })$

   c. $-4 = (-1)(\underline{\ ?\ })$

3. a. $(-2)(3) = (-1)(\underline{\ ?\ })(3)$

   $\qquad = (-1)(\underline{\ ?\ })$

   $\qquad = \underline{\ ?\ }$

   b. $(-7)(4) = (-1)(\underline{\ ?\ })(\underline{\ ?\ })$

   $\qquad = (-1)(\underline{\ ?\ })$

   $\qquad = \underline{\ ?\ }$

   c. $(-3)(4) = \underline{\ ?\ }$

4. If you multiply a negative number by a positive number, is the product positive or negative? Use the multiplicative property of -1 to explain your reasoning.

For Question 5, use the multiplicative property of −1 to complete each simplification. Then answer Question 6.

5. a. $(-3)(-5) = (-1)(\underline{\,?\,})(-5)$

$= (-1)(\underline{\,?\,})$

$= \underline{\,?\,}$

b. $(-7)(-4) = (-1)(\underline{\,?\,})(\underline{\,?\,})$

$= (-1)(\underline{\,?\,})$

$= \underline{\,?\,}$

c. $(-3)(-4) = \underline{\,?\,}$

6. If you multiply two negative numbers, is the product positive or negative? Use the multiplicative property of −1 to explain your reasoning.

The rules below summarize what you saw in Think and Communicate Questions 1–6.

The product or quotient of two numbers is **positive** if **both** numbers are positive or if **both** are negative.

Same sign → Positive product

$6 \cdot 2 = 12$

$(-6)(-2) = 12$

The product or quotient of two nonzero numbers is **negative** if only **one** of the numbers is negative.

Different sign → Negative product

$(-6)(2) = -12$

$(6)(-2) = -12$

Same sign → Positive quotient

$\dfrac{6}{2} = 3 \qquad \dfrac{-6}{-2} = 3$

Different sign → Negative quotient

$\dfrac{-6}{2} = -3 \qquad \dfrac{6}{-2} = -3$

In general, for all numbers **a** and all nonzero numbers **b:**

$$\frac{-a}{b} = \frac{a}{-b} = -\frac{a}{b}$$

EXAMPLE 1

Evaluate each expression when $m = -3$.

a. $-5m + 4$

b. $\dfrac{7}{m}$

SOLUTION

a. Substitute −3 for m. Then follow the order of operations.

$-5m + 4 = (-5)(-3) + 4$

$= 15 + 4$

−5 and −3 are **both** negative, so the product is **positive**.

$= 19$

b. Substitute −3 for m.

$\dfrac{7}{m} = \dfrac{7}{-3}$

7 and −3 have **different** signs, so the quotient is **negative**.

$= -\dfrac{7}{3} \approx -2.3$

◀ **WATCH OUT!**

$-5m$ means $(-5) \cdot m$. Use parentheses when you substitute −3 for m to avoid confusion: $(-5)(-3)$.

Properties of Addition and Multiplication

There often are many ways to simplify an expression. You can use the *commutative* and *associative properties* to choose the easiest way for you.

Commutative Property

You can add or multiply numbers in any order.

Addition:	**Multiplication:**
$-3 + 7 = 7 + (-3)$	$(-3)(7) = (7)(-3)$

Associative Property

When you add or multiply three or more numbers, you can group the numbers without affecting the result.

Addition:	**Multiplication:**
$(7 + 8) + 2 = 7 + (8 + 2)$	$(7 \cdot 8) \cdot 2 = 7 \cdot (8 \cdot 2)$

EXAMPLE 2

Simplify each expression.

a. $\left(\frac{2}{3} \cdot 7\right) \cdot 3$

b. $-9 + 7 + 2 \cdot 5$

SOLUTION

a. It may be easier to find $\frac{2}{3} \cdot 3$ than to find $\frac{2}{3} \cdot 7$.

$$\left(\frac{2}{3} \cdot 7\right) \cdot 3 = \left(7 \cdot \frac{2}{3}\right) \cdot 3$$

Use the commutative property.

$$= 7 \cdot \left(\frac{2}{3} \cdot 3\right)$$

Use the associative property.

$$= 7 \cdot 2$$

$$= 14$$

b. $-9 + 7 + 2 \cdot 5 = -9 + 7 + 10$

Do the **multiplication** first.

$$= [10 + (-9)] + 7$$

$$= 1 + 7$$

$$= 8$$

Choose which numbers to add first. Brackets [ ] are grouping symbols like parentheses.

THINK AND COMMUNICATE

7. Describe at least two other ways to simplify the expression in part (b) of Example 2. Which way do you find easiest? Explain your choice.

8. Do you think division is commutative? associative? Give some examples to support your answer.

☑ CHECKING KEY CONCEPTS

Evaluate each expression when $t = -2$.

1. $3t$

2. $-1 \cdot t$

3. $\dfrac{12}{t}$

4. $\dfrac{t}{-4}$

5. $6(t + 5)$

6. $|4 \cdot 2 + 5t|$

Rewrite each fraction with at most one negative sign. Explain each answer.

7. $\dfrac{-13}{-16}$

8. $\dfrac{4}{-15}$

9. $-\left(\dfrac{-12}{-7}\right)$

Describe two ways to simplify each expression. Which do you prefer? Why?

10. $(0.12)(-3)(100)$

11. $7 \cdot 8 \cdot 0$

12. $3 + 5 - 10 + 4 - 5 - 9$

13. Open-Ended Problem Write a numerical expression that you might want to use the associative property to simplify. Simplify the expression.

1.5 **Exercises and Applications**

1. Writing If you owe a friend $2, you can think of yourself as having -2 dollars. Think of a situation when you might have $(5)(-2)$ dollars. Explain why it makes sense that $(5)(-2)$ is negative for the situation.

Extra Practice exercises on page 558

Simplify each expression.

2. $2(-5)$

3. $(-1)(4)$

4. $(-3)(3)$

5. $(-7)(-10)$

6. $\dfrac{-24}{6}$

7. $-35 \div (-5)$

8. $2 - 4 + (-2) + 4$

9. $|-18 \div 9 + 9|$

10. $\dfrac{|(4)(-3)|}{-2}$

11. Writing Peter says that $-a \cdot -b$ is always positive because the signs are the same. Explain why you agree or disagree. Use examples to support your answer.

12. Writing Explain the difference between a subtraction sign $(13 - 4)$ and a negative sign (-4).

Pure water freezes and ice melts at 32°F. When you dissolve something in the water, however, the freezing temperature of the water is lowered. Most scientists measure temperature in degrees Celsius (°C). If you are more familiar with degrees Fahrenheit (°F), you can use the conversion formula below. Use the formula to convert the temperatures in Exercises 13–15.

Let F = the temperature in degrees Fahrenheit.

$$F = \frac{9}{5}C + 32$$

Let C = the temperature in degrees Celsius.

13. During the winter, salt is used to melt ice on the streets. A solution of 30 g salt and 70 g water freezes at −21°C.

14. Because salt can harm plants, the substance $(NH_2)_2CO$ is sometimes used to melt ice. A solution of 79 g $(NH_2)_2CO$ and 100 g water freezes at −24°C.

15. Ethylene glycol is used as antifreeze in car radiators to lower the freezing point of the coolant. A solution of 1 kg ethylene glycol and 2 kg water freezes at −15°C.

16. Research Ocean water does not freeze at 32°F. At what temperature does ocean water freeze?

Evaluate each expression when $x = -6$.

17. $2x$

18. $-2x$

19. $3x - 12$

20. $16 + x - 4$

21. $-3x - 12$

22. $2 \cdot x \cdot 5$

23. $|2x|$

24. $\frac{5}{x}(-10)(12)$

25. Open-ended Problem Karen explained how she remembers how to multiply with negative numbers.

  a. Write your own positive statement that can be made negative by using the word *not*.

  b. Write your own negative statement that can be made positive by using the word *not*.

Positive statement:
If the sky is clear,
that's good. (+)

Negative statement:
If the sky is cloudy,
that's bad. (−)

If it's not clear,
that's bad.
(−) · (+) = (−)

And if it's not cloudy,
that's good.
(−) · (−) = (+)

Tell whether the expressions are *equal* or *not equal*. If they are equal, tell whether they demonstrate the *associative property*, the *commutative property*, or *both*.

26. $7 \cdot 10 \stackrel{?}{=} 10 \cdot 7$

27. $-0.2 \cdot (4 \cdot 15) \stackrel{?}{=} (-0.2 \cdot 4) \cdot 15$

28. $6 + (3 \cdot 5) \stackrel{?}{=} (6 + 3) \cdot 5$

29. $(5 + 0.8) + 0.2 \stackrel{?}{=} 5 + (0.8 + 0.2)$

30. $(20 \div 4) \div 2 \stackrel{?}{=} 20 \div (4 \div 2)$

31. $6 - 2 \stackrel{?}{=} 2 - 6$

32. $-1.7 + 13 \stackrel{?}{=} 13 + (-1.7)$

33. $(0.01)(3.42)(100) \stackrel{?}{=} (3.42)(0.01)(100)$

34. **Open-ended Problem** Write and simplify expressions to show that the associative property of addition works for positive and negative numbers.

35. **Challenge** Are the commutative and associative properties true for subtraction? Give examples to support your answers.

36. **SAT/ACT Preview** Which expression has a positive value?

   A. $-3|4 - 9|$ **B.** $3 \cdot 4 - 7 \cdot 9$ **C.** $9(-72)$

   D. $-19 + |8 - 3|$ **E.** $|12 - 6 \cdot 3|$

ONGOING ASSESSMENT

37. **Open-ended Problem** Choose the easiest way to simplify $(2)(-4.7)(-5)$ and simplify it. Explain how you made your choice. What properties did you use?

SPIRAL REVIEW

For each pair of numbers, find the greatest common factor and the least common multiple. *(Toolbox, page 584)*

38. 9 and 16   **39.** 24 and 40   **40.** 17 and 51

Write each fraction as a decimal and as a percent. *(Toolbox, page 587)*

41. $\dfrac{6}{5}$ **42.** $\dfrac{4}{3}$ **43.** $\dfrac{8}{9}$

ASSESS YOUR PROGRESS

Vocabulary

integer (p. 19)   **opposite** (p. 20)
absolute value (p. 19)

Evaluate each variable expression when $s = -5$ and $t = 7$. *(Section 1.4)*

1. $11 - s - t$ 2. $2 - (t + s)$ 3. $t - 5 + s$

4. $|s|$ 5. $|-t| - s$ 6. $20 - |t - s|$

7. **Writing** A friend tells you that $-x$ is less than x for all values of x. Explain why you agree or disagree.

Simplify each expression. *(Section 1.5)*

8. $(-7)(-6)$ 9. $(-4)(2)$ 10. $-15 \div (-3)$

Evaluate each variable expression when $x = -2$ and $y = 8$. *(Section 1.5)*

11. $x + y + 1$ 12. $y \cdot \dfrac{1}{6} \cdot x$ 13. $-2y + y + 20$

14. **Journal** Describe some real-world situations in which you might need to perform operations on negative numbers. Explain how to do the operations.

1.6

Exploring Variable Expressions

How many toothpicks do you need to make 10 triangles in a row? 100 triangles? 1,000,000? Patterns can help you answer questions like these and write variable expressions.

EXPLORATION
COOPERATIVE LEARNING

Patterns and Variable Expressions

Work with a partner.
You will need:
- about 30 toothpicks

SET UP Make as many triangles in a row as you can. Copy and extend the table. Then answer the questions below.

Number of triangles	Number of toothpicks
1	3
2	5
3	7
4	?
5	?
?	?

1 triangle

2 triangles

3 triangles

Questions

1. Describe any number patterns you see. What is the relationship between the number of triangles and the number of toothpicks?

2. How many toothpicks do you need to make 10 triangles? 100 triangles? 1,000,000 triangles? How did you get your answers?

3. How many toothpicks do you need to make *n* triangles? Write your answer as a variable expression. Compare your answer to the answers of other groups in the class.

EXAMPLE 1 Application: Personal Finance

Jason is going dancing. The dance club charges **$8** at the door and **$1.25** for each soft drink. Write a variable expression for the amount of money Jason should bring, based on the number of soft drinks he buys.

SOLUTION

Problem Solving Strategy: Identify a pattern.

Number of soft drinks Jason buys	Money needed (dollars)
0	$8 + 1.25 \cdot 0 = 8.00$
1	$8 + 1.25 \cdot 1 = 9.25$
2	$8 + 1.25 \cdot 2 = 10.50$
3	$8 + 1.25 \cdot 3 = 11.75$
4	$8 + 1.25 \cdot 4 = 13.00$
10	$8 + 1.25 \cdot 10 = 20.50$

Toolbox p. 600
Problem Solving

Step 1 Try different numbers to see if you can find a pattern.

Step 2 After a pattern appears, test a larger number.

Step 3 Use the pattern to write a variable expression.

Let d = the number of soft drinks Jason buys.

$$8 + 1.25d$$

Always tell what the variable stands for.

If Jason wants to buy d soft drinks, then he needs to bring $8 + 1.25d$ dollars.

THINK AND COMMUNICATE

1. In Example 1, suppose the dance costs $10 and each soft drink costs $1. How would the variable expression change?

EXAMPLE 2 Application: Recreation

Suppose you are a new customer at the Cactus Video Arcade. Write a variable expression for the amount you pay, based on the number of games you play.

SOLUTION

Problem Solving Strategy: Identify a pattern.

Each game costs **$.50**.

$0.50(3 - 2)$ You play **3 games** but **2 are free**.

$0.50(4 - 2)$ You play **4 games** but **2 are free**.

Let g = the number of games you play.

$0.50(g - 2)$ You play **g games** but **2 are free**.

You pay $0.50(g - 2)$ dollars to play g games.

2. Does the variable expression in Example 2 make sense if you play only one or two games? Explain.

3. a. You pay a library fine of $.05 a day every day a book is overdue. What is your fine *d* days after the book is due?

   b. Suppose you pay $.05 sales tax for every dollar you spend in a store. How much sales tax do you pay when you spend *d* dollars?

   c. How are the variable expressions in parts (a) and (b) alike? How are they different?

☑ CHECKING KEY CONCEPTS

1. In Example 2, suppose each game costs $.75 and the first five games are free. How would the variable expression change?

2. Tim buys a water pitcher for $12 and matching glasses for $2.50 each. Write a variable expression for the amount he pays, based on the number of glasses he buys.

3. Jill invites her five family members to her birthday party. She also invites her friends. Each friend brings a guest. Use this information to write a variable expression. Tell what the variable stands for. Explain what the expression tells you about the situation.

1.6 | **Exercises and Applications**

Extra Practice exercises on page 559

1. Use the pizza menu. Write a variable expression for the cost of a cheese pizza with *t* toppings. Based on your own experience, what values of *t* make sense for your variable expression?

Pizza

Plain Cheese $10.00
1 Topping. $11.50
2 Toppings $13.00

Extra Toppings $1.50 ea.

2. Visual Thinking The table below shows the height of a corn plant each day after it first appeared above ground. Write a variable expression for the height of the corn plant *d* days after it first appeared above ground.

1.5 cm	3 cm	4.5 cm	...	?
1 day	2 days	3 days	...	*d* days

Read the article about ballet shoes. Then answer Exercises 3–5.

That's a Lot of Shoes!

Ballet dancing is so strenuous that a professional ballerina's toe shoes usually wear out after about 15 minutes of performing. According to one ballet shoe manufacturer, the New York City Ballet Company orders about 50 pairs of toe shoes for each of its ballerinas four times per year. The American Ballet Theater orders about 60 pairs for each of its ballerinas twice a year.

3. Write a variable expression for the number of pairs of toe shoes the New York City Ballet Company orders in one year, based on the number of ballerinas.

4. Write a variable expression for the number of pairs of toe shoes the American Ballet Theater orders in one year, based on the number of ballerinas.

5. Writing Can you tell which company orders more toe shoes? Explain.

6. PERSONAL FINANCE Use the health club ad. Write a variable expression for the amount you would pay to join Northside Fitness, based on the number of months you belong to the health club.

7. Chris says, "I read a lot when I'm on vacation. I bet I read about two books a day. I also like to take a couple of extra books along." Use this information to write a variable expression. Tell what the variable stands for. Explain what the expression tells you about the situation.

north SIDE

The Only Health Club That Meets Your Needs!

just $**65** per month!

No initiation fee, and your first month is free!

Open-ended Problem **The same variable expression can represent many situations, depending on what the variable stands for. For example:**

Tickets cost $10. The price of n tickets is $10n$ dollars.

One van holds 10 students. The number of students who fit in n vans is $10n$.

Describe a situation that each variable expression can represent.

8. $50m$

9. $16 + d$

10. $1.50m + 3.25n$

11. $\frac{p}{25}$

12. $20 + 0.15m$

13. $0.15(m - 60)$

14. NUTRITION Calories in food come from fat, protein, and carbohydrates. There are nine calories in each gram of fat, four calories in each gram of protein, and four calories in each gram of carbohydrates.

a. Write a variable expression for the total calories in a food, based on its fat, protein, and carbohydrate content.

b. Research Find a food package that gives information about the fat, protein, and carbohydrate content of the food. Show how you can use your variable expression to find the number of calories in the food.

ROAST TURKEY

How long to cook?

You should cook a turkey 20 min for each pound. If the turkey is stuffed, add a half hour.

How many servings per pound?

For a whole turkey, allow 1.5 lb per person. This amount is roughly based on 4 oz servings of cooked turkey with some leftovers for sandwiches the next day.

For Exercises 15–17, refer to the cookbook excerpt.

15. Write a variable expression for the weight of the turkey you should buy, based on the number of people who will be eating it.

16. a. Write a variable expression for the amount of time in minutes you should cook a stuffed turkey, based on its weight.

b. When should you begin cooking a 16 lb stuffed turkey if you want to take it out of the oven at noon?

17. a. Challenge In a pressure cooker, food cooks in about one third of the time it would take otherwise. Write a variable expression for the amount of time it takes to cook a dish in a pressure cooker, based on the normal cooking time.

b. Food in a pressure cooker is usually cooked at a pressure of 15 lb/in.^{2}. You should increase the pressure by 0.5 lb/in.^{2} for every 1000 ft above sea level. What pressure should you use h ft above sea level?

ONGOING ASSESSMENT

18. Cooperative Learning Work with a partner. You each need about 30 toothpicks. One of you should answer part (a) and the other should answer part (b). Work together on part (c).

a. Write a variable expression for the number of toothpicks needed to make a row of n squares. Use a table to support your answer.

b. Write a variable expression for the number of toothpicks needed to make a row of n pentagons. Use a table to support your answer.

c. Writing Write a paragraph comparing the variable expressions you wrote for squares and for pentagons. Compare these expressions with the ones you wrote for triangles in the Exploration on page 30.

SPIRAL REVIEW

Evaluate each variable expression when $x = 4$. *(Section 1.2)*

19. $0.25x$
20. $12(3 + x)$
21. $15(2x + 5)$
22. $3x + 12$

23. $3(x + 12)$
24. $x^2 - 4x$
25. $\frac{1}{3}(9x - 9)$
26. $\frac{x + 3}{x - 3}$

1.7 Applying Variable Expressions

Learn how to...
- simplify variable expressions
- use the distributive property

So you can...
- apply variable expressions more easily in problem solving situations

The expressions $3(x + 1)$ and $3x + 3$ are called **equivalent expressions**. No matter what value you substitute for x, the expressions are equal. Algebra tiles show why the expression $3(x + 1)$ is equivalent to $3x + 3$.

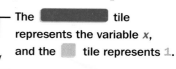 The ▮ tile represents the variable x, and the ▯ tile represents 1.

3 groups of $x + 1$ $3(x + 1)$ = $3x + 3(1)$ 3 x's and 3 ones

You can use the *distributive property* to write equivalent expressions.

Distributive Property

For all numbers a, b, and c:
$$a(b + c) = ab + ac$$

For example:
$$4(23 + 3) = 4 \cdot 23 + 4 \cdot 3$$

EXAMPLE 1

For each expression, use the distributive property to write an equivalent expression.

a. $2(2x + 5)$ **b.** $-(3t - 2)$

SOLUTION

a. $2(2x + 5) = (2 \cdot 2x) + (2 \cdot 5)$

$= 4x + 10$

Use the distributive property. Multiply both terms inside the parentheses by 2.

b. $-(3t - 2) = -1(3t - 2)$

$= -1[3t + (-2)]$

$= (-1)(3t) + (-1)(-2)$

$= -3t + 2$

Use the multiplicative property of -1.

Remember: Subtracting 2 is the same as adding -2. Use the distributive property.

Simplifying Variable Expressions

Algebra tiles can help you see how to *simplify* a variable expression.

Use tiles to represent the expression $3x + 2 + 2x + 1$.

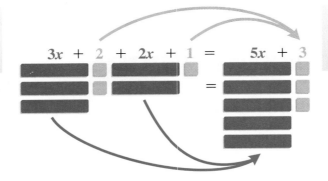

$3x + 2 + 2x + 1 = 5x + 3$

Simplify the expression by grouping the ▢ tiles and the ▬ tiles.

When you group algebra tiles, you are combining *like terms*. **Terms** are the parts of a variable expression that are added together. When terms have identical variable parts, they are called **like terms**.

$$2x + 4 + x^2 + 3x + 6x^2 - 7$$

like terms

This term is -7 because you are adding -7.

To combine like terms, add their *coefficients*. The numerical part of a variable term is called the **coefficient**. For example, the coefficient of $6x^2$ is **6** and the coefficient of x^2 is **1**.

$$x^2 + 6x^2 = 1 \cdot x^2 + 6 \cdot x^2 = (1 + 6)x^2 = 7x^2$$

An expression is in **simplest form** when there are no parentheses and all like terms are combined.

EXAMPLE 2

Simplify each variable expression by combining like terms.

a. $2x + 4 + x^2 + 3x + 6x^2 - 7$ b. $3x + 6xy + 4y$

SOLUTION

a. Group like terms.

$$2x + 4 + x^2 + 3x + 6x^2 - 7 = (2x + 3x) + (x^2 + 6x^2) + (4 - 7)$$

$$= 5x + 7x^2 + (-3)$$

$$= 5x + 7x^2 - 3$$

b. $3x + 6xy + 4y$

There are no like terms. The expression cannot be simplified.

EXAMPLE 3

Simplify the variable expression $5(2x - 4) - (8x + 3)$.

SOLUTION

Use the distributive property.

$$5(2x - 4) - (8x + 3) = 5 \cdot 2x - 5 \cdot 4 - 8x - 3$$
$$= 10x - 20 - 8x - 3$$
$$= (10x - 8x) + (-20 - 3) \quad \longleftarrow \text{Combine like terms.}$$
$$= 2x - 23$$

CHECKING KEY CONCEPTS

Write equivalent expressions for each tile diagram.

1.

2.

Simplify each variable expression.

3. $2(x - 4)$

4. $-(2t + 1)$

5. $8(5x - 2)$

6. $7 + y - 10$

7. $8 + 7x - 4 + 9 + x$

8. $x^2 + x - 2$

9. $2(x - 1) - 5$

10. $12t - (t - 1)$

11. $-5(x + 4) + 2x$

1.7 Exercises and Applications

Simplify each variable expression.

Extra Practice
exercises on
page 559

1. $2(x + y)$

2. $4(4 - 4d)$

3. $2(-5t - 7)$

4. $-8(2u + 3v)$

5. $\frac{1}{3}(-5a + 6b)$

6. $\frac{1}{2}(6x + 8y)$

7. $-6(-4 + 2y)$

8. $-6(z - 6)$

9. $-5(-3z - 4)$

10. $0.25(2x + 10)$

11. $-(1.75 - x)$

12. $-(2.5x + 8)$

13. A teacher distributes a textbook and a calculator to each algebra student. How is this situation like using the distributive property?

14. Joe used the distributive property to rewrite two variable expressions. Decide whether he rewrote the expressions correctly. If not, correct his work.

a. $-6(-3x + 10) = -18x - 60$

b. $-4(x + 4) = -4x - 16$

15. Cloth shopping bags cost $3.95. Liam used mental math to find the total cost of five bags:

$$5(3.95) = 5(4.00 - 0.05) = 20.00 - 0.25 = 19.75$$

a. **Writing** Explain how Liam used the distributive property.

b. **Open-ended Problem** Describe a situation in which you might use the distributive property to do mental math.

Simplify each variable expression.

16. $3m + 5m$

17. $10m - 7m + 4$

18. $3l - 2w$

19. $4x + 4y^2 - x$

20. $8t - 5 - 16t + 12$

21. $20 + 6b^2 - 5 - 6b^2$

22. $5x^2 - 13 + 6x + 4$

23. $-21u + 21uv - 2v$

24. $-n^2 + 2n^2 - 4n^2$

25. **FUNDRAISING** Dave is raising money in a walk for charity. His friends pay him for each mile he walks. Write two equivalent variable expressions for the amount of money Dave raises, based on the number of miles he walks. Which expression do you prefer to evaluate? Why?

Sponsor's name	Address	City, State, Zip	Telephone number	Pledge per mile	T ple
1. Vera Santos	3 River Street	Westbridge	555-8355	$1.00	
2. Bruce Simms	10 River Street	Westbridge	555-8788	$1.50	
Paul and Sarah Wu	88 Center Street	Westbridge	555-0208	$2.00	
Danny Wu	88 Center Street	Westbridge	555-0208	$1.00	
Sam Peterson	1500 8th Street	Westbridge	555-9600	$1.00	

26. Use the book sale sign at the right.

a. A student wrote the variable expression shown for the total cost of buying some hardcover books and some softcover books. Explain why this variable expression does not give the correct cost. Then write a variable expression that does give the correct cost.

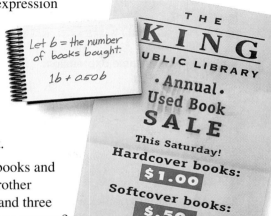

Let b = the number of books bought.

$1b + 0.50b$

THE KING UBLIC LIBRARY

· Annual ·
Used Book
SALE
This Saturday!
Hardcover books:
$1.00
Softcover books:
$.50

b. Sheila bought five softcover books and three hardcover books. Her brother bought five hardcover books and three softcover books. Who paid more money?

Simplify each variable expression.

27. $16 + 4(x + 5)$

28. $-11(2 - 2x) + 4$

29. $10 - 3(x - 1)$

30. $4 + 2(6y - 2) + 5y$

31. $2(3x - 2) + 3(5y - 1)$

32. $5 - (n + 3) + 3n$

33. $-3(6x - 6) + 2(9x - 9)$

34. $3(d + 5) - (3 - d)$

35. a. Cooperative Learning Work with a partner to evaluate the variable expression $-5(6 + 12x) + 3(20x + 10)$ when $x = 5$. One of you should evaluate the expression as is. The other should simplify the expression first. Which of you do you think did more work? Why?

b. Work together to write a variable expression with two variables that equals 0 no matter what values you choose for each variable. Simplify the expression to show why it equals 0.

36. Look back at Example 2 on page 31 in Section 1.6. Describe another way you can write the amount you pay for g games at the Cactus Arcade.

37. Challenge Evaluate each expression when $x = 1$ and when $x = -1$.

a. $x^2(x - 2)$

b. $x^3 - 2$ ←——— *Hint:* x^3 means $x \cdot x \cdot x$.

c. Writing Explain why $x^2(x - 2)$ and $x^3 - 2$ are not equivalent expressions. Support your answer with an example.

38. GEOMETRY The perimeter of a rectangle is the distance around the rectangle. You can use a variable expression to represent the perimeter.

$$l + w + l + w \quad \longleftarrow \quad \begin{array}{l} \text{Let } l = \text{length.} \\ \text{Let } w = \text{width.} \end{array}$$

a. Write two other variable expressions that are equivalent to $l + w + l + w$.

b. Measure the length and the width of a rectangular calculator. Then use the calculator to find its perimeter. Which of the three perimeter expressions describes the key sequence you used? Explain.

ONGOING ASSESSMENT

39. Writing Explain how to simplify the variable expression $10(d - 6) - 3(2d + 8)$. Explain why the original expression and the simplified expression are equivalent.

SPIRAL REVIEW

Evaluate each variable expression when $n = -5$ and $p = 10$. *(Section 1.5)*

40. $8p + 3p - 20$

41. $n - 5n$

42. $\dfrac{-3n - 2p}{5}$

43. $-4(n + 3) + 2n$

44. $2n + 2(6 - n) + 3n$

45. $-11n + 11(n + 2)$

1.8

Organizing Data

Learn how to...
- organize data using spreadsheets and matrices

So you can...
- work conveniently with large amounts of data, such as the win-loss histories of baseball teams

Are you a baseball fan? The tables below compare the win-loss history of each team in the American League East before and after the 1992 All-Star Game.

AMERICAN LEAGUE

Games Before the All-Star Game

EAST	W	L
Toronto	53	34
Baltimore	49	38
Milwaukee	45	41
Boston	42	43
New York	42	45
Detroit	41	48
Cleveland	36	52
WEST	W	

AMERICAN LEAGUE

Games After the All-Star Game

EAST	W	L
Toronto	43	32
Baltimore	40	35
Milwaukee	47	29
Boston	31	46
New York	34	41
Detroit	34	39
Cleveland	40	34
WEST	W	

Each table can be written as a *matrix*. A **matrix** is a group of numbers arranged in rows and columns. The plural of matrix is *matrices*.

Matrices are usually named by capital letters and enclosed in brackets.

$$B = \begin{bmatrix} 53 & 34 \\ 49 & 38 \\ 45 & 41 \\ 42 & 43 \\ 42 & 45 \\ 41 & 48 \\ 36 & 52 \end{bmatrix} \quad A = \begin{bmatrix} 43 & 32 \\ 40 & 35 \\ 47 & 29 \\ 31 & 46 \\ 34 & 41 \\ 34 & 39 \\ 40 & \mathbf{34} \end{bmatrix}$$

Each number is called an **element** of the matrix. This element shows Cleveland's losses after the All-Star Game.

EXAMPLE 1   Application: Sports

Find each team's total wins and losses for the season.

SOLUTION

Add matrices B and A. Add each element in B to the element in the same position in matrix A.

$$53 + 43 = 96$$

$$B + A = \begin{bmatrix} 53 & 34 \\ 49 & 38 \\ 45 & 41 \\ 42 & 43 \\ 42 & 45 \\ 41 & 48 \\ 36 & 52 \end{bmatrix} + \begin{bmatrix} 43 & 32 \\ 40 & 35 \\ 47 & 29 \\ 31 & 46 \\ 34 & 41 \\ 34 & 39 \\ 40 & 34 \end{bmatrix} = \begin{bmatrix} 96 & 66 \\ 89 & 73 \\ 92 & 70 \\ 73 & 89 \\ 76 & 86 \\ 75 & 87 \\ 76 & 86 \end{bmatrix}$$

Matrix $B + A$ has 7 rows and 2 columns. Its **dimensions** are 7×2. You can add matrices only when they have the same dimensions.

1. What is represented by the element in the top left-hand corner of matrix B? of matrix A? of matrix $B + A$?

2. Can the matrices $\begin{bmatrix} 5 & 4 \\ 9 & -1 \end{bmatrix}$ and $\begin{bmatrix} 0 & 7 & 5 \\ 3 & 6 & 1 \end{bmatrix}$ be added? If they can be added, show how. If they cannot, explain why not.

Spreadsheets

Who was the top team in the American League East at the end of the 1992 season? To find out, you can compare the winning percentages of the teams. A computer **spreadsheet** can help you perform the calculations. The data in a spreadsheet are arranged in rows and columns. The spreadsheet can also include row and column headings. Each box in the spreadsheet is called a **cell**.

For more information about spreadsheets, see the *Technology Handbook*, p. 610.

When you highlight a cell, its contents appear in the formula bar.

Wins/Losses

C2		66	
	A	**B**	**C**	**D**
1		Wins	Losses	
2	Toronto	96	66	
3	Baltimore	89	73	
4	Milwaukee	92	70	
5	Boston	73	89	

Cell C2 shows Toronto's losses for the season.

BY THE WAY

The origin of baseball is an English game called *rounders*. People in the United States played rounders as early as 1834 under the name *base ball* or *goal ball*.

EXAMPLE 2 **Application: Sports**

Use a spreadsheet to calculate the winning percentage for each team.

SOLUTION

Step 1 Enter a formula to tell the computer how to find the winning percentage for Toronto. The formula for winning percentage is:

$$\text{Winning percentage} = \frac{\text{Wins}}{\text{Wins} + \text{Losses}}$$

Put parentheses around the denominator and use / for division.

Wins/Losses

D2		=B2/(B2+C2)	
	A	**B**	**C**	**D**
1		Wins	Losses	Winning percentage
2	Toronto	96	66	0.593
3	Baltimore	89	73	
4	Milwaukee	92	70	
5	Boston	73	89	
6	New York	76	86	
7	Detroit	75	87	
8	Cleveland	76	86	

The spreadsheet finds Toronto's winning percentage:

$$\frac{96}{96 + 66} \approx 0.593$$

Solution continued on next page.

SOLUTION *continued*

Step 2 Highlight cells D2 through D8. Then use the *fill down* command.

For more information about the fill down command, see the *Technology Handbook*, p. 610.

	Wins/Losses		
D2	=B2/(B2+C2)		

	A	B	C	D
1		Wins	Losses	Winning percentage
2	Toronto	96	66	0.593
3	Baltimore	89	73	0.549
4	Milwaukee	92	70	0.568
5	Boston	73	89	0.451
6	New York	76	86	0.469
7	Detroit	75	87	0.463
8	Cleveland	76	86	0.469

The spreadsheet calculates the winning percentages for all the teams.

THINK AND COMMUNICATE

3. Which team was the American League East's top team at the end of the 1992 season? In what order should the teams be ranked? Explain.

4. In Example 2, the formula for cell D4 is B4/(B4 + C4). Compare this formula to the formula for cell D2. What does the fill down command do?

☑ CHECKING KEY CONCEPTS

For Questions 1 and 2, use matrices *A*, *B*, *C*, and *D*.

$$A = [-5 \ \ 0.5 \ \ 0] \qquad B = \begin{bmatrix} 5 \\ 3.5 \\ 2 \end{bmatrix} \qquad C = [1 \ \ 0 \ \ 0] \qquad D = \begin{bmatrix} 3 \\ 0 \\ 3 \end{bmatrix}$$

1. What are the dimensions of each matrix?

2. Add each pair of matrices. If they cannot be added, explain why.

   a. *A* and *C*   **b.** *A* and *B*   **c.** *B* and *D*

3. **Spreadsheets** Use the spreadsheet below showing heights of NBA basketball players.

   a. What formula can you enter to find Rik Smits's height in inches?

   b. Suppose you fill down to find each athlete's height in inches. Then you enter the formula (D3 + D4 + D5 + D6)/4 in cell D7. What number will you get? How is the number related to the height data?

	A	B	C	D
1		Height		
2		Feet	Inches	Height in inches
3	Rik Smits	7	4	
4	David Robinson	7	1	
5	Grant Hill	6	8	
6	Jason Kidd	6	4	

1.8 Exercises and Applications

For Exercises 1 and 2, use matrices A, B, C, and D.

Extra Practice exercises on page 559

$$A = \begin{bmatrix} 12 & 19 & 34 \\ -11 & 25 & 16 \end{bmatrix} \quad B = \begin{bmatrix} 3 & 6 & -4 \\ -4 & 5 & 9 \end{bmatrix} \quad C = \begin{bmatrix} 1 & 4 \\ 9 & 3 \\ -5 & 7 \end{bmatrix} \quad D = \begin{bmatrix} 3 & 4 \\ 2 & -1 \\ 5 & -8 \end{bmatrix}$$

1. Give the dimensions of each matrix.

2. Add each pair of matrices. If the matrices cannot be added, explain why.

   a. C and D **b.** A and B **c.** A and C

3. **Open-ended Problem** Give an example of two matrices that cannot be added. What are the dimensions of each matrix?

Spreadsheets A company asks its traveling salespeople to keep a monthly record of mileage and gasoline expenses. For Exercises 4–6, use the monthly records shown in the spreadsheet below.

Monthly Travel Expenses							
D3		=C3/B3					
A	**B**	**C**	**D**	**E**	**F**	**G**	**H**
		June Travel				July Travel	
Name	Mileage	Gas expenses (dollars)	Cost per mile		Mileage	Gas expenses (dollars)	Cost per mile
Milner	3000	120	0.04		2600	110	
Dubois	2000	90	0.05		2200	100	
Perez	2600	110	0.04		2800	120	
Chen	2400	100	0.04		1800	75	

4. Write a matrix showing the mileage and gas expenses for each time period. Give the dimensions of each matrix.

   a. June   **b.** July   **c.** total for June and July

5. **a.** The sales manager of the company used the spreadsheet to find the cost per mile for each salesperson during the month of June. Explain what the sales manager did.

   b. Use a spreadsheet to find the cost per mile for each salesperson during the month of July.

6. Describe formulas you could use to find the mean mileage and the mean gas expenses for June.

7. **Technology** You can use a graphing calculator to add matrices. Before you enter a matrix, you need to give its dimensions and name it with a letter.

   a. Give the dimensions of matrices A and B.

   b. Use a graphing calculator to add matrices A and B. If you do not have a graphing calculator, add them by hand.

$$A = \begin{bmatrix} 2 & 4.5 \\ 0 & 9.1 \\ 40 & 7 \end{bmatrix} \quad B = \begin{bmatrix} -5.4 & 37 \\ -4.8 & 16.6 \\ -4.9 & 23 \end{bmatrix}$$

For more information on adding matrices with a graphing calculator, see the *Technology Handbook*, p. 608.

Look back at the article on pages xxx–2.

One of the statistics Carrie Teegardin looked at in her report on "Growing Up Southern" was teenage employment. Teens who are working or looking for jobs are part of the work force.

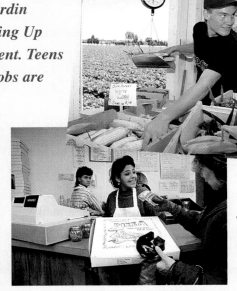

Number of 16- and 17-year-olds in the Work Force (millions)

Year	Males	Females	Total
1960	1.3	0.8	2.1
1970	1.8	1.3	3.1
1980	2.1	1.8	3.9
1990	1.5	1.4	2.9

Total Number of 16- and 17-year-olds (millions)

Year	Males	Females	Total
1960	2.9	2.8	5.7
1970	3.9	3.8	7.7
1980	4.3	4.1	8.4
1990	3.4	3.3	6.7

In 1990, about 44% of working teens in the United States held jobs in wholesale or retail trade.

8. In which year was the number of males in the work force the highest? In which year was the number of females in the work force the highest?

9. Describe in general what happened to the number of males in the work force and the number of females in the work force from 1960 to 1990.

10. **Spreadsheets** What percent of 16- and 17-year-olds were in the work force during each of the four years? Describe how to use a spreadsheet to make the calculations.

11. Look at the numbers of 16- and 17-year-olds in the work force from year to year and the percents in the work force from year to year. Which figures do you think give you a better sense of the changes in the teen work force from 1960 to 1990? Explain.

12. **Spreadsheets** Find the percent of males in the work force and the percent of females in the work force for the years 1960, 1970, 1980, and 1990. Organize your results in a table.

13. **Writing** Look back at what Carrie Teegardin said about using percents in her work. How does your analysis of teens in the work force support her statements?

14. **Challenge** Write a matrix for each table. Write a matrix showing the number of 16- and 17-year-olds who were not in the work force from 1960 to 1990. How did you get your answer? How can you use matrix addition to check your answer?

The Atlanta Journal / THE ATLANTA CONSTITUTION

SUNDAY, JUNE 6, 1993

Hutchison win in Texas is big loss for Clinton

Republican ran against budget plan

Growing up Southern SPECIAL REPORT

For most kids, ...'s getting better

Source: 1990 census

Teens working

Southern teens between 16 and 19 are less likely to be working than teens outside the South.

SOUTH
▶ 16 to 19-year olds working or looking for jobs: **48.2%**
▶ Unemployment rate: **20.9%**

NON-SOUTH
▶ 16 to 19-year olds working or looking for jobs: **52.7%**
▶ Percent of the population actually employed: **45.2%**
▶ Unemployment rate: **18.0%**

Minority teens have a much ...

Teegardin used data from the 1990 Census to support her report.

15. Writing Describe a situation in which you might want to add matrices. Describe a situation in which you might want to use a spreadsheet to perform calculations.

Evaluate each expression when $a = 4$, $b = \frac{1}{2}$, and $c = -3$. *(Section 1.5)*

16. $2c + a$ **17.** $6b + c$ **18.** ab

19. $a - c$ **20.** $-c$ **21.** $0.25a - b$

ASSESS YOUR PROGRESS

VOCABULARY

equivalent expressions (p. 35)   **matrix** (p. 40)
terms (p. 36)   **element** (p. 40)
like terms (p. 36)   **dimensions** (p. 40)
coefficient (p. 36)   **spreadsheet** (p. 41)
simplest form (p. 36)

For Exercises 1 and 2, write a variable expression. *(Section 1.6)*

1. the number of minutes in h hours

2. the current mileage of a car that had m mi before a 250 mi road trip

Simplify each variable expression. *(Section 1.7)*

3. $3(x - 2)$ **4.** $5x^2 + 5y - 3y - x$ **5.** $2\left(x - \frac{1}{2}\right) - x$

6. Give the dimensions of each matrix. Tell which matrices can be added. Then add them. *(Section 1.8)*

$$X = \begin{bmatrix} 5 & 0.25 & -2 \\ 4 & 7 & 0 \\ 0.5 & -1 & 7 \end{bmatrix} \quad Y = \begin{bmatrix} 1.6 & 2 & 0 \\ -6 & 11 & 3 \\ 4.5 & 4 & 4 \end{bmatrix} \quad Z = \begin{bmatrix} -2 & -2 \\ -1.4 & 2.5 \end{bmatrix}$$

7. **Spreadsheets** Use the spreadsheet below. What formula can you enter to find the total number of T-shirts sold from January through April? *(Section 1.8)*

	A	B	C	D
1		T-shirts sold	Sweatshirts sold	Polo shirts sold
2	January	96	129	51
3	February	88	108	46
4	March	72	115	62
5	April	104	92	57

8. Journal Describe some advantages of using a spreadsheet to organize data. How does knowing about variable expressions help you work with spreadsheet formulas?

PORTFOLIO PROJECT

Analyzing Change

Would your school feel more crowded if enrollment increased by 50 students? It depends on the size of your school. If your school has 1000 students, enrollment would increase by 5%. If your school has 100 students, enrollment would increase by 50%. Knowing a percent change often gives you more information than a numerical change.

PROJECT GOAL Your goal is to investigate the percent change in something that interests you.

Collecting Data

Work with a partner. Decide on a topic for your investigation. You can collect data yourself, or you can find statistics in a newspaper, a magazine, or an almanac. Your data should include at least four numerical values that change from one time to another.

Some possibilities are:

- the populations of your state and four nearby states in two different years

- the number of cars passing through a green light in the morning and afternoon at various intersections

Use a table or spreadsheet to record and analyze your data. For example, suppose you want to explore a surge in the popularity of in-line skating.

1. USE A SPREADSHEET to record the number of in-line skaters in four regions of the United States for two different years.

2. USE A FORMULA to calculate the percent change from 1992 to 1993 in the Northeast.

$$\text{Percent change} = \frac{\text{New value} - \text{Old value}}{\text{Old value}} \cdot 100$$

Edit   Format

In-Line Skating

D2		=((C2–B2)/B2)*100		
	A	**B**	**C**	**D**
1		1992	1993	Percent change
2	Northeast	1,720,000	2,889,000	67.97
3	North central	2,646,000	3,963,000	49.77
4	South	2,306,000	2,899,000	25.72
5	West	2,692,000	2,808,000	4.31
6				
7				
8				

3. FILL DOWN to find the percent change in the other three regions of the country.

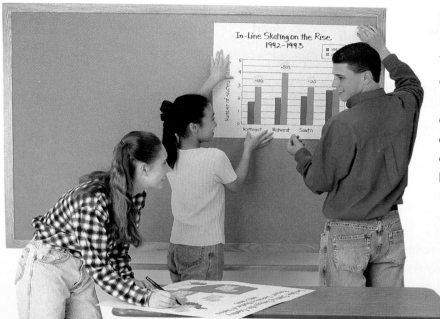

Making a Graph

Present your results visually. Your display should make it clear which quantities increased and which decreased. You can choose any of these types of displays, or you may have an idea of your own.

- a bar graph
- a colored map
- a histogram

Presenting Your Analysis

Write a report, make a poster, prepare a video, or give a presentation explaining your results.

Remember to:

- Include your original data. Explain how you measured and recorded the data.

- Describe how you did your calculations. (If you used a spreadsheet, include a copy and explain how you used it.)

- Explain your visual display.

- Explain the difference between a positive percent change and a negative percent change.

- Try to draw conclusions from the percent change or from the original data.

- Discuss reasons why these changes may have occurred.

Self-Assessment

What have you and your partner learned about percent change? Describe what you think went well in your project. Describe what you think did not work as smoothly as you would have liked. If you could do the project again, what would you do differently? Why?

1 Review

There are many ways to review what you have learned. Complete the following list of study ideas with some that have worked well for you.

To study for a math test, I usually...

- **Make an outline of the chapter with worked out sample problems.**
- **Redo the homework problems that were difficult for me.**
- **...**

Compare ideas with other students. As you review this and other chapters, you may want to try new ways of studying.

VOCABULARY

data (p. 3)
variable (p. 3)
value of a variable (p. 3)
variable expression (p. 9)
evaluate (p. 9)
exponent (p. 10)
mean (p. 14)
median (p. 14)
mode (p. 14)
integer (p. 19)
absolute value (p. 19)

opposite (p. 20)
equivalent expressions (p. 35)
terms (p. 36)
like terms (p. 36)
coefficient (p. 36)
simplest form (p. 36)
matrix, matrices (p. 40)
element (p. 40)
dimensions (p. 40)
spreadsheet (p. 41)

SECTIONS | 1.1, 1.2, *and* 1.3

To **evaluate** a **variable expression**, substitute a value for each **variable**. Then simplify the numerical expression using the order of operations.

Evaluate the expression $3(a + 2) - 4b + 6$ when $a = 5$ and $b = 2$.

1. Do all work inside parentheses.
2. Then do all multiplications and divisions from left to right.
3. Then do all additions and subtractions from left to right.

$$3(a + 2) - 4b + 6 = 3(5 + 2) - 4 \cdot 2 + 6$$
$$= 3 \cdot 7 - 4 \cdot 2 + 6$$
$$= 21 - 8 + 6$$
$$= 19$$

There are three kinds of averages, as shown in the example below.

The **modes** are **9** and **18** because they are the most frequent values.

$$9 \ 9 \ 10 \ 11 \ 13 \ \underset{\downarrow}{14} \ \underset{\downarrow}{16} \ 17 \ 18 \ 18 \ 19 \ 22$$
$$15$$

There is an even number of values, so the **median** is the mean of the two middle values.

The **mean** is the sum of the data items divided by the number of data items.

$$\frac{9 + 9 + 10 + 11 + 13 + 14 + 16 + 17 + 18 + 18 + 19 + 22}{12} \approx 14.67$$

SECTIONS 1.4 *and* 1.5

Subtracting a number is the same as adding its **opposite**.

$$2 - 3 = 2 + (-3) = -1$$

You can use these rules to multiply and divide negative numbers:

Same sign $\rightarrow$ **Positive**

$3 \cdot 9 = 27$ and $(-5)(-4) = 20$

$\dfrac{16}{2} = 8$ and $\dfrac{-15}{-5} = 3$

Different sign $\rightarrow$ **Negative**

$(-3)(9) = -27$ and $(5)(-4) = -20$

$\dfrac{-16}{2} = -8$ and $\dfrac{15}{-5} = -3$

SECTIONS $1.6, 1.7,$ *and* 1.8

You can use patterns to write variable expressions.

To simplify a variable expression:
1. Use the distributive property.
2. Combine like terms.

$$2(5x + 3) - 2x = 10x + 6 - 2x$$
$$= 10x - 2x + 6$$
$$= 8x + 6$$

A **matrix** is a group of numbers arranged in rows and columns. You can add matrices that have the same **dimensions**. The dimensions of each of these matrices are 3×2.

$$\begin{bmatrix} 4 & 9 \\ 7 & 1 \\ 0 & -3 \end{bmatrix} + \begin{bmatrix} 1 & 2 \\ 2 & 1 \\ 6 & 3 \end{bmatrix} = \begin{bmatrix} 5 & 11 \\ 9 & 2 \\ 6 & 0 \end{bmatrix}$$

Computer **spreadsheets** contain data arranged in rows and columns. Spreadsheets are useful for performing calculations within large sets of data.

	Miles Traveled		
	A	B	C
1		Week 1	Week 2
2	Alexa Maglio	75	160
3	Rick Steele	130	95

Assessment

VOCABULARY QUESTIONS

1. Give an example of a variable expression that is not in simplest form. Then write the same expression in simplest form.

2. Explain the meaning of *absolute value*.

3. Explain the difference between finding the mean and finding the median of a data set.

SECTIONS 1.1, 1.2, *and* 1.3

For Questions 4–6, use the histogram. Let g = each student's test grade.

4. For how many students is $g \geq 91$?

5. How many students are represented?

6. Write two inequalities that describe the grades of the students.

Evaluate each variable expression when $a = 3$, $b = 4$, and $c = 5$.

7. $4a + 3b$ **8.** $6a - 2c$ **9.** $\dfrac{a^2 - b}{c}$

10. a. The ages of students in a college class are 18, 19, 17, 22, 19, 19, 19, 18, 23, 22. Find the mean, the median, and the mode(s) of the ages.

  b. Writing Which of the averages in part (a) will be affected if a 68-year-old person joins the class? Explain.

SECTIONS 1.4 *and* 1.5

Simplify each expression.

11. $7 - 10$ **12.** $(-3) - (-4)$ **13.** $(-3) + (-4)$ **14.** $(-6)(-5)$

15. $-15 \div 3$ **16.** $|-2| + |3|$ **17.** $|-2 + 3|$ **18.** $-12 + \dfrac{|-3|}{3(3)}$

Evaluate each variable expression when $x = -5$ and $y = 2$.

19. $x + 7y + 2x - 5y$ **20.** $-8y + \dfrac{20}{x}$ **21.** $y^2 + xy - 4$

22. There are 24 h in one day. Write a variable expression for the number of hours in d days. Use your variable expression to find the number of hours in a week.

23. John is treating his friends to lunch. He pays $2.25 for each sandwich and $.65 for each soft drink. Write a variable expression for John's cost if he and his friends each have a sandwich and a soft drink. What is John's cost if he and 7 friends have lunch?

Simplify each variable expression.

24. $2x - 6 + 9x + 17$

25. $5 + 6x + 4y - x + 3y$

26. $-3(2x - y) + 2(3x - y)$

27. $4x^2 - 2x - 10x^2 + 8x$

28. A secretary used a spreadsheet to find the cost of ordering some supplies, as shown. What information is contained in columns B, C, and D? Explain what spreadsheet formula was used for cell D2.

	Office Supplies			
	A	**B**	**C**	**D**
1	Item	Qty	Price	Cost
2	Paper	6	4.40	26.40
3	Pencils	25	0.69	
4	Envelopes	2	22.90	
5	Disks	5	19.10	

For Questions 29 and 30, use matrices *A*, *B*, *C*, and *D*.

$$A = \begin{bmatrix} 5 & 6 \\ 2 & 8 \\ 7 & 4 \end{bmatrix} \quad B = \begin{bmatrix} 4 & -3 \\ 7 & 0.5 \end{bmatrix} \quad C = \begin{bmatrix} 3 \\ 6 \\ -4 \end{bmatrix} \quad D = \begin{bmatrix} -3 & 2 \\ 5 & -8 \\ -7 & 2 \end{bmatrix}$$

29. What are the dimensions of each matrix?

30. Tell which matrices can be added. Then add them.

PERFORMANCE TASK

31. Work in a group of 2–4 students. Follow the steps below to create a game using variable expressions.

Each group should make a set of 20 cards. Half of the cards contain variable expressions. The other half describe matching situations.

Exchange cards with another group. Shuffle the cards and place them face down. Turn two cards face up. If they are a matching pair, take both cards and go again. If they don't match, put them back and the next person takes a turn. The person with the most pairs of cards wins.

Give comments and suggestions for improvement to the group whose game you played. Make changes to your game if needed.

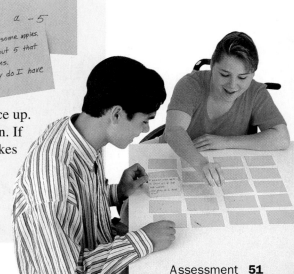

$a - 5$

I picked some apples.
I threw out 5 that
had worms.
How many do I have
now?

2 Equations and Functions

Life in the fast lane

INTERVIEW **Bob Hall**

*B*ob Hall is not slowed down because he uses a wheelchair. In 1975, he became the first wheelchair athlete to win the Boston Marathon. He won again in 1977, setting national records in both races. Hall is also a wheelchair designer. His designs have helped transform the sport of wheelchair racing. "I always thought wheelchairs were the most disabling aspect of being disabled," he says. "They weren't made for speed or comfort. They were made for keeping you still. But I like to keep moving. I needed a chair that would let me do that."

"**All of a sudden, you can take a negative and turn it into a positive.**"

The Sky's the Limit

When Hall first began racing, wheelchairs weighed as much as 50 pounds. In 1980, he designed a wheelchair for himself that weighed only 22 pounds. Today Hall owns a company that makes advanced racing wheelchairs for wheelchair athletes all over the world. As chairs have become faster and safer, the races have grown more competitive. "A whole group of people who didn't have an outlet in sports can now compete on an international basis," Hall says.

Revolutionizing a Sport

Probably the most important advances in wheelchair racing come from weight reduction. Through the use of lightweight materials like titanium, Hall's competition chairs now weigh as little as 12.5 pounds. Racers sit lower in their chairs to reduce wind resistance.

Hall also invented a new steering system that makes turning easier. To change course, racers push a "handlebar" connected to the front wheel. After completing the turn, they let go. A powerful spring automatically snaps the wheel back to a centered position. This mechanism keeps the chair straight, so that racers can push with both arms and not worry about steering.

"With the evolution of equipment, suddenly the sky's the limit!"

In 1975, Bob Hall was the only wheelchair entrant in the Boston Marathon. The evolution of equipment has made the race more competitive. In 1994, 90 men and women participated in the race's wheelchair division.

Turn Angle and Turning Radius

Racers learn to judge just how much the wheel has to be nudged to make the desired turn. However, mathematics can also provide the answer. There is a precise relationship between the number of degrees the wheel is turned and the turning radius. (The turning radius is the radius of the circular path the wheelchair follows when you turn it.)

When the wheel is turned by one degree, the turning radius is 206 feet. In other words, the wheelchair would need a road wider than the width of a football field to make a U-turn. When the wheel is turned by 15 degrees, the turning radius is 14 feet. The wheelchair could make a U-turn in a small garage. The relationship between the number of degrees and the turning radius is called a *function*.

Turn angle

Turn angle	1°	2°	3°	6°	9°	12°	15°
Turning radius	206 ft	103 ft	69 ft	35 ft	23 ft	17 ft	14 ft

EXPLORE AND CONNECT

Bob Hall is shown here in his workshop, modifying one of his racing wheelchairs.

1. Research Find out more about wheelchair sports. For example, how are wheelchairs designed for tennis and basketball players different from racing wheelchairs? Why do the chairs have different designs?

2. Writing Does it make sense to talk about the turning radius of the chair when the turn angle is 0°? Explain.

3. Project Visit a race track in your town and measure its dimensions. Draw a sketch showing your measurements. What turn angle do you think Bob Hall would have to use to round a turn on this track? Explain your thinking.

Mathematics & Bob Hall

In this chapter, you will learn more about how mathematics is related to the sport of wheelchair racing.

Related Examples and Exercises

Section 2.1
• **Example 4**
• **Exercises 31 and 32**

Section 2.5
• **Exercises 11–14**

2.1 Solving One-Step Equations

Learn how to...
- **write and solve one-step equations**

So you can...
- **solve real-world problems about distance, rate, and time, for example**

It's not every day you see algebra in a newspaper headline. As you can see from this article in the *New York Times*, zoo officials used an *equation* to find the weight of a stubborn baby elephant.

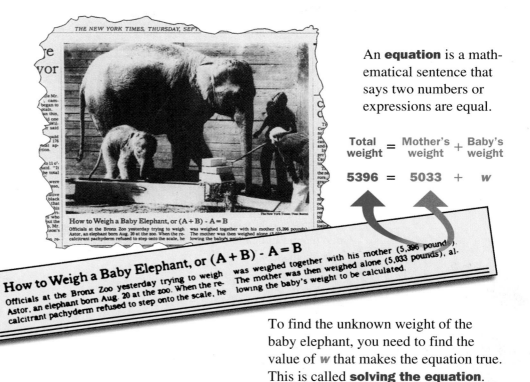

THE NEW YORK TIMES, THURSDAY, SEPT

How to Weigh a Baby Elephant, or (A + B) - A = B
Officials at the Bronx Zoo yesterday trying to weigh Astor, an elephant born Aug. 20 at the zoo. When the recalcitrant pachyderm refused to step onto the scale, he was weighed together with his mother (5,396 pounds). The mother was then weighed alone (5,033 pounds), allowing the baby's weight to be calculated.

An **equation** is a mathematical sentence that says two numbers or expressions are equal.

Total weight	=	Mother's weight	+	Baby's weight
5396	=	5033	+	*w*

To find the unknown weight of the baby elephant, you need to find the value of *w* that makes the equation true. This is called **solving the equation**.

EXAMPLE 1   Application: Zoology

Find the weight of the baby elephant.

SOLUTION

Undo the addition. Subtract 5033 from both sides of the equation.

Solve the equation $5396 = 5033 + w$.

$$
\begin{aligned}
5396 &= 5033 + w \\
-5033 &\quad -5033 \\
\hline
363 &= \qquad\quad w
\end{aligned}
$$

The weight of the baby elephant is 363 lb.

Check
Substitute 363 for w.
$5396 = 5033 + w$
$5396 \overset{?}{=} 5033 + 363$
$5396 = 5396$ ✔

Inverse Operations

Any value of a variable that makes an equation true is called a **solution** of the equation. To find a solution, you need to get the variable alone on one side of the equation. You can do this by using *inverse operations*. Addition and subtraction are called **inverse operations** because they undo each other. Multiplication and division are also inverse operations.

EXAMPLE 2

Solve each equation.

a. $\dfrac{t}{60} = 12.5$ b. $-15x = -37$

SOLUTION

a. $$\dfrac{t}{60} = 12.5$$

Undo the division. Multiply both sides of the equation by **60**.

$$60\left(\dfrac{t}{60}\right) = 60(12.5)$$

$$\dfrac{60}{60} \cdot t = 60(12.5)$$

$$t = 750$$

Check
Substitute **750** for t.
$$\dfrac{t}{60} = 12.5$$
$$\dfrac{750}{60} \overset{?}{=} 12.5$$
$$12.5 = 12.5 \;\checkmark$$

b. $$-15x = -37$$

Undo the multiplication. Divide both sides by **–15**.

$$\dfrac{-15x}{-15} = \dfrac{-37}{-15}$$

$$x = \dfrac{37}{15}$$

or $x \approx 2.47$

Check
Substitute $\dfrac{37}{15}$ for x.
$$-15x = -37$$
$$-15\left(\dfrac{37}{15}\right) \overset{?}{=} -37$$
$$-37 = -37 \;\checkmark$$

THINK AND COMMUNICATE

1. What inverse operations were used to solve the equation in Example 1 and the equations in Example 2?

2. Explain why you divide by -15 in the solution of part (b) of Example 2.

3. The solution of part (b) of Example 2 says that $x \approx 2.47$. If you try to check this solution, you get $-15(2.47) = -37.05$. Does this mean the solution is incorrect? Explain your reasoning.

Sometimes it helps to think about opposites when you solve equations. Remember that $-x$ means "the opposite of x."

$$-x = 8 \quad \Longleftarrow \quad \text{The opposite of } x \text{ is 8.}$$

$$x = -8 \quad \Longleftarrow \quad \text{So } x \text{ is the opposite of 8.}$$

EXAMPLE 3

Solve the equation $-x - 13 = 9$.

SOLUTION

Undo the subtraction.
Add **13** to both sides.

$$
\begin{array}{rcr}
-x - 13 &=& 9 \\
+ 13 && + 13 \\
\hline
-x &=& 22
\end{array}
$$

$$x = -22$$

The opposite of x is 22, so $x = -22$.

Check

Substitute -22 for x.

$$-x - 13 = 9$$
$$-(-22) - 13 \stackrel{?}{=} 9$$
$$22 - 13 \stackrel{?}{=} 9$$
$$9 = 9 \checkmark$$

EXAMPLE 4 Interview: Bob Hall

In 1975, Bob Hall became the first wheelchair athlete to win the Boston Marathon. He finished the 26.2 mi race in about 178 min. Use this formula to find his average speed, or rate, in miles per minute:

$$\textbf{Distance} = \textbf{Rate} \times \textbf{Time}$$
$$d = rt$$

SOLUTION

The distance *d* is 26.2 mi. The time *t* is 178 min. Find the rate *r*.

Substitute the values you know in the formula.

$$d = rt$$
$$26.2 = r \cdot 178$$
$$\frac{26.2}{178} = \frac{178r}{178}$$

Divide both sides by **178**. Round to the nearest hundredth.

$$0.15 \approx r$$

Bob Hall's average speed was about 0.15 mi/min.

BY THE WAY

The back wheels of a racing wheelchair are tilted in at the top. This makes them easier for the rider to reach and also makes the chair more stable.

☑ CHECKING KEY CONCEPTS

Solve each equation.

1. $27 + x = 4$ **2.** $8 = y - 4$ **3.** $v + 0.25 = 1$ **4.** $s - \frac{1}{3} = \frac{2}{3}$

5. $16 = -m$ **6.** $-r = -10$ **7.** $28 = \frac{t}{5}$ **8.** $\frac{m}{3} = 11$

9. $-60 = 6t$ **10.** $-2p = 18$ **11.** $13m = 53.3$ **12.** $1.5 = -6y$

13. **SPORTS**   At the 1994 winter Olympics in Lillehammer, Norway, Lee-Kyung Chun of South Korea won the women's 1000 m short-track speed skating race with a time of about 97 seconds. Find her average speed in meters per second.

2.1 **Exercises and Applications**

Extra Practice exercises on page 559

Solve each equation.

1. $520 + m = 105$ **2.** $-21 = p - 13$ **3.** $-0.5 + n = 6$

4. $-8 = 14 - p$ **5.** $-c + \frac{1}{8} = \frac{1}{2}$ **6.** $-6 - v = 3\frac{1}{2}$

7. $-10 = x - 5$ **8.** $-7 = -x$ **9.** $-s = 0$

10. $10 = \frac{n}{5}$ **11.** $0 = \frac{t}{16}$ **12.** $\frac{n}{12} = -12$

13. $16x = 100$ **14.** $3n = -27$ **15.** $3 = -6h$

16. $3x = 7.5$ **17.** $13.5 = -3z$ **18.** $\frac{x}{2} = 2.5$

19. **Challenge**   Solve for x. Tell what inverse operations you used.

    a. $\sqrt{x} = 4$ **b.** $x^2 = 9$

LANGUAGE ARTS   The sentences in Exercises 20–22 are like equations because only certain names make them true. Match each name with a sentence.

    A. Mae Jemison   **B.** Daniel Galvez   **C.** Bob Hall

(*Hint:* Each person is described in an interview in one of the chapters of this book.)

20. $\underline{?}$ invented a steering mechanism for racing wheelchairs.

21. $\underline{?}$ is the first African-American woman astronaut.

22. $\underline{?}$ is an artist who painted a mural called *Crosswinds*.

23. **Open-ended Problem**   Write a word sentence like the ones in Exercises 20–22.

24. **SAT/ACT Preview**   How many positive integers less than 20 are equal to 3 times an even integer?

    A. one   **B.** two   **C.** three   **D.** four   **E.** five

On May 21, 1932, Amelia Earhart became the first woman to complete a solo west-east flight across the Atlantic Ocean. On September 5, 1936, Beryl Markham became the first woman to complete a solo east-west flight across the Atlantic. The map shows their routes.

Amelia Earhart — 2026 miles

Beryl Markham — 2212 miles

Baleine, Nova Scotia

Harbor Grace, Newfoundland

Londonderry, Northern Ireland

Abingdon, England

25. Amelia Earhart made her flight in about 14.9 h. Find her average speed in miles per hour.

26. Beryl Markham made her flight in about 21.6 h. Find her average speed in miles per hour.

27. About 100 acres of pizza are eaten each day in the United States. San Francisco, California, has an area of about 29,887 acres. Write and solve an equation to find how many days it would take people in the United States to eat a pizza the size of San Francisco.

28. MEDICINE Health officials predict that by the year 2020, there will be 292,000 women doctors in the United States. These women will represent one third of all doctors. Which equation can you use to find the total number of doctors d in 2020? Explain your choice and solve the equation.

    A. $3d = 292{,}000$ **B.** $\frac{292{,}000}{3} = d$ **C.** $\frac{d}{3} = 292{,}000$

29. HISTORY Covered wagons traveled about 15 miles per day on the Oregon Trail. Which equation can you use to find about how many days d the 2000 mi journey took? Explain your choice and solve the equation.

    A. $15d = 2000$ **B.** $\frac{d}{15} = 2000$ **C.** $d + 15 = 2000$

30. Investigation How thick is a piece of paper? Make a stack of 100 pieces of paper. Then measure the height h of the stack. Use the value you find for h and one of the formulas below to find the approximate thickness t of one piece of paper. Explain why you chose this formula.

    A. $\frac{t}{100} = h$ **B.** $100t = h$ **C.** $100h = t$

Look back at the
article on pages
52–54.

*In 1975, Bob Hall reached the finish line
of the Boston Marathon in the wheelchair
shown at the right. His average speed was
about 0.15 mi/min. Today Hall uses a racing
wheelchair that he designed and built himself.*

31. In 1990, Bob Hall finished a marathon (26.2 mi) in a little over
117 min. He used a racing wheelchair like the one shown at the left.
Find his average speed in miles per minute.

32. Writing Compare Bob Hall's average speed in the 1990
marathon to his average speed in the 1975 marathon.
Write a paragraph describing how improvements in
wheelchair design may have affected Hall's speed.

ONGOING ASSESSMENT

33. a. Open-ended Problem Write three word problems that can be
solved using the formula $d = rt$. Write one problem that asks for a rate,
one that asks for a time, and one that asks for a distance. Explain how
solving for a distance is different than solving for a rate or a time.

b. Cooperative Learning Work with several other students to read
and evaluate each other's problems from part (a). Which are the most
interesting? Which are the most challenging? Why? Based on your
conclusions, rewrite your problems to make them more interesting and
more challenging.

SPIRAL REVIEW

Evaluate each expression when $p = -3$. *(Sections 1.4 and 1.5)*

34. $6p + 5$

35. $2 - 8p$

36. $\dfrac{p}{10 + 1}$

37. $\dfrac{|p + 1|}{3}$

38. $|-p + p|$

39. $|-p| + |p|$

**GEOMETRY Write a variable expression for the perimeter of each figure.
Evaluate each expression when $x = 10$.** *(Section 1.6)*

40.

41.

2.2

Solving Two-Step Equations

Learn how to...

- write and solve two-step equations

So you can...

- make purchasing decisions in real-world situations, such as at a state fair
- solve geometry problems, such as finding a perimeter

Sometimes it takes more than one step to solve an equation. You may need to add or subtract, and then multiply or divide. Algebra tiles help you see the steps involved.

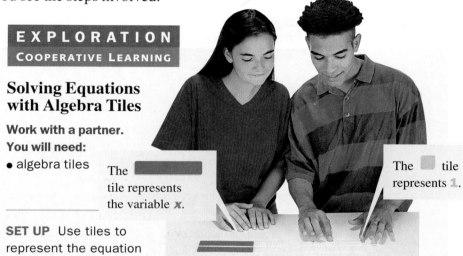

EXPLORATION
COOPERATIVE LEARNING

Solving Equations with Algebra Tiles

Work with a partner.
You will need:
- algebra tiles

The ▬ tile represents the variable x.

The ▢ tile represents 1.

SET UP Use tiles to represent the equation $2x + 3 = 11$.

1 **Subtract 3** from both sides of the equation.

2 **Divide** both sides of the equation into **two** equal parts.

You can see that each ▬ tile equals four ▢ tiles. So $x = 4$.

Show how to solve each equation using algebra tiles.

1. $5x + 2 = 7$ **2.** $2x + 5 = 15$ **3.** $3x + 5 = 11$ **4.** $13 = 4 + 3x$

EXAMPLE 1

Solve the equation 8 + 2x = 14.

SOLUTION

First undo the
addition. Subtract
8 from both sides.

$$8 + 2x = 14$$
$$\underline{-8 \qquad -8}$$
$$2x = 6$$

$$2x = 6$$

Then undo the
multiplication.
Divide both sides
by **2**.

$$\frac{2x}{2} = \frac{6}{2}$$

$$x = 3$$

Example 2 shows how to solve a two-step equation using symbols instead of
algebra tiles. This method is called *solving algebraically*.

EXAMPLE 2

Solve the equation $\frac{x}{5} - 7 = -3$.

SOLUTION

Undo the subtraction.
Add **7** to both sides.

$$\frac{x}{5} - 7 = -3$$
$$\underline{+7 \qquad +7}$$
$$\frac{x}{5} = 4$$

Undo the division.
Multiply both sides
by **5**.

$$5\left(\frac{x}{5}\right) = 5(4)$$
$$x = 20$$

Check
Substitute 20 for x.

$$\frac{x}{5} - 7 = -3$$
$$\frac{20}{5} - 7 \stackrel{?}{=} -3$$
$$4 - 7 \stackrel{?}{=} -3$$
$$-3 = -3 \checkmark$$

THINK AND COMMUNICATE

1. Describe the steps you would use to solve the equation $8 - 2x = 14$.

2. Does it make sense to solve Example 2 with algebra tiles? Why or
why not?

3. Jean tried to solve the equation $2x + 3 = 4$, as shown at the left.
What did she do wrong? How would you solve the equation?

EXAMPLE 3   Application: Personal Finance

The state fair charges $6 admission plus $2.25 for each ride. Surajit has $25 to spend. How many rides can he take?

SOLUTION

Problem Solving Strategy: Use an equation.

Let r = the number of rides Surajit can take.

Admission		Price per ride		Number of rides		Amount Surajit has to spend
$6	+	$2.25	·	r	=	$25

How many rides can Surajit take with $25?
Solve the equation for r.

$$6 + 2.25r = 25$$

Subtract **6** from both sides of the equation.

$$\underline{-\ 6 \qquad\qquad -\ 6}$$
$$2.25r = 19$$

Divide both sides by **2.25**.

$$\frac{2.25r}{2.25} = \frac{19}{2.25}$$
$$r \approx 8.4$$

The solution of the equation is about 8.4. But Surajit must take a whole number of rides. Surajit can take eight rides.

THINK AND COMMUNICATE

4. In Example 3, suppose Surajit has $40 to spend. How does the equation change? How does the solution change?

✓ CHECKING KEY CONCEPTS

Each diagram represents an equation. Write and solve the equation.

1.   **2.**

Solve each equation.

3. $5x + 7 = 14$ **4.** $12t - 7 = 19$ **5.** $24 = 3x - 3$

6. $16 - 4c = 20$ **7.** $89 = 11 - 11p$ **8.** $-3 - 9x = -21$

9. $\frac{m}{3} + 6 = 3$ **10.** $20 + \frac{x}{4} = 25$ **11.** $0 = \frac{m}{10} - 1$

Exercises and Applications

Extra Practice
exercises on
page 560

Solve each equation.

1. $1200 + 50x = 2750$ **2.** $6 + 3h = -3$ **3.** $10 = 5 - 5m$

4. $12 = 15n - 8$ **5.** $-4 - 2p = 8$ **6.** $-17 - 6t = -17$

7. $2m - 2.5 = 1.5$ **8.** $37 = -4s - 1$ **9.** $9 + \frac{x}{2} = 7$

10. $\frac{b}{2} - 7 = -11$ **11.** $10 = \frac{t}{7} + 16$ **12.** $\frac{n}{4} - 3 = 4$

13. $4 = \frac{p}{8} - 5$ **14.** $-9 + \frac{w}{4} = 18$ **15.** $-12 = 10.5 + \frac{r}{3}$

16. Eli tried to solve the equation $3 = -7 - 2m$, as shown at the left. What mistake did he make? Show how to solve the equation correctly.

$$3 = -7 - 2m$$
$$\underline{+7 \qquad +7}$$
$$\frac{10}{\quad} \qquad \frac{2m}{\quad}$$
$$\frac{10}{2} = \frac{2m}{2}$$
$$5 = m$$

17. **Technology** You can use a calculator to solve equations. For example, you can solve the equation $\frac{x}{6} - 13 = 5$ using the key sequence 5 [+] 13 [ENTER] [×] 6 [ENTER] .

   a. What operation do you undo when you enter 5 [+] 13 [ENTER] ? What operation do you undo next?

   b. **Writing** Describe a key sequence you can use to solve the equation $7x + 14 = 42$. How can you use a calculator to check your solution?

18. **Challenge** How many values of x make each statement true? Explain your reasoning.

   a. $3x + 3 = 3(x + 1)$ **b.** $x - 4 - x = 0$

Connection   CARS

When your car breaks down, you may have to call a tow truck. Towing companies usually charge one fee to hook up the car. They also may charge for each mile the car is towed. The table shows towing charges for two companies.

19. Suppose Company A charges you $70. About how far was your car towed?

20. A friend tells you that Company B would have been a better bargain. Do you agree or disagree? Explain.

21. **Open-ended Problem** Under what circumstances would you recommend using Company B? What other information would you like to know before hiring a towing company?

Company	Charge to hook up car	Charge per mile
A	$34	$2
B	$28	$3

Solve each equation.

22. $5x + 150 = -100$　　**23.** $-6x - 3 = -18$　　**24.** $-20 = 4 - 12m$

25. $10 - 5x = 25$　　**26.** $0 = 12 - 4x$　　**27.** $10 = \frac{n}{5} + 4$

28. $17 = \frac{p}{5} - 13$　　**29.** $\frac{t}{7} + 3 = -4$　　**30.** $-3 + \frac{z}{3} = 3$

31. $\frac{x}{10} - 13 = 0$　　**32.** $-3 = \frac{w}{9} - 9$　　**33.** $-28 = -8 + \frac{t}{5}$

34. Open-ended Problem Write a two-step equation for which -5 is the solution. Explain how you found your equation.

GEOMETRY **For Exercises 35–38, find the value of x. Then find the unknown dimension(s).**

35.

4 in.
5x
4 in.
7 in.
Perimeter = 20 in.

36.

5 cm　8 cm
4x
Perimeter = 23 cm

37.

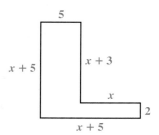

5
x + 5
x + 3
x
2
x + 5
Perimeter = 50 mm

38.

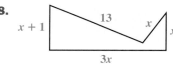

x + 1　13　x　x
3x
Perimeter = 44 ft

ONGOING ASSESSMENT

39. Open-ended Problem Make a list of the most important steps you should use to solve an equation. Include two examples and solutions that illustrate all four operations.

SPIRAL REVIEW

Write a variable expression to describe each situation. Tell what the variable stands for. *(Section 1.6)*

40. There are about 13 calories in each ounce of a glass of apple juice.

41. Rob is 5 in. taller than his sister.

42. Sandra took $50 out of her savings account.

Simplify each variable expression. *(Section 1.7)*

43. $12t + 13t - 7$　　**44.** $3(-6 - p)$　　**45.** $-4(2m - 7)$

46. $3 - 5(6 + 2m)$　　**47.** $7 - (2x + 7)$　　**48.** $4(6t + 3) - 3(t + 3)$

2.3

Applying Functions

Many portrait photographers offer you a deal. You order a basic package at a set cost and pay extra for additional photos. The relationship between the number of additional photos you order and your total cost is an example of a *function*.

A **function** is a relationship between input and output. For each input, there is exactly one output. Output depends on input.

For example, the cost of buying photos at Galante Studios is a function of the number of additional wallet photos you order.

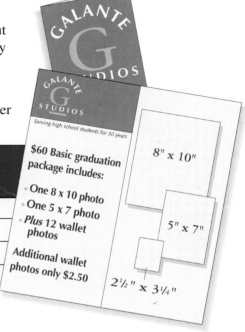

Input					Output
Additional wallet photos					**Total cost (dollars)**
0	$60	+	$2.50(0)	=	$60.00
1	60	+	2.50(1)	=	62.50
2	60	+	2.50(2)	=	65.00
3	60	+	2.50(3)	=	67.50

You can represent functions in many ways, including tables and equations.

$60 Basic graduation package includes:

- One 8 x 10 photo
- One 5 x 7 photo
- Plus 12 wallet photos

Additional wallet photos only $2.50

8" x 10"

5" x 7"

2½" x 3¼"

EXAMPLE 1   **Application: Personal Finance**

a. Write an equation that describes total cost as a function of the number of additional wallet photos ordered.

b. You have $87 to spend on graduation photos at Galante Studios. How many additional wallet photos can you order?

SOLUTION

a. Let n = the number of additional wallet photos ordered.
Let C = the total cost in dollars.

$$60 + 2.50n = C$$

Cost of basic package   Cost of additional wallet photos   Total cost

b. Use the equation from part (a). To find how many photos you can order, substitute $87 for C and solve for n.

$$60 + 2.50n = C$$

Subtract **60** from both sides.

$$
\begin{array}{rcr}
60 + 2.50n &=& 87 \\
-60 & & -60 \\
\hline
2.50n &=& 27
\end{array}
$$

Divide both sides by **2.50**.

$$\frac{2.50n}{2.50} = \frac{27}{2.50}$$

$$n = 10.8$$

You can order 10 additional wallet photos.

THINK AND COMMUNICATE

1. Explain how making a table like the one showing the costs at Galante Studios can help you write an equation for a function.

2. In part (b) of Example 1, the solution of the equation $60 + 2.50n = 87$ is 10.8. Why can you order only 10 additional wallet photos?

Domain and Range

The table below shows only positive whole numbers for n, the number of additional wallet photos you can order from Galante Studios. It doesn't make sense to order -12 photos or $\frac{1}{3}$ of a photo.

The **domain** of a function consists of all input values that make sense for the function.

n	C
0	60.00
1	62.50
2	65.00
3	67.50
...	...

The **range** of a function consists of all possible output values. The range depends on the domain.

These three dots show that the pattern of numbers continues.

EXAMPLE 2

Describe the domain and the range of the cost function in Example 1.

SOLUTION

The domain consists of the whole numbers 0, 1, 2, 3, ...

The range consists of these amounts in dollars and cents:
60.00, 62.50, 65.00, 67.50, ...

EXAMPLE 3

Decide whether each table represents a function. Explain your reasoning.

a.

Price (dollars)	Sales tax (dollars)
1	0.05
2	0.10
3	0.15
4	0.20
5	0.25
...	...

b.

x	x^2
−2	4
−1	1
0	0
1	1
2	4
...	...

c.

x	Any numbers that equal x when squared
0	0
1	1 and −1
4	2 and −2
9	3 and −3
...	...

SOLUTION

a. Yes. For each value in the domain, there is exactly one value in the range. The table represents a function.

b. Yes. For each value in the domain, there is exactly one value in the range. The table represents a function.

c. No. For some values in the domain, there is more than one value in the range. The table does not represent a function.

THINK AND COMMUNICATE

3. In Example 2 on page 67, how can you find the next three values in the range?

4. a. Use the movie theater sign. Explain why the price of a movie ticket is a function of age. Describe the domain and the range of the function.

b. Can you write an equation for the function? Why or why not?

☑ CHECKING KEY CONCEPTS

1. When you turn the steering wheel of a car to the left, you expect the car to turn to the left. How is the steering wheel like a function? Can you think of another machine that is like a function?

2. a. A soft drink at Cafe Jamaica costs $.80. If you go to the salad bar, you pay $.20 per ounce of salad. Write an equation for the cost of a soft drink and a salad at Cafe Jamaica.

b. Explain why the equation you wrote for part (a) represents a function. Describe the domain and the range of the function.

c. You have $6 to spend on a soft drink and a salad. How many ounces of salad can you buy?

2.3 Exercises and Applications

Decide whether each table represents a function. Explain your reasoning.

Extra Practice exercises on page 560

1.

Age (years)	Height (in.)
13	60
13	63
14	64
14	68
16	68

2.

n	$|n|$
0	0
−3	3
3	3
15	15
−15	15

3.

x	$x - 1$
−1	−2
0	−1
1	0
2	1
3	2

4. LANGUAGE ARTS A doctor says, "Good health is a function of good nutrition." What does the doctor mean? How is this use of the word *function* like the use of the word *function* in mathematics? How is it different? In what other ways can you use the word *function*?

For Exercises 5–8, match each equation with a situation. Tell what the variables stand for.

    A. $C = p - 10$ **B.** $p = \dfrac{10}{n}$ **C.** $C = 10 - p$ **D.** $C = 10d$

5. Rita buys a book with a $10 bill. The change she gets from the cashier is a function of the price of the book.

6. Matt rents a bicycle for $10 a day. His total cost is a function of the number of days he rents the bicycle.

7. Some friends want to share equally the cost of a $10 pizza. The price each friend pays is a function of the number of friends.

8. Tony buys a slightly damaged pair of shoes. The store clerk subtracts $10 from the original price. Tony's cost is a function of the original price of the shoes.

9. Describe the domain and the range of each function in Exercises 5–8. Explain your reasoning.

PERSONAL FINANCE The cost of going to a concert depends on the number of tickets you buy and the number of cars you take, as shown in the notebook at the right. Use the information in the notebook for Exercises 10 and 11.

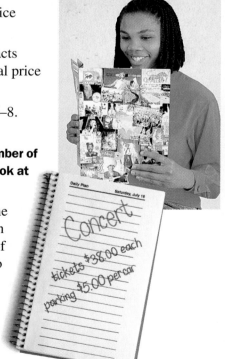

10. One car seats up to five people. Suppose you take one car and some friends to the concert. Write an equation for your cost as a function of the number of tickets you buy. Give the domain and the range of the function. How does the equation change if you take two cars to the concert? How do the domain and the range change?

11. Janet Neville wants to take some of her cousins to the concert. She has $160 to spend on tickets and parking for one car. How many tickets can she buy?

The United States government requires that workers receive a minimum hourly wage. The circle graph shows that most teenagers earn more than the minimum wage.

12. What percent of working teenagers earn more than the minimum wage?

13. a. Write a formula for converting the numbers on the circle graph to percents.

  b. Does your formula from part (a) represent a function? Explain.

14. a. In 1994, the minimum wage was $4.25 per hour. Write an equation for earnings as a function of the number of hours worked at this wage.

  b. Describe the domain and the range of the function from part (a).

15. Ana Martinez earns $500 for working a 40 h week. She also earns time and a half for working overtime. That means her overtime hourly wage is 1.5 times her regular hourly wage.

  a. What is Ana's regular hourly wage?

  b. What is her overtime hourly wage?

  c. How many hours of overtime must she work to earn a total of $650 in one week?

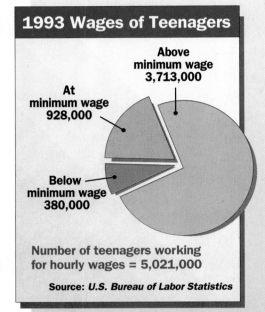

1993 Wages of Teenagers

Above minimum wage 3,713,000

At minimum wage 928,000

Below minimum wage 380,000

Number of teenagers working for hourly wages = 5,021,000

Source: *U.S. Bureau of Labor Statistics*

GEOMETRY For Exercises 16–18, write an equation for the perimeter *P* of each figure as a function of the side length *s*. Describe the domain and range of each function.

16.   **17.**   **18.**

      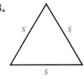

19. Open-ended Problem Draw a figure like the ones in Exercises 16–18. Express the perimeter of the figure as a function of the side length. Give the domain and the range of your function.

20. Challenge Write an equation for the perimeter *P* of the triangle at the right as a function of the side length *s*. The Triangle Inequality says that the sum of the lengths of any two sides of a triangle is greater than the length of the third side. What does this tell you about the domain and the range of the perimeter function for this triangle?

21. **Writing** A hiking club raises money by collecting returnable bottles. The club earns $.05 for each bottle. Tell how this situation represents a function. Describe the domain and the range of the function.

22. **SAT/ACT Preview** If $2x - 15 = -5$, then $x = \underline{\ ?\ }$.

   A. 25   **B.** −10   **C.** 10   **D.** 5   **E.** 7.5

ONGOING ASSESSMENT

23. **Writing** Give five examples of functions in everyday life. Describe the domain and the range of each one. Then describe a relationship between input and output that is not a function. Explain why it is not a function.

SPIRAL REVIEW

Solve each equation. *(Sections 2.1 and 2.2)*

24. $42 = -b$

25. $-64 + p = -12$

26. $\frac{n}{27} = 3$

27. $57d = 378.5$

28. $72 = 45 + 12h$

29. $-21 = -2w - 16$

30. $2r - 17 = 27$

31. $35 - 2x = 5$

32. $22 - p = -15$

33. $4 + \frac{s}{3} = -13$

34. $\frac{h}{3} - 7 = -19$

35. $\frac{q}{7} + 12 = 15$

ASSESS YOUR PROGRESS

VOCABULARY

equation (p. 55)   **function** (p. 66)
solve an equation (p. 55)   **domain** (p. 67)
solution (p. 56)   **range** (p. 67)
inverse operations (p. 56)

Solve each equation. *(Sections 2.1 and 2.2)*

1. $-3 = 18x$

2. $-r - 7 = 1$

3. $-4 = 37 + s$

4. $24 - 3d = 0$

5. $-n = 20$

6. $20 = 18 + \frac{m}{7}$

7. $\frac{p}{4} = -12$

8. $\frac{t}{6} - 4 = 9$

9. **AVIATION** A plane is flying at an altitude of 10,000 ft. The pilot wants to climb to a higher altitude at a rate of about 1000 ft/min. About how many minutes will it take the pilot to climb to 14,500 ft? *(Section 2.2)*

10. **a.** Ben buys plant food for $3 and plants for $2 each. Write an equation for his total cost as a function of the number of plants he buys. Describe the domain and the range of the function. *(Section 2.3)*

    b. Ben has $45. How many plants can he buy?

11. **Journal** Evaluate $2x + 8$ when $x = 5$. Then solve $2x + 8 = 16$. How is evaluating an expression different from solving an equation?

2.4 Coordinate Graphs

Learn how to...

- **read and create coordinate graphs**

So you can...

- **analyze biological and geographical data, for example**

- **read and interpret maps**

Winters in Antarctica are so cold that several feet of ice cover the ocean for most of the year. This icy barrier does not stop the Weddell seal from diving beneath the ice to hunt for fish. The graph below shows data collected by a biologist studying Weddell seals.

EXAMPLE 1 | Application: Biology

The graph compares the dives of two seals.

a. Where was Seal A 5 min after it dove under the ice?

b. About how deep did Seal A dive?

c. How long did it take Seal A to return to the surface?

SOLUTION

The **horizontal axis** shows the amount of **time t** that has passed. The **vertical axis** shows the **depth** of each dive at different times.

a. Seal A dove into the water at $t = 0$ min. After **5 min** it was about **600 ft** below the surface.

c. Seal A returned to the surface after about **11 min**.

b. Seal A dove about **1000 ft** below the surface.

Seal A ········
Seal B ───────

THINK AND COMMUNICATE

1. Which seal stayed under water longer? How can you tell from the graph?

2. How deep was Seal B's dive? About how many minutes did it take for Seal B to reach the deepest point of its dive?

Graphing in the Coordinate Plane

You can graph points in a **coordinate plane** like the one shown. You can use an **ordered pair** such as **(−4, 2)** to identify each point.

horizontal coordinate, or **x-coordinate**

vertical coordinate, or **y-coordinate**

The point with coordinates (0, 0) is called the **origin**.

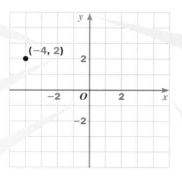

The vertical axis is usually called the **y-axis**.

The horizontal axis is usually called the **x-axis**.

The **x**- and **y**-axes divide the coordinate plane into four regions called **quadrants**.

This point is in Quadrant III.

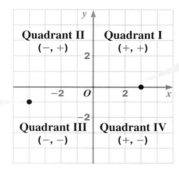

This point is not in any quadrant because it is on an axis.

EXAMPLE 2

Graph each point in a coordinate plane. Name the quadrant (if any) in which the point lies.

    a. (−3, 2)   **b.** (0, −4)

SOLUTION

a. Start at the origin. Move **3** units left. Then move **2** units up. Label the point. The point is in Quadrant II.

b. Start at the origin. Move **0** units on the x-axis. Move **4** units down. Label the point. The point is not in any quadrant.

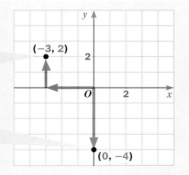

> ◀ **WATCH OUT!**
>
> The horizontal coordinate always comes first in an ordered pair. So (−3, 2) is *not* the same as (2, −3).

THINK AND COMMUNICATE

  3. Which direction is positive on the horizontal axis? on the vertical axis?

  4. Which quadrant does the graph on page 72 show? How can you tell?

☑ CHECKING KEY CONCEPTS

For Questions 1 and 2, use the graph in Example 1 on page 72.

1. Where was Seal A 10 min after it dove under the ice?

2. About how many minutes did it take Seal A to swim from the deepest point of its dive to the surface?

Match each ordered pair with a point on the graph. Name the quadrant (if any) in which the point lies.

3. $(1, 1)$

4. $(4, 4)$

5. $(-4, 3)$

6. $(3, -4)$

7. $(0, -3)$

8. $(4, 0)$

9. $(-3.5, -1)$

10. $\left(\frac{7}{3}, -2\right)$

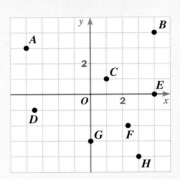

2.4 │ Exercises and Applications

Extra Practice exercises on page 560

Graph each point in a coordinate plane. Label each point with its letter. Name the quadrant (if any) in which the point lies.

1. $A(4, 1)$

2. $B(4, -1)$

3. $C(-3, -5)$

4. $D(-4, 2)$

5. $E(4, 5)$

6. $F(-5, 4)$

7. $G(0, -2)$

8. $H(-2, 3)$

9. $I(5, -5)$

10. $J(0, 0)$

11. $K(3, 0)$

12. $L(-4, -4)$

13. $M(1.5, 0)$

14. $N\left(-2, 4\frac{1}{2}\right)$

15. $P(2.75, 2.75)$

16. Challenge Point (a, b) is in the second quadrant and point (c, d) is in the fourth quadrant. Complete each blank with < or >. Explain your reasoning.

    a. $a \underline{\;?\;} b$ **b.** $c \underline{\;?\;} d$ **c.** $a \underline{\;?\;} c$ **d.** $b \underline{\;?\;} d$

PERSONAL FINANCE Some students decide to raise money by having a car wash. The graph at the left shows how much money they can raise, depending on how many cars they wash.

17. What does the horizontal axis of the graph show? What does the vertical axis show?

18. How much do the students raise per car? How do you know?

19. The students want to raise $200. How many cars do they have to wash?

Depending on your location on the globe and the time of year, the number of daylight hours varies, as shown by the graph below. Use the graph for Exercises 20–22.

Daylight Hours in Two Cities

- Dakar, Senegal
- Chicago, United States

Hours of daylight

16
15
14
13
12
11
10
9
8

Jan Feb Mar Apr May Jun Jul Aug Sep Oct Nov Dec

Time of year

BY THE WAY

Dakar and Chicago are both in the Northern Hemisphere. In the Northern and Southern Hemispheres, the day with the greatest number of daylight hours is called the summer solstice. The day with the least number of daylight hours is called the winter solstice.

20. a. When does each city receive the fewest hours of daylight? About how many hours of daylight per day does each city receive at this time?

b. When does each city receive the most hours of daylight? About how many hours of daylight per day does each city receive at this time?

21. During which months are the hours of daylight in Chicago and Dakar about equal? About how many hours of daylight per day does each city receive during these months?

22. Compare the shapes of the two graphs. How are they the same? How are they different?

For Exercises 23 and 24, use the graph in Example 1 on page 72.

23. Writing About how long do you think a seal would stay under water if it dove to a depth of 700 ft? Explain your answer.

24. Open-ended Problem About how long do you think you can hold your breath under water? How deep do you think you can dive? Answer these questions by drawing a graph. How is your graph different from the seal graph? Give some possible reasons for these differences.

25. PERSONAL FINANCE Suppose you want to borrow money from your parents for a new pair of in-line skates. You plan to pay your parents $10 a week until the loan is paid off.

a. According to the graph at the right, how much money do you borrow? How do you know?

b. How much do you owe after six weeks?

c. How long does it take you to pay back the loan? How can you tell from the graph?

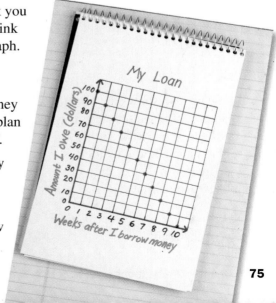

My Loan

Amount I owe (dollars)

100
90
80
70
60
50
40
30
20
10
0

0 1 2 3 4 5 6 7 8 9 10

Weeks after I borrow money

Around the second century A.D., Chinese geographers began using grids on maps. In 1311, the geographer Chu Ssu-Pen created a map of China. The map was printed in book form in about 1555. This section of the map shows part of China's coast.

26. About how many square miles does the map represent? How do you know?

27. The black band in the upper left corner of the map represents the Gobi Desert. According to the map, about how long from east to west is this section of the Gobi Desert? How can you tell?

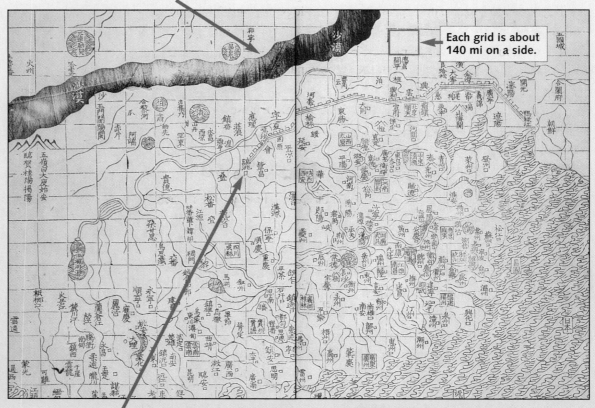

Each grid is about 140 mi on a side.

28. The map shows the Great Wall of China with these special lines (▱▱▱). According to the map, about how long is the Great Wall? How did you make your estimate?

29. Research Use an encyclopedia to find out more about the Great Wall. How long is it? When and why was it built?

30. Research Find a modern map showing roughly the same area as the map above. Does the modern map use a coordinate grid? Can you use the modern map to identify any bodies of water on Chu Ssu-Pen's map? any land features?

For Exercises 31–35, use the coordinate graph.

31. Write an ordered pair for each point.

32. Name two points for which the *x*-coordinate is less than the *y*-coordinate.

33. Name two points for which the *x*-coordinate is greater than the *y*-coordinate.

34. Name two points for which the *x*-coordinate equals the *y*-coordinate.

35. Name two points for which the *x*-coordinate is the opposite of the *y*-coordinate.

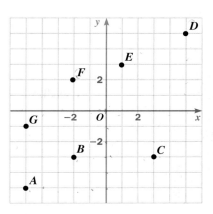

ONGOING ASSESSMENT

36. a. Draw a coordinate plane. Graph at least six points for which the *x*-coordinate equals the *y*-coordinate. In which quadrant(s) (if any) do the points lie?

b. Draw another coordinate plane. Graph at least six points for which the *x*-coordinate is the opposite of the *y*-coordinate. In which quadrant(s) (if any) do the points lie?

c. Writing Compare your graphs. Write a paragraph describing any patterns you see.

SPIRAL REVIEW

For Exercises 37 and 38, find the mean, the median, and the mode(s) of each set of data. *(Section 1.3)*

37. Movie prices: $6.00, $3.50, $5.50, $5.75, $4.50, $6.75, $6.75, $6.50

38. Overtime hours: 7, 1, 10, 5, 0, 3, 1, 1, 2, 2, 0, 0, 0, 9, 0, 4, 2

39. **Technology** Give the dimensions of each matrix. Tell which matrices can be added. Then add them. Use a graphing calculator if you have one. *(Section 1.8)*

$$A = \begin{bmatrix} 2 & 4 \\ 7 & -2 \\ 1 & 16 \end{bmatrix} \qquad B = \begin{bmatrix} 3 & 7 & 11 \\ 3 & 2 & 13 \end{bmatrix} \qquad C = \begin{bmatrix} 5 & 3 & 2 \\ 1 & -1 & -3 \\ 0 & 0 & 3 \end{bmatrix}$$

$$D = \begin{bmatrix} 6 & 0 & 0 & 1 \end{bmatrix} \qquad E = \begin{bmatrix} 1 & 2 \\ 2 & 1 \end{bmatrix} \qquad F = \begin{bmatrix} 1 & 0 & -7 \end{bmatrix}$$

$$G = \begin{bmatrix} 3 & -1 \\ 6 & 17 \\ -12 & 3 \end{bmatrix} \qquad H = \begin{bmatrix} 10 & 5 & 3 & 6 \end{bmatrix} \qquad I = \begin{bmatrix} -3 & -2 & 11 \\ 0 & 1 & -1 \\ 5 & 7 & 1 \end{bmatrix}$$

2.5 Representing Functions

Learn how to...
- graph equations

So you can...
- visually represent functions, such as costs at a swimming pool

Take any number and subtract 1 from it. You can think of this command as a function. The number is the input. Your answer is the output. One way to represent this function is to use a graph.

EXPLORATION
COOPERATIVE LEARNING

Graphing Equations

Work with your class.

1 Turn your class into a coordinate grid, as shown. What are your coordinates?

2 Stand up if your coordinates make the equation $y = x - 1$ true. What do you notice about the people standing up?

3 Sit down again. Your teacher will give you other equations. For each equation given, stand up if your coordinates make the equation true. In each case, describe what you observe.

This student is standing because her coordinates, (1, 0), make the equation true.

$$y = x - 1$$
$$0 = 1 - 1$$

You can represent the same function in several different ways. As shown in Example 1, you can use an equation, a table of values, or a graph.

EXAMPLE 1

Graph the equation $y = 2x - 1$.

SOLUTION

Step 1 Make a table of values. Choose at least three values of x. Be sure to include both positive and negative values of x.

$y = 2x - 1$	
x	y
-2	-5
0	-1
2	3

Step 2 Draw a coordinate grid. Use the table of values to plot points.

Step 3 Connect the points to make a line. Put an arrowhead on each end to show that the line extends in both directions.

The coordinates of every point on the line make the equation $y = 2x - 1$ true, not just the points plotted in Step 2. For example, the point $\left(-\frac{3}{2}, -4\right)$ is on the graph of the equation, and its coordinates make the equation true:

$$-4 = 2\left(-\frac{3}{2}\right) - 1$$

An ordered pair that makes an equation such as $y = 2x - 1$ true is called a **solution of the equation**. The graph of the equation $y = 2x - 1$ contains all possible solutions of the equation.

THINK AND COMMUNICATE

1. Which of these ordered pairs are solutions of the equation $y = 2x - 1$? How do you know?

   A. $(3, 4)$ **B.** $\left(\frac{1}{2}, 0\right)$ **C.** $\left(\frac{3}{2}, 2\right)$ **D.** $\left(-\frac{5}{2}, -6\right)$

2. Based on what you now know about the graph of an equation, explain what you observed in the Exploration on page 78.

EXAMPLE 2   **Application: Consumer Economics**

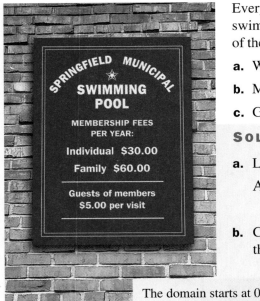

Every year, the Takedas buy a family membership at the Springfield swimming pool. The sign shows that their annual cost is a function of the number of guests they take to the pool.

a. Write an equation for the function.

b. Make a table of values for the function.

c. Graph the function.

SOLUTION

a. Let C = the annual cost. Let g = the number of guests.

Annual cost = $60 + $5 per guest × number of guests

$$C = 60 + 5g$$

b. Choose values for g, the number of guests. It helps to think about the domain of the function.

The domain starts at 0 and includes only whole numbers.

Number of guests	Annual cost (dollars)
0	60
1	65
2	70
3	75
...	...

Use the equation $C = 60 + 5g$.

$60 + 5(0)$
$60 + 5(1)$
$60 + 5(2)$
$60 + 5(3)$

c. Draw a pair of axes. Put the domain values on the horizontal axis and the range values on the vertical axis.

Use the table to plot points. The Takedas can take only a whole number of guests. Do not connect the points because a line would include numbers that are not whole numbers.

Swimming Pool Costs

THINK AND COMMUNICATE

3. In Example 2, suppose the points were connected. Then the point (4.5, 82.5) would be on the graph. Explain why these values do not make sense for the function.

4. The cost of an individual membership at the Springfield swimming pool is a function of the number of guests the member takes. How is this function different from the one in Example 2?

☑ CHECKING KEY CONCEPTS

Match each equation with a graph. Explain how you got your answers.

1. $y = \frac{1}{2}x$

2. $y = -x - 2$

3. $y = -x$

A.

B.

C.

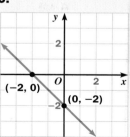

For Questions 4 and 5, show three different ways to represent each function.

4. Think of a number and triple it.

5. **PERSONAL FINANCE** Chandra spends $8 for 15 graduation party invitations. It costs $.32 to mail each invitation. Her total cost is a function of the number of invitations she mails.

2.5 Exercises and Applications

Graph each equation and each point. Tell whether the ordered pair is a solution of the equation.

1. $y = -3x$; $(-1, 3)$

2. $y = 2 - 2x$; $(2, 4)$

3. $y = 4x - 3$; $(-3, 0)$

4. $y = -4x + 1$; $(6, 21)$

5. $y = 4 - 5x$; $(4, 0)$

6. $y = 2x - 5$; $(3, 4)$

7. $y = -5x - 1$; $(0, -1)$

8. $y = \frac{1}{3}x$; $(3, 1)$

9. **Writing** When you graph an equation, two points are enough to draw a line. How does plotting three points help you graph the equation?

10. **ENERGY USE** Suppose a new refrigerator costs $1000. Electricity to run the refrigerator costs about $68 per year. The total cost of the refrigerator is a function of the number of years it is used.

  a. Write an equation for the function.

  b. Make a table of values for the function.

  c. Graph the function.

BY THE WAY

There are an estimated 110 million refrigerators in the United States. They use more than 20% of all household energy.

2.5 Representing Functions **81**

Bob Hall's invention of a new steering system for racing wheel-chairs has made it possible for many more athletes to compete in national and international races.

11. The steering mechanism on Bob Hall's racing wheelchair is designed to allow a racer to turn the wheelchair by a maximum angle of 15°.

a. Use the table on page 54 to draw a graph showing turning radius as a function of turn angle. Connect the points on your graph to form a smooth curve.

b. Use your graph to estimate the turning radius when the turn angle is 4°.

Look back at the article on pages 52–54.

c. Describe the domain and the range of the turning radius function.

The table and graph show the number of male wheelchair athletes in the Boston Marathon as a function of the year and the number of female wheelchair athletes as a function of the year. Use the table and graph for Exercises 12–14.

Wheelchair Athletes in the Boston Marathon, 1975–1993

Year	Males	Females
1975	1	0
1976	0	0
1977	6	1
1978	18	2
1979–1983	No official records	
1984	16	3
1985	21	3
1986	30	4
1987	37	4
1988	45	3
1989	34	8
1990	41	5
1991	62	8
1992	52	4
1993	75	9

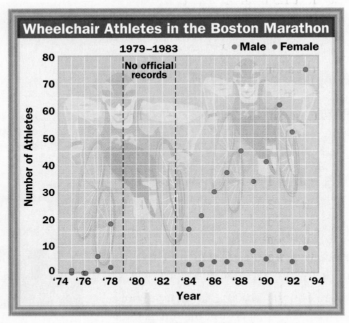

Wheelchair Athletes in the Boston Marathon

12. Describe the domain and the range of each function.

13. Official records were not kept for the years 1979–1983. Estimate the number of male wheelchair athletes in 1983. Estimate the number of female wheelchair athletes in 1983. How did you make your estimates?

14. Writing What are some of the advantages and disadvantages of seeing these functions displayed in a table? in a graph?

For Exercises 15–18, match each sentence with an equation. For each equation, find the value of y when $x = -6$.

15. Multiply any number by 10 and then add 3.   **A.** $y = 3x + 10$

16. Multiply any number by 3 and then add 10.   **B.** $y = 3x - 10$

17. Subtract 10 from the product of any number and 3.   **C.** $y = 10 - 3x$

18. Subtract the product of any number and 3 from 10.   **D.** $y = 10x + 3$

 Spreadsheets You can use a spreadsheet to make a table of values. You need to enter a formula and use the fill down command. Use the spreadsheet at the right for Exercises 19 and 20.

19. The spreadsheet shows a table of values for the function *Add 7 to a number.* Which column, A or B, contains the domain values? the range values?

20. What spreadsheet formula is used to represent the function? What would a spreadsheet formula for the function $y = x - 7$ look like?

Worksheet 1		
B3		=A3+7
	A	**B**
1	−5	2
2	−4	3
3	−3	4
4	−0.5	6.5
5	0	7
6	1	8
7	2	9
8	5	12

 Spreadsheets Make a table of values for each function. Use a spreadsheet if you have one.

21. Subtract 19 from any number.

22. Multiply any number by 3 and then subtract 1.5 from the result.

23. Divide any positive integer by 2.

24. **SAT/ACT Preview** If $-3x + 3 = y$, and $y = 15$, then $x = \underline{\ ?\ }$.

25. **Challenge** Describe the graph of each equation.

     a. $y = 3$ **b.** $x = -8$

26. **Open-ended Problem** Describe a situation in which a graph would be more helpful than a table. Explain why.

ONGOING ASSESSMENT

27. **Cooperative Learning** Work with a partner. Each of you should make up a function that models something from everyday life, such as buying lunch for a number of friends. Exchange functions and show three ways to represent your partner's function. Check your work together.

SPIRAL REVIEW

Simplify each expression. *(Sections 1.2, 1.4, and 1.5)*

28. $\dfrac{-8(1 - 8)}{7}$ 29. $\dfrac{5}{8} \cdot \dfrac{4}{3} + \dfrac{7}{12}$ 30. $\dfrac{3}{-4} - 5 + 6(5)$

Simplify each variable expression. *(Section 1.7)*

31. $25s + 3t - 7 - 6s + t$ 32. $10 - 3(2t + 8)$

33. Show that $-6(2m - 3n) + 12m$ is equivalent to $18n$.

2.6

Using Graphs to Solve Problems

Learn how to...

- use a graph to solve a problem

So you can...

- analyze the costs and profits of running a business, for example

Have you ever thought about putting your own design on a T-shirt and selling it? Graphing an equation helps you analyze the costs and profits of making and selling T-shirts.

Suppose **silk-screening supplies** cost **$70**. Plain **T-shirts** cost **$3** each. You can write an equation for the cost of making T-shirts.

Cost = $3 × number of shirts + $70
$$C = 3n + 70$$

As shown below, the graph of this equation can help you analyze the costs of running your own T-shirt business.

EXAMPLE 1   Application: Business

You receive 87 orders for T-shirts. You have $300 to spend on production costs. Decide whether you have enough money to fill all 87 orders.

SOLUTION

Graph the equation $C = 3n + 70$.

Find $300 on the vertical axis. Move across to the graph of the equation.

T-Shirt Production Costs

Move down to the horizontal axis. You can make about 75 T-shirts.

You do not have enough money to fill all 87 orders.

The graph in Example 1 helps you estimate production costs. But a graphing calculator can help you make even closer estimates, as shown in Example 2.

EXAMPLE 2 Application: Business

a. You decide to charge customers $10 per T-shirt. Your production costs are the same as in Example 1. Write an equation for your profit P from making and selling n T-shirts.

b. What is your profit from making and selling 75 T-shirts?

SOLUTION

a. Profit = Sales Income − Production Cost

$10 × number of shirts − $3 × number of shirts + $70

P = $10n$ − $(3n + 70)$

$P = 10n - (3n + 70)$

You can leave the equation in this form. Or you can simplify it.

$P = 10n - 3n - 70$

$P = 7n - 70$

b. **Problem Solving Strategy:** Use a graph.
First enter the equation $P = 7n - 70$ as $y = 7x - 70$.

Then set the viewing window by choosing *scales* for the x- and y-axes.

For more information about using a graphing calculator, see the *Technology Handbook*, p. 603.

Then graph the equation. Use the *trace* feature to find your profit from selling 75 T-shirts.

The *trace* feature shows you x- and y-values on the graph.

Use the *zoom* feature to get more accurate values for x and y.

When the x-value is close to 75, the y-value is close to 455.

Your profit from selling 75 T-shirts is about $455.

THINK AND COMMUNICATE

Use the information in Examples 1 and 2 on pages 84 and 85.

1. Suppose you want to make a profit of $1000. Use the graph shown at the right to estimate the number of T-shirts you must sell.

2. Explain how you could use the graph in Example 1 to estimate the cost of making 87 T-shirts. Why is this cost an estimate?

Graph of $y = 7x - 70$

X=153.19149 Y=1002.3404

☑ CHECKING KEY CONCEPTS

Cooperative Learning Work with a partner. One of you should do part (a) of each question. The other should do part (b). Compare your work. Discuss which way of solving each problem you prefer.

1. Mandy has to read a 126-page book in Spanish. She has already read 19 pages. She estimates she can read about 9 pages per day.

   a. Write and solve an equation to find how many days it will take Mandy to finish the book.

   b. Use a graph to estimate how many days it will take Mandy to finish the book.

2. **BUSINESS** Jon Li buys sand from a building supply company. The sand costs $19 per ton. There is a $20 delivery charge. Jon needs 2 tons of sand delivered to a construction site.

   a. Use a graph to estimate Jon's total cost.

   b. Write and solve an equation to find Jon's total cost.

2.6 | Exercises and Applications

Extra Practice exercises on page 561

BUSINESS For Exercises 1 and 2, use the information in Examples 1 and 2 on pages 84 and 85.

1. Suppose you charge $10 per T-shirt. How many T-shirts do you have to sell to break even? (To break even means that you sell just enough T-shirts to cover your production costs.)

2. a. Suppose you charge $15 per T-shirt. Write an equation for your profit P from selling n T-shirts.

   b. How many T-shirts do you have to sell to make $500?

RECREATION Exercises 3 and 4 are about Central Park in New York City. Solve at least one of the exercises by writing and graphing an equation.

3. Central Park is rectangular and has a perimeter of about 6 mi. The length of one side is about 2.5 mi. Find the lengths of the other three sides.

4. To celebrate each new year, runners participate in a "Midnight Run" through Central Park. The women's record for this 5 mi race is about 26.2 min. The men's record is about 23.3 min. Estimate the running speed in miles per minute of these record-holding runners.

Central Park West

CENTRAL PARK

Central Park South

Central Park North

Fifth Avenue

2.5 mi

SCIENCE The graph shows the relationship between the Fahrenheit and Celsius temperature scales. Use the graph for Exercises 5–9.

5. Estimate the temperature in degrees Fahrenheit when the temperature is:

   a. 0°C   **b.** 20°C   **c.** −20°C   **d.** −30°C

6. Estimate the temperature in degrees Celsius when the temperature is:

   a. 32°F   **b.** 90°F   **c.** 0°F   **d.** −40°F

7. You can use the formulas below to convert Celsius temperatures to Fahrenheit temperatures and Fahrenheit temperatures to Celsius. Use the formulas to find more exact answers to Exercises 5 and 6.

   Let F = the temperature in degrees Fahrenheit.

   Let C = the temperature in degrees Celsius.

   $$\begin{cases} F = \dfrac{9}{5}C + 32 \\ C = \dfrac{5}{9}(F - 32) \end{cases}$$

   Temperature Scales

   Fahrenheit temperature (°F)

   Celsius temperature (°C)

8. **Writing** A friend on a trip to Africa writes that the temperature in Nairobi, Kenya, is 25°. Which temperature scale do you think your friend is using? Why?

9. **Open-ended Problem** A friend from a country that uses the Celsius scale wants to visit you. Write a letter to your friend describing the range of temperatures in your area. Give all temperatures in degrees Celsius.

10. The film club pays $500 to rent *Casablanca*. Members sell tickets for $3.50 each. Write and graph an equation for the club's profit P from selling n tickets. How many tickets must be sold to make a $150 profit?

Use the article below for Exercises 11–15.

SPECIAL ISSUE

Can crickets tell the temperature?

Many can, if you know how to listen. For example, the snowy tree cricket chirps faster as the temperature rises. Just count the number of chirps you hear in 15 seconds. Add 40 to get the approximate temperature in degrees Fahrenheit.

Each of a cricket's wings has a vein called the file and a hard edged portion called the scraper. The cricket chirps by rubbing the file on one wing against the scraper on the other.

scraper and file

Snowy tree cricket

(partial text in left margin and bottom of article, partially obscured)

hoppers, to the ra. The cket fam- ets have ince the riod ears ago). well- United cket. Field black or crickets, ree cricket, r pale-

wn and he world for ng." Cricket pecies to cies may pro- ills; others eries of chirps. ng or trilling uced primarily . Crickets pro- ubbing their The chirping ed close to the re greater force . In most e male cricket chirps. In some ver, the female capable of chirp-

s sound-sensing cated on its front have shown that receptors can hear rceptible to

use their antennae as sensory organs. On most types of crickets, the antennae are slender and

A cricket can determine from which direction a sound is coming by positioning itself so that the sound is minimized. At that

11. Suppose you hear 23 cricket chirps in 15 seconds. What is the approximate temperature?

12. Write and graph an equation for the temperature as a function of the number of chirps you hear in 15 seconds.

13. Suppose you count the number of chirps in 15 seconds. According to your graph, what is the temperature when you hear:
   a. 5 chirps   **b.** 10 chirps
   c. 20 chirps   **d.** 25 chirps

14. How many chirps would you expect to hear in 15 seconds when the temperature is:
   a. 50°F   **b.** 65°F
   c. 0°F   **d.** 95°F

15. a. Challenge According to your graph, what is a lower limit for the temperature at which a snowy tree cricket starts chirping? Explain.

   b. Do you think there is an upper limit for the number of chirps you might hear in 15 seconds? Explain your thinking.

 Technology Graph each equation. Use a graphing calculator if you have one. Use the graph to estimate the unknown value to the nearest tenth.

16. $y = 2 - 0.3x$; find the value of x when $y = 0$.

17. $y = 8 - 0.4x$; find the value of x when $y = 6$.

18. $y = -5 + 2.7x$; find the value of x when $y = 1$.

19. $y = 0.5x - 7$; find the value of x when $y = 2$.

20. $y = -9.2 + 3.7x$; find the value of y when $x = 5$.

21. SAT/ACT Preview Which ordered pair is a solution of the equation $y = 3x - 4$?

   A. $(-1, 1)$ **B.** $(1, -1)$ **C.** $(-4, 0)$ **D.** $\left(0, \frac{4}{3}\right)$ **E.** $\left(0, \frac{3}{4}\right)$

22. a. Suppose you read that you can exchange one United States dollar for 100 Japanese yen. Make a graph that shows the number of yen you receive as a function of the number of dollars you exchange.

    b. Writing Suppose you are going to Japan. You think you will need between 100,000 and 150,000 yen. Explain how can you use the graph from part (a) to find how many dollars you should exchange.

SPIRAL REVIEW

Use the graph. Tell whether each ordered pair is a solution of the equation $y = 6 - 2x$. *(Section 2.5)*

Graph of $y = 6 - 2x$

23. $(6, 6)$ **24.** $(3, 0)$ **25.** $(-2, 4)$ **26.** $(4, -2)$

Show three ways to represent each function. *(Section 2.5)*

27. Take any number, double it, and add 3.

28. Take any number and multiply it by 5.

29. Subtract any number from 4.

ASSESS YOUR PROGRESS

VOCABULARY

horizontal axis (p. 72)	**y-coordinate** (p. 73)
vertical axis (p. 72)	**origin** (p. 73)
coordinate plane (p. 73)	**x-axis** (p. 73)
ordered pair (p. 73)	**y-axis** (p. 73)
horizontal coordinate (p. 73)	**quadrant** (p. 73)
x-coordinate (p. 73)	**solution of an equation**
vertical coordinate (p. 73)	**in two variables** (p. 79)

Graph each point in a coordinate plane. Name the quadrant (if any) in which the point lies. *(Section 2.4)*

1. $A(-5, 7)$ **2.** $B(0, 0)$ **3.** $C(-1, -1)$ **4.** $D(-6, 0)$

5. $E(2, -3)$ **6.** $F(0, 7)$ **7.** $G(4, 1)$ **8.** $H(-3, 3)$

Graph each equation. *(Section 2.5)*

9. $y = -4x - 1$ **10.** $y = 3x + 2$ **11.** $y = -x + 5$

12. The student council pays $780 for a band and refreshments for a dance. They sell dance tickets for $8 each. Write and graph an equation for their profit P from selling n tickets. About how many tickets must be sold to make a profit of $500? *(Section 2.6)*

13. Journal Show four different ways to represent the same function. Which do you think is easiest to create? Which do you think is easiest to use? Explain your answers.

Functions of Time

Many of the changes you see are examples of functions. A graph of a function can help you see a pattern in the change. For example, the graph on page 82 shows how the number of wheelchair athletes in the Boston Marathon has grown over time.

PROJECT GOAL Collect and graph data about something that changes over time. You will look for patterns in your graph.

Collecting Data

Work with a partner. Choose an event that occurs in the world around you. Whatever you choose should be easy to measure and should change noticeably within minutes, hours, or days.

Some possibilities:

- the height of a pedal on a moving bicycle

- the temperature at each hour of the day

- the number of cars in a fast food restaurant's parking lot at different times of the day

- the number of people going into a mall on different days or at different hours

Investigate and record what you observe. For example, suppose you decide to investigate how the height of a bicycle pedal changes over time.

1. MEASURE the height of the pedal in different positions.

2. RECORD when the pedal is in each position. Making a video can help you see a change that happens quickly.

3. ORGANIZE your data in a table or spreadsheet.

Making a Graph

Use your data to make a graph of the function. Do you see a pattern? What do you think happens between the points that you plot? Does it make sense to connect the points?

Presenting Your Ideas

Write a report, make a poster, or give an oral presentation explaining your graph. Describe what you observed and discuss any patterns you see. Describe the domain and the range of your function.

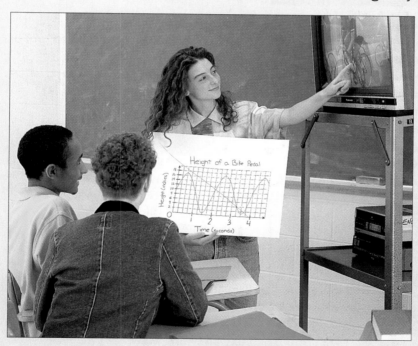

You may also want to look through books, magazines, and other sources at the library. Can you find data about functions like the one you investigated? How do the data compare with yours?

Before giving your presentation, you may want to show your classmates your graph without the title or the labels to see if they can guess what you observed.

Self-Assessment

Describe what you and your partner learned about collecting and graphing data. For example, how did you decide what to investigate? How did you decide on values to use for the axes of your graph? Did you discuss ideas together? Did you divide up some of the work? What might you do differently next time?

2 Review

Summarize what you have learned in the chapter by following these steps:

- **Look back through the sections in the chapter.**
- **Read each section title and each subtitle.**
- **Write a sample question or a short summary for each one.**

In your summary be sure to include the topics that gave you the most difficulty. Use your summary as a study sheet.

VOCABULARY

equation (p. 55)
solve an equation (p. 55)
solution (p. 56)
inverse operations (p. 56)
function (p. 66)
domain (p. 67)
range (p. 67)
horizontal axis (p. 72)
vertical axis (p. 72)
coordinate plane (p. 73)

ordered pair (p. 73)
horizontal coordinate (p. 73)
x-coordinate (p. 73)
vertical coordinate (p. 73)
y-coordinate (p. 73)
origin (p. 73)
x-axis (p. 73)
y-axis (p. 73)
quadrant (p. 73)
**solution of an equation
    in two variables** (p. 79)

SECTIONS 2.1 *and* 2.2

To **solve** an **equation**, find the value of the variable that makes the equation true. Use **inverse operations** and keep both sides of the equation equal.

First **undo** any addition or subtraction.

Then **undo** any multiplication or division.

$$3x + 29 = 8$$
$$\underline{-29 \qquad -29}$$
$$3x = -21$$
$$\frac{3x}{3} = \frac{-21}{3}$$
$$x = -7$$

SECTIONS 2.3, 2.4, and 2.5

A **function** is a relationship between input and output. Each input has only one output. You can use equations, tables, and graphs to represent functions.

Dollar value V of n quarters

Equation

$V = 0.25n$

Table

domain values	**range** values
Number of quarters	**Value (dollars)**
0	0
1	0.25
2	0.50
3	0.75
...	...

Graph

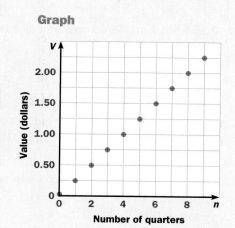

You can graph equations in a **coordinate plane**.

	$y = -2x - 3$	
	x	**y**
	−2	1
	−1	−1
	0	−3
	1	−5
	2	−7

Graph of $y = -2x - 3$

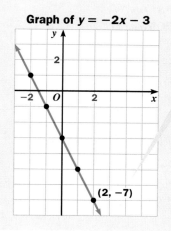

The **ordered pair** $(2, -7)$ is a **solution** of the equation $y = -2x - 3$. The graph contains all solutions of the equation.

SECTION 2.6

You can use an equation or a graph to solve a problem. For example, to find how many hours h it will take you to drive 500 miles at 55 mi/h, you can solve an equation.

$$500 = 55h$$

$$9.1 \approx h$$

Graph of $y = 55x$

The graph at the right gives you the same answer.

When $y \approx 500$, $x \approx 9.1$.

Assessment

VOCABULARY QUESTIONS

For Questions 1 and 2, complete each paragraph.

1. The graph shows the cost of buying pretzels as a function of the number of pretzels. The _?_ of the function consists of the numbers 0, 1, 2, 3, … . The _?_ of the function consists of these amounts in dollars and cents: 0, 0.75, 1.50, 2.25, … .

Cost of Pretzels

2. When you solve an equation, first undo any addition or subtraction, then undo any _?_. Addition and subtraction are called _?_ because they undo each other.

SECTIONS 2.1 *and* 2.2

Solve each equation.

3. $6t + 122 = 278$ 4. $3 = 18 + 16w$ 5. $-x = 23$

6. $-w = -32$ 7. $11 - m = 15$ 8. $-h - 21 = -26$

9. $\frac{x}{13} - 7 = 1$ 10. $-4 + \frac{n}{3} = 14$ 11. $-3p + 1.8 = 10.8$

12. **HISTORY** On February 20, 1962, John Glenn became the first person from the United States to orbit Earth. He traveled 80,966 mi in about 4.92 h. Find his average speed in miles per hour.

SECTIONS 2.3, 2.4, *and* 2.5

13. **a. PERSONAL FINANCE** Sequina Jackson pays $50 to join a golf club. She pays an additional $12 every time she plays golf. Write an equation that describes her total cost for playing golf as a function of the number of times she plays.

   b. Describe the domain and the range of the function.

   c. Sequina does not want to spend more than $250 to play golf during the summer. Find the maximum number of times she can play for this amount.

14. ASTRONOMY The graph shows the temperature on Mars in degrees Celsius during a 24-hour period.

a. Estimate the highest and lowest temperatures during this 24-hour period. At what recorded hours did these temperatures occur?

b. In which quadrant of the coordinate plane does the graph lie?

Graph each point. Name the quadrant (if any) in which the point lies.

15. $A(0, 4)$ **16.** $B(-3, -3)$ **17.** $C(2, -7)$ **18.** $D(-4, 5)$

Graph each equation in a coordinate plane.

19. $y = -5x$ **20.** $y = 5 - 2x$ **21.** $y = 4x - 3$

22. $y = -x - 1$ **23.** $y = \frac{x}{2} + 3$ **24.** $y = 2.5 + x$

25. For which equation in Exercises 19–24 is $(-1, -7)$ a solution?

26. Open-ended Problem Describe the steps you took to graph the equation in Exercise 24. Describe another way to graph the equation.

SECTION 2.6

27. a. CONSUMER ECONOMICS Ted and Elaine Kemp are planning an anniversary party. The disc jockey costs $450. The caterer charges $20 per person. Write an equation describing the Kemps' cost for the disc jockey and catering as a function of the number of people at the party.

b. Graph the equation.

c. The Kemps do not want to spend more than $2000 for a disc jockey and catering. At most, how many people can attend the party?

d. **Writing** How did you solve part (c)? Describe one other way to solve the problem.

PERFORMANCE TASK

28. Make a poster of a function. Represent your function in at least three ways. Be sure to describe the domain and the range of your function.

3 Graphing Linear Equations

Taking science into SPACE

INTERVIEW Mae Jemison

On the morning of September 12, 1992, Dr. Mae Jemison was aboard the space shuttle *Endeavour* at Kennedy Space Center, awaiting blastoff. The giant booster rockets fired. As the shuttle accelerated, reaching a speed of over 17,000 miles per hour in a little more than 8 minutes, Jemison felt 3 times the force of gravity pressing on her. But that wasn't her strongest impression. "The biggest thing I remember was that I was smiling," she says. "I always knew I would go into space—I didn't know how— I just knew I would go."

> " I always knew I would go into space, to the stars. "

Music of the spheres

When **Mae Jemison** went into space in September 1992, she took a few items for different organizations: an Alvin Ailey dance poster, a flag for the Organization for African Unity, A Spelman College flag. . . . "I had all kinds of music in space – **Stevie Wonder, Olatunji, Hiroshima, Aretha Franklin** and **Nancy Wilson**. . . . I was jammin'."

This article appeared in the *Boston Globe* on Saturday, July 10, 1993.

A Love of Science

Jemison achieved her seat on *Endeavour* by following her love of science. "From as long ago as I can remember, I wanted to be involved in science," Jemison says. In her teens, she wanted to design spacecraft. Later, she discovered biomedical engineering, which "allowed you to do engineering but bring biology into it—both areas that I liked." Jemison earned degrees in chemical engineering and medicine, achievements that helped open the door to NASA's astronaut program.

Hands, Eyes, and Ears

Aboard *Endeavour*, Jemison worked as a science mission specialist, overseeing scientific experiments sent along by earthbound researchers. "The scientists had worked on these studies for years, and they entrusted their work to us," Jemison says. "We were their hands, eyes, and ears on the shuttle." One experiment used weightlessness to grow large protein crystals. Protein crystals are important body substances that do many jobs, such as fighting disease. "In weightlessness, you can grow any size crystals and they won't break. Then you can look at them under microscopes and study their exact structures." By studying protein structures, scientists can build new molecules that copy the protein, Jemison explains. "We took up some proteins that we think are responsible for the immune system responding to AIDS. If we can grow large crystals and see their structure, we might be able to help make drugs to fight the disease."

Experiencing Weightlessness

Some people think there is no gravity outside Earth's atmosphere. In fact, the shuttle is affected by Earth's gravity, though less so than on the planet's surface. So why do astronauts experience weightlessness? The answer has to do with the rate at which the shuttle travels. To maintain its orbit, the shuttle has to travel about 17,400 miles per hour. Even when traveling at this great speed, the shuttle and the astronauts are falling back to Earth. It's just that the shuttle is moving fast enough so that the path of its fall matches the curvature of Earth. The astronauts inside the shuttle are also falling, so gravity does not press them against the shuttle. Because the astronauts are in a constant state of falling, they feel weightless.

Stages of a Shuttle Mission

ORBIT

Maneuvering for atmospheric entry

Maneuvering for orbit

External tank separation
Main engine cut off

Atmospheric entry

Solid rocket booster separation

External tank impact

Solid rocket booster landing

LIFT-OFF

LANDING

EXPLORE AND CONNECT

Mae Jemison floating weightless during the *Endeavour* mission.

1. Project Create a poster that gives an idea of what it would be like to travel at 17,000 miles per hour. For example, you may want to compare this speed to speeds of other fast-moving objects.

2. Research Find out the requirements for becoming an astronaut. What are some reasons astronauts need to study mathematics? What types of science experiments have astronauts performed in space?

3. Writing Mae Jemison began to study Russian when she was in high school. Several events in the 1950s and 1960s may have convinced her that Russian would become an important language for space scientists. Create a time line of events in space travel that support this conclusion.

Mathematics & Mae Jemison

In this chapter, you will learn more about how mathematics is related to space science.

Related Examples and Exercises

Section 3.1
• Examples 1 and 2
• Exercises 19–22

Section 3.2
• Exercises 7 and 8

3.1

Learn how to...

- find unit rates from words and graphs

So you can...

- compare real-world rates, such as those of the space shuttle and the Concorde jet

Applying Rates

It did not take long for astronaut Mae Jemison to leave Earth far behind as she rocketed into space aboard the space shuttle *Endeavour*. The shuttle climbed about 71 mi in about 8.5 min. You can use this information to describe the shuttle's average *rate* of ascent.

A **rate** is a ratio that compares two units. $\dfrac{71 \text{ mi}}{8.5 \text{ min}}$

> *Endeavour*'s rate of ascent is 71 mi in 8.5 min.

Rates are often expressed as *unit rates.*
A **unit rate** is a rate per one given unit. $\dfrac{8.4 \text{ mi}}{1 \text{ min}}$

> Find the unit rate in miles per minute. Divide 71 by 8.5 and round to the nearest tenth.

To get an idea of how fast 8.4 mi/min really is, you can convert it to units that are more familiar, such as miles per hour.

Toolbox p. 586
Ratios and Rates

EXAMPLE 1   Interview: Mae Jemison

Express *Endeavour*'s average rate of ascent, 8.4 mi/min, in miles per hour.

SOLUTION

$$\text{Rate of ascent} = 8.4 \text{ mi/min}$$

$$= \frac{8.4 \text{ mi}}{1 \text{ min}} \cdot \frac{60 \text{ min}}{1 \text{ h}}$$

> Use the fact that 60 min = 1 h. Multiply by $\frac{60 \text{ min}}{1 \text{ h}}$, which is equal to 1.

$$= \frac{8.4 \text{ mi} \cdot 60 \text{ min}}{1 \text{ min} \cdot 1 \text{ h}}$$

> The minute units divide out.

$$= \frac{8.4 \text{ mi} \cdot 60}{1 \text{ h}}$$

$$= \frac{504 \text{ mi}}{1 \text{ h}}$$

> Simplify.

Endeavour's average rate of ascent was about 504 mi/h.

THINK AND COMMUNICATE

1. Describe how to convert *Endeavour*'s average rate of ascent to miles per second. Which units give you the best idea of the speed of the shuttle?

2. Give some examples of rates you use in everyday life. Are any of these rates unit rates?

EXAMPLE 2   Interview: Mae Jemison

The interview with Mae Jemison gives the speed of the space shuttle in orbit as about 17,400 mi/h. How fast is this in feet per second?

SOLUTION

First change miles per *hour* to miles per *second.*

$$\frac{17{,}400 \text{ mi}}{1 \text{ h}} \cdot \frac{1 \text{ h}}{3600 \text{ s}} \approx \frac{4.83 \text{ mi}}{1 \text{ s}} \qquad 1 \text{ h} = 3600 \text{ s}$$

Then change *miles* per second to *feet* per second.

$$\frac{4.83 \text{ mi}}{1 \text{ s}} \cdot \frac{5280 \text{ ft}}{1 \text{ mi}} \approx \frac{25{,}502 \text{ ft}}{1 \text{ s}} \qquad 5280 \text{ ft} = 1 \text{ mi}$$

The speed of the space shuttle in orbit is about 25,502 ft/s.

Finding Rates from Graphs

The graph below shows the altitude of the supersonic Concorde jet during a flight from New York to London. You can use the graph to find the Concorde's rate of change in altitude.

EXAMPLE 3   Application: Aviation

During its first ten minutes of flight, the Concorde climbs from ground level to the altitude shown by point *B* on the graph. Estimate the Concorde's rate of change in altitude during this time. Round your answer to the nearest hundred feet per minute.

Altitude of the Concorde During a Transatlantic Flight

The coordinates of point *A* are (**0**, **0**). At takeoff, 0 min have elapsed and the Concorde's altitude is 0 ft.

SOLUTION

Estimate the coordinates of point *B* and point *A*. Use the coordinates to find the Concorde's rate of change in altitude.

$$B \approx (11 \text{ min}, 27{,}000 \text{ ft}) \text{ and } A = (0 \text{ min}, 0 \text{ ft})$$

$$\text{Rate of change} = \frac{\text{Change in altitude}}{\text{Change in time}}$$

$$= \frac{\text{Altitude at point } B - \text{Altitude at point } A}{\text{Time at point } B - \text{Time at point } A}$$

$$\approx \frac{27{,}000 \text{ ft} - 0 \text{ ft}}{11 \text{ min} - 0 \text{ min}}$$

$$\approx 2455 \text{ ft/min}$$

The Concorde's rate of change in altitude is about 2500 ft/min.

EXAMPLE 4   Application: Aviation

Between point *G* and point *H* on the graph, the Concorde is preparing to descend into London. Estimate the rate of change in altitude during this time.

SOLUTION

Estimate the coordinates of point *H* and point *G*.

$$H \approx (183 \text{ min}, 39{,}000 \text{ ft}) \text{ and } G \approx (175 \text{ min}, 53{,}000 \text{ ft})$$

$$\text{Rate of change} = \frac{\text{Altitude at point } H - \text{Altitude at point } G}{\text{Time at point } H - \text{Time at point } G}$$

$$\approx \frac{39{,}000 \text{ ft} - 53{,}000 \text{ ft}}{183 \text{ min} - 175 \text{ min}}$$

$$\approx \frac{-14{,}000 \text{ ft}}{8 \text{ min}}$$

$$\approx -1750 \text{ ft/min}$$

The Concorde's rate of change in altitude is about −1800 ft/min.

BY THE WAY

One reason the Concorde flies at such a high altitude is to limit the noise pollution caused by sonic booms. Sonic booms occur when jets first cross the sound barrier by traveling faster than the speed of sound. The Concorde does not cross the sound barrier until it is about 100 mi offshore.

THINK AND COMMUNICATE

Use the graph on page 100.

3. Why is the rate in Example 4 negative? Explain how the graph shows whether a rate of change is positive or negative.

4. Find the rate of change in altitude between points *E* and *F* on the graph. Is this rate greater than or less than the rate between points *A* and *B*? How does the graph visually show that one rate is greater than the other?

5. Can the rate of change between any two points ever be 0? Explain. If possible, give an example from the graph.

Copy and complete each equation.

1. $\dfrac{\$6.75}{1 \text{ h}} \cdot \dfrac{8 \text{ h}}{1 \text{ day}} = \underline{?}$ **2.** $\dfrac{\$7}{1 \text{ h}} \cdot \dfrac{40 \text{ h}}{1 \text{ week}} \cdot \dfrac{52 \text{ weeks}}{1 \text{ year}} = \underline{?}$

3. $\dfrac{21 \text{ mi}}{1 \text{ gal}} \cdot 15 \text{ gal} = \underline{?}$ **4.** $\dfrac{24 \text{ h}}{1 \text{ day}} \cdot \underline{?} = \dfrac{168 \text{ h}}{1 \text{ week}}$

5. **SPORTS** In 1988, Florence Griffith-Joyner set an Olympic record of 10.54 s in the women's 100 m dash. Find her average speed in kilometers per hour. Then use the fact that 1 km ≈ 0.62 mi to give her average speed in miles per hour.

Use the graph in Example 3 on page 100. Find the rate of change in altitude between the altitudes indicated by each pair of points on the graph.

6. *B* and *C*   **7.** *C* and *D*   **8.** *I* and *J*

3.1 **Exercises and Applications**

**Extra Practice
exercises on
page 561**

Copy and complete each equation.

1. $\dfrac{18 \text{ rooms}}{1 \text{ dormitory}} \cdot \dfrac{2 \text{ students}}{1 \text{ room}} = \underline{?}$ **2.** $\dfrac{\$2.12}{8.2 \text{ min}} \cdot \underline{?} \approx \dfrac{\$15.51}{1 \text{ h}}$

3. $\dfrac{1500 \text{ mi}}{1 \text{ s}} \cdot \underline{?} = \dfrac{90,000 \text{ mi}}{1 \text{ min}}$ **4.** $\dfrac{50 \text{ km}}{1 \text{ h}} \cdot \dfrac{1 \text{ h}}{3600 \text{ s}} \cdot \underline{?} \approx \dfrac{13.9 \text{ m}}{1 \text{ s}}$

Express each rate in the given units. For conversion rates, see the Table of Measures on page 611.

5. 25 cm/s = $\underline{?}$ km/h   **6.** 186,282 mi/s = $\underline{?}$ mi/h

7. 5280 ft/mi = $\underline{?}$ yd/mi   **8.** $15,900 per year = $\underline{?}$ per week

9. 4.2 mi/h = $\underline{?}$ ft/min   **10.** $13 per hour = $\underline{?}$ cents per minute

11. **ASTRONOMY** During the Perseid meteor shower, Lance counted 9 shooting stars in 15 min. As shown below, he tried to estimate the number of shooting stars per hour.

$$\dfrac{9 \text{ shooting stars}}{15 \text{ min}} \cdot \dfrac{1 \text{ h}}{60 \text{ min}} = \dfrac{0.01 \text{ shooting stars}}{1 \text{ h}}$$

What is wrong with his work? Show how to find the correct rate.

BY THE WAY

The Perseid meteor shower lasts from mid-July through mid-August every year, as Earth passes through an area of space with a large concentration of dust particles. These particles ignite when they pass through Earth's upper atmosphere.

12. Use the graph at the right to estimate the average rate of change in the number of job openings for each time period. Round your answers to the nearest ten thousand.

   a. 1987–1988

   b. 1988–1989

   c. 1989–1990

   d. 1990–1991

Job Openings per Year

SPORTS For Exercises 13–16, use the news clipping.

13. Give two rates that indicate the team's scoring ability. Be sure to include units in your answers.

14. Express home game attendance as a unit rate.

15. Suppose the fullback carries the ball an average of 20 times per game. About how many yards does he gain per game?

16. A quarterback's pass completion rate is the ratio of passes completed to passes attempted. Find Fred Sorenson's pass completion rate. Give your answer as a percent.

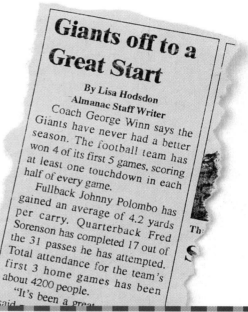

Giants off to a Great Start

By Lisa Hodsdon
Almanac Staff Writer

Coach George Winn says the Giants have never had a better season. The football team has won 4 of its first 5 games, scoring at least one touchdown in each half of every game.

Fullback Johnny Polombo has gained an average of 4.2 yards per carry. Quarterback Fred Sorenson has completed 17 out of the 31 passes he has attempted. Total attendance for the team's first 3 home games has been about 4200 people.

"It's been a great...

Connection | **MUSIC**

The compact disk (CD) was first sold to the public in 1982. During the years that followed, music buyers could choose to buy long-playing records (LPs) or CDs. Use the graph for Exercises 17 and 18.

Sales of LPs and CDs in the United States

- CD Sales
- LP Sales

17. For each time period, find the average rate of change for LP sales and the average rate of change for CD sales. Organize your results in a table. Round your answers to the nearest ten million.

   a. 1984–1985   **b.** 1985–1986

   c. 1986–1987   **d.** 1987–1988

18. Writing Take yourself back in time to the year 1988. A music producer wants to know how many CDs and LPs to produce in 1989. Write a letter describing a good business strategy. Use rates of change to support your suggestions.

Look back at the article on pages 96–98.

Acceleration is the rate of change in velocity over time. The graph shows how **Endeavour**'s *velocity changed during its first eight minutes after takeoff. The velocity is measured in feet per second, so acceleration is measured in feet per second per second. (The units for acceleration are written ft/s².)*

Velocity of *Endeavour*

Acceleration Between Pairs of Points	
Pairs of points	**Acceleration (ft/s²)**
A and *B*	?
B and *C*	?
C and *D*	?
D and *E*	?
E and *F*	?
F and *G*	?
G and *H*	?
H and *I*	?

19. Copy and complete the table. Round each acceleration to the nearest ten.

20. Describe *Endeavour*'s acceleration over time. When did *Endeavour* accelerate the most? When did it accelerate the least? Describe any trends you see.

21. *Endeavour* became lighter as it rose because it used up more and more fuel. How does this factor seem to affect *Endeavour*'s velocity and acceleration over time? Explain your reasoning.

Not drawn to scale

188 mi

3960 mi

diameter of orbit

22. Challenge The diagram shows that *Endeavour*'s orbit was about 188 mi above Earth. While in orbit, *Endeavour* traveled at a rate of about 17,400 mi/h.

  a. Find the distance traveled by *Endeavour* during one orbit around Earth. Explain how you got your answer.

  b. Use what you know about converting rates to estimate the number of orbits *Endeavour* made during its eight days in space. Explain how you made your estimate.

WEATHER For Exercises 23–25, use the table showing rainfall amounts.

• • • • World Record Rainfall Amounts • • • •		
Location	Amount	Duration
Unionville, Maryland (U.S.)	3.1 cm	1 min
Curtea-de-Arges (Romania)	20.5 cm	20 min
Holt, Missouri (U.S.)	30.5 cm	42 min
Cilaos, La Réunion	188 cm	24 h
Cherrapunji (India)	930 cm	1 month

23. Write a rate and a unit rate for each location. Use the same units for all the unit rates.

24. In Exercise 23, what information does each rate give you that the unit rate does not? What information does the unit rate give you that the rate does not?

25. Suppose rain continued to fall at a rate of 3.1 cm/min in Unionville. How long would it take to break Cherrapunji's record? Explain.

26. **Open-ended Problem** A child's height is an example of a variable showing a positive rate of change over time. Give two examples of a variable showing a negative rate of change over time. For one of these variables, estimate a rate of change. Explain your estimate.

27. **SAT/ACT Preview** Vera has $20. The cost of a pair of socks ranges from $1.75 to $2.60. What is the greatest number of pairs she can buy?

   A. 7   **B.** 8   **C.** 11   **D.** 9   **E.** 10

ONGOING ASSESSMENT

28. **Writing** Write a story that relates the graph and the picture shown at the right. Be sure to include rates in your story.

SPIRAL REVIEW

Write each fraction as a decimal. *(Toolbox, page 587)*

29. $\dfrac{3}{8}$ 30. $\dfrac{1}{2}$ 31. $\dfrac{5}{4}$ 32. $\dfrac{7}{11}$

Graph each point in a coordinate plane. Label each point with its letter. Name the quadrant (if any) in which the point is located. *(Section 2.4)*

33. $A(4, 7)$ 34. $B(3, 23)$ 35. $C(28, 25)$ 36. $D(-12, 6)$

3.2 Exploring Direct Variation

The artist Leonardo da Vinci (1452–1519) studied human proportions in order to make more accurate drawings. He observed that the kneeling height of a person is $\frac{3}{4}$ of the person's standing height. In the Exploration, you will test Leonardo da Vinci's observation.

EXPLORATION
COOPERATIVE LEARNING

Patterns and Direct Variation

Work in a group of at least 6 students.
You will need:
- a meterstick
- graph paper or a graphing calculator

1 Measure and record the kneeling and standing heights of each person in your group.

2 For each person, find the ratio of kneeling height to standing height. Compare your group's results with results from other groups.

Name	Standing height (cm)	Kneeling height (cm)	Kneeling height / Standing height
Tim	156	119	0.76
?	?	?	?

Standing and Kneeling Heights

3 Make a graph of your data. Plot a point for each student. Do not connect the points. This **scatter plot** shows the relationship between the two sets of data.

4 Based on Leonardo da Vinci's observation, write an equation for kneeling height as a function of standing height. Graph the equation on your scatter plot.

5 Describe any patterns you see. Do you agree with Leonardo da Vinci's observation? Explain.

Whenever two variables have a constant ratio, they are said to show **direct variation**. The constant ratio is called the **constant of variation**.

Let x = standing height.

Let y = kneeling height.

$$\frac{y}{x} \approx 0.75$$

The constant of variation is 0.75.

You can rewrite the equation $\frac{y}{x} = 0.75$ to show that kneeling height y is a function of standing height x.

$$x \cdot \frac{y}{x} = 0.75x$$

Multiply both sides of the equation by x.

$$y = 0.75x$$

Kneeling height is a function of standing height.

Direct Variation

A direct variation can be described by an equation in this form:

$$y = kx, \text{ or } \frac{y}{x} = k, \text{ where } k \neq 0$$

The letter k represents the constant of variation.

You say that y *varies directly* with x.

EXAMPLE 1   **Application: Employment**

Do earnings vary directly with the number of hours worked? If so, give the constant of variation and write an equation that describes the situation.

Hours worked	2	3	4	5
Earnings (dollars)	11.50	17.25	23.00	28.75

SOLUTION

Check to see whether the data pairs have a constant ratio. The data show direct variation because the data pairs have a constant ratio.

The constant of variation is 5.75.

Let E = earnings in dollars, and h = hours worked.

$$E = 5.75h$$

$\dfrac{\text{Earnings}}{\text{Hours worked}}$	
$\dfrac{11.50}{2} = 5.75$	
$\dfrac{17.25}{3} = 5.75$	
$\dfrac{23.00}{4} = 5.75$	
$\dfrac{28.75}{5} = 5.75$	

Application: Consumer Economics

Do the prices of different containers of juice vary directly with their sizes? If so, give the constant of variation and write an equation.

Size (oz)	10	48	64	64
Price (dollars)	0.40	1.99	1.69	2.99

SOLUTION

The data pairs do not have a constant ratio. The data do not show direct variation.

Juice Prices			
C2		=B2/A2	
	A	**B**	**C**
1	Size (oz)	Price (dollars)	Ratio
2	10	0.40	0.04
3	48	1.99	0.04
4	64	1.69	0.03
5	64	2.99	0.05

THINK AND COMMUNICATE

1. In Example 1, what real-world information does the constant of variation give you? Explain why the constant of variation is a unit rate.

2. A person buys bananas for $.63 per pound. How does this situation show direct variation? What are the variables? What is the constant of variation? Describe another situation that shows direct variation.

Direct Variation Graphs

The graph of a direct variation equation is always a line through the origin. You can use this fact to see whether data show direct variation. For example, the scatter plot below shows data collected by one group of students in the Exploration on page 106. The data points on their scatter plot lie on or close to the graph of the equation $y = 0.75x$, so the data show direct variation.

Standing and Kneeling Heights

The points on the scatter plot lie on or close to a line through the origin.

THINK AND COMMUNICATE

3. The graph at the right shows data from one of the Examples in this section. Which data does the graph represent?

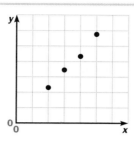

☑ CHECKING KEY CONCEPTS

Tell whether the data show direct variation. If they do, give the constant of variation and write an equation that describes the situation.

1.

2.

PHYSICS A group of students dropped a ball from different heights and measured the bounce heights. The table shows their data.

3. Bounce height varies directly with drop height. Give the constant of variation and write an equation that describes the situation.

4. Make a scatter plot of the data. Graph your equation from Question 3 in the same coordinate plane. What does your scatter plot show?

Drop height (in.)	Bounce height (in.)
20	15.0
30	24.0
40	32.5
50	41.0
60	48.5

3.2 | **Exercises and Applications**

Tell whether the data show direct variation. If they do, give the constant of variation and write an equation that describes the situation.

Extra Practice exercises on page 561

1.

Time (h)	Distance (mi)
0.5	15.0
1.5	24.0
3.0	32.5
4.5	41.0

2.

Weight (lb)	Weight (kg)
50	22.7
75	34.0
120	54.4
180	81.6

3.

Year	Cable subscribers
1970	5,100,000
1980	17,500,000
1985	35,430,000
1990	50,500,000

For Exercises 4 and 5, decide whether each scatter plot suggests that the data show direct variation. Give reasons for your answers.

4.

5.
Overdue Charges

6. **Open-Ended Problem** Suppose you buy cans of juice from a vending machine. The cost varies directly with the number of cans you buy. Describe three other real-world situations that show direct variation. For one of the situations, write an equation and draw a graph.

INTERVIEW **Mae Jemison**

Look back at the article on pages 96–98.

The force of gravity affects how much you weigh. Astronauts use the letter g to stand for the pull of Earth's gravity on an object at sea level. Accelerating away from Earth increases the "g-force," making you weigh more than you would ordinarily.

BY THE WAY

NASA used this centrifuge to give astronauts the experience of high g-forces. A force of 3 g's is considered the maximum force under which astronauts can function easily. During liftoff for the Apollo missions to the Moon, astronauts experienced g-forces as high as 8 g's.

7. During *Endeavour*'s launch, the astronauts experienced a maximum force of 3 g's. This made them weigh 3 times their normal weight. Use a graph to show that maximum weight varied directly with normal weight.

8. The weight of an object on another planet varies directly with its weight on Earth. Use the table to answer the following questions.

   a. Let w = the weight in pounds of an object on Earth. Write an equation for the weight of the object on each planet.

   b. **Writing** Explain how weight varies on different planets. Be sure to use equations and talk about direct variation in your answer. Include graphs.

Weight of a 100 lb Object on Various Planets	
Mercury	36.5 lb
Venus	90.5 lb
Mars	38 lb
Jupiter	266 lb
Saturn	114 lb
Uranus	107 lb
Neptune	135 lb
Pluto	22.5 lb

In his notebooks, Leonardo da Vinci described how proportions can be used to draw figures accurately. For example, he observed that "... from the top to the bottom of the chin ... is the sixth part of the face, and it is the fifty-fourth part of the man."

9. Write an equation describing chin length l as a function of face length f.

10. Write an equation describing chin length l as a function of height h.

11. Challenge Based on Leonardo's observation, do you think a person's face length varies directly with the person's height? If so, find the constant of variation. Explain how you got your answer.

12. Cooperative Learning Leonardo da Vinci wrote "... the fourth part of a man's height ... is equal to the greatest width of the shoulders." Work in a group of 4–6 students to test this observation.

   a. Measure and record the distance across each person's shoulders. Measure and record each person's height.

   b. Make a scatter plot of your group's data. Put height on the horizontal axis and shoulder width on the vertical axis.

   c. Do your data show direct variation? Do your data support Leonardo da Vinci's observation? Explain why or why not.

BY THE WAY

Leonardo da Vinci often used mirror writing in his notebooks. He was left-handed and may have found writing backward as easy as writing forward.

BUSINESS Use the graphs at the right for Exercises 13 and 14.

13. In her new sales job, Mina Joshi will not make any commissions until she has sold $1000 of merchandise. Then she will receive a 5% commission on every sale. Which graph shows the amount Mina will receive in commissions? Explain how you know.

14. Does the amount Mina receives in commissions vary directly with her sales? Give a reason for your answer.

A. Mina's Commissions

B. Mina's Commissions

15. GEOMETRY Write an equation for the perimeter of each of the figures shown at the right. For each figure, decide whether the perimeter varies directly with *x*. Give reasons for your answers.

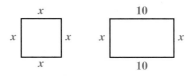

16. Writing Explain why the graph of an equation in the form $y = kx$ always goes through the origin. Give an example of a graph that shows direct variation and one that does not show direct variation.

17. Find the mean, the median, and the mode(s) of the data shown in the table, if each of the three secretaries has a $25,000 salary. *(Section 1.3)*

Salaries at a Small Company	
President	$135,000
Vice-president	$90,000
Office manager	$60,000
3 Secretaries	$25,000

Graph each equation and each point.
Tell whether the ordered pair is a solution of the equation. *(Section 2.5)*

18. $y = 2x + 2$; (2, 6)

19. $y = \frac{2}{3}x + 1$; (3, 4)

20. $y = 5 - 2x$; (−1, 3)

21. $y = 6x - 2$; (1, 4)

ASSESS YOUR PROGRESS

VOCABULARY

rate (p. 99)
unit rate (p. 99)
scatter plot (p. 106)

direct variation (p. 107)
constant of variation (p. 107)

Express each rate as a unit rate in kilometers per hour. *(Section 3.1)*

1. 0.7 km in 2 min   **2.** 500 km in 5 days   **3.** 1 m/s

Tell whether the data show direct variation. If they do, give the constant of variation and write an equation that describes the situation. *(Section 3.2)*

4.

Age of car (years)	Mileage
1	10,020
2	30,023
3	50,072
4	90,365

5.

Tickets sold	Profit (dollars)
50	150
100	300
150	450
200	600

6. Journal Give some examples of direct variation that can help you remember what direct variation means.

3.3

Finding Slope

Learn how to...
- **find the slope of a line**

So you can...
- analyze a real-world graph showing fuel consumption, for example

The cheetah is the fastest animal on land. Race horses are among the fastest domestic animals. The graph shows the distances these animals can cover when traveling at top speeds.

The steepness of a line is called its **slope**. The slope of a line is a rate of change.

$$\text{Slope} = \frac{\text{Vertical change}}{\text{Horizontal change}}$$

$$= \frac{27}{1}$$

The slope of the cheetah's line is 27.

Animal Speeds

THINK AND COMMUNICATE

1. Estimate the slope of the horse's line. What information does the slope of the line give you?

2. Which line is steeper? What does this tell you about the speeds?

EXAMPLE 1

Graph the equation $y = 2x + 3$. Find the slope of the line.

SOLUTION

Step 1 To graph the equation on a coordinate plane, make a table of values and plot the points.

Step 2 To find the slope of the line, choose two points and count the vertical change and the horizontal change between them.

$$\text{Slope} = \frac{\text{Vertical change}}{\text{Horizontal change}}$$

$$= \frac{4}{2}$$

$$= 2$$

The slope of the line is 2.

x	y
1	5
0	3
−1	1

BY THE WAY

Cheetahs can maintain top speeds for about 300 m. They are an endangered species because of the destruction of their habitat and the desirability of their fur.

Positive and Negative Slope

A line that rises as you move from left to right has a **positive slope**. A line that falls as you move from left to right has a **negative slope**.

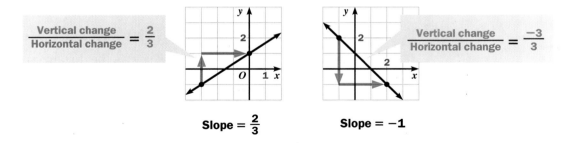

$$\frac{\text{Vertical change}}{\text{Horizontal change}} = \frac{2}{3}$$

$$\frac{\text{Vertical change}}{\text{Horizontal change}} = \frac{-3}{3}$$

Slope = $\frac{2}{3}$

Slope = -1

EXAMPLE 2 **Application: Energy Use**

Rob Mueller had 12 gal of gasoline in his car. After he drove 160 mi on the highway, there were 6 gal left. The graph shows the amount of gasoline left in his car as a function of the number of miles driven. Find the slope of the line. What does the slope tell you about the situation?

SOLUTION

$$\text{Slope} = \frac{\text{Vertical change}}{\text{Horizontal change}}$$

$$= \frac{6 - 12}{160 - 0}$$

In many real-world situations, it may make sense to write the slope as a decimal.

$$= \frac{-6}{160}$$

$$= -0.0375$$

The slope of the line is -0.0375. For every mile Rob drives, the car burns about 0.04 gal of gasoline. The slope is negative because the amount of gasoline in the car is decreasing.

Rob Mueller's Car

The amount of gasoline in the tank decreases by 6 gal when the car is driven 160 mi.

Amount of gasoline in tank (gal)

Distance driven (mi)

THINK AND COMMUNICATE

Use the information in Example 2.

3. Suppose you graphed the distance Rob's car travels as a function of time. Would the slope of the graph be positive or negative? Explain.

4. Rob's car uses more gasoline when he drives on city streets than it does on the highway. Suppose you graphed the amount of gasoline left in Rob's car after he drove 160 mi in the city. How would the graph be different from the graph shown in Example 2? How would the slope be different?

Example 2 shows that you can calculate slope if you know the coordinates of two points on a line.

$$\textbf{Slope} = \frac{\textbf{Vertical change}}{\textbf{Horizontal change}} = \frac{y_2 - y_1}{x_2 - x_1}$$

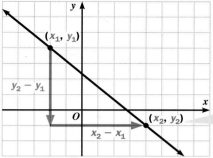

Subscripts are used to designate the coordinates of the **first** point (x_1, y_1) and the coordinates of the **second** point (x_2, y_2).

If you use this formula to find the slope of a horizontal line, you will see that the slope is zero. If you use this formula with a vertical line, you will see that the slope of a vertical line is **undefined**.

Horizontal line

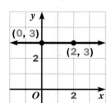

$$\frac{\textbf{Vertical change}}{\textbf{Horizontal change}} = \frac{3 - 3}{2 - 0}$$

$$= \frac{0}{2}$$

The slope of a horizontal line is always zero.

Vertical line

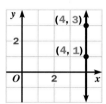

$$\frac{\textbf{Vertical change}}{\textbf{Horizontal change}} = \frac{3 - 1}{4 - 4}$$

$$= \frac{2}{0}$$

Division by zero is not possible. The slope of a vertical line is always undefined.

☑ CHECKING KEY CONCEPTS

For Questions 1–5, use the coordinate grid.

1. Which lines have a positive slope?

2. Which lines have a negative slope?

3. Which line has a slope of zero?

4. Which line has an undefined slope?

5. Which line has the greatest slope?

6. Calculate the slope of the line through the points $(-4, 5)$ and $(1, -4)$.

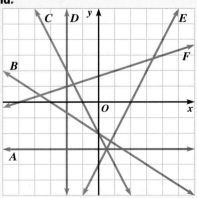

3.3 Exercises and Applications

Extra Practice
exercises on
page 562

Find the slope of each line.

1.

2.

3.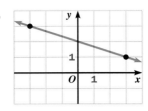

Graph each equation. Find the slope of the line.

4. $y = 3x - 2$

5. $y = 10 - x$

6. $y = -5$

7. $y = \frac{1}{2}x + 7$

8. $y = \frac{1}{4}x$

9. $y = -2x - \frac{1}{2}$

Find the slope of the line through each pair of points.

10. $(1, 3)$ and $(-2, 1)$

11. $(0, 0)$ and $(-5, 3)$

12. $(5, -1)$ and $(1, 6)$

13. $(2, 12)$ and $(2, -9)$

14. $(3, 3)$ and $(-7, -7)$

15. $(45, 3)$ and $(30, 35)$

16. $(-12, 5)$ and $(23, 5)$

17. $(5, 22)$ and $(126, 7)$

18. $(7, -3)$ and $(-11, 22)$

19. Mona Jhaveri asked two students
to find the slope of the line through
the points $(3, -1)$ and $(-2, 2)$.

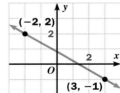

a. Writing Describe each
student's method. Explain
why both methods work.

b. Look back at the formula
$\frac{y_2 - y_1}{x_2 - x_1}$ on page 115.
Does it matter which
coordinates are labeled
(x_1, y_1) and which are
labeled (x_2, y_2)? Explain.
Give examples to support your answer.

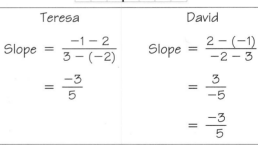

20. CIVIL ENGINEERING A road's slope, or grade, is usually expressed as a
percent. For example, a section of road that rises 5 ft for every 100 ft of
horizontal distance has a 5% grade. The United States Department of
Transportation says that a sign like the one shown should be used when a
section of road has an 8% grade or greater and is more than 750 ft long.
Find the grade of each road and tell whether the sign should be used.

a. A road 803 ft long rises 65 ft over a horizontal distance of 800 ft, as
shown in the diagram.

b. A road 975 ft long rises 75 ft over a horizontal distance of 972 ft.

c. A road 650 ft long rises 57 ft over a horizontal distance of 647 ft.

Connection ARCHITECTURE

The tread of a stair is the flat part that you step on. The riser is the vertical part of the stair. The ratio of the height of the riser to the depth of the tread can affect the safety of the staircase. The diagram shows a traditional staircase design and a new design. Building code officials prefer the new design because they say it is safer. Builders prefer the old design because it is less costly.

21. Open-ended Problem Why do you think the old design costs less to build? Why might the new design be safer? Give reasons for your answers.

TRADITIONAL DESIGN
9 INCH TREAD
8.25 INCH RISER

LANDING   LANDING

NEW DESIGN
11 INCH TREAD
7 INCH RISER

22. a. An architect is designing a house with a staircase 9 ft high and 14 ft long, as shown at the left. Which design is the architect using? How do you know?

b. How long would a 9 ft high staircase be if the architect were using the other design? Explain your reasoning.

9 ft

14 ft

23. Research Find the slopes of at least three different staircases near your home or school. Are they the old design, the new design, or neither? Which do you think is safest? Why?

24. Challenge The American National Standards Institute (ANSI) specifications state that the maximum slope of a wheelchair ramp should be $\frac{1}{12}$.

a. Does a ramp 7 in. high and 94 in. long meet the ANSI specifications?

b. Does a ramp with slope $\frac{1}{6}$ meet the ANSI specifications? Explain.

c. A 14 in. high ramp needs to be built. Find the length l of a ramp that meets the ANSI specifications.

25. **Visual Thinking** Tanya filled two jars with water. She poured the water into each jar at the same rate. Then she made a graph showing the depth of the water in each jar over time. On her graph, which line represents the water level in which jar? Explain your reasoning.

26. **Cooperative Learning** Graph a line that has a slope of −2. Compare your graph to other students' graphs. What patterns do you see?

27. **a.** Give the constant of variation for each direct variation equation.

$$y = 4x \qquad y = -7x \qquad y = \frac{3}{5}x$$

   b. Graph each equation in part (a). Find the slope of the line.

   c. For each equation, what relationship do you see between the slope of the line and the constant of variation?

28. **SAT/ACT Preview** What is the slope of a line through the points (0, 3) and (3, 3)?

   A. 3   **B.** −3   **C.** 1   **D.** 0   **E.** undefined

Typing Speed

29. **a.** Use the graph at the left. Give the slope of each line. What information do the slopes give you about the typists?

   b. **Open-ended Problem** The amount each typist charges varies directly with the person's typing speed. Suppose you want to hire someone to type a paper for you. What is a good reason for hiring **Bob**? for hiring **Paul**? Be sure to talk about slopes and rates.

SPIRAL REVIEW

Decide whether each table represents a function. Explain your reasoning.
(Section 2.3)

30.

n	Number with absolute value equal to n
2	2 and −2
3	3 and −3
5	5 and −5

31.

Height (in.)	Shoe size
69	9
69	9.5
70	9.5
70	10.5

32.

x	$5x$
−2	−10
−1	−5
0	0
1	5
2	10

Express each rate in the given units. For conversion rates, see the Table of Measures on page 611. *(Section 3.1)*

33. 50 cm/h = _?_ km/h

34. 6160 ft/min = _?_ mi/h

35. 2 km/min = _?_ m/h

36. $20 per hour ≈ _?_ cents per minute

3.4

Finding Equations of Lines

An equation whose graph is a line is called a **linear equation**. One way to write a linear equation is to use the form $y = mx + b$. As you will see in the Exploration, the numbers m and b give you information about the graph of a linear equation.

Learn how to...

- **write linear equations in slope-intercept form**

So you can...

- **analyze real-world problems represented by graphs**

For more information on setting the viewing window, see the *Technology Handbook*, p. 603.

For more information on setting the viewing window, see the *Technology Handbook*, p. 603.

E X P L O R A T I O N

COOPERATIVE LEARNING

Exploring Linear Equations and Graphs

Work with a partner.
You will need:
- a graphing calculator
- graph paper

GETTING STARTED Set the viewing window on your calculator. The *standard viewing window* uses values from −10 to 10 on both the *x*-axis and the *y*-axis.

1 Graph each equation. For each equation give the slope of the line and write the coordinates of the point where the graph crosses the *y*-axis.

a. $y = 2x$ **b.** $y = 2x + 3$ **c.** $y = 2x - 3$

2 Sketch the graphs of the three equations. Describe any patterns you see.

3 Predict what the graph of $y = 2x + 6$ will look like. Predict the slope. Where will the graph cross the *y*-axis? Explain your prediction. Then check it.

4 One way to write linear equations is in the form $y = mx + b$. Based on your observations, what information do the numbers m and b give you about a graph? Use graphs to support your answer.

Slope-Intercept Form

As you can see, the 3 in the linear equation $y = 3x - 2$ is the slope of the graph. The -2 is the y-value of the point where the graph crosses the y-axis. This value is called the **y-intercept**, or the **vertical intercept**.

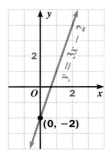

One way to write a linear equation is in **slope-intercept form**.

$$y = mx + b$$

The letter m represents the slope of the line.

The letter b represents the vertical intercept.

EXAMPLE 1

The graph shows the height of a candle as it burns. Find the slope and the vertical intercept of the graph. Write an equation for the height of the candle in slope-intercept form.

SOLUTION

Choose two convenient points to find the slope.

$$\text{Slope} = \frac{9 - 10}{30 - 0}$$

$$= \frac{-1}{30}$$

$$\approx -0.03$$

The slope is about -0.03.

The graph crosses the vertical axis at (0, **10**). The vertical intercept is **10**.

Let h = the height of the candle in centimeters after it has been burning t minutes.

An equation for the height of the candle is $h = -0.03t + 10$.

THINK AND COMMUNICATE

1. In Example 1, what information does the vertical intercept give you about the candle? What information does the slope give you? Why is the slope negative?

2. Estimate the vertical intercept of a graph of the height of a burning birthday candle. How would the line compare to the graph in Example 1? Would the slope be greater or less? Explain.

EXAMPLE 2

Graph the equation $y = 6 - \frac{3}{2}x$.

SOLUTION

Method 1 Use graph paper.

Step 1 Rewrite the equation in slope-intercept form to find the slope and y-intercept.

$$y = 6 - \frac{3}{2}x$$

$$y = -\frac{3}{2}x + 6$$

Step 2 The y-intercept is 6. Graph the point $(0, 6)$.

Step 3 Start at the y-intercept. Count 3 units down and 2 units to the right. Graph a second point.

Step 4 Count down 3 and 2 to the right again to plot a third point. Connect the points with a line.

Method 2 Use a graphing calculator.

Enter the equation $y = 6 - \frac{3}{2}x$ as $y = 6 - (3/2)x$.

WATCH OUT!

Use parentheses to enter $y = 6 - (3/2)x$, so your calculator graphs $y = 6 - \frac{3}{2}x$, not $y = 6 - \frac{3}{2x}$.

THINK AND COMMUNICATE

3. Check that the three points plotted in method 1 of Example 2 are solutions of the equation $y = 6 - \frac{3}{2}x$. How can checking help you make sure a graph is accurate? How can you check a graph on a graphing calculator?

4. In Section 2.5, you learned to graph equations by creating a table of values and plotting points. Compare that method with method 1 of Example 2. What are some advantages and disadvantages of each of the two methods?

☑ CHECKING KEY CONCEPTS

For Questions 1–3, give the slope and vertical intercept of the graph of each equation. Then graph the equation.

1. $y = x + 4$ **2.** $y = \frac{1}{2}x - 3$ **3.** $y = x$

4. Some students performed an experiment by dropping marbles into a jar of water. They measured the height of the water after every five marbles. Then they graphed their data.

   a. Write an equation for the height of the water as a function of the number of marbles.

   b. What do the vertical intercept and the slope of the graph of your equation tell you about the experiment?

3.4 | Exercises and Applications

Extra Practice exercises on page 562

Write an equation in slope-intercept form for each graph.

1.

2.

3.

4.

5.

6.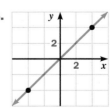

Graph each equation.

7. $y = x - 3$ **8.** $y = 3x + 1$ **9.** $y = 5x - 2$

10. $y = -x$ **11.** $y = -5$ **12.** $x = 3$

13. $x = -\frac{1}{2}$ **14.** $y = -\frac{2}{3}x + 4$ **15.** $y = \frac{5}{3}x + 1$

16. **Open-ended Problem** Describe two different ways to graph the equation $y = \frac{1}{4}x + 2$ by hand. Which method do you prefer? Why?

17. **a.** Explain why the equations in Exercises 10 and 11 are in slope-intercept form. Give the values of m and b for the equation.

    b. **Writing** What is the y-intercept of the graph of a direct variation? Explain your answer, using both graphs and equations.

18. **Cooperative Learning** Work with a partner.

    a. Write three linear equations in slope-intercept form. On a separate sheet of paper, graph each of your equations. Keep your graphs secret. Trade equations with your partner.

    b. Graph the three equations that you received from your partner.

    c. Compare your graph of each equation with your partner's. If they are different, work together to make an accurate graph of the equation.

19. **Challenge** What is true about an equation of any line through the point $(0, -2)$? What is true about any line with a slope of 3? How many lines have slope 3 and pass through the point $(0, -2)$? Explain.

Connection   HISTORY

Alfred the Great (849–899) was king of Wessex in south-western England. He is said to have invented a candle clock. A 12 in. candle was divided into twelve equal parts colored alternately black and white. Three parts burned each hour.

20. Graph the height h of one candle as a function of the time t it has burned.

21. Give the slope and vertical intercept of the graph. Explain what each tells you about the situation.

22. How long did it take for one candle to burn? How do you know?

23. Write an equation of the graph in slope-intercept form.

Egyptian water clock

24. **Research** Another type of clock used before modern mechanical clocks was the clepsydra, or water clock. Find out more about the clepsydra. Write a report and draw a graph that illustrates how the clepsydra worked.

Modern reconstruction of King Alfred's candle clock

BY THE WAY

King Alfred also is said to have invented the "lant-horn" or lantern to protect the candles from drafts that would affect the rate at which the candle burned.

Bracelet Sales

Profit (dollars)

Bracelets sold

25. **a. Writing** Alicia Hartwell owns a jewelry business. Every week she buys materials and sells the bracelets she makes at a crafts market. The graph shows her weekly profit from bracelet sales. What does the slope of the line tell you about Alicia's business? What does the vertical intercept of the line tell you?

  b. The crafts market has decided to charge a weekly membership fee of $15. How does this affect the graph of Alicia's weekly profit? Do you think she should increase the price of her bracelets? Explain your suggestion using graphs and equations.

SPIRAL REVIEW

Solve each equation. *(Sections 2.1 and 2.2)*

26. $4x - 2 = 10$

27. $6t + 1 = 17$

28. $-2m - 9 = 11$

29. $-s + 4s = 45$

30. $2x + 18 = 38$

31. $35y + 17y = 104$

32. $\frac{1}{3}x + 2 = 5$

33. $\frac{5}{6}r + 9 = 4$

34. $\frac{2}{7}z + 5 = 9$

ASSESS YOUR PROGRESS

VOCABULARY

slope (pp. 113, 115)
positive slope (p. 114)
negative slope (p. 114)
undefined slope (p. 115)

linear equation (p. 119)
y-intercept, vertical intercept (p. 120)
slope-intercept form (p. 120)

Find the slope of the line through each pair of points. *(Section 3.3)*

1. (2, 7) and (1, 5)

2. (−1, 0) and (−1, 3)

3. (12, 9) and (9, 12)

4. Sue Hill drove at a constant speed from her home to Tampa. The graph at the right shows her distance from Tampa as a function of time.
(Sections 3.3 and 3.4)

  a. Write an equation for the line in slope-intercept form.

  b. What information do the slope and the vertical intercept give you about the situation?

Sue Hill's Distance from Tampa

Distance (mi)

Time (hours)

Graph each equation. Give the slope and y-intercept of the graph. *(Section 3.4)*

5. $y = 3x - 4$

6. $y = -x + 2$

7. $y = \frac{1}{3}x - \frac{4}{3}$

8. **Journal** Describe two ways to find the slope of the graph of a linear equation. Which do you prefer? Why?

3.5

Writing an Equation of a Line

Learn how to...

- **find the *y*-intercept of a graph**

So you can...

- **model real-world situations, such as the altitude of a hot air balloon**

Every year at the International Balloon Fiesta in Albuquerque, New Mexico, thousands of hot air balloons take to the sky. The graph below shows the altitude of one balloon as it climbs.

This break in the scale means that the *y*-axis is not shown for values of *y* between 0 and 6000.

EXAMPLE 1 **Application: Recreation**

As shown by the slope of the graph, the balloon rises at a rate of 110 ft/min. Write an equation for the balloon's altitude *a* as a function of time *t*.

SOLUTION

Step 1 Write the equation in slope-intercept form.

Altitude = Slope × Time + Vertical intercept

$$a = mt + b$$

Step 2 Substitute the slope of the line for *m* in the equation.

$$a = 110t + b$$

Step 3 To find the vertical intercept, substitute the coordinates of any point on the graph for *t* and *a* in the equation. Then solve for *b*.

$$6135 = 110(1) + b \qquad \text{Choose the point } (1, 6135).$$
$$6135 = 110 + b$$
$$6025 = b \qquad \text{Solve for } b.$$

Step 4 Substitute **6025** for *b* in the equation $a = 110t + b$.

An equation for the altitude of the balloon as a function of time is $a = 110t + 6025$.

THINK AND COMMUNICATE

1. In Example 1, why can you substitute the coordinates of the point (1, 6135) into the equation and solve for b? Can you use another point? Explain why or why not.

2. Federal aviation regulations require that balloons at the International Balloon Fiesta fly at an altitude of at least 6500 ft while over the Fiesta grounds. How long will it take the balloon in Example 1 to reach this altitude?

EXAMPLE 2

Find an equation of the line through the points (2, 6) and (3, 1).

SOLUTION

Draw a graph.

Step 1 The equation will be in slope-intercept form:

$$y = mx + b$$

Step 2 Find the slope.

$$\text{slope} = \frac{1-6}{3-2} = \frac{-5}{1} = -5$$

Substitute -5 for m in the equation.

$$y = -5x + b$$

Step 3 Find the *y*-intercept. Substitute the coordinates of any point on the line for *x* and *y* in the equation. Then solve for *b*.

$$6 = -5(2) + b \qquad \text{Choose the point (2, 6).}$$
$$6 = -10 + b$$
$$16 = b \qquad \text{Solve for } b.$$

Step 4 Substitute **16** for *b* in the equation.

An equation of the line is $y = -5x + 16$.

THINK AND COMMUNICATE

3. In Example 2, substitute the coordinates of the second point for x and y in the equation $y = -5x + b$. Why do you get the same value for b?

4. a. Graph the horizontal line through (1, 2) and (5, 2). Write an equation for this line.

b. What is the slope of any horizontal line?

c. What do all equations of horizontal lines have in common?

EXAMPLE 3

Writing Find an equation of the vertical line through the point (−2, 3). Explain your reasoning.

SOLUTION

Here is one student's solution.

The student included a diagram.

First I drew a sketch of the line.

Next I picked a second point on the line (-2, 1), and tried to find the slope.

There was a zero in the denominator, so I knew the slope was undefined. I started looking at some other points. The coordinates of every point I chose looked like (-2, y). If I plot every point with x = -2, I get the vertical line, so an equation of the line is x = -2.

The student explained each step in reaching a solution.

✓ CHECKING KEY CONCEPTS

Find an equation of the line with the given slope and through the given point.

1. slope = 2; (0, −3)   **2.** slope = −2; (5, 7)

3. slope = 0; (3, 4)   **4.** slope is undefined; (4, 6)

Find an equation of each line.

5.   **6.**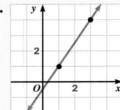

3.5 | **Exercises and Applications**

Find an equation of the line with the given slope and through the given point.

Extra Practice exercises on page 563

1. slope = 1; (5, 8)   **2.** slope = 4; (5, 9)   **3.** slope = −3; (7, 10)

4. slope = $\frac{1}{3}$; (3, 9)   **5.** slope = −2; (−4, 8)   **6.** slope = −$\frac{1}{2}$; (1, 6)

7. slope = 0; (−2, 5)   **8.** slope = −0.5; (3, −2)   **9.** undefined slope; (1, 1)

10. A balloon at the International Balloon Fiesta lifted off from a different location from the one in Example 1. It reached an altitude of 6125 ft above sea level 1 min into its flight. After 3 min, it reached an altitude of 6245 ft. Write an equation for the balloon's altitude a after t min.

11. A line passes through the points (0, 4) and (5, 0).

   a. Graph the line and find an equation for it.

   b. A line passes through the point (0, 6) and has the same slope as the line in part (a). Graph this line and find an equation for it.

   c. **Open-ended Problem** Write an equation for another line with the same slope as the lines in parts (a) and (b). Explain how you found this line. What can you say about these lines and their equations?

Find an equation of the line through the given points.

12. (8, 4), (10, 2)   **13.** (12, 18), (−22, 18)   **14.** (7, 1), (7, 4)

15. (0, −3), (6, 9)   **16.** (0.5, −2), (2.5, −2)   **17.** (−2, 5), (4, −5)

18. (15, −3), (15, 2)   **19.** (−1, 12), (6, −7)   **20.** (−5, −7), (7, 6)

21. Challenge In the graph at the right, an equation of line j is $y = x$.

   a. Write an equation that could be an equation of line l. Write another for line k.

   b. Write an inequality that describes all possible slopes for line k. Write inequalities that describe all possible slopes for line l.

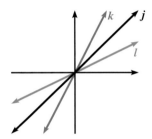

Connection   GEOGRAPHY

The Grand Canyon has a large range in temperatures. Park rangers suggest that if you visit the park in June, you should plan on temperatures like those in the table.

South Rim 7050 ft

North Rim 8340 ft

Inner Gorge 2450 ft

8000 ft
6000 ft
4000 ft
2000 ft
Sea Level

	Temperature (°F)	
	High	Low
South Rim	81	47
Inner Gorge	101	72
North Rim	73	40

22. Make a scatter plot with elevation on the horizontal axis and high temperature on the vertical axis.

23. Use the temperatures at the South Rim and Inner Gorge to find an equation for high temperature as a function of elevation. Use your equation to find the expected temperature at the North Rim.

24. Repeat Exercise 23 using the low temperature data.

25. What high and low temperatures would you expect to find at an elevation of 4600 ft on the trail down to the Inner Gorge? at 5200 ft?

The Colorado River runs through the Inner Gorge.

Ruth Emerson thought that her oven was not heating to the temperature on the dial. She recorded the oven temperature at several dial settings. For Exercises 26 and 27, use her results as shown in the table below.

26. Make a scatter plot of Ruth's results. Write an equation for the temperature of Ruth's oven as a function of the dial setting. What assumptions did you make?

27. **Writing** Write instructions that explain how to find the correct dial setting for Ruth's oven.

Dial Setting (°F)	Oven Temperature (°F)
300	325
320	350
400	450

28. **Writing** Cynthia Wong frequently visits the science museum with her two children. She has three options for paying admission:

    (a) Pay admission for one adult and two children each time she visits.

    (b) Buy a family membership.

    (c) Buy an individual membership, and pay admission for two children.

    For each of the three options, write an equation for cost as a function of the number of visits. Graph each equation. What do the slope and the vertical intercept of each line tell you about that option?

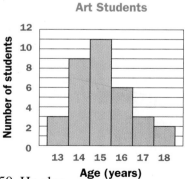

THE MUSEUM OF SCIENCE

■ General admission fees
  Adults $8.00
  Children (3–14) $6.00
  Seniors (65+) $6.00
      (Members free)

■ Annual membership fees
  Individual $50.00
    Free admission for member presenting card

  Family $80.00
    Free admission for 2 adults and their children aged 18 and under

For Exercises 29–32, use the histogram at the right. *(Section 1.1)*

Art Students

29. How many students are at least 15 years old?

30. How many students are younger than 18?

31. What percent of the students are over 16 years old?

32. Write two inequalities that describe the ages of the students.

33. Nikki buys 15 post card stamps and 5 airmail stamps. Write a variable expression for her total cost, based on the cost of each type of stamp. *(Section 1.6)*

34. Bryan Noland uses his car for his job. His weekly salary is $550. He also gets paid $.29 per mile. Write a variable expression for his total weekly pay, based on the number of miles he drives. *(Section 1.6)*

3.6

Learn how to...

- fit a line to data

So you can...

- make predictions about natural occurrences and the life expectancy of future Americans

Modeling Linear Data

Old Faithful is a geyser in Yellowstone National Park in Wyoming. For more than 100 years, Old Faithful has erupted every day at intervals of less than two hours. The data in the table suggest that the time between eruptions may be related to the length of the previous eruption.

The table is an example of a *mathematical model*. A mathematical model can be a table, a graph, an equation, a function, or an inequality that describes a real-world situation. Using mathematical models is called modeling.

Length of eruption (min)	Time until next eruption (min)
2.0	57
2.5	62
3.0	68
3.5	75
4.0	83
4.5	89
5.0	92

EXAMPLE   Application: Geology

Find an equation that models the data about Old Faithful.

SOLUTION

Make a scatter plot of the data. Draw a line on the scatter plot that is as close as possible to all of the points on the plot. This is called a **line of fit**.

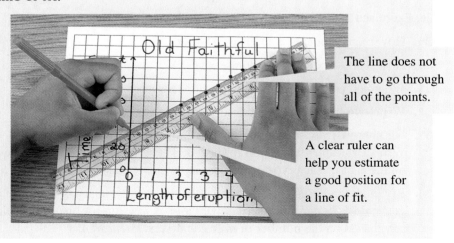

The line does not have to go through all of the points.

A clear ruler can help you estimate a good position for a line of fit.

BY THE WAY

A geyser is a natural hot spring that periodically ejects a column of water and steam into the air. Most geysers are less predictable than Old Faithful.

Write an equation for the line of fit by choosing two points on the line.

Step 1 Let l = the length of the eruption in minutes.
Let t = the number of minutes until the next eruption.
The equation will be in the form

$$t = ml + b$$

Step 2 Find the **slope** m of the line of fit.

Choose the points $(2, 57)$ and $(5, 92)$.

$$m = \frac{92 - 57}{5 - 2} = \frac{35}{3} \approx 12$$

The slope of the line is 12.

Step 3 Find the **vertical intercept** b of the line.

$$t = 12l + b$$
$$57 = 12(2) + b$$
$$57 = 24 + b$$
$$33 = b$$

Choose the point $(2, 57)$.

Old Faithful

The vertical intercept of the line is 33.

An equation for the line of fit is $t = 12l + 33$.

THINK AND COMMUNICATE

1. In the Example, what information does the slope of the line of fit give you about the situation?

2. Suppose an eruption of Old Faithful lasts 4.2 min. Use the equation for the line of fit to predict the time until the next eruption.

3. In the Example, why does it make sense in this situation to round the slope to 12?

When the points on a scatter plot come close to forming a line, you can say that the data are linear. You can use a line of fit to model linear data.

Even though a few of the points are far from the line, you can say the data are linear.

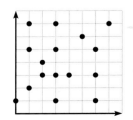

These data are not linear.

Creating a Linear Model

Work in a group of 3–4 students:

You will need:
- a paper cup
- a paper clip
- 30 marbles
- 25 pieces of uncooked spaghetti
- 2 chairs of the same height

SET UP Arrange the chairs, spaghetti, and cup as shown.

1 Place marbles in the cup, one at a time, until the spaghetti breaks. Copy and complete the table.

Number of pieces of spaghetti	Number of marbles needed to break spaghetti
1	?
2	?
3	?
4	?
5	?

2 Make a scatter plot of the data. Draw a line of fit. Find an equation for your line of fit.

3 Use your equation to predict how many marbles are needed to break six pieces of spaghetti. Check your prediction using spaghetti.

4 Do you think that moving the chairs farther apart or closer together would affect your results? If so, how? Describe and carry out an experiment to test your theory.

☑ CHECKING KEY CONCEPTS

Are the data linear? If so, use an equation to model the data.

1.

2.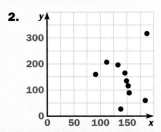

For Questions 3–5, use the equation from the Example to predict the time until the next eruption for the given length of an eruption.

3. 5.8 min   **4.** 2.3 min   **5.** 6.1 min

3.6 | Exercises and Applications

Find an equation that models the data in each scatter plot.

*Extra Practice
exercises on
page 563*

1.

2.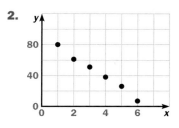

3. **Writing** Comment on the trends suggested by the data shown below. Support your comments with graphs, equations, and predictions.

Garbage Generated and Recycled in the United States							
(pounds per person per day)							
Year	1960	1965	1970	1975	1980	1985	1990
Garbage generated	2.66	3.00	3.27	3.26	3.65	3.77	4.30
Garbage recycled	0.18	0.19	0.23	0.25	0.35	0.38	0.70

Connection ▸ STATISTICS

The table shows the life expectancy of people born in the United States of America between 1900 and 1990. Use these data for Exercises 4–7.

4. Make a scatter plot of the data. Draw a line of fit on your scatter plot. Find an equation for your line of fit.

5. Predict the life expectancy of a person born in each year.

   a. 1978   b. 1994   c. 1942

6. a. Use your equation to predict the life expectancy of a person born in the United States in the year 2000.

   b. **Writing** The United States Census Bureau predicts that the average life expectancy of a person born in the United States in the year 2000 will be 76.7 years. Compare this with your prediction. Why do you think the predictions might be different?

7. **Challenge** Use your equation to predict the life expectancies of people born in 1800 and in 2100. How accurate do you think your predictions are? What do your answers say about the accuracy of using a linear equation to model the data?

Life Expectancy	
Year of birth	Life expectancy (years)
1900	47.3
1910	50.0
1920	54.1
1930	59.7
1940	62.9
1950	68.2
1960	69.7
1970	70.8
1980	73.7
1990	75.4

8. **Technology** Choose a data set from the Example on page 130 or from the Exercises in this lesson.

a. Follow the steps below to use a graphing calculator to find a line of fit for the data. Your graphing calculator manual gives more complete instructions.

Step 1 Clear the calculator's statistical memory. Enter the data as *xy*-pairs.

Step 2 Choose a good viewing window. Then draw a scatter plot.

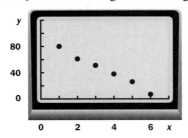

These numbers are the result of performing a linear regression on the data from Exercise 2. The calculator's line of fit is $y = -13.8x + 92.13$.

Step 3 Have the calculator perform a linear regression. (Many calculators use letters different from those shown.) The number *r* is called the correlation coefficient. The closer $|r|$ is to 1, the more linear the data.

Step 4 Graph the line of fit on the scatter plot from Step 2.

b. Does the calculator give a line of fit that is close to the one you found?

9. **SAT/ACT Preview** If $2x + 3 = 9$, then $x - 2 = \underline{\ ?\ }$.

A. 4   **B.** -1 **C.** 3   **D.** 1   **E.** -2

ONGOING ASSESSMENT

10. a. Model the data with a scatter plot, a line of fit, and an equation.

b. **Open-ended Problem** What predictions can be made from each of your models? Who might be interested in using these predictions?

Energy Consumption		
State	**1990 Population (thousands)**	**Energy consumption (trillion Btu)**
Alaska	550	582
Iowa	2,777	899
Tennessee	4,877	1753
Massachusetts	6,016	1341
Ohio	10,847	3698
Florida	12,938	3059
New York	17,990	3584
California	29,760	7307

Simplify each variable expression. *(Section 1.7)*

11. $17b + 5 - 3b - 2$ **12.** $0.25(8 + 20b)$ **13.** $3(3 - 3z)$

14. $9x - 2(x + 7)$ **15.** $-6(2n - 4) + 5n$ **16.** $16(0.25a + 2)$

For Exercises 17–19, use a graph to find the solution. *(Section 2.6)*

17. The drama club spent $1000 on their production of *Fiddler on the Roof*. The club's goal is to make a profit of at least $300. If tickets are sold at $5 each, how many does the club need to sell to reach this goal?

18. The club is considering increasing the ticket price to $10. How many tickets must be sold at this price to reach the club's goal of $300?

19. Suppose the club earns $200 by selling advertising space in their program. To reach their goal of $300, how many tickets must the club sell at $5 each? at $10 each?

ASSESS YOUR PROGRESS

VOCABULARY

line of fit (p. 130)

Find an equation of the line through the given points. *(Section 3.5)*

1. $(7, 2), (4, -1)$ **2.** $(2, 3), (-1, 3)$ **3.** $(6, 1), (6, -3)$

4. $(-4, 6), (1, -4)$ **5.** $(3, 7), (-9, -1)$ **6.** $(5, -3), (7, 8)$

7. a. Graph the line through the points $(-3, 0)$ and $(0, 2)$. Write an equation of this line. *(Section 3.5)*

b. A line through the point $(2, 0)$ has the same slope as the line in part (a). Graph this line and find an equation for it.

The table lists the number of cable-television systems in the United States from 1980 to 1990. (The number of cable systems has been rounded to the nearest hundred.) Use the table for Exercises 8–10. *(Section 3.6)*

8. Make a scatter plot of the data.

9. Draw a line of fit for the data. Find an equation for your line.

10. Predict the number of cable systems in the United States in 2005.

11. Journal How are mathematical models different from other kinds of models you have seen or used? How are they the same?

Year	Number of cable systems
1980	4200
1982	4800
1984	6200
1986	7600
1988	8500
1990	9600

Creating Linear Puzzles

Computer graphics programs use equations to determine each shape in a picture. Displaying a line on a computer starts with finding an equation of the line and specifying its endpoints. You can use this idea to create a puzzle with linear equations.

PROJECT GOAL Design and create a book of linear puzzles for other students in your class to solve.

Designing the Puzzles

Work in groups of five or six to create a book of linear puzzles. Here are some hints for designing your puzzles:

1. CHOOSE A SHAPE OR LETTER for your puzzle that you can draw using straight lines. **DRAW** your design on a coordinate plane. Make sure that the endpoints of the line segments are on grid intersections. Keep your design secret from the other members of your group.

Choose a shape that has line segments with at least four different slopes.

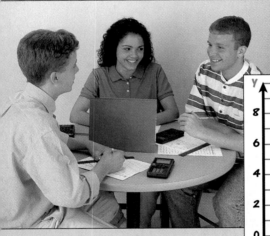

Solution

2. FIND an equation in slope-intercept form of each line in your design. Write an equation for each line segment on a separate sheet of paper.

3. PROVIDE INFORMATION about the endpoints of each line segment. Write inequalities that describe the smallest and largest x- or y-values for each line. Add this information to your set of equations. This list is your puzzle. Your drawing is the solution.

Puzzle

Equation	Endpoints	
$y = \frac{3}{2}x + \frac{1}{2}$	$x \geq 1$	$x \leq 5$
$y = -2x + 18$	$x \geq 5$	$x \leq 8$
$y = -\frac{1}{2}x + \frac{5}{2}$	$x \geq 1$	$x \leq 5$
$y = \frac{2}{3}x - \frac{10}{3}$	$x \geq 5$	$x \leq 8$
$x = 5$	$y \geq 0$	$y \leq 8$

Checking Your Work

• Exchange puzzles within your group and solve them by graphing the lines described by the equations and inequalities.

• Check to see if your drawing matches the solution.

• Write an evaluation of the puzzle you solve. Give suggestions for improvements if you have any.

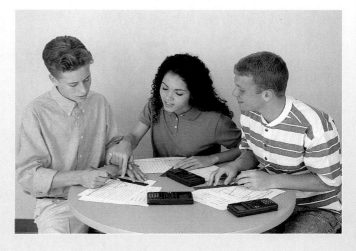

Assembling Your Puzzle Book

Collect a copy of the puzzle and solution made by each member of your group. Use these puzzles to make a book with instructions for solving and creating puzzles.

Remember to include:

• names of puzzle designers

• instructions for solving

• an answer key

You may want to extend your report to explore some of the ideas below:

• Images on a computer monitor are made up of tiny dots called pixels. Each pixel has a coordinate based on its distance from one corner of the monitor. How do computer programmers use pixel coordinates to generate images on a computer screen?

• Chess players move pieces according to a specific set of rules. Find out what these rules are. Then write a set of equations to describe the paths of different pieces as they are moved from one end of the chess board to the other.

Self-Assessment

In your report, describe which puzzles were the most challenging to solve. Why were they challenging? How could you have made your puzzle more challenging? What general advice would you give someone who is trying to create a puzzle?

3 Review

STUDY TECHNIQUE

What did you just study? Without looking at the book, take five or ten minutes and list all the words and important concepts related to this chapter. When you are finished, compare your list with the vocabulary and with the lesson titles in the table of contents. Did you miss anything important?

VOCABULARY

rate (p. 99)
unit rate (p. 99)
scatter plot (p. 106)
direct variation (p. 107)
constant of variation (p. 107)
slope (pp. 113, 115)
positive slope (p. 114)

negative slope (p. 114)
undefined slope (p. 115)
linear equation (p. 119)
y-intercept, vertical intercept (p. 120)
slope-intercept form (p. 120)
line of fit (p. 130)

Fuel Economy

The **constant of variation** is 30.

SECTIONS 3.1 *and* 3.2

A **unit rate** is a rate per one given unit.

A **rate** is a ratio that compares two units.

$$\frac{220 \text{ miles}}{4 \text{ hours}} = \frac{55 \text{ miles}}{1 \text{ hour}} = 55 \text{ mi/h}$$

The points on the **scatter plot** shown at the left lie on a line through the origin. An equation for this **direct variation** is $y = 30x$.

SECTIONS 3.3 *and* 3.4

The **slope** or steepness of a line is a rate of change.

$$\text{Slope} = \frac{y_2 - y_1}{x_2 - x_1}$$

$$= \frac{5 - 1}{4 - 1}$$

$$= \frac{4}{3}$$

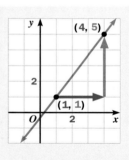

A line that rises to the right has **positive slope**.

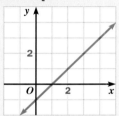

A line that falls to the right has **negative slope**.

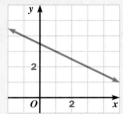

A horizontal line has **0 slope**.

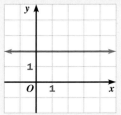

A vertical line has **undefined slope**.

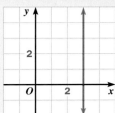

A linear equation can be written in **slope-intercept form**: $y = mx + b$, where m is the slope and b is the y-intercept.

SECTIONS 3.5 *and* 3.6

Mathematical models are used to describe real-world situations.

A **line of fit** is a line drawn as close as possible to all the points on a scatter plot. You can write an equation for the line in slope-intercept form using two of the points on the line.

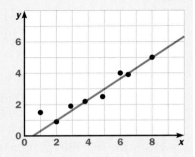

Find the slope using the points $(2, 1)$ and $(8, 5)$.

$$m = \frac{5 - 1}{8 - 2} = \frac{4}{6} = \frac{2}{3}$$

Find the vertical intercept by substituting the slope and the coordinates of one point in the equation $y = mx + b$. Then solve for b.

$$1 = \frac{2}{3}(2) + b \qquad \text{Choose the point (2, 1).}$$

$$-\frac{1}{3} = b$$

An equation for the line of fit is $y = \frac{2}{3}x - \frac{1}{3}$.

Assessment

VOCABULARY QUESTIONS

1. Draw a scatter plot of data that show direct variation. Give the constant of variation for the data.

2. Draw a graph of a line with each type of slope.

    a. positive   **b.** negative   **c.** zero   **d.** undefined

SECTIONS 3.1 *and* 3.2

For Questions 3 and 4, copy and complete each equation.

3. $\dfrac{420 \text{ mi}}{1 \text{ day}} \cdot \dfrac{?}{} = \dfrac{2940 \text{ mi}}{1 \text{ week}}$ **4.** $\dfrac{\$7.25}{1 \text{ h}} \cdot \dfrac{8 \text{ h}}{\text{day}} = \dfrac{?}{}$

Express each rate in the given units.

5. $\dfrac{\$30.82}{23 \text{ gal}} = \underline{?} \text{ cents/gal}$ **6.** $\dfrac{0.75 \text{ mi}}{2 \text{ min}} = \underline{?} \text{ mi/h}$

7. Which graph represents the direct variation $C = 1.25g$?

A.

B.

C.

8. The amount of garbage produced in the United States varies directly with the number of people who produce it. It is estimated that 200 people produce 50 tons of garbage annually. How many tons of garbage are produced each year in a city with 100,000 inhabitants?

9. Decide whether the data in the table show direct variation. If they do, give the constant of variation and write an equation that describes the situation.

Circumference (cm)	Radius (cm)
12.56	2
31.4	5
43.96	7
62.8	10

Sections 3.3 and 3.4

10. **Writing** Explain how the constant of variation in a direct variation equation is related to the slope of the graph of the equation. Explain why a direct variation equation is also a linear equation.

11. Write an equation in slope-intercept form for the graph shown at the right. What information do the slope and vertical intercept give you about the situation?

Find the slope of the line through each pair of points.

12. $(7, 5)$ and $(8, 4)$

13. $(-6, 3)$ and $(2, -3)$

Write an equation in slope-intercept form for each graph.

14.

15.

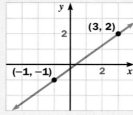

Graph each equation.

16. $y = -3x - 3$

17. $y = \frac{2}{5}x$

18. $y = \frac{3}{2}x - 4$

Towing Costs

Sections 3.5 and 3.6

Find an equation of the line through the given points.

19. $(2, 3), (4, 6)$

20. $(-5, -6), (6, 5)$

21. $(2, 7), (-3, 7)$

22. Find an equation of a line with slope 3 and through the point $(4, 7)$.

23. **a.** Make a scatter plot of the data in the table at the right.

   b. Draw a line of fit on your scatter plot.

   c. Write an equation for the line of fit.

   d. Use your equation to predict a weight for a 120 cm tall male who is nine years old.

Average Weight and Height of American Males Aged 6 to 10 Years	
Weight (kg)	Height (cm)
20.9	114.2
23.1	119.3
25.9	127.0
28.6	132.0
31.3	137.1

Performance Task

24. Write a story about a trip you have taken or might want to take. Include at least two rates and one linear equation. Illustrate your story with at least one graph. Include as many vocabulary words from the chapter as you can.

Cumulative Assessment

C H A P T E R S $1–3$

CHAPTER 1

Use the histogram at the right.

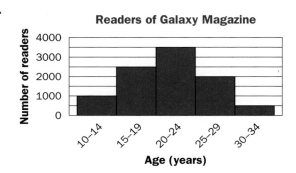

Readers of Galaxy Magazine

1. How many readers of *Galaxy Magazine* are less than 20 years old?

2. How many readers are at least 25 years old?

3. What interval is the mode of the readers' ages?

Simplify each expression.

4. $6 + 2 \cdot 8$

5. $-8 - (-3)$

6. $-44 \div (-4)$

Evaluate each expression when $a = 3$ and $b = -5$.

7. $5a - b^2$

8. $\dfrac{a^2 + 3b}{6}$

9. $\dfrac{a - b^2}{b}$

For Questions 10 and 11, find the mean, the median, and the mode(s) of the data.

10. Hours studied
    3, 3, 3, 4, 4, 5, 8, 16

11. Meals served
    20, 21, 21, 25, 30

12. **Writing** Uma says that $\left|n^2\right| = n^2$ for all values of n. Do you agree or disagree? Explain.

Open-ended Problem **Describe two real-world situations that each variable expression can represent.**

13. $t + 2$

14. $\dfrac{p}{4}$

15. $5n$

CHAPTER 2

Solve each equation.

16. $-7p = 105$

17. $\dfrac{t}{4} = -52$

18. $15 = 7 + x$

19. $2n - 6 = 16$

20. $3 - 6x = -19$

21. $7 + \dfrac{x}{4} = 15$

Graph each point on a coordinate grid. Name the quadrant (if any) in which the point lies.

22. $A(0, 1)$ **23.** $B(1, 2)$ **24.** $C(-1, 2)$ **25.** $D(-1, -2)$

The graph below shows the velocity v of a car undergoing constant acceleration for t seconds. Use the graph for Questions 26 and 27.

26. Estimate the velocity v when the time is 4 s.

27. Estimate the time t when the velocity is 15 mi/h.

28. Writing Each pizza ordered for the Ski Club gathering costs $7. Tell how this situation represents a function. Describe the domain and the range of the function.

Car Acceleration

29. a. Open-ended Problem Describe a situation that can be represented by the equation $y = 4x$. Graph the equation.

    b. Tell how your graph can be used to solve a problem relating to the situation in part (a).

CHAPTER 3

For Questions 30 and 31, copy and complete each equation.

30. $\dfrac{\$4.25}{1\ h} \cdot \dfrac{20\ h}{1\ week} = \underline{\ ?\ }$ **31.** $\dfrac{55\ mi}{1\ h} \cdot \underline{\ ?\ } = \dfrac{440\ mi}{day}$

32. Decide whether the data in the table at the right show direct variation. If they do, give the constant of variation and write an equation that describes the situation.

Size (oz)	Price (dollars)
2.4	2.88
3.0	3.60
3.5	4.20

33. A line contains the two points $A(-2, 3)$ and $B(4, 12)$.

    a. Find the slope of the line.

    b. Write an equation of the line in slope-intercept form.

    c. What is the y-coordinate when $x = 0$?

Use the scatter plot at the right.

34. Which equation is a better model for the data?

    A. $y = -\dfrac{1}{2}x + 1$ **B.** $y = \dfrac{1}{2}x + 1$

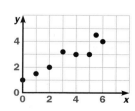

35. Writing Can a line with a steep slope have a negative slope? Write a few sentences to explain your answer.

36. Open-ended Problem Choose two points in different quadrants of a coordinate plane. Find an equation of the line through the points.

4 Solving Equations and Inequalities

Shaking up SCIENCE

INTERVIEW **Barbara Romanowicz**

When the ground in Northern California starts to rumble, Barbara Romanowicz goes to work. A seismologist at the University of California at Berkeley, Romanowicz studies the sudden, violent upheavals of the land called earthquakes. "Although we can't predict earthquakes, we can save lives and money by promptly telling people where an earthquake is," she says. "Armed with this information, officials will be able to direct emergency aid much more effectively."

"It's important to know where earthquakes are,..."

Romanowicz is interested not only in earthquakes, but also in the information that earthquakes convey about the structure of the planet. Her studies have led her to propose a new theory, that stresses caused by a large earthquake can trigger other quakes, decades later, far from the initial site.

Collecting Data

Romanowicz and other seismologists collect data about earthquakes from seismology stations located throughout the country. The stations are equipped with seismographs, machines that measure the movement of the ground.

In an earthquake, the ground moves in waves that spread out from the epicenter. (The epicenter is the point on Earth's surface closest to the source of the earthquake. Large earthquakes originate at great depths, sometimes 10 miles below the surface.) Earthquakes produce different kinds of waves. For example, pressure waves (called *P waves*) move the ground back and forth as they travel away from the epicenter. Shear waves (called *S waves*) move the ground sideways as they travel away from the epicenter. P waves and S waves show up as different patterns when recorded by a seismograph. P waves appear first on the seismograph record, because P waves travel faster than S waves.

Locating the Epicenter

To find out how far a particular seismology station is from an earthquake's epicenter, seismologists find the difference between the arrival times of

"...but it's also important to know how big they are and the time at which they occurred."

Rescuers on the roof of a collapsed house try to free people trapped inside following a major earthquake in western Japan, January, 1995.

In January, 1995, this section of the elevated Hanshin Expressway collapsed during an earthquake in western Japan.

A modern model of Chang Heng's seismograph of 132 A.D.

P waves and S waves. They use this value to solve an equation that gives the distance of the station from the epicenter. (You will learn more about this equation in Section 4.4.)

Predicting Earthquakes

"It's important to know where earthquakes are, but it's also important to know how big they are and the time at which they occurred," Romanowicz says. This information is collected in catalogs that record the location, size, and date of major earthquakes all over the world. After reviewing catalogs that cover the past 80 years, Romanowicz was able to propose her theory about the long-term effects of earth-

quakes. "If I'm correct," she says, "soon we might be able to state with confidence that there will be a large earthquake in a certain region sometime in the next 20 years. Of course, that doesn't help people living there much, but it's a step towards our ultimate goal—reliable, long-range earthquake predictions."

EXPLORE AND CONNECT

Despite Japan's strict building codes, the earthquake caused severe damage.

1. Research The map on page 144 shows that some regions of the world are more susceptible to earthquakes than others. Find out why earthquakes tend to occur in these regions. Are earthquakes common in the area where you live? Report your findings to your class.

2. Writing To measure the size of an earthquake, seismologists commonly use the *Richter scale*. Write a report explaining how the Richter scale works and how seismologists use it.

3. Project Find out how modern seismographs work. Draw a diagram or build a model of one. Show how it works.

Mathematics
& Barbara Romanowicz

In this chapter, you will learn more about how mathematics is related to seismology.

Related Examples and Exercises

Section 4.4
• Example 1
• Exercises 28–31

4.1

Solving Problems Using Tables and Graphs

Learn how to...

- model situations with tables and graphs

So you can...

- compare options, such as buying CDs from a store or from a CD club

Is it less expensive to buy compact discs (CDs) at a discount store or through a CD club? You can compare options using tables or graphs.

EXPLORATION
COOPERATIVE LEARNING

Comparing Tables and Graphs

Work in 2 teams of 2 students each.
You will need:

- a graphing calculator or graph paper

The CD club minimum is 8 + 6, or 14.

One team should do Steps 1–4 below. The other team should do Steps 5–8 on the next page. The teams should work together on Step 9.

1 Copy and complete the table.

Number of CDs	Cost at store ($)	CD club cost ($)
14	?	?
15	?	?
16	?	?

2 For the values in the table, which way of buying CDs is less expensive? Will this always be the case? Explain.

3 Expand your table. Include rows for 20, 30, 40, and 50 CDs.

4 When will the total cost through the CD club equal the total cost at the discount store? How can you tell? When would you choose the CD club? the discount store? Why?

Exploration continued on next page.

5 Write an equation for the cost *y* of buying *x* CDs from the discount store.

6 Repeat Step 5 for the CD club.

7 Graph the two equations in the same coordinate plane. Set a good viewing window, like the one shown below.

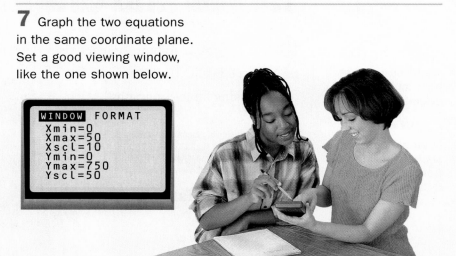

```
WINDOW FORMAT
Xmin=0
Xmax=50
Xscl=10
Ymin=0
Ymax=750
Yscl=50
```

8 When will the total cost through the CD club equal the total cost at the discount store? How can you tell? When would you choose the CD club? the discount store? Why?

9 Discuss your answers to Steps 4 and 8 with the other team. Compare the table method to the graphing method. List some advantages and disadvantages of each method.

EXAMPLE   Application: Consumer Economics

At Service Bank customers pay a monthly fee of $6.25 plus $.30 per transaction. At Slate Bank customers pay $3.00 per month plus $.40 per transaction. (Checks written, deposits, and visits to automatic teller machines count as transactions.) When will the total monthly cost at Slate Bank be greater than at Service Bank?

SOLUTION

Problem Solving Strategy: Make a table.

You can use a spreadsheet or a graphing calculator, or make a table by hand. Include a column for the number of transactions and a column for the cost at each bank. Use this general formula.

Cost = Monthly fee + Transaction fee × Number of transactions

Step 1 Enter an initial value for the number of transactions in cell A2. Enter a formula in cell A3 to add a fixed amount to the initial value. Fill down to get a series of transaction values.

Bank Options		
A3	=A2+5	

	A	B	C
1	Transactions	Service Bank	Slate Bank
2	5	7.75	5.00
3	10	9.25	7.00
4	15	10.75	9.00
5	20	12.25	11.00
6	25	13.75	13.00
7	30	15.25	15.00
8	35	16.75	17.00
9	40	18.25	19.00

Step 2 Enter a formula and fill down to find the **cost** at Service Bank.
For example, **18.25 = 6.25 + (0.3)(40)**.

Step 3 Enter a formula and fill down to find the **cost** at Slate Bank.
For example, **19 = 3 + (0.4)(40)**.

The total monthly cost at Slate Bank becomes greater than at Service Bank somewhere between 30 and 35 transactions. Try more values between 30 and 35 transactions until you find more precisely when the cost at Slate Bank becomes greater.

Bank Options		
A8	=A7+1	

	A	B	C
7	30	15.25	15.00
8	31	15.55	15.40
9	32	15.85	15.80
10	33	16.15	16.20
11	34	16.45	16.60
12	35	16.75	17.00

When the number of transactions in a month reaches 33, the total monthly cost at Slate Bank becomes greater than at Service Bank.

THINK AND COMMUNICATE

1. a. For each bank in the Example, write an equation for the monthly cost of a checking account as a function of the number of transactions.

  b. Describe how you could use a graph of the equations to solve the problem.

2. Suppose Jennifer expects to write fewer than 30 checks in a month. Based on the Example, what advice would you give her? Why?

People who want to use a community pool during the summer have the two options modeled by the graph shown. Use the graph for Questions 1–3.

1. Write an equation for the red line. Describe the relationship between the cost of a season pass and the number of visits to the pool.

2. Write an equation for the green line. What is the cost without a season pass of one visit to the pool?

3. Is it worth buying a season pass if you expect to use the pool only 10 times during the summer? Explain.

Cost of Using a Pool

4.1 : **Exercises and Applications**

Extra Practice exercises on page 563

Cooperative Learning For Exercises 1–4, work with a partner. Use the table below.

1. Which club has a greater initiation fee? a greater monthly fee? How can you tell?

Fitness Club Membership Costs		
Number of months	Cost at Club A ($)	Cost at Club B ($)
0	100	0
1	140	50
2	180	100
3	220	150

2. One person should copy and expand the table to cover a total of 18 months. The other person should write and graph two equations to model the membership costs at the two clubs.

3. When is the membership cost the same for both fitness clubs?

4. **Open-ended Problem** Suppose you decide to join a fitness club. What factors will you consider besides cost? Explain.

PERSONAL FINANCE Exercises 5–8 should be done together.

5. Use both a table and a graph to model the earnings from these two jobs as a function of the number of overtime hours worked.

Job A: $30,000 per year for full-time work, with no pay for overtime

Job B: $24,000 per year for full-time work, plus $12 per hour for each hour of overtime

6. How many hours of overtime will you have to work to earn as much in a year from Job B as from Job A? How can you tell from the table? from the graph?

7. Based on earnings alone, under what circumstances would you be better off with Job A? with Job B?

8. Choose the letter of the equation you could use to solve Exercise 6.

   A. $12h = 30{,}000$

   B. $24{,}000 + 12h = 30{,}000$

   C. $24{,}000 + 12h = 30{,}000 + 12h$

Connection   ENERGY USE

An advertisement for a fluorescent light bulb says that it lasts 9 times longer than a comparable standard light bulb—up to 9000 hours. It also uses less energy. The fluorescent bulb uses only 15 watts. A typical standard bulb costs about $.75 and uses 60 watts. This graph models the cost of buying and using each kind of bulb.

Costs of Using Standard and Fluorescent Light Bulbs

9. How does the graph show that after using a fluorescent light bulb for 9000 hours, you must buy a new one? Estimate the price of one fluorescent bulb.

10. Why aren't there noticeable jumps in the graph for standard bulbs?

11. How many hours each month do you think you have the light on in your bedroom? How many hours do you think the lights are on in your school each month?

12. Writing Do you think you should use fluorescent lights at home? at school? Explain.

13. **RECREATION** A small movie theater has two admissions options.

   Members: $40 annual membership fee, plus $4.50 per show
   Nonmembers: $7.00 per show

   Under what circumstances do you recommend buying a membership?

14. **Challenge** Montreal and Quebec City are about 260 km apart along Route 20 in Canada. Sylvie Potvin leaves Montreal and drives toward Quebec City on Route 20 at the same time that Anne Durand leaves Quebec City headed for Montreal. Sylvie drives at a steady speed of 90 km/h and Anne drives at a steady speed of 105 km/h.

   Quebec City

   Montreal

   a. When and where do Sylvie and Anne pass each other? (Assume that neither one stops along the way.) Include a table or graph with your answer. (*Hint:* When are they the same distance from Montreal?)

   b. Write an equation to model this situation. Then show how to solve the equation.

ONGOING ASSESSMENT

15. **Writing** Every time Mr. and Mrs. Dalton go to the movies, it costs them $13. They wonder whether they should buy a VCR for $350 and rent videotapes for $3 each instead. Write a letter to the Daltons to give them your advice. Include a table or graph in your letter.

SPIRAL REVIEW

Solve each equation. (*Sections 2.1 and 2.2*)

16. $m - 17 = -3$ 17. $\frac{n}{5} = 2$ 18. $15 = \frac{c}{2} + 1$

19. $5a - 3 = 7$ 20. $-6 = 3x + 21$ 21. $13 - c = 18$

Tell whether the data show direct variation. If they do, give the constant of variation and write an equation. (*Section 3.2*)

22.

Time (minutes)	Distance (meters)
3	5.25
8	14
14	24.5
21	36.75

23.

Age (years)	Height (inches)
2	32
10	60
25	66
45	66

4.2

Using Reciprocals

Learn how to...
- solve one-step and two-step equations using reciprocals

So you can...
- solve problems involving fractions

How would you solve an equation like $\frac{3}{4}x = 12$? You could divide both sides by $\frac{3}{4}$ or multiply both sides by the reciprocal of $\frac{3}{4}$, which is $\frac{4}{3}$. Two numbers are **reciprocals** if their product is 1.

$$\frac{3}{4} \cdot \frac{4}{3} = \frac{3 \cdot 4}{4 \cdot 3} = \frac{12}{12} = 1$$

THINK AND COMMUNICATE

1. What is the reciprocal of $\frac{2}{3}$? of $\frac{1}{3}$? of 3? of 1? Give some other examples of reciprocals.

2. Do you think 0 has a reciprocal? Explain.

Property of Reciprocals

For every nonzero number a, there is exactly one number $\frac{1}{a}$ such that

$$a \cdot \frac{1}{a} = 1 \text{ and } \frac{1}{a} \cdot a = 1.$$

EXAMPLE 1

Solve the equation $\frac{3}{4}x = 12$.

SOLUTION

In order to solve an equation algebraically, you must get the variable alone on one side of the equation.

Multiply both sides of the equation by the **reciprocal** of $\frac{3}{4}$.

$$\frac{3}{4}x = 12$$
$$\frac{4}{3} \cdot \frac{3}{4}x = \frac{4}{3} \cdot 12$$
$$1 \cdot x = \frac{4 \cdot 12}{3}$$
$$x = 16$$

When you solve an equation, remember to keep it balanced. If you apply an operation like multiplication to one side of the equation, then you must apply the same operation to the other side of the equation.

BY THE WAY

The Babylonians used reciprocals for division. Instead of dividing by a number, they multiplied by its reciprocal. They created tables of reciprocal values to simplify this work.

Reciprocals and Negative Numbers

The reciprocal of a negative number is negative.

For example, the reciprocal of $-\frac{5}{4}$ is $-\frac{4}{5}$.

The product of two negative numbers is positive.

$$-\frac{5}{4} \cdot -\frac{4}{5} = +\frac{5 \cdot 4}{4 \cdot 5} = \frac{20}{20} = 1$$

EXAMPLE 2

Solve each equation.

a. $-\frac{1}{6}y = -18$

b. $\frac{-3n}{10} = 21$

SOLUTION

a.
$$-\frac{1}{6}y = -18$$
$$(-6)\left(-\frac{1}{6}\right)y = (-6)(-18)$$
$$1 \cdot y = 108$$
$$y = 108$$

Multiply both sides by the **reciprocal**.

b.
$$\frac{-3n}{10} = 21$$
$$-\frac{3}{10}n = 21$$
$$\left(-\frac{10}{3}\right)\left(-\frac{3}{10}\right)n = \left(-\frac{10}{3}\right)(21)$$
$$n = -70$$

Rewrite $\frac{-3n}{10}$ with a fraction as the coefficient.

Multiply both sides by the **reciprocal**.

WATCH OUT!

When you work with negative fractions, remember these rules:

$$-\frac{a}{b} \neq \frac{-a}{-b}$$

$$-\frac{a}{b} = \frac{-a}{b} = \frac{a}{-b}$$

THINK AND COMMUNICATE

3. Explain how a reciprocal was used in part (a) of Example 2 to get y alone on one side of the equation.

4. How could you solve part (b) of Example 2 using inverse operations, as you learned in Chapter 2?

5. Describe how you would solve the equation $\frac{-a}{3} = 5$.

EXAMPLE 3

Sheila has $20 available to spend on materials to make a vest. The pattern, thread, and buttons cost $10 altogether. The pattern calls for $\frac{7}{8}$ yd of fabric. What is the most Sheila can afford to pay per yard of fabric?

SOLUTION

Problem Solving Strategy: Use an equation.

Let p = the maximum price per yard.

$$\underset{\text{fabric}}{\text{Cost of}} + \underset{\text{other materials}}{\text{Cost of}} = \underset{\text{available}}{\text{Money}}$$

$$\frac{7}{8}p + 10 = 20$$

$$\frac{7}{8}p + 10 - 10 = 20 - 10$$

$$\frac{7}{8}p = 10$$

Multiply both sides by the **reciprocal**.

$$\frac{8}{7} \cdot \frac{7}{8}p = \frac{8}{7} \cdot 10$$

$$p = \frac{80}{7} \approx 11.43$$

Thread: $1.90

Pattern: $3.50

Buttons: $4.60

The most Sheila can afford to pay per yard of fabric is $11.43.

THINK AND COMMUNICATE

6. Suppose someone wrote the equation in Example 3 as $20 - \frac{7}{8}p = 10$. Describe the steps you would take to solve the equation.

7. Solve Example 3 assuming that the pattern calls for only $\frac{3}{4}$ yd of fabric.

☑ CHECKING KEY CONCEPTS

Find the reciprocal of each number.

1. $\frac{9}{10}$ **2.** $\frac{1}{11}$ **3.** 8 **4.** $-\frac{7}{12}$

Rewrite each expression with a fraction as the coefficient.

5. $\frac{2n}{3}$ **6.** $\frac{-4a}{7}$ **7.** $\frac{6c}{-5}$ **8.** $\frac{-x}{2}$

Solve each equation.

9. $\frac{4}{5}x = 16$ **10.** $\frac{-4x}{3} = 5$ **11.** $\frac{3}{5}n - 11 = 19$

4.2 Exercises and Applications

Extra Practice
exercises on
page 564

Identify the reciprocal you would use to solve each equation. Then solve.

1. $\frac{2}{9}x = 8$ **2.** $\frac{1}{3}n = 21$ **3.** $12 = \frac{5}{6}a$ **4.** $\frac{2}{5}b = -3$

Rewrite each expression with a fraction as the coefficient.

5. $\frac{7c}{16}$ **6.** $\frac{n}{15}$ **7.** $\frac{-5x}{11}$ **8.** $\frac{-b}{20}$

Solve each equation.

9. $-\frac{4}{9}c = 20$ **10.** $2 = -\frac{1}{15}n$ **11.** $-\frac{2}{3}x = -7$ **12.** $\frac{10c}{3} = -20$

13. $9 = \frac{5a}{2}$ **14.** $\frac{-12x}{5} = 36$ **15.** $10 = \frac{-n}{8}$ **16.** $\frac{-4a}{5} = 30$

17. SAT/ACT Preview Which number is the solution of $-\frac{3}{5}x = -15$?

A. -25 **B.** -9 **C.** $-\frac{5}{3}$ **D.** 9 **E.** 25

Connection ▷ MUSIC

Musicians generate sound waves when they sing or play an instrument. The rate
at which a sound wave vibrates determines the musical note, or *pitch*, that you
hear. For example, the A above middle C has a frequency of
440 vibrations per second and is referred to as *international*,
or *concert*, pitch.

middle C

18. When two notes are a "fifth" apart, the frequency of the lower note

is $\frac{2}{3}$ the frequency of the higher note.

a. Write an equation to model this relationship.

b. Find the frequency of the note that is a fifth above concert pitch.

c. Find the frequency of the note that is a fifth below concert pitch.

19. When two notes are a "fourth" apart, the frequency of the lower note

is $\frac{3}{4}$ the frequency of the higher note. Repeat Exercise 18 using a

"fourth" instead of a "fifth."

20. Research What is the letter of the note a fifth above concert pitch?
a fourth below concert pitch? Compare the notes and their frequencies.

BY THE WAY

Orchestra concerts
begin with the
musicians tuning their
instruments to a single
note. That note is the A
above middle C.

GEOMETRY **Find each unknown measurement.**

Toolbox p. 594
Area and Volume

21. parallelogram
Area = 2 yd^2

$\frac{3}{4}$ yd
b

22. triangle
Area = 10 in.^{2}

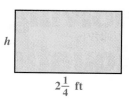
h
5 in.

23. rectangle
Area = 3 ft^2

h

$2\frac{1}{4}$ ft

24. Writing Explain how you found the unknown measurement for one of the figures in Exercises 21–23.

Solve each equation.

25. $\frac{2}{3}x + 7 = 15$

26. $\frac{1}{10}n - 5 = 15$

27. $60 = \frac{4}{5}a + 16$

28. $\frac{9}{10}c - 8 = 4$

29. $-\frac{3}{4}x - 12 = 15$

30. $14 - \frac{7}{8}n = 56$

31. $\frac{-2a}{5} + 9 = 5$

32. $32 = \frac{-n}{8} + 8$

33. $-2 - \frac{5c}{6} = 7$

34. Open-ended Problem Write an equation that can be solved using inverse operations or using reciprocals. Solve the equation both ways.

35. FUEL ECONOMY The gasoline tank on Lorenzo Shepard's car holds 13 gal. The car uses $\frac{1}{35}$ gal of gasoline for every mile that Lorenzo drives it. The car's fuel light comes on when there are 3 gal of gasoline left in the tank. How many miles can Lorenzo drive the car before the fuel light comes on?

36. Challenge How is the equation $\frac{3}{5x} + 2 = 8$ different from the others in this section? How would you solve this equation?

ONGOING ASSESSMENT

37. Open-ended Problem Write a quiz with five questions that could be used to check your understanding of the material in this section. Then make a solution key for your quiz.

SPIRAL REVIEW

Simplify each expression. *(Section 1.2)*

38. $4 + 10 \div 2$

39. $18 - 2(12 - 4)$

40. $35 \cdot 4 + 20 \div 4$

41. $\frac{5 + 3(5)}{4}$

42. $\frac{16 + 8}{2 + 4}$

43. $\frac{15 - 3}{5} + 7(2)$

Simplify each variable expression. *(Section 1.7)*

44. $12x + 1 - 3x$

45. $6a^2 - 2 + 5a - a^2$

46. $0.35(3q - 4)$

47. $-4(2m - 3n)$

48. $3(c + 5) + c - 7$

49. $5(p + 7) + 3(p + 7)$

4.3 Solving Multi-Step Equations

Some equations that represent real-world situations have variables on both sides of the equals sign. Algebra tiles can help you understand how to solve this type of equation.

EXPLORATION
COOPERATIVE LEARNING

Getting Variables onto One Side

Work with a partner.
You will need:
- algebra tiles

SET UP Use tiles to represent the equation $3x + 1 = x + 5$.

1 Remove an x-tile from each side of your tile equation. What mathematical operation does this model?

2 Finish solving the equation using the algebra tile model. What is the value of one x-tile? Check this result in the original equation.

3 Take turns using the algebra tile method introduced in the previous steps to solve each of these equations.

   a. $4x + 2 = x + 5$ **b.** $x + 9 = 5x + 1$
   c. $2x + 8 = 4x + 2$ **d.** $5x + 3 = 2x + 6$

4 Work together to make a generalization about the first step you should take to solve an equation with variables on both sides of it.

In Section 4.1 you used a table and a graph to analyze options for buying CDs. The example below shows how to solve the same problem using an equation with variables on both sides.

EXAMPLE 1 **Application: Personal Finance**

A discount store charges $11 per CD. A CD club gives members 8 free CDs, then charges $15 per CD. When does the cost at the discount store equal the cost through the CD club?

SOLUTION

Let x = the number of CDs you buy.

Write the cost for each option as a function of the number of CDs you buy.

$$\text{Discount store: Cost} = 11x$$
$$\text{CD club: Cost} = 15(x - 8)$$

To find when the two costs are equal, write and solve an equation.

Cost at discount store = Cost through CD club

Use the distributive property to simplify $15(x - 8)$.

Get the x-terms on just one side by subtracting $15x$ from both sides.

$$11x = 15(x - 8)$$
$$11x = 15x - 120$$
$$11x - 15x = 15x - 120 - 15x$$
$$-4x = -120$$
$$\frac{-4x}{-4} = \frac{-120}{-4}$$
$$x = 30$$

WATCH OUT!
The second function does not apply to the first 8 CDs.

The cost at the discount store equals the cost through the CD club when you buy 30 CDs.

THINK AND COMMUNICATE

1. What are the three key steps used to solve the equation in Example 1? In what order are the steps done?

2. How can you check the solution of Example 1?

3. What would you do first to solve the equation $5x = 3(x + 10)$? What would you do second? What would you do third?

4. **a.** What would you do first to solve the equation $5x = 3x + 10$? What would you do second?

   b. Are your answers to part (a) the same as your answers to Question 3? Why or why not?

EXAMPLE 2

Al is a sprinter on a track team. He can run at a rate of 9 m/s. His brother Bill can run only 6 m/s, so Al gives him a 25 m head start in a 100 m race.

a. After how many seconds will Al catch up to Bill?

b. How far will they be from the starting line when Al catches up to Bill?

SOLUTION

Problem Solving Strategy: Use a diagram.

Toolbox p. 600
Problem Solving

Let t = the number of seconds they run.

You know each runner's rate r. Use the formula $d = rt$ to find the distance d each person can run in t seconds.

Al's distance from the starting line after t seconds

the point where Al catches up to Bill

$9t$

$6t$

0 m   25 m   ? m   100 m

$25 + 6t$

Bill's distance from the starting line after t seconds

a. To find out when the distances are equal, write and solve an equation.

$$\textbf{Al's distance} = \textbf{Bill's distance}$$

$$9t = 25 + 6t$$

Subtract **6t** from both sides.

$$9t - 6t = 25 + 6t - 6t$$

$$3t = 25$$

$$\frac{3t}{3} = \frac{25}{3}$$

$$t = 8\frac{1}{3} = 8.\overline{3}$$

Al catches up to Bill after running about 8.3 seconds.

b. Al's distance $= 9t$

$$= 9 \cdot \frac{25}{3} \qquad \text{Substitute } 8\frac{1}{3}, \text{ or } \frac{25}{3}, \text{ for } t.$$

$$= 75$$

Al catches up to Bill when they are 75 m from the starting line.

You can follow these steps to solve an equation with a variable term on both sides of the equals sign.

1. Simplify each side of the equation.

2. Get all the variable terms on one side of the equation. Remember to keep it balanced by doing the same operation on both sides.

3. Get the variable alone on one side of the equation.

Sometimes an equation has *no solution*, because no value of the variable will make the equation true. Sometimes *all numbers* are solutions of an equation. An equation that is true for all numbers is called an **identity**. The following questions give examples of these types of equations.

THINK AND COMMUNICATE

Identify the steps used to solve each equation and interpret the final statement. If all numbers are solutions, write *identity*. If there is no solution, write *no solution*. Explain your reasoning.

5.
$$7x - 4 = 7(x - 4)$$
$$7x - 4 = 7x - 28$$
$$7x - 4 - 7x = 7x - 28 - 7x$$
$$-4 = -28$$

This statement is never true.

6. $8x - 12 = 2x + 2(3x - 6)$
$$8x - 12 = 2x + 6x - 12$$
$$8x - 12 = 8x - 12$$

This statement is always true.

☑ CHECKING KEY CONCEPTS

Copy and complete each solution by replacing each _?_ with the correct expression.

1.
$$3x + 3 = 2x + 5$$
$$3x + 3 - \underline{?} = 2x + 5 - 2x$$
$$x + 3 = 5$$
$$x + 3 - \underline{?} = 5 - \underline{?}$$
$$x = \underline{?}$$

2.
$$3(2x - 4) = 5x - 6$$
$$\underline{?} - 12 = 5x - 6$$
$$6x - 12 - 5x = 5x - 6 - \underline{?}$$
$$x - 12 = -6$$
$$x - 12 + \underline{?} = -6 + \underline{?}$$
$$x = \underline{?}$$

Tell what you would do first to solve each equation. If an equation is an *identity* or there is *no solution*, say so.

3. $4x = 2x + 9$

4. $3y = 20 - y$

5. $5n - 17 = 7n + 3$

6. $9a - 60 = 12a + 3$

7. $2(6x + 4) = 12x - 6$

8. $2(5c + 3) = 6 + 10c$

4.3 Exercises and Applications

Extra Practice exercises on page 564

1. **RECREATION** Look back at Exercises 1–4 on page 150. You can use the equations below to model the membership costs for the two fitness clubs.

   Let n = the number of months you belong to the club.

   Club A: Cost = $100 + 40n$ Club B: Cost = $50n$

   Write and solve an equation to answer this question: When is the membership cost the same for both fitness clubs?

Solve each equation.

2. $7x = 3x + 12$

3. $3n = n - 2$

4. $8b - 4 = 5b + 8$

5. $4a - 5 = 6a - 11$

6. $12c - 5 = 7c + 30$

7. $3n - 11 = 29 - 5n$

8. $2y - 1 = 5y$

9. $4n + 4 = 16n$

10. $-7x + 5 = 4x + 5$

11. $3y - 9 = 9y - 3$

12. $7x + 8 = -18 - 6x$

13. $2r - 15 = -9r + 18$

14. **Writing** Write a letter to a friend explaining how to get a variable term from one side of an equation onto the other side. Include examples.

15. **TRANSPORTATION** Suppose the McCall family wants to rent a truck for a day. They have a choice of two truck rental companies.

    Company A: $30 per day plus $.60 per mile
    Company B: $55 per day plus $.35 per mile

    Write and solve an equation to answer this question: After how many miles are the total costs at the two companies equal?

Visual Thinking For Exercises 16–19, use the graph below.

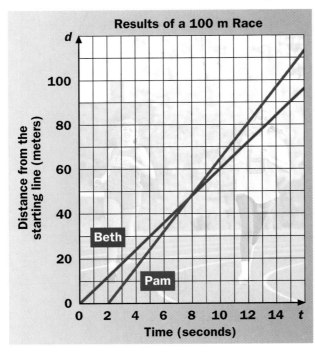

Results of a 100 m Race

In Example 2, Al and Bill started running at the same time, but Bill had a *distance* advantage. Another way to give someone a head start is to give a *time* advantage. Both runners leave from the starting line, but the slower runner leaves earlier. The graph gives information about another 100 m race in which one person was given a time advantage.

16. Who is the slower runner? How can you tell?

17. Estimate how far the runners were from the starting line when one of them caught up to the other.

18. Write an equation for the distance covered by each runner as a function of time t.

19. Write and solve an equation to find the time t at which the faster runner caught up to the slower runner. Calculate how far the runners were from the starting line at that time.

This may not look like a math problem to you, but it is. It was written in Africa over 3000 years ago by a scribe named Ahmes. Scholars have studied the Ahmes papyrus to understand how the Egyptians used mathematics.

The diagram at the right shows the reasoning used in the Ahmes papyrus to solve this problem:

A quantity and one quarter of the quantity added together equal 15. What is the quantity?

This method of reasoning is often called *false assumption*.

For Exercises 20–22, use the method of false assumption to find each unknown number.

Assume the quantity is 4.   **Choose 4 because $\frac{1}{4}$ of 4 is an integer.**   **The result is 5. It should be 15. The assumption is false.**

▲▲▲▲ + △ = ▲▲▲▲▲△   **Notice that 3·5 = 15.**

▲▲▲▲
▲▲▲▲ + △ = ▲▲▲▲△
▲▲▲▲   △   ▲▲▲▲△   △   **Work backward. Multiply each term by 3.**

3·4 = 12   3·1 = 3   3·5 = 15

20. A number plus $\frac{1}{4}$ of the number is 25.

The unknown quantity is 12.

21. A number plus $\frac{1}{3}$ of the number is 36.

22. A number plus $\frac{1}{12}$ of the number is 65.

23. Use an equation to find each unknown number in Exercises 20–22.

24. **Research** Find out about the Rosetta stone. How did it enable scholars to read Egyptian writing?

Solve. If an equation is an *identity* or there is *no solution*, say so.

25. $4(x - 3) = 7x$

26. $4(x - 3) = 4x - 3$

27. $4(x - 3) = 2(2x - 6)$

28. $7(-6 - 2t) = 9t + 4$

29. $10q - 36 = 2(3q + 4)$

30. $6n - 5 = 2(3n - 4)$

31. $2(4a - 2) = \frac{2}{3}(9 - 48a)$

32. $12x - (6 + 8x) = \frac{1}{2}(8x - 12)$

33. **Open-ended Problem** Write an equation that has one solution, another equation that has no solution, and another equation that is an identity.

34. SAT/ACT Preview Which of these equations is not an identity?

A. $5(n + 2) = 5n + 10$

B. $8(n - 2) = 8n - 16$

C. $3n + 8 = \frac{1}{4}(12n + 32)$

D. $2(n + 6) - 5n = 14 - 3n$

35. a. Challenge Graph the equations $y = x + 2$ and $y = x - 2$ in the same coordinate plane. How does the graph help you to see that there is no solution to the equation $x + 2 = x - 2$?

b. When will an equation in the form $ax + b = cx + d$ have no solution? only one solution? When will all numbers be solutions?

ONGOING ASSESSMENT

36. Cooperative Learning Work with a partner. Take turns using algebra tiles to solve each equation. Record the steps with algebraic symbols.

a. $x + 6 = 6x + 1$ b. $4x + 8 = 5x + 5$ c. $6x + 2 = 4x + 10$

SPIRAL REVIEW

Find the least common multiple of each set of numbers. *(Toolbox, page 584)*

37. 2, 3, 5   **38.** 2, 6, 12   **39.** 3, 12, 15   **40.** 4, 14, 21

41. Gerry makes a phone call at a rate of $.20 per minute or fraction thereof. Her total cost is a function of the call's length. Find the domain and range of the function. *(Section 2.3)*

ASSESS YOUR PROGRESS

VOCABULARY

reciprocals (p. 153)   **identity** (p. 161)

1. A bank has two basic checking accounts. For the *Minimum Account* the monthly fee is $2.50 and the first ten checks you write each month are free. After that you pay $.75 per check. The *Maximum Account* fee is $7.00 no matter how many checks you write each month. *(Section 4.1)*

   a. Model this situation with a table.

   b. When will the monthly cost be the same for both accounts?

Solve. If the equation is an *identity* or there is *no solution*, say so.
(Sections 4.2 and 4.3)

2. $\frac{2}{3}a = 6$ **3.** $-\frac{2}{7}b + 4 = 6$ **4.** $\frac{5t}{3} = -20$

5. $15 = -\frac{1}{5}b + 2$ **6.** $6x + 72 = 15x$ **7.** $-9c - 8 = 6c + 7$

8. $4n - 21 = 15 - 8n$ **9.** $2(2a + 4) = 4a + 1$ **10.** $8x + 6 = 2(4x + 3)$

11. Journal Compare solving problems by using equations to solving problems by using tables and graphs. Which method do you prefer? Why?

4.4

Equations with Fractions or Decimals

Learn how to...

- **solve equations that involve more than one fraction or decimal**

So you can...

- **solve real-world problems, such as finding the distance to the epicenter of an earthquake**

A catastrophic earthquake struck Kobe, Japan, in January, 1995. A seismograph in Western Samoa recorded the wave patterns shown.

For the same earthquake, a seismograph near Tulsa, Oklahoma, recorded a difference of 700 seconds between the arrival times of pressure waves (P waves) and shear waves (S waves). Both waves traveled the same distance from the epicenter to Tulsa, but the average speed of the P waves was about 13 km/s, while the average speed of the S waves was only about 7 km/s.

EXAMPLE 1 Interview: Barbara Romanowicz

Seismologists like Barbara Romanowicz calculate distances to epicenters. Estimate the distance from the epicenter to the seismograph near Tulsa, to the nearest ten thousand kilometers.

SOLUTION

Let d = the distance in kilometers from the epicenter to the seismograph.

$$\underset{\text{the S waves}}{\text{Travel time for}} - \underset{\text{the P waves}}{\text{Travel time for}} = \underset{\text{S waves and P waves}}{\text{Time lapse between}}$$

$$\frac{d}{7} - \frac{d}{13} = 700 \qquad \text{Remember: } d = rt, \text{ or } t = \frac{d}{r}.$$

$$91\left(\frac{d}{7} - \frac{d}{13}\right) = 91 \cdot 700 \qquad \text{Multiply both sides by } 91, \text{ the least common multiple of 7 and 13.}$$

$$\frac{91d}{7} - \frac{91d}{13} = 63{,}700$$

$$13d - 7d = 63{,}700 \qquad \text{Divide both sides of the equation by 6.}$$

$$6d = 63{,}700$$

$$d = \frac{63{,}700}{6} \approx 10{,}617$$

The epicenter was about 10,000 km from the seismograph near Tulsa.

Toolbox p. 584
Factors and Multiples

You can make an equation with fractions easier to solve. You multiply both sides of the equation by the least common denominator of the fractions. As you saw in Example 1, the least common denominator of a group of fractions is the least common multiple (LCM) of the denominators.

$$91\left(\frac{d}{7} - \frac{d}{13}\right) = 91 \cdot 700$$

The least common denominator is the LCM of 7 and 13:

$7 \cdot 13 = 91$

EXAMPLE 2

Solve the equation $\frac{3}{2}x - \frac{1}{5} = \frac{1}{6}x$.

SOLUTION

$$\frac{3}{2}x - \frac{1}{5} = \frac{1}{6}x$$

$$30\left(\frac{3}{2}x - \frac{1}{5}\right) = 30 \cdot \frac{1}{6}x$$

Multiply both sides of the equation by **30**, the LCM of 2, 5, and 6.

$$\frac{90}{2}x - \frac{30}{5} = \frac{30}{6}x$$

$$45x - 6 = 5x$$

$$45x - 6 - 45x = 5x - 45x$$

Subtract **45x** from both sides of the equation.

$$-6 = -40x$$

$$\frac{-6}{-40} = \frac{-40x}{-40}$$

$$\frac{3}{20} = x$$

THINK AND COMMUNICATE

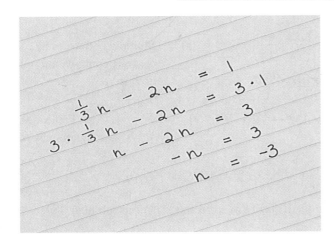

1. A student tried to solve the equation $\frac{1}{3}n - 2n = 1$ as shown. What mistake did the student make? What is the correct solution?

2. In Example 2, why can't you just multiply the right side of the equation by 6 and the left side of the equation by the least common denominator of $\frac{3}{2}$ and $\frac{1}{5}$?

3. Why might someone try to solve the equation in Example 2 by multiplying both sides by 60 instead of 30? Will that method work?

Equations with Decimals

You can also make an equation with decimals easier to solve. You multiply both sides of the equation by a power of ten that will eliminate the decimals.

BY THE WAY

EXAMPLE 3 Application: Personal Finance

It costs $6.50 to see a movie at the theater near where Nina Ramirez lives. It costs only $2.99 to rent a video at the local video store. Nina wants to buy a VCR for $325.95. How many movies will Nina have to watch at home instead of at the theater before she recovers the cost of the VCR?

SOLUTION

Let n = the number of movies Nina watches.

Write and solve an equation to find the number of movies that will make the costs equal.

$$\underset{\text{at the theater}}{\text{Cost of seeing } n \text{ movies}} = \text{Cost of VCR} + \underset{n \text{ videos}}{\text{Cost of renting}}$$

$$6.50n = 325.95 + 2.99n$$

$$100(6.50n) = 100(325.95 + 2.99n)$$

$$650n = 32{,}595 + 299n$$

$$650n - 299n = 32{,}595 + 299n - 299n$$

$$351n = 32{,}595$$

$$n \approx 92.86$$

Multiply both sides of the equation by **100** to eliminate the decimals.

Divide both sides of the equation by 351.

Nina will have to watch 93 movies at home instead of at the theater before she recovers the cost of the VCR.

Hollywood studios earn almost half of their revenues from video rentals and sales.

☑ CHECKING KEY CONCEPTS

Find the least common denominator of each group of fractions.

1. $\frac{1}{2}, \frac{3}{5}, \frac{1}{3}$

2. $\frac{1}{10}, \frac{1}{5}, \frac{7}{2}$

3. $\frac{5}{9}, \frac{5}{6}, \frac{3}{8}$

Tell what power of ten you would multiply by to make each equation easier to solve.

4. $10.5x + 1.1 = 43.1$ **5.** $1.5n = 11.25 - 6n$ **6.** $5.15x - 29 = -2.1x$

Solve each equation.

7. $\frac{x}{2} + \frac{x}{3} = 1$ **8.** $\frac{d}{8} + \frac{d}{12} = \frac{5}{6}$ **9.** $\frac{3}{8}n = 10 - \frac{1}{4}n$

10. $8y + 40.8 = 1.2y$ **11.** $0.34c + 1.2 = 0.74c$ **12.** $1.5a = 15 - 2.25a$

Toolbox p. 585
Powers of Ten

4.4 Exercises and Applications

Extra Practice
exercises on
page 564

Find the least common denominator of each group of numbers.

1. $\frac{1}{3}, \frac{5}{6}, \frac{1}{2}$

2. $\frac{3}{2}, \frac{3}{4}, \frac{1}{3}$

3. $\frac{2}{18}, \frac{4}{15}, 8$

Solve each equation.

4. $\frac{m}{3} + \frac{m}{4} = 14$

5. $\frac{x}{2} + \frac{x}{3} = \frac{25}{3}$

6. $\frac{7y}{9} + 2 = \frac{2y}{3}$

7. $\frac{2}{3}a - \frac{1}{2}a = \frac{7}{2}$

8. $\frac{3}{4}a - 6 = \frac{1}{4}$

9. $\frac{3}{5}p - \frac{7}{10} = \frac{7}{10}p$

10. $\frac{2y}{3} + \frac{1}{3} = \frac{y}{2} + \frac{1}{6}$

11. $\frac{3x}{4} - \frac{5}{2} = \frac{x}{8} + 10$

12. $\frac{1}{8}c - \frac{1}{5}c + 1 = \frac{5}{2}$

 Technology For Exercises 13–16, use a graphing calculator to graph and compare the equations.

13. Graph the equations $y = \frac{x}{2} + \frac{x}{3}$ and $y = \frac{25}{3}$ in the same coordinate plane.

14. Graph the equations $y = 3x + 2x$ and $y = 50$ in the same coordinate plane.

15. How are the equations in Exercise 14 related to the equations in Exercise 13? What do their graphs have in common?

16. How is the solution of the equation in Exercise 5 related to the graphs?

17. SPORTS With the wind at his back, Russell can skate across a lake at a speed of 14 km/h. When skating back into the wind, he can skate only 12 km/h. It takes Russell half an hour to skate across the lake and back.

a. Copy and complete the table.

b. Write and solve an equation to find the distance d across the lake.

Skating Across a Lake and Back			
	Distance across (km)	Rate (km/h)	Time (hours)
Wind at his back	d	14	$\frac{d}{14}$
Into the wind	?	?	?

Solve each equation.

18. $0.2x + 5.3 = 6.1$

19. $0.05a + 2.33 = 1.08$

20. $4.8 - 0.3n = 0.3n$

21. $2.54a - 0.008 = 0.5$

22. $3.25c = 6 + 2.5c$

23. $464 - 0.018n = -0.25n$

24. $3k - 0.75 = k - 0.5$

25. $8 - 0.02x = 2.23x + 3.5$

26. Writing Write a letter to a friend explaining how you decide what power of ten to multiply by when you solve equations like those in Exercises 18–25. Include examples.

27. SAT/ACT Preview If $0.3x - 14 = 0.02x$, what is the value of $2x + 9$?

A. 0.5 **B.** 14 **C.** 50 **D.** 59 **E.** 109

Look back at the article on pages 144–146.

Barbara Romanowicz is shown here reading a seismogram.

Seismologists like Barbara Romanowicz need data from at least three seismographs at three different locations in order to locate the epicenter of an earthquake.

• • • • Readings for the Northridge Earthquake • • • • January 17, 1994			
	Average S wave speed (km/s)	Average P wave speed (km/s)	Difference in arrival time (seconds)
Topopah Spring, Nevada	3.3	6.0	49
San Andreas Observatory, California	3.75	7.5	51
Albuquerque, New Mexico	3.5	7.7	169

For Exercises 28 and 29, round each answer to the tens' place.
For Exercise 30, round the answer to the hundreds' place.

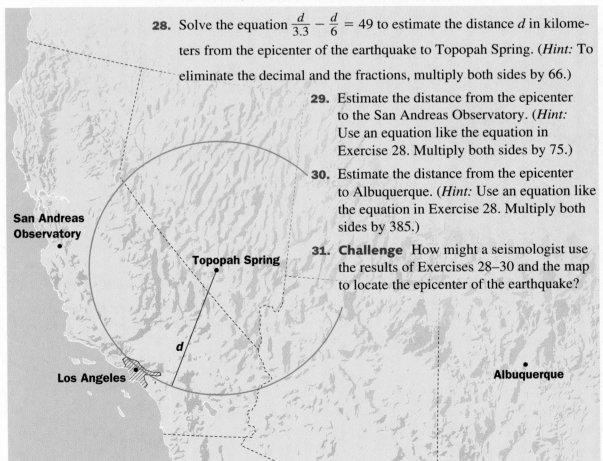

28. Solve the equation $\frac{d}{3.3} - \frac{d}{6} = 49$ to estimate the distance d in kilometers from the epicenter of the earthquake to Topopah Spring. (*Hint:* To eliminate the decimal and the fractions, multiply both sides by 66.)

29. Estimate the distance from the epicenter to the San Andreas Observatory. (*Hint:* Use an equation like the equation in Exercise 28. Multiply both sides by 75.)

30. Estimate the distance from the epicenter to Albuquerque. (*Hint:* Use an equation like the equation in Exercise 28. Multiply both sides by 385.)

31. Challenge How might a seismologist use the results of Exercises 28–30 and the map to locate the epicenter of the earthquake?

Spreadsheets **For Exercises 32–35, use a spreadsheet and the information below.**

Kellem Company wants to buy a fax machine. The company will buy either a machine that uses ordinary paper or one that uses special paper.

Cost (ordinary paper): $1274 plus $2.41 per package of 500 sheets of paper

Cost (special paper): $622 plus $8.45 per roll of 350 sheets of paper

Kellem Company is strongly considering buying the fax machine for $622, but faxes on ordinary paper are easier to read.

32. Explain how to find the cost per page of each type of paper.

33. Complete the spreadsheet shown.

Fax Machine Costs		
B2	=1274+(2.41/500)*A2	
	A	**B**	**C**
1	Number of pages	Cost (ordinary)	Cost (special)
2	1,000	1278.82	
3	10,000		
4	100,000		

34. Write the total cost of each type of fax machine as a function of the number of pages. Then write and solve an equation to find the number of pages for which the costs are equal.

35. **Open-ended Problem** Suppose the manager of Kellem Company asks for your advice. What questions would you ask the manager before giving advice? Under what conditions would you recommend the machine that uses ordinary paper? the machine that uses special paper?

ONGOING ASSESSMENT

36. **Cooperative Learning** Work with a partner. Together make a list of the important things to remember about solving equations with fractions. Make a similar list for equations with decimals. Write a brief summary of each list and include examples.

SPIRAL REVIEW

Technology **Graph each equation. Use a graphing calculator if you have one. Use the graph to estimate the unknown value to the nearest tenth.** *(Section 2.6)*

37. $y = 2x + 3$; find the value of x when $y = 7$.

38. $y = -x + 4$; find the value of x when $y = 6$.

39. $y = \frac{2}{3}x + 1$; find the value of x when $y = -1$.

40. Find the mean, the median, and the mode(s) of the data. *(Section 1.3)*

Ages of employees in an office: 21, 25, 30, 45, 52, 24, 38, 27, 35, 48

4.5

Writing Inequalities from Graphs

Learn how to...

- **use inequalities to represent intervals on a graph**

So you can...

- **represent a group of solutions, such as times of day when temperatures are more than 100°F**

You can be comfortable inside an adobe house even when the outside temperature is more than 100°F. The clay building materials absorb heat during the day and give off heat during the night. The graph below models this situation.

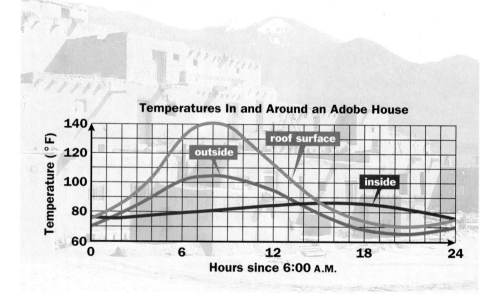

Temperatures In and Around an Adobe House

THINK AND COMMUNICATE

Use the graph above.

1. Where will you find the hottest temperature: *inside*, *outside*, or *on the roof*? Where will you find the coolest temperature?

2. What is the range of inside temperatures? of outside temperatures?

3. Where do the temperatures vary least: *inside*, *outside*, or *on the roof*?

4. When is the outside temperature hotter than the inside temperature? When is the outside temperature cooler than the inside temperature?

5. How do the outside temperature and the temperature of the roof surface compare?

An **inequality** is a mathematical statement formed by placing an inequality symbol between numerical or variable expressions. You can use an inequality to represent the solution of a problem.

EXAMPLE 1   Application: Architecture

Use the graph on page 171. When is the outside temperature more than 100°F? Write your answer as an inequality.

SOLUTION

Here the **outside** temperature is more than 100°F.

Temperatures In and Around an Adobe House

Let t = hours since 6:00 A.M. Then t is between 5.5 and 10.5, or $5.5 < t < 10.5$.

EXAMPLE 2   Application: Recreation

A dance organization sponsors a few dances every week. The admission price to each dance is $4 for members and $6 for nonmembers. It costs $50 for an annual membership. When is the total cost for a nonmember more than the total cost for a member? Write your answer as an inequality.

SOLUTION

Problem Solving Strategy: Use a graph.

Let x = the number of dances a person attends in a year.

Step 1 Write and graph two equations that model this situation.

Cost for a member: $y = 4x + 50$

Cost for a nonmember: $y = 6x$

Step 2 Find the coordinates of the intersection point.

Nonmember costs are more than **member** costs beyond this point.

For more information about finding where graphs intersect, see the *Technology Handbook*, p. 607.

The total cost for a nonmember is more than the total cost for a member when a person attends more than 25 dances, or when $x > 25$.

THINK AND COMMUNICATE

6. In Example 2, why is the inequality $x > 25$ and not $x \geq 25$?

7. In Example 2, when is the total cost for a nonmember *less than* the total cost for a member? Write your answer as an inequality.

For Questions 1–4, use the graph. Write an inequality to represent the x-values that correspond to each range of y-values.

$$y = -\frac{1}{2}x + 4$$

1. $y > 0$ **2.** $y \leq 0$

3. $y < 3$ **4.** $0 < y < 3$

5. In Example 2, suppose the annual membership is $30.00, the dance admission price for members is $4.50, and the dance admission price for nonmembers is $5.00. When is the total cost for nonmembers greater than the total cost for members? Write your answer as an inequality.

Extra Practice
exercises on page 565

4.5 Exercises and Applications

1. FUEL ECONOMY The graph shows the fuel economy of an average car as a function of speed.

a. At what speeds should the car be driven so that it gets more than 25 mi/gal? Write your answer as an inequality.

b. Writing What does the graph tell you about fuel economy at different speeds?

c. Open-ended Problem Write three problems that can be answered using the graph and an inequality.

Speed and Fuel Economy

For Exercises 2–4:

a. Graph each equation.

b. Write an inequality to represent the x-values when $y > 1$.

2. $y = 2x + 1$ **3.** $y = -2x + 1$ **4.** $y = 2x$

Earth's atmosphere is divided into layers. Temperatures vary from layer to layer. The black jagged line on the graph shows a representative temperature for each altitude. For Exercises 5–7, write each answer as an inequality. Use the variable *a* to represent altitude, and the variable *t* to represent temperature.

5. What is the range of temperatures in the troposphere? in the mesosphere?

6. At which altitudes is the temperature above 30°C?

7. At which altitudes is the temperature below −70°C?

8. Which layer has the greatest range of temperatures? the smallest range?

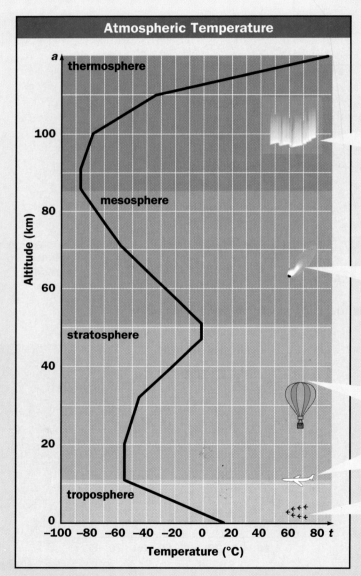

Atmospheric Temperature

thermosphere

mesosphere

stratosphere

troposphere

Altitude (km)

Temperature (°C)

Auroras are bright displays of light in the sky. They usually occur at least 97 km above Earth's surface.

Meteors burn up at heights around 65 km.

A record-setting balloon flight in 1961 reached 34.668 km.

Airplanes generally fly between 8.8 km and 12.5 km.

Pilots have observed geese in migratory flight at 1.05 km.

9. Challenge Between which two altitudes is the rate of decrease in temperature the greatest? How can you tell?

10. Research Temperatures do not follow the pattern in the graph during an *inversion*. Find out what an *inversion* is.

11. COMMUNICATIONS The graph models the cost of using two different long distance companies. Cost is a function of the total number of minutes of your long distance calls each month.

a. Describe each company's fees in words.

b. When does it cost less to use Company A? Write your answer as an inequality.

12. **Technology** A ten-wheeler truck weighs 25,000 lb when empty. It can carry up to 13 yd^3 of sand. Each cubic yard weighs 2750 lb.

a. Write an equation for the weight of the truck as a function of the number of cubic yards of sand it carries.

b. Graph the equation from part (a). Set a good viewing window.

c. For what amounts of sand is it safe for the truck to cross a bridge with a 20 ton load limit? Write your answer as an inequality. (*Hint:* Use the fact that 1 ton = 2000 lb.)

13. PERSONAL FINANCE The camcorder that Yoshi Minami is considering costs $750 to buy. A similar model costs $40 per day to rent. When is the cost to rent a camcorder greater than the cost to buy a camcorder? Write your answer as an inequality.

ONGOING ASSESSMENT

14. Open-ended Problem
The graph models car trips for Theresa Jones and Jason Wills between Dallas and Houston. Write a story based on the graph and the point where the lines intersect. Be sure to include speeds and inequalities in your story.

SPIRAL REVIEW

Find an equation of the line with the given slope and through the given point. (*Section 3.5*)

15. slope = 2; (5, 3)

16. slope = $-\frac{3}{5}$; (−15, 9)

17. slope = 0; (12, −12)

18. undefined slope; (2, 2)

Graph each inequality on a number line. (*Toolbox, page 591*)

19. $x > -4$ **20.** $a \le 3$ **21.** $n \ge 0$ **22.** $y < -1$

4.6

Learn how to...

- solve inequalities

So you can...

- tell when one quantity is greater than another, such as when renting is more expensive than buying

Solving Inequalities

Suppose you start with an inequality that is true. What happens when you add the same number to both sides? The example below should give you some idea.

$$2 < 3$$
$$2 + 4 < 3 + 4$$
$$6 < 7$$

EXPLORATION
COOPERATIVE LEARNING

Working with Inequalities

Work with a partner.

For each of Steps 1–4, draw pairs of number lines like those shown above. One of you should work with the inequality 2 < 3 and one of you should work with the inequality 1 < 5. Use these numbers for each step:

a. 2   **b. 10**   **c. −2**   **d. −10**

---

1 **Add** each number to both sides of your inequality. Is each new inequality *True* or *False*?

2 **Subtract** each number from both sides of your inequality. Is each new inequality *True* or *False*?

3 **Multiply** both sides of your inequality by each number. Is each new inequality *True* or *False*?

4 **Divide** both sides of your inequality by each number. Is each new inequality *True* or *False*?

Questions

1. Work together to generalize how an inequality is affected by each of the four basic operations: addition, subtraction, multiplication, and division. Test your generalization with some inequalities.

2. Assume that $x > 0$, $y > 0$, and $x > y$. Replace each $\underline{?}$ with > or <.

    a. $x - 3 \underline{} y - 3$ **b.** $3x \underline{} 3y$ **c.** $\dfrac{x}{-3} \underline{} \dfrac{y}{-3}$

When you add, subtract, multiply by, or divide by the same positive number on both sides of an inequality, the inequality symbol stays the same. It also stays the same when you add or subtract the same negative number. When you multiply by or divide by the same **negative** number on both sides, the inequality symbol is **reversed**. For example:

$$1 < 2$$
$$1 - 5 < 2 - 5$$
$$-4 < -3$$

$$1 < 2$$
$$5 \cdot 1 < 5 \cdot 2$$
$$5 < 10$$

$$1 < 2$$
$$-5 \cdot 1 > -5 \cdot 2$$
$$-5 > -10$$

When you solve an inequality that involves a variable, the **solution of the inequality** is the set of numbers that makes the inequality true.

EXAMPLE 1 Application: Recreation

Brian pays $4 to rent skates every time he goes roller skating. He thinks he can save money if he buys his own skates for $45. He wonders, "When is the cost of renting skates more expensive than buying a new pair?"

SOLUTION

Here is how Brian solved the problem.

Let n = the number of times I rent skates

Cost of renting skates : $4n$

Cost of buying skates : 45

Cost of renting > Cost of buying

$$4n > 45$$

$$\frac{4n}{4} > \frac{45}{4}$$

$$n > 11.25$$

If I rent skates more than 11 times, I will pay more than if I buy new skates.

Check

Substitute a number **greater** than 11 **for** n.

$$4n > 45$$

$$4(12) \overset{?}{>} 45$$

$$48 > 45 \checkmark$$

EXAMPLE 2

Solve the inequality $-2x \le 10$. Graph the solution on a number line.

SOLUTION

Divide both sides by -2 and **reverse** the inequality symbol.

$$-2x \le 10$$

$$\frac{-2x}{-2} \ge \frac{10}{-2}$$

$$x \ge -5$$

Graph -5 and all numbers greater than -5.

Toolbox p. 591
Graphing Inequalities

THINK AND COMMUNICATE

1. Describe how Brian could have used a graph to solve the problem in Example 1 on page 177.

2. In Example 2 on page 177, can you use -5 to check the solution? Why is it better to use a number greater than -5 to check the solution? Check the solution.

EXAMPLE 3

Solve the inequality $6 + x < 3 - 4x$. Graph the solution on a number line.

SOLUTION

$$6 + x < 3 - 4x$$
$$6 + x - 6 < 3 - 4x - 6$$
$$x < -3 - 4x$$
$$x + 4x < -3 - 4x + 4x$$
$$5x < -3$$
$$\frac{5x}{5} < \frac{-3}{5}$$
$$x < -\frac{3}{5}$$

Graph all numbers less than $-\frac{3}{5}$.

Check

Substitute a number *less than* $-\frac{3}{5}$ for x, such as -1.

$$6 + x < 3 - 4x$$
$$6 + (-1) \overset{?}{<} 3 - 4(-1)$$
$$5 \overset{?}{<} 3 + 4$$
$$5 < 7 ✓$$

WATCH OUT!

The number you are dividing by is positive 5, so you do not reverse the inequality symbol.

x is less than but not equal to $-\frac{3}{5}$, so you use an open circle.

☑ CHECKING KEY CONCEPTS

Tell whether you would reverse the inequality symbol to solve each inequality. Then solve each inequality.

1. $-3x > 6$

2. $4n < -20$

3. $a + 4 \geq 5$

4. $\frac{m}{2} \leq 3$

5. $\frac{x}{-4} \geq 5$

6. $-\frac{c}{3} > 1$

Solve each inequality. Graph each solution on a number line.

7. $x + 9 < 4$

8. $\frac{n}{-5} \leq 7$

9. $-5t + 6 > 46$

10. $15 \geq 3a - 6$

11. $-3x > x - 9$

12. $7n + 4 < 9n - 3$

13. In Example 1 on page 177, suppose that Brian pays only $3 each time he rents skates. When is the cost of renting skates more expensive than buying a new pair?

4.6 Exercises and Applications

Solve each inequality. Graph each solution on a number line.

Extra Practice exercises on page 565

1. $x - 5 > 0$

2. $-4n \leq 16$

3. $10 > k + 9$

4. $10x \geq 7$

5. $\dfrac{c}{-3} > 2$

6. $\dfrac{3}{4}a \leq 21$

7. $8 - x > 8$

8. $-7k + 4 < 18$

9. $3n + 4 \leq -8$

10. $6 + 4c > -10$

11. $15 < 3a - 5$

12. $1 - 4a > 21$

13. Malcolm has already invited some people to a party. Now he wants to invite 8 more. His father says, "If you add 8 more people to the guest list, that's over 40 people." Write and solve an inequality based on this situation.

14. **GEOMETRY** A rectangle is 7 cm longer than it is wide. The perimeter is between 30 cm and 50 cm. Find the range of possible widths. Write your answer as an inequality.

Connection ▶ TRANSPORTATION

Many people work in areas where public transportation is available. They often have a number of commuting options to choose from. The options shown are for a 36 mi commute.

For Exercises 15–17, use the information with each photograph. Explain your reasoning.

15. For what number of days is it less expensive to commute by train than by subway?

16. When is it less expensive to commute by train than by bus?

17. When is it less expensive to commute by subway than by bus?

18. **Open-ended Problem** Is the least expensive option always the best option? What other things besides cost should you consider when you choose a commuting option?

Walk to the commuter bus stop and pay $7 per day.

Drive to the subway and pay $13.50 per day for subway, parking, and car operating costs.

Walk to the train station and pay $120 per month.

Match each inequality with its solution.

19. $3x + 4 < 7$
20. $3x + 4 < 3x$
21. $3x + 4 < 3(x + 4)$

A. all numbers
B. $x < 1$
C. no solution

Solve each inequality. Graph each solution on a number line. If there is no solution, write _no solution_.

22. $12x - 36 < 3x$
23. $7n > 12n - 5$
24. $2.5n - 14 < -12.75$

25. $-\dfrac{2}{5}c \geq \dfrac{3}{10}$
26. $\dfrac{5}{2}x - 1 > 14$
27. $-\dfrac{3}{2}t - 1 \leq 5$

28. $-6a - 16 \leq 6a + 8$
29. $5a < 2(a - 9)$
30. $3(5x - 4) < 15x$

31. $2(1 - x) \geq 2x$
32. $5(n + 1) < 5n$
33. $-6k > -2(3k + 3)$

34. Writing How is solving an inequality like solving an equation? How is it different?

35. BUSINESS Veronica Kelly is a salesperson in a large department store. She receives a yearly salary of $19,500. In addition, she makes a 5% commission on all sales. Last year, Veronica made more than $43,000. Write and solve an inequality based on this situation.

36. Challenge Solve the inequalities $x + 3 \leq 5$ and $x + 3 \geq -5$. Explain how you can use the solutions to solve the inequality $|x + 3| \leq 5$.

ONGOING ASSESSMENT

37. Cooperative Learning Work with a partner. Discuss how you could solve the inequality $5 - \dfrac{1}{2}x > 4 - \dfrac{1}{4}x$ by graphing the equations $y = 5 - \dfrac{1}{2}x$ and $y = 4 - \dfrac{1}{4}x$ in the same coordinate plane. Also discuss how you would solve the inequality algebraically. Then solve the inequality both ways.

SPIRAL REVIEW

Copy and complete each equation. *(Section 3.1)*

38. $\dfrac{\$10.25}{1 \text{ page}} \cdot \dfrac{5 \text{ pages}}{1 \text{ h}} = \underline{\ ?\ }$

39. $\dfrac{2 \text{ tokens}}{1 \text{ trip}} \cdot \underline{\ ?\ } = 56 \text{ tokens}$

40. $\dfrac{4 \text{ quarters}}{1 \text{ wash}} \cdot \dfrac{\$1}{4 \text{ quarters}} \cdot \dfrac{6 \text{ washes}}{1 \text{ day}} = \underline{\ ?\ }$

41. $\dfrac{60 \text{ mi}}{1 \text{ h}} \cdot \underline{\ ?\ } \cdot \dfrac{1 \text{ h}}{3600 \text{ s}} = \underline{\ ?\ } \text{ ft/s}$

Find the slope of the line through each pair of points. *(Section 3.3)*

42. $(5, 1)$ and $(4, 4)$
43. $(-2, 5)$ and $(7, 1)$
44. $(1, 8)$ and $(1, 0)$

ASSESS YOUR PROGRESS

VOCABULARY

inequality (p. 171)   **solution of an inequality** (p. 177)

Solve each equation. *(Section 4.4)*

1. $\dfrac{x}{5} = \dfrac{3}{2}$ **2.** $\dfrac{3}{4}a - \dfrac{2}{3} = \dfrac{5}{6}$ **3.** $\dfrac{3}{8} - \dfrac{n}{2} = 1$

4. $0.75t + 0.15 = 1.65$ **5.** $5x - 0.5 = 5.75$ **6.** $2.208 - 0.09n = 0.6n$

7. SPORTS When you hit a baseball near either end of a bat, it vibrates. It stings your hands and the ball does not travel far. But if you hit the ball near the bat's *node*, the bat vibrates less and the ball travels farther.

Suppose a pitcher throws a ball at 85 mi/h and you swing the bat at about 70 mi/h. The graph shows the distance the ball will travel, based on where the bat hits the ball. *(Section 4.5)*

BY THE WAY

Aluminum bats, used almost exclusively in amateur baseball, were first made in the early 1970s. They weigh 2–4 oz less than wooden bats and are usually filled with a sound–deadening insert such as cork.

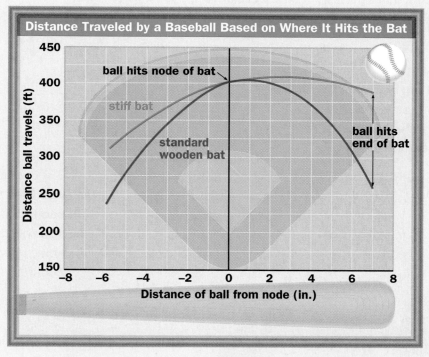

Distance Traveled by a Baseball Based on Where It Hits the Bat

ball hits node of bat

stiff bat

standard wooden bat

ball hits end of bat

Distance ball travels (ft)

Distance of ball from node (in.)

a. With a stiff bat, how far from the node should you hit the ball so that it travels at least 350 ft? Write your answer as an inequality.

b. Repeat part (a) for a standard wooden bat.

Solve each inequality. Graph each solution on a number line. *(Section 4.6)*

8. $n + 8 > 5$ **9.** $\dfrac{x}{-5} \ge 8$ **10.** $3 - 2a < 7$

11. $10t - 4 \le 2t$ **12.** $6k + 9 \ge k - 1$ **13.** $4(c + 2) < 4c + 10$

14. Journal Write a summary of the various methods you have learned for solving equations and inequalities.

Making a Profit

Have you ever wondered how prices are determined? People who make a product must determine how much money to charge so that they make a *profit*. Profit is the money remaining after expenses are subtracted from income.

PROJECT GOAL Determine the expenses of making cookies and decide on a selling price so that you earn a profit.

Financial Plans
Work with a partner. Choose a cookie recipe you want to make.

1. CALCULATE EXPENSES

• **INGREDIENTS**
Calculate or estimate how much the ingredients in one batch of cookies cost. For example:

Cost of 1 cup of sugar =

$$\frac{\text{Cost of a bag}}{\text{Cups per bag}} = \frac{\$1.89}{10} \approx \$.19$$

• **TIME**
Calculate the cost of your time to make one batch.

$$\begin{array}{c}\text{Labor}\\\text{costs}\end{array} = \begin{array}{c}\text{Hours}\\\text{worked}\end{array} \times \begin{array}{c}\text{Hourly}\\\text{pay}\end{array}$$

• **FIXED EXPENSES**
Assume that you must pay \$25 for a permit to sell cookies.

2. CALCULATE INCOME Research a reasonable price to charge for selling your cookies individually. How much income can you make per batch?

Analyzing Your Data

Use a table and a graph to model expenses and income as functions of the number of batches sold. What is the point at which you make a profit? Write and solve an inequality to show another method of finding the point at which you make a profit.

Making a Report

Write a report about your cookie-making venture. Describe the decisions that you made concerning ingredients, labor costs, and pricing. Include your table, your graph, and a discussion of income and expenses.

You may want to extend your report by examining some of the ideas below:

• Survey friends and neighbors to find the price they would be willing to pay for a cookie. Display that data and any conclusions you can make from the data.

• How do cookie manufacturers' prices differ from your price? Why do you think this is so? How are their expenses different from yours?

Self-Assessment

Write a paragraph explaining your decisions. Which method, making a table, making a graph, or writing an inequality, works best for you?

Describe how you did research. For example, how did you find out what the minimum wage is or what local businesses pay workers for similar work?

What was the most difficult aspect of the project? the easiest? Why?

4 Review

STUDY TECHNIQUE

The diagram at the right is the start of a **concept map**. A map like this can help you review new words and ideas. Keep these things in mind when you make a concept map:

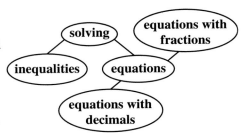

- **A group of ideas can be summarized in more than one way.**

- **Lines should connect related concepts.**

- **Concepts should be arranged to emphasize relationships.**

Expand this concept map to review the key ideas in this chapter.

VOCABULARY

reciprocals (p. 153)
identity (p. 161)

inequality (p. 171)
solution of an inequality (p. 177)

SECTIONS · 4.1, 4.2, *and* 4.3

Tables, graphs, and equations can be used to compare different situations.

For example, should you pay $7 for each visit to a museum or pay $50 for an annual membership? The models below show that 8 or more visits in a year cost more than the annual $50 fee.

Number of visits	Total cost ($)	
	With membership	Without membership
1	50	7
2	50	14
3	50	21
4	50	28
5	50	35
6	50	42
7	50	49
8	50	56

Cost of Visiting a Museum

Let v = the number of visits.

$$7v = 50$$
$$v \approx 7.14$$

You can use *reciprocals* to solve equations with fractional coefficients.

The numbers x and $\frac{1}{x}$ are **reciprocals** because their product is one. Zero does not have a reciprocal.

$$\frac{5}{8}x = 10$$

$$\frac{8}{5} \cdot \frac{5}{8}x = \frac{8}{5} \cdot 10$$

$$x = 16$$

Multiply both sides of the equation by the **reciprocal** of $\frac{5}{8}$.

To solve multi-step equations with parentheses:

1. Simplify each side of the equation.

2. Get all the variable terms on one side.

3. Get the variable alone on one side.

$$5(x + 3) = 3 + 7x - 14$$

$$5x + 15 = -11 + 7x$$

$$5x + 15 - 5x = -11 + 7x - 5x$$

$$15 = -11 + 2x$$

$$15 + 11 = -11 + 2x + 11$$

$$26 = 2x$$

$$13 = x$$

An **identity** is an equation that is true for all values of the variable. An equation has no solution when no value of the variable will make the equation true.

SECTIONS 4.4, 4.5, *and* 4.6

You can solve equations with fractions by **multiplying both sides by the least common denominator**.

$$\frac{3}{4}x + 5 = \frac{2}{3}$$

$$12\left(\frac{3}{4}x + 5\right) = 12\left(\frac{2}{3}\right)$$

$$9x + 60 = 8$$

$$9x = -52$$

$$x = -5\frac{7}{9}$$

You can eliminate decimals in an equation by **multiplying both sides by a power of ten**.

$$3.65x - 0.6 = 3.45x$$

$$100(3.65x - 0.6) = 100(3.45x)$$

$$365x - 60 = 345x$$

$$-60 = -20x$$

$$3 = x$$

Subtract $365x$ from both sides.

An **inequality** is a statement formed by placing an inequality symbol between mathematical expressions. For the museum example on page 184, you can use the inequality $v \geq 8$ to represent the interval where the total cost for a nonmember is more than the cost for a member.

When you find the **solution of an inequality**, remember to **reverse** the inequality symbol when you multiply or divide both sides by a **negative** number.

$$-3x < 12$$

$$\frac{-3x}{-3} > \frac{12}{-3}$$

$$x > -4$$

Assessment

VOCABULARY

Complete each statement.

1. The product of a number and its _?_ is 1.

2. All numbers are solutions of a(n) _?_.

SECTIONS 4.1, 4.2, *and* 4.3

3. A public golf course charges $25 for an 18-hole round of golf. Renting clubs costs an additional $10 per round. Suppose you can buy a set of golf clubs for $525. Make a graph that models this situation. When will it begin to cost more if you rent clubs each time than if you buy clubs?

4. Write an equation to model the point at which the two options in Question 3 cost the same amount.

Find the reciprocal of each number.

5. $\dfrac{3}{5}$

6. $-\dfrac{7}{8}$

7. $\dfrac{3}{2}$

Solve each equation.

8. $\dfrac{3}{5}m = 15$

9. $\dfrac{2x}{3} - 8 = 4$

10. $-17 = 19 - \dfrac{6}{7}a$

11. $8a = 2a + 54$

12. $4y - 3 = 6y + 7$

13. $2 - 7n = 8n - 3$

14. Which equation is an identity? Which equation has no solution?

   A. $2(6a + 3) = 3(4a + 2)$ **B.** $5(x + 3) + 4 = 9x + 15$

   C. $8x - 10 = \dfrac{2}{3}(15x + 3)$ **D.** $6(c + 3) - 15 = 6c + 18$

15. Solve the equation $3(x + 2) = 5(x - 3) + 13$. Explain each step.

Solve. If an equation is an *identity* or there is *no solution*, say so.

16. $-30x + 15 = -6x + 31$ **17.** $5t - 6 - 8t = -3(t + 2)$

18. $\dfrac{1}{2}(6x - 8) = 3x - 2$ **19.** $4a - 2 = \dfrac{4}{5}(10a - 5) - 6$

20. A season ticket for an orchestra seat at a theater costs $140 for six performances. The cost of a ticket for the same seat for one performance is $35. Write and solve an equation to find out how many performances Rafael Santos will have to attend to recover the cost of a season ticket.

21. What power of ten would you use to eliminate the decimal numbers in the equation $0.02n + 3.5 = 8.067$ most easily?

22. What is the smallest number that you can multiply by to eliminate the fractions in the equation $\frac{4}{5}x + \frac{2}{9} = \frac{1}{6} - \frac{4}{15}x$?

Solve each equation.

23. $\frac{2}{5}x + 2 = \frac{2}{3} - 2x$

24. $\frac{4}{5}m + \frac{3}{4} = \frac{7}{10}m + \frac{1}{4}$

25. $0.2n + 3 = 5.4 - n$

26. $0.7 - 0.8x = 0.18 + 0.24x$

The graph shows the fuel economy of an average car as a function of speed. Use the graph for Exercises 27 and 28.

27. At what speeds is a car's fuel economy less than 25 mi/gal? Write your answer as two inequalities.

28. At what speeds is a car's fuel economy more than 30 mi/gal? Write your answer as an inequality.

29. The statement $10 > 2$ is a true inequality. Is the inequality still true if both sides of the inequality are multiplied by 7? by -3? by $\frac{1}{2}$? by $-\frac{5}{2}$? Explain.

30. If $m < r$, which of the following inequalities is not true?

   A. $m + 7 < r + 7$ **B.** $-3m < -3r$ **C.** $m - 3 < r - 3$ **D.** $\frac{m}{5} < \frac{r}{5}$

Speed and Fuel Economy

Fuel economy (mi/gal) vs. Speed (mi/h)

Solve each inequality. Graph each solution on a number line.

31. $c - 3 > 4$

32. $2x \leq -8$

33. $-3f \leq -27$

34. $7 - 4x \leq -5$

35. $-\frac{2}{5}x + 2 \geq 8$

36. $5x + 3 < 2x - 6$

PERFORMANCE TASK

37. Find something that can be paid for in two different ways, such as admission to a museum. Compare the costs. Explain the circumstances under which each option is better. Justify your explanation with a table, a graph, or an equation. Use inequalities to express your findings.

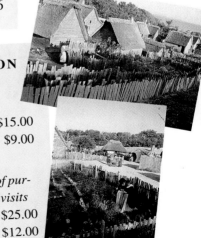

PLIMOTH PLANTATION RATES

INDIVIDUALS:

Adult	$15.00
Child 5–12	$9.00

ANNUAL PASS:
Valid for one full year from date of purchase for an unlimited number of visits

Adult	$25.00
Child 5–12	$12.00

1 YEAR MEMBERSHIP:

Family	$70.00
Individual	$35.00

5 Connecting Algebra and Geometry

Painting a neighborhood

Daniel Galvez

Daniel Galvez always wanted to be an artist, but he didn't know what kind of artist until he learned in college about the great Mexican muralists of the 1900s. Galvez admired the way these artists used colorful paintings on public walls to reach all types of people. "I liked the idea of making a painting that you don't have to go see in a gallery," he says, "the idea that it was part of people's lives whether or not they understood art. It would be something they would see on their way to school or work."

> " I liked the idea of making a painting that you don't have to go see in a gallery."

Viva la Raza

Galvez began to look for places to paint murals and for money to pay for his time and supplies. When he was 24, he painted *Viva la Raza*, a history of Mexican Americans, on the wall of a building in Berkeley, California. Over the next 16 years, he painted more than 15 murals in California and Massachusetts.

Painting a Neighborhood

In 1992, Galvez was hired to paint a mural on the wall of a restaurant in Cambridge, Massachusetts. He wanted the mural, called *Crosswinds*, to show the diverse cultures of people living near the restaurant. "The more I can make people aware of their differences, the more they become aware of their similarities," he says. "The more you know of your neighbors, the less threatening they are." He hired a photographer who took hundreds of photos of people from the neighborhood. Galvez chose his favorite photos to include in the mural.

Using a Scale Drawing

The wall of the restaurant was 24 ft high and 41 ft long. To create his mural, Galvez first made a scale drawing. He used a scale of 1 in. to 1 ft, which made the drawing 24 in. high and 41 in. long. Using slides, he projected the photos onto his scale drawing to find the right size and position for each figure. Then he took the same slides and projected them onto sheets of paper at the actual size he wanted the figures to appear on the wall. After drawing the figures on the sheets of paper, Galvez coated the back of each sheet with chalk and traced the outlines of the figures onto the wall of the restaurant.

He estimated he needed at least a half gallon of each color of paint for every 1000 ft^2 of wall. He multiplied the height of the restaurant wall by the length and got 984 ft^2, so he bought a half gallon of each color and painted the figures using the outlines on the wall. By using paint, mathematics, and a lot of talent, Galvez was able to capture the face of a neighborhood.

EXPLORE AND CONNECT

Cambridge residents stand in front of their portraits with artist Daniel Galvez.

1. Project Measure a bulletin board or wall in your school or community and create a scale drawing of a mural for the space. Include the scale with your drawing. Write a letter to the school board or the owner of the space proposing that your mural be painted.

2. Research José Clemente Orozco, Diego Rivera, David Alfaro Siqueiros, and Rufino Tamayo are four Mexican artists who painted murals. Find out more about one of these artists and write a report about his work.

3. Writing Find a scale model of an object, such as a toy car or doll. Estimate the scale that was used to create the model. How accurate is the model? Be sure to explain your reasoning.

Mathematics & Daniel Galvez

In this chapter, you will learn more about how mathematics is related to art.

Related Exercises

Section 5.1
• **Exercises 17–19**

Section 5.2
• **Exercises 13–16**

5.1

Ratios and Proportions

How do wildlife biologists find the number of fish in a lake or birds in a forest? They don't count the animals one by one. Instead, they use an estimation method called *capture-recapture.*

EXPLORATION
COOPERATIVE LEARNING

The Capture-Recapture Method

Work with a partner.
You will need:
- a large coffee can
- a bag of dried beans
- a marker

Follow Steps 1–4. Then answer the questions below.

1 Pour the bag of beans into the can.

2 Take out a handful of beans, mark them, and record how many you marked.

3 Put the marked beans back into the can. Mix the beans thoroughly.

4 Take out another handful, or *sample*, of beans. Record the number of marked beans and the total number of beans in your sample.

Questions

1. Let B = the unknown number of beans in the can. What is the fraction of beans that are marked? (Your answer will involve B.)

2. What fraction of beans in your sample were marked? How do you think this fraction relates to the fraction from Question 1?

3. Ayita marked 23 beans. Then she took a sample of 12 beans and found 3 marked. Use Ayita's method and the fractions you found in Questions 1 and 2 to estimate the number of beans in your coffee can.

$$\frac{3}{12} = \frac{23}{B}$$

$$(12B)\frac{3}{12} = (12B)\frac{23}{B}$$

$$(12B)\frac{3}{12} = (12B)\frac{23}{B}$$

$$3B = 276$$

$$B = 92$$

Ratios

A **ratio** is a quotient that compares two quantities. For example, in the Exploration on page 191, there were 3 marked beans in a sample of 12 beans. The ratio of marked beans to all beans in the sample is 3 to 12, or 1 to 4. You can write this ratio in three ways:

$$\frac{1}{4} \qquad 1:4 \qquad 1 \text{ to } 4$$

EXAMPLE 1 Application: History

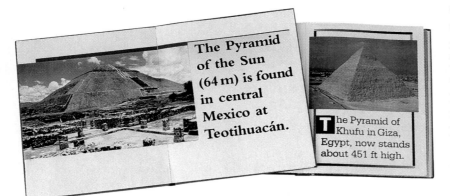

The Pyramid of the Sun (64 m) is found in central Mexico at Teotihuacán.

The Pyramid of Khufu in Giza, Egypt, now stands about 451 ft high.

Grace wants to compare the heights of two pyramids, but the books she has use different units of measure. Use the information shown to find the ratio of the height of the Pyramid of Khufu to the height of the Pyramid of the Sun.

SOLUTION

Convert the height of the Pyramid of Khufu to meters.

WATCH OUT!

When you write a ratio, the two quantities must be measured in the same units.

$$451 \text{ ft} \cdot \frac{0.305 \text{ m}}{1 \text{ ft}} \approx 138 \text{ m}$$

Use the fact that 1 ft = 0.305 m.

$$\frac{\text{Height of the Pyramid of Khufu}}{\text{Height of the Pyramid of the Sun}} = \frac{138 \text{ m}}{64 \text{ m}}$$

$$= \frac{69}{32}$$

Write the fraction in lowest terms.

The ratio of the height of the Pyramid of Khufu to the height of the Pyramid of the Sun is $\frac{69}{32}$.

THINK AND COMMUNICATE

1. Convert $\frac{69}{32}$ to a decimal rounded to the nearest hundredth. What does this number tell you about the heights of the pyramids in Example 1?

2. Suppose you found the ratio in Example 1 by first converting the height of the Pyramid of the Sun to feet. What do you think the ratio would be?

3. Find the ratio of the height of the Pyramid of the Sun to the height of the Pyramid of Khufu. How is this ratio related to the ratio found in Example 1?

Proportions

A **proportion** is an equation that shows two ratios to be equal. Each of the proportions below can be read as "*a* is to *b* as *c* is to *d*."

You call *a* and *d* the **extremes** of the proportion.

$$\frac{a}{b} = \frac{c}{d} \qquad a : b = c : d \qquad a \text{ to } b = c \text{ to } d$$

You call *b* and *c* the **means** of the proportion.

Means-Extremes Property

For any proportion, the product of the **extremes** is equal to the product of the **means**.

$$\text{If } \frac{a}{b} = \frac{c}{d}, \text{ then } ad = bc.$$

For example, you know that $\frac{4}{6} = \frac{2}{3}$.

$$\frac{4}{6} = \frac{2}{3} \qquad\qquad 4 \cdot 3 = 6 \cdot 2$$

EXAMPLE 2 Application: Ecology

At Dryden Lake in New York, researchers caught, marked, and released 213 largemouth bass. Later, they took a sample of 104 bass and found that 13 were marked. Estimate the number of largemouth bass in Dryden Lake.

SOLUTION

Let x = the total number of largemouth bass in the lake.

$$\frac{\text{Marked bass in sample}}{\text{Total bass in sample}} = \frac{\text{Marked bass in lake}}{\text{Total bass in lake}}$$

$$\frac{13}{104} = \frac{213}{x} \qquad \text{Use the means-extremes property.}$$

$$13x = 104 \cdot 213$$

$$13x = 22{,}152$$

$$x = 1704$$

There are about 1700 largemouth bass in Dryden Lake.

EXAMPLE 3 Application: Forestry

In a large state forest, the ratio of white pine trees to red pine trees is 7 to 4. A forester estimates that the total number of white and red pine trees is 5000. How many white pine trees are in the forest?

SOLUTION

Let w = the number of white pine trees. Then $5000 - w$ = the number of red pine trees.

$$\frac{7}{4} = \frac{w}{5000 - w}$$

$$7(5000 - w) = 4w \qquad \text{Use the distributive property.}$$

$$35{,}000 - 7w = 4w$$

$$35{,}000 = 11w$$

$$3182 \approx w$$

There are about 3200 white pine trees in the state forest.

☑ **CHECKING KEY CONCEPTS**

Find the ratio of the first quantity to the second quantity. Write your answer as a fraction in lowest terms.

1. $4\frac{1}{2}$ lb; $13\frac{1}{2}$ lb   **2.** 6 days; 3 weeks   **3.** 5 ft; 15 in.

Solve each proportion.

4. $\dfrac{12}{35} = \dfrac{p}{100}$ **5.** $\dfrac{x}{x + 12} = \dfrac{2}{3}$ **6.** $9{:}11 = 6{:}m$

7. **NUTRITION** A 10.5 oz can of clam chowder contains 220 Cal. Find the number of calories in an 18.5 oz can of the same brand of clam chowder.

5.1 | Exercises and Applications

Extra Practice exercises on page 566

Find the ratio of the first quantity to the second quantity. Write your answer as a fraction in lowest terms.

1. length of a football field: 360 ft; width of a football field: 160 ft

2. number of gray whales in 1970: 21,000; number of gray whales in 1946: 2000

3. weight of a mouse: 20 g; weight of an African elephant: 7500 kg

4. speed of light: 186,282 mi/s; speed of sound: 1100 ft/s

Solve each proportion.

5. $\dfrac{x}{5} = \dfrac{2}{5}$

6. $\dfrac{5}{1} = \dfrac{d}{7}$

7. $\dfrac{9}{y} = \dfrac{3}{4}$

8. $\dfrac{7}{4} = \dfrac{v}{6}$

9. $\dfrac{1.5}{18} = \dfrac{2}{k}$

10. $\dfrac{2}{5.08} = \dfrac{h}{8.89}$

11. 144 to 18 = 32 to a **12.** $6:1 = w:12$ **13.** $3:n = 7:11$

14. ECOLOGY In 1974, researchers at the Montezuma Wildlife Refuge in New York captured, marked, and released 364 young wood ducks. A few weeks later, the researchers took a sample of 196 young wood ducks and found that 60 were marked. Estimate the number of young wood ducks in the refuge. Round your answer to the nearest ten.

15. Open-ended Problem Write a real-life problem that can be solved using the proportion $\dfrac{6.5}{180} = \dfrac{g}{325}$.

16. LANGUAGE ARTS The statement "Bat is to baseball as racquet is to tennis" is called an *analogy*. Complete the following analogies.

   a. Kitten is to cat as ___?___ is to dog.

   b. Red is to ___?___ as green is to go.

   c. Driver is to car as pilot is to ___?___ .

   d. ___?___ is to Texas as Sacramento is to California.

   e. Writing Explain how analogies are like proportions.

INTERVIEW Daniel Galvez

Look back at the article on pages 188–190.

The photo below at the right shows a part of Daniel Galvez's mural, Crosswinds, in Cambridge, Massachusetts. The girl is standing in front of her portrait that Galvez painted as part of the mural.

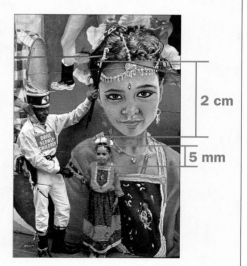

17. Use the measurements in the photo to estimate the ratio of the size of the girl's portrait to the size of the girl in real life.

18. The face of the girl in the photo is about 3 mm wide. How wide do you expect the face of the portrait in the photo to be? Explain your reasoning.

2 cm

5 mm

19. How many times as large as the girl is the portrait? How can you tell from the photo?

There are 435 seats in the United States House of Representatives. Every 10 years, these seats are apportioned, or divided, among the states. You can find a state's quota, or share, of House seats by solving the proportion below.

$$\frac{\text{State's quota of House seats}}{\text{Total number of House seats}} = \frac{\text{Population of the state}}{\text{Population of United States}}$$

20. Copy and complete the table at the right. Use 248,709,873 for the population of the United States in 1990. Round your answers to the nearest hundredth.

State	Population in 1990	Quota
Arizona	3,665,228	?
Massachusetts	6,016,425	?
New Jersey	7,730,188	?
Oklahoma	3,145,585	?

21. Research Find the number of House seats each state in the table actually had in 1990. How do these numbers compare to the states' quotas?

22. Alexander Hamilton proposed one method for apportioning the House seats among the states. He suggested the government give each state the whole-number part of its quota, and then distribute the extra seats to the states whose decimal parts of their quotas are largest. (For example, 8 is the whole-number part of 8.53.)

District	Population	Quota	Seats
A	35,000	?	?
B	27,000	?	?
C	12,000	?	?
D	36,000	?	?

a. Consider a city having 14 seats on its city council. Copy and complete the table at the left using the Hamilton method.

b. Open-ended Problem Suppose the city council increases the number of seats to 15. How does the number of seats change for each district? Do you think the new apportionment is fair? Explain.

Solve each proportion.

23. $\dfrac{2}{9} = \dfrac{4a}{13}$

24. $\dfrac{3}{g} = \dfrac{g}{12}$

25. $\dfrac{y + 1}{12} = \dfrac{2}{3}$

26. $\dfrac{3}{7} = \dfrac{t - 4}{21}$

27. $\dfrac{11}{8} = \dfrac{1}{3r + 2}$

28. $40 = \dfrac{80}{x}$

29. $\dfrac{b}{6} = \dfrac{2 + b}{12}$

30. $\dfrac{p}{36 - p} = \dfrac{4}{5}$

31. $\dfrac{6x - 2}{7} = \dfrac{5x + 7}{8}$

32. Challenge One third of an unknown number of beans is marked. Twenty more unmarked beans are added. Then a sample of 25 beans is taken. Six beans have marks. Estimate the total number of beans there were before the twenty beans were added.

33. SPORTS Major league baseball teams play 162 games during a typical season. Due to the baseball strike in 1994, the San Francisco Giants played only 115 games during that season. In these 115 games, Matt Williams hit 43 home runs for the Giants.

    a. Predict how many home runs Matt Williams would have hit during the 1994 season if it had been a typical season.

    b. Research Find out the record for the number of home runs hit in a single season by one player. Do you think Matt Williams would have broken this record if it had been a typical season? Explain your answer.

34. SAT/ACT Preview Suppose $0 < x < y$. Which of the following is true?

    A. $\dfrac{x}{y} > 1$ **B.** $\dfrac{x}{y} < 1$ **C.** $\dfrac{x}{y} = 1$

    D. relationship cannot be determined

35. LANGUAGES The bar graph shows the six most widely spoken languages in the world in 1993. To put these numbers in perspective, scale the numbers in the graph to the number of students in your school. Use a 1993 world population estimate of 5.5 billion people. How many students would speak each of the six languages?

O N G O I N G   A S S E S S M E N T

36. Writing Explain how to solve a proportion without using the means-extremes property. Illustrate your method with an example.

S P I R A L   R E V I E W

Find an equation of the line through the given points. *(Section 3.5)*

37. $(-3, -16)$ and $(4, 5)$ **38.** $(-6, 3)$ and $(12, -15)$

39. $(0, 0)$ and $\left(\dfrac{1}{3}, \dfrac{1}{4}\right)$ **40.** $\left(-\dfrac{1}{2}, -2\right)$ and $(1.5, -6)$

41. $(0, 1)$ and $(0, -1)$ **42.** $(0, b)$ and $(1, b + m)$

Solve each equation. Round answers to the nearest hundredth. *(Section 4.4)*

43. $3.57u - 4.73 = -8.07$ **44.** $\dfrac{13.42(t - 7.53)}{12.56} = -4.79$

45. $12.71v - 13.07 = 1.12v$ **46.** $0.1w + 0.7w = 0.24$

5.2

Scale Measurements

Learn how to...

- use scales in drawings and on graphs
- use scale factors to compare sizes of objects and drawings

So you can...

- model large objects with scale drawings
- find large dimensions, such as the size of a whale
- compare maps of different sizes

The blue whale is the largest animal on Earth. To appreciate its size, compare the pictures below. The drawing of the whale has the same *scale* as the photograph of the elephant.

The drawing of the whale and the photograph of the elephant both have a **scale** of 1 in. = 24 ft. This means that each inch in the pictures represents 24 feet in real life.

EXAMPLE 1   Application: Zoology

The drawing of the whale is about 4 in. long. Find the whale's real length.

SOLUTION

Let w = the whale's real length in feet.

$$\frac{\text{Whale's length in drawing in inches}}{\text{Whale's real length in feet}} = \frac{1 \text{ in.}}{24 \text{ ft}}$$

$$\frac{4}{w} = \frac{1}{24}$$

Use the means-extremes property.

$$4 \cdot 24 = w \cdot 1$$

$$96 = w$$

The whale is about 96 ft long.

THINK AND COMMUNICATE

1. The photograph of the elephant is about $\frac{3}{8}$ in. high. Explain how to estimate the real height of the elephant.

2. Suppose a drawing of a whale has a scale of 1 cm = 2 m. To find the whale's real length, what units must you use to measure the drawing? In what units will the whale's length be? How do you know?

Scale Factors

The drawing here has a different scale from the one on page 198. In this drawing the whale is 2 in. long.

To describe how the drawing has changed, you can use a ratio called the *scale factor*.

1 in. = 48 ft

$$\text{Scale factor} = \frac{\text{Whale's length in second drawing}}{\text{Whale's length in first drawing}} = \frac{2 \text{ in.}}{4 \text{ in.}} = \frac{1}{2}$$

The second drawing is half the size of the original drawing. You can write the scale factor as a fraction $\left(\frac{1}{2}\right)$, as a decimal (0.5), or as a percent (50%).

◄ **WATCH OUT!**

The scale factor is the ratio of the new size to the old size, not the old size to the new size. Be sure to write the ratio in the correct order.

EXAMPLE 2

Nathan plans to include a map in his report on Africa. The map he has is 6 in. wide and 8 in. long. He wants to use a photocopier to enlarge the map so that it fills the width of the page.

a. What scale factor should he use to make the map $7\frac{1}{2}$ in. wide? Give your answer as a percent.

b. How long will the enlarged map be?

SOLUTION

a. Let p = the scale factor at which Nathan should copy the map.

$$\frac{\text{Width of enlargement}}{\text{Width of original}} = p$$

$$\frac{7.5}{6} = p$$

$$1.25 = p$$

Nathan should copy the map at 125%.

b. Let l = the length of the enlarged map in inches.

$$\frac{\text{Length of enlarged map}}{\text{Length of original map}} = 1.25$$

Set the ratio of the maps' lengths equal to the scale factor.

$$\frac{l}{8} = 1.25$$

$$l = 8(1.25)$$

$$l = 10$$

The new map will be 10 in. long.

3. In Example 2, the original map has a scale of 1 in. = 1000 mi. Nathan did the calculation shown to find the scale of the enlarged map. Explain why he multiplied 1 in. by 125% to find the new scale.

4. How many times as small as the original is a picture changed by a scale factor of $\frac{1}{20}$? How many times as small as the building is a blueprint drawn to a scale of 1 in. = 20 ft? Use these examples to describe the difference between a scale and a scale factor.

5. How does the size of a copy compare to the size of the original when the scale factor is greater than 1? less than 1? equal to 1?

Scales on Graphs

The table shows an increase in tourism in the United States from 1970 through 1990.

Even though the graphs below both represent these data, they give different impressions of the increase in tourism. The choice of the scale on the vertical axis of each graph affects the appearance of the graph.

Year	Visitors from other countries
1970	12,362,000
1975	15,698,000
1980	22,326,000
1985	25,399,000
1990	39,539,000

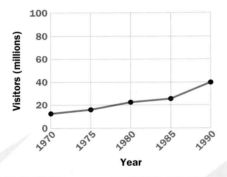

Each interval on the vertical axis represents 20 million visitors.

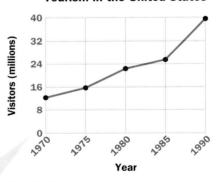

Each interval on the vertical axis represents 8 million visitors.

6. Compare and contrast the graphs. How does the scale on the vertical axis of each graph affect your impression of the data?

7. Suppose you are trying to encourage a hotel manager to hire more bilingual employees. Which graph above would you show him? How might you change the graph to make your point even more dramatic?

☑ CHECKING KEY CONCEPTS

1. The floor plan of a 90 ft by 200 ft building has a scale of 1 in. = 8 ft. Give the dimensions of the floor plan.

2. A portrait of a man about 6 ft tall is drawn 2.5 times life size in one of Daniel Galvez's murals. About how tall is the portrait?

For Questions 3 and 4, find the scale factor used to create each object.

3. a 5 in. by $7\frac{1}{2}$ in. postcard that pictures a 2 ft by 3 ft painting

4. a photograph 15 cm wide made from a negative 35 mm wide

5. Look back at the graphs on page 200. How can you change the scale on the *horizontal* axis to make the graph seem even more dramatic?

5.2 **Exercises and Applications**

1. This photograph of the space shuttle *Columbia* has a scale of 1 in. = 48 ft. Use a ruler to measure the photograph. Then use a proportion to find the real length of *Columbia*.

Extra Practice exercises on page 566

Tonya, the editor of the school yearbook, has to scale each picture in Exercises 2–6 to fill a column 2 in. wide. Give the scale factor she should use for each picture. Then find the height of each scaled picture.

2. $1\frac{1}{4}$ in. ↕   ← $1\frac{1}{4}$ in. →

3. $1\frac{1}{2}$ in. ↕   ← 4 in. →

4. $2\frac{1}{2}$ in. ↕   ← $1\frac{1}{2}$ in. →

5. 2 in. ↕   ← 3 in. →

6. $1\frac{7}{8}$ in. ↕   ← 2 in. →

7. **Challenge** Kathryn Tate is designing a building in Germany. Her drawing has a scale of 1 cm = 1200 cm. She wants to compare the drawing to the blueprint of a building in the United States, which has a scale of 1 in. = 200 ft. At what percent should she copy the drawing of the building in Germany so that its scale becomes 1 in. = 200 ft? (*Hint:* First convert 200 ft to inches. Then use the fact that the scale 1 cm = 1200 cm is equivalent to the scale 1 in. = 1200 in.)

8. **BUSINESS** The graph shows estimated circulation figures for two competing newspapers, the *City News* and the *City Daily*.

   a. What does the graph suggest will happen in 1995?

   b. Find the average rate of change per year for each paper from 1990 to 1994. Use this rate to predict each paper's circulation for 1995.

   c. **Writing** What is deceptive about this graph? Which newspaper do you think is more likely to have created the graph? Why?

9. **Open-ended Problem** Draw a floor plan of a room in your home. Show the scale you used and explain how you calculated each measurement.

10. Troy wants to reduce a 12 in. by 12 in. drawing to 3 in. by 3 in. Unfortunately, the smallest scale factor the photocopier allows for is 50%. What scale factor can he use twice to reduce the drawing to the size he wants?

11. A map of Alabama has a scale of 1 in. = 20 mi. The close-up of Montgomery has a scale of 1 in. = 4 mi. What scale factor was used to change the size of the map of Alabama? Explain your reasoning.

Circulation of the City News and the City Daily

Circulation (vertical axis): 170,000; 160,000; 150,000; 140,000; 130,000; 120,000; 110,000; 50,000; 40,000; 30,000; 20,000

City News: 161,000; 141,500; 126,200; 117,500; 108,600; ?

City Daily: 22,600; 31,900; 34,700; 36,100; 44,900; ?

Year (horizontal axis): 1990, 1991, 1992, 1993, 1994, 1995

Map from the Road Atlas,
© 1995 by Rand McNally, R.L. 95–S–161

12. **Technology** The table gives the average number of gallons of milk consumed per person in the United States for various years from 1970–1990.

a. Use a graphing calculator to make a scatter plot of the data in the table. Set the viewing window as shown below. Make a sketch of the graph.

b. Change the scale on the vertical axis so that $0 \le y \le 100$. Make a sketch of the new graph.

c. Compare the two graphs. How does the scale affect the visual impression given by the data?

d. **Open-ended Problem** Find a viewing window that you think gives the "best" graph. Sketch your graph and show the scale you used. Explain why you think your graph is better than the graphs in parts (a) and (b).

Milk Consumption in the U.S. (per capita)	
Year	Gallons consumed
1970	31.2
1975	29.5
1980	27.6
1985	26.7
1990	25.7

```
WINDOW FORMAT
Xmin=1965
Xmax=1995
Xscl=5
Ymin=25
Ymax=32
Yscl=1
```

INTERVIEW **Daniel Galvez**

Look back at the article on pages 188–190.

Before Daniel Galvez painted the mural Crosswinds, he made a drawing of it by projecting slides onto a piece of paper. In this activity, your class will create a poster using a similar method.

Cooperative Learning You will need an overhead projector, overhead transparencies, markers, a ruler, and a large piece of paper.

13. Draw a picture of a person on a transparency. Measure the height of the person. What scale factor will make the person appear 12 in. tall when projected on a piece of paper taped to the wall?

14. Put your transparency on the overhead projector and measure the height of the person as it appears on the paper. By what scale factor did the image change? Without moving the projector, project the other students' transparencies onto the paper. Did they all change by the same scale factor? How do you know?

15. Will you have to move the projector forward or backward to make the person appear 12 in. tall? How do you know? Find this position and trace the outline of the person on the piece of paper. Then draw the details of your picture to the new scale. Repeat these steps for each student's picture.

16. Writing Describe a way to change this activity so that each person in the poster appears a different size.

17. **SAT/ACT Preview** A person spent $1\frac{3}{4}$ min filling a car's empty gasoline tank completely. If the person had stopped after 1 min, what part of the tank would have been filled?

   A. $\frac{2}{7}$ **B.** $\frac{3}{7}$ **C.** $\frac{4}{7}$ **D.** $\frac{3}{4}$ **E.** $\frac{5}{7}$

ONGOING ASSESSMENT

18. **Writing** Describe the difference between a scale and a scale factor. Explain why one specifies units and the other does not.

SPIRAL REVIEW

Solve each inequality. Graph the solution on a number line. *(Section 4.6)*

19. $2x + 1 < 7$ 20. $5(x - 4) > 0$ 21. $3 - 5x \leq 5(x + 3)$

22. $\frac{3}{5}x \geq 8$ 23. $4(-x + 1) \geq 3x + 9$ 24. $3x - 5 \leq 6\left(\frac{1}{2}x + 1\right)$

ASSESS YOUR PROGRESS

VOCABULARY

ratio (p. 192)   **extremes** (p. 193)
proportion (p. 193)   **scale** (p. 198)
means (p. 193)   **scale factor** (p. 199)

Find the ratio of the first quantity to the second quantity. Write your answer as a fraction in lowest terms. *(Section 5.1)*

1. 5 min; 2 h   2. 3 kg; 200 g

Solve each proportion. *(Section 5.1)*

3. $4:x = 6:15$ 4. $\dfrac{3.6}{10.8} = \dfrac{p}{100}$ 5. $\dfrac{x}{20 - x} = \dfrac{2}{3}$

6. A week before Election Day, 276 out of 450 people said that they would vote for Jean Ferreira. If 15,240 registered voters plan to vote, how many votes should Jean Ferreira expect to get? *(Section 5.1)*

7. Gary Leeds wants to build a basement playroom for his children. He makes a drawing using a scale of $\frac{3}{4}$ in. = 1 ft. *(Section 5.2)*

   a. What width in the scale drawing should Gary use to represent a doorway 30 in. wide?

   b. The room in the scale drawing is 15 in. by 22 in. Find the actual dimensions of the playroom.

8. **Journal** Explain the difference between a ratio and a scale. Which was easier for you to understand? Why?

5.3

Learn how to...

- recognize and draw similar figures

So you can...

- measure large distances in real-life situations

Working with Similarity

What do triangles have to do with totem poles, trees, and other tall objects? *Similar* triangles can help you measure the heights of tall objects indirectly.

EXPLORATION

COOPERATIVE LEARNING

Constructing Similar Triangles

Work with a partner.
You will need:

- a rectangular sheet of paper
- scissors
- a protractor
- a ruler

SET UP Cut along the diagonal of a piece of paper to make two triangles. Label the corners of one triangle *A*, *B*, and *C* as shown. Fold the unlabeled triangle to make a line parallel to the shortest side. Cut along this line and label the corners of the smaller triangle *D*, *E*, and *F* as shown. Then answer the questions below.

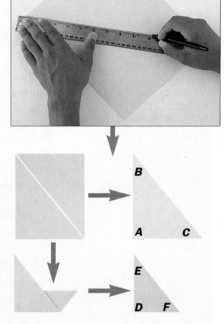

Questions

1. Copy and complete the tables. Give lengths to the nearest millimeter.

2. What do you notice about the angles of △*ABC* and △*DEF*?

△*ABC*		△*DEF*	
Measure of ∠ *A*	?	Measure of ∠ *D*	?
Measure of ∠ *B*	?	Measure of ∠ *E*	?
Measure of ∠ *C*	?	Measure of ∠ *F*	?
Length of $\overline{AB}$	?	Length of $\overline{DE}$	?
Length of $\overline{BC}$	?	Length of $\overline{EF}$	?
Length of $\overline{AC}$	?	Length of $\overline{DF}$	?

3. Find each ratio. What do you notice? (*AB* stands for the length of $\overline{AB}$.)

   a. $\dfrac{AB}{DE}$ **b.** $\dfrac{BC}{EF}$ **c.** $\dfrac{AC}{DF}$

4. The triangles you made are called *similar* triangles. What do you think it means for two triangles to be similar?

Similar Figures

The triangles you made in the Exploration are called *similar figures.* Two figures are **similar** if corresponding angles have equal measure and the ratios of the lengths of corresponding sides are equal.

WATCH OUT!

When you name similar figures, always name the corresponding angles in the same order. That is, if $\angle A = \angle D$ and $\angle B = \angle E$ and $\angle C = \angle F$, then write $\triangle ABC \sim \triangle DEF$.

The figures below are similar. To show this you write $\triangle KLM \sim \triangle RST$, which you read as "triangle *KLM* is similar to triangle *RST*."

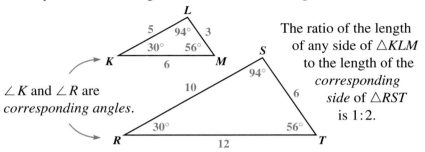

$\angle K$ and $\angle R$ are *corresponding angles.*

The ratio of the length of any side of $\triangle KLM$ to the length of the *corresponding* side of $\triangle RST$ is $1:2$.

Similar figures can be the same size or different sizes, but they are always the same shape.

EXAMPLE 1

Figure *ABCD* ~ figure *EFGH*.

a. Find the measure of $\angle F$.

b. Find *FG*.

SOLUTION

a. $\angle F$ and $\angle B$ are corresponding angles, so $\angle F = \angle B = 100°$.

b. Find the ratio of the lengths of corresponding sides.

$\overline{BC}$ and $\overline{FG}$ are also corresponding sides. Set the ratio of their lengths equal to $\frac{AB}{EF}$.

$$\frac{AB}{EF} = \frac{20}{12}$$

$$\frac{BC}{FG} = \frac{20}{12}$$

$$\frac{50}{FG} = \frac{20}{12} \qquad \text{Substitute 50 for } BC.$$
$$\text{Solve the proportion.}$$

$$600 = FG \cdot 20$$

$$30 = FG$$

THINK AND COMMUNICATE

Questions 1 and 2 refer to figures *ABCD* and *EFGH* in Example 1.

1. How would you use similarity to find *GH*?

2. Can you use similarity to find *EH*? Explain.

Similar Triangles

To see whether two figures are similar, you need to check that all corresponding angles are equal in measure and that the ratios of lengths of corresponding sides are equal. There is a simpler test for similar triangles.

A Test for Similar Triangles

Two triangles are similar if two angles of one triangle have the same measures as two angles of the other triangle.

EXAMPLE 2 | **Application: Measuring Height Indirectly**

Jane wants to estimate the height of the Thunderbird House Post totem pole. She stands so that the end of her shadow meets the end of the totem pole's shadow (A). Use similar triangles and the measurements given to find the height of the totem pole.

SOLUTION

First, find a pair of similar triangles. $\triangle ABC$ and $\triangle ADE$ share $\angle A$. Since Jane and the totem pole are both perpendicular to the ground, $\angle C = \angle E = 90°$. Two angles of $\triangle ABC$ have the same measures as two angles of $\triangle ADE$, so the triangles are similar.

Let x = the height of the totem pole. Write and solve a proportion involving x.

$$\frac{BC}{DE} = \frac{AC}{AE}$$

Since $\triangle ABC \sim \triangle ADE$, the ratios of the lengths of corresponding sides are equal.

$$\frac{5}{x} = \frac{7}{7 + 14}$$

$$5(7 + 14) = x \cdot 7$$

$$105 = 7x$$

$$15 = x$$

The totem pole is about 15 ft tall.

BY THE WAY

Totem poles are posts carved and painted by Native Americans of the Northwest Coast. These poles support houses, serve as memorials, and preserve family histories.

☑ CHECKING KEY CONCEPTS

For Questions 1–5, use the fact that _ABCD_ ∼ _EFGH_.

1. Complete each statement.

   a. ∠_B_ and ___?___ are corresponding angles.

   b. ___?___ and $\overline{FG}$ are corresponding sides.

2. Find the measure of ∠_A_.

3. Find the measure of ∠_H_.

4. Find _AB_.

5. Find _HG_.

5.3 | Exercises and Applications

Extra Practice
exercises on
page 567

1. Use the fact that △_MNO_ ∼ △_PQR_ to complete each statement.

   a. ∠_N_ = ___?___

   b. $\overline{PR}$ and ___?___ are corresponding sides.

   c. $\dfrac{MN}{?} = \dfrac{NO}{?}$

Tell whether the figures are similar. If they are, list each pair of corresponding angles and sides. If they are not similar, explain why.

2.

3.

4.

5.

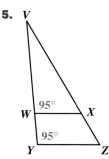

For Exercises 6 and 7, use the quadrilaterals *JKLM* and *NOPQ*.

6. Figure *JKLM* ~ figure *NOPQ*.
Find each missing side length
and angle measure.

7. Tell whether each statement
is *True* or *False*. Explain
your reasoning.

   a. *KLMJ* ~ *OPQN*   **b.** *MLKJ* ~ *QPON*   **c.** *JKLM* ~ *QPON*

8. **HISTORY** The photo shows a patchwork coat that belonged to Uesugi
Kenshin, a famous Japanese general. The coat is over 400 years old.

   a. Look carefully at the shapes of the
patches. Find a pair of quadrilaterals
that appear to be similar. Explain
your reasoning.

   b. **Writing** Explain precisely what you
would do to decide whether two patches
in the coat are actually similar.

9. **GEOMETRY** A *regular polygon* is a figure
whose angles are all of equal measure and
whose side lengths are all equal.

   a. Are all squares similar to each other?
How do you know?

   b. An equilateral triangle has three sides
of equal length and three 60° angles. Are all equilateral triangles
similar to each other? How do you know?

   c. Draw a regular pentagon whose angles all measure 108°. Draw
another one the same way, but use a different length for the sides.
Are these pentagons similar? Are all regular pentagons similar to
each other? Explain.

   d. Make a conjecture about regular polygons and similarity based on
your answers to parts (a)–(c).

10. **Open-ended Problem** Draw and label two similar triangles so that the
ratio of the lengths of corresponding sides is 3:5.

11. **Writing** Is a figure similar to itself? Explain.

12. **Visual Thinking** Draw a triangle and label its corners *A*, *B*, and *C*. Use
a ruler to find the midpoint of each side. Draw △*DEF* where *D* is the
midpoint of $\overline{AB}$, *E* is the midpoint of $\overline{BC}$, and *F* is the midpoint of $\overline{AC}$.

   a. How many triangles are inside the original triangle?

   b. Name all the similar triangles. Give the ratio of corresponding sides.

   c. Draw another triangle and repeat the exercise. Did you get the same
answers?

One of the earliest known mathematics textbooks is the *Nine Chapters on the Mathematical Art*, the *Jiŭzhāng suànshù* of China. The following problem is from this 2000-year-old book.

A hill lies west of a tree whose height is 95 ch'ih. The distance between the hill and tree is known to be 53 li. A man 7 ch'ih tall stands 3 li east of the tree. If the tops of the hill and tree are aligned in the path of his vision, what is the height of the hill?

Not drawn to scale

95 ch'ih   7 ch'ih

53 li   3 li   y

13. Explain why $\triangle AFG \sim \triangle ADE$.

14. Solve the proportion below:

$$\frac{y \text{ li}}{y + 3 \text{ li}} = \frac{7 \text{ ch'ih}}{95 \text{ ch'ih}}$$

15. Challenge Explain why you can use the proportion in Exercise 14 to find AF, even though both li and ch'ih are used.

16. Explain why $\triangle AFG \sim \triangle ABC$.

17. Find the length of $\overline{BC}$. Explain how you got your answer.

18. Writing Based on the measurements in the illustration, about how long do you think a li is? How long do you think a ch'ih is? Explain your reasoning.

19. SAT/ACT Preview For which of the following values of x is rectangle *ABCD* similar to rectangle *EFGH?*

I. 5   **II.** 17   **III.** 7

A. I only

B. II only

C. III only

D. I and II only

E. I, II, and III

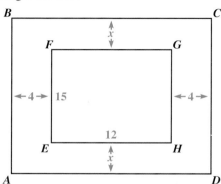

20. **Cooperative Learning** Work with a partner to follow the steps below. You will need a yardstick or a tape measure and a small mirror.

- Place the mirror some distance from a wall. Measure the distance between the mirror and the wall.

- Facing the wall, back up along the line perpendicular to the wall and passing through the mirror. Stop when you can see the top of the wall in the mirror. Have your partner measure the distance between the mirror and your feet.

- Have your partner measure your height.

a. Draw a diagram like the one shown. Include all measurements.

b. The two triangles in your drawing are similar. Use this fact to calculate the height of the wall.

c. Exchange roles with your partner and repeat the activity with the mirror in a different position. Do you get the same answer for the height of the wall?

ONGOING ASSESSMENT

21. **Writing** If $\triangle ABC \sim \triangle DEF$ and $\triangle DEF \sim \triangle GHI$, is it true that $\triangle ABC \sim \triangle GHI$? Explain your reasoning.

SPIRAL REVIEW

Express each rate in the given units. Write each answer as a unit rate. *(Section 3.1)*

22. $21,000 per year = _?_ per month

23. 160 pages per hour = _?_ pages per day

24. 186,000 miles per second = _?_ feet per minute

Tell whether the data show direct variation. If they do, give the constant of variation and write an equation. *(Section 3.2)*

25.

Capacity (liters)	Distance (meters)
2.5	13.125
2.6	14.976
2.7	16.983
2.8	19.152
2.9	21.489
3.0	24.000

26.

Time (min)	Distance (cm)
1.3	5.59
2.5	10.75
4.0	17.20
6.1	26.23
7.2	30.96
7.7	33.11

27.

Age (years)	Weight (lb)
5	7.5
10	15.0
15	22.5
30	45.0
50	75.0
70	112.5

5.4 Perimeters and Areas of Similar Figures

Learn how to...

- **find the ratios of the perimeters and areas of similar figures**

So you can...

- **use scale drawings to find the perimeters and areas of houses, for example**

- **create accurate pictographs that represent statistics such as the price of stamps**

What does it mean for a square to double in size? Look at the sets of tiles below and see whether you can find a pair of figures in which one set appears to be twice as large as the other.

Figure *A*   Figure *B*   Figure *C*   Figure *D*

THINK AND COMMUNICATE

Use the figures above to answer Questions 1–3.

1. Explain how you know that figures *A, B, C,* and *D* are similar.

2. For each pair of figures, find the ratio of the perimeters. Then find the ratio of the areas. Describe any patterns you see.

   a. *A* and *B*   **b.** *B* and *D*   **c.** *A* and *C*   **d.** *B* and *C*

3. How do the perimeter and area of a square change when its side length is doubled? when it is tripled? Explain your answers.

4. Look at the diagrams below. All the triangles shown are similar equilateral triangles. How do the perimeter and area of a triangle change when its side length is doubled? when it is tripled?

   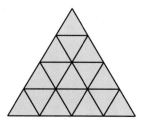

As you saw in Questions 1–4, there is a relationship between the side lengths, perimeters, and areas of similar figures. One way to express this relationship is to use ratios.

Suppose the ratio of the lengths of corresponding sides of two similar figures is $a:b$. The ratio of the perimeters is $a:b$, and the ratio of the areas is $a^2:b^2$.

For example, the rectangles shown are similar.

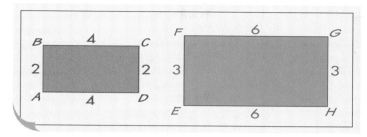

Side length of **ABCD** : Side length of **EFGH** = **2:3**

Perimeter of **ABCD** : Perimeter of **EFGH** = **12:18 = 2:3**

Area of **ABCD** : Area of **EFGH** = **8:18 = 4:9**

EXAMPLE 1   **Application: Architecture**

The drawing below shows the floor plan of the Manuél Atencio house in Trampas, New Mexico. The ratio of the length of a side of the house in the diagram to the length of the corresponding side of the actual house is 1 cm: 300 cm. Estimate the perimeter of the actual house.

SOLUTION

Find the perimeter of the house in the diagram by adding the side lengths.

$$1.7 + 7.4 + 1.9 + 2 + 3.6 + 9.4 = 26 \text{ cm}$$

Then use what you know about the perimeters of similar figures to estimate the perimeter of the actual house. Let P = the perimeter of the house.

$$\frac{\text{Side length of house in diagram}}{\text{Side length of actual house}} = \frac{\text{Perimeter of house in diagram}}{\text{Perimeter of actual house}}$$

$$\frac{1}{300} = \frac{26}{P} \qquad \text{Use the means-extremes property.}$$

$$P = 7800$$

The perimeter of the house is about 7800 cm, or 78 m.

EXAMPLE 2 **Application: Architecture**

Find the area of the Manuél Atencio house.

SOLUTION

First find the area of the house in the diagram by dividing the house into two rectangles. Add the areas.

$$(1.7)(7.4) + (3.6)(2) = 19.78 \text{ cm}^2$$

Now use what you know about the areas of similar figures to find the area of the actual house.

> The ratio of the squares of the side lengths is equal to the ratio of the areas.

$$\frac{(\text{Side length of house in diagram})^2}{(\text{Side length of actual house})^2} = \frac{\text{Area of house in diagram}}{\text{Area of actual house}}$$

$$\frac{1^2}{300^2} = \frac{19.78}{A}$$

$$\frac{1}{90,000} = \frac{19.78}{A}$$

$$A = 1,780,200$$

The area of the house is about 1,780,200 cm^2. Convert this to square meters by using the fact that $(1 \text{ m})^2 = (100 \text{ cm})^2 = 10,000 \text{ cm}^2$.

$$1,780,200 \text{ cm}^2 \cdot \frac{1 \text{ m}^2}{10,000 \text{ cm}^2} \approx 178 \text{ m}^2$$

The area of the house is about 178 m^2.

☑ CHECKING KEY CONCEPTS

Find the missing perimeter and area for each pair of similar figures.

1.

10   8

P = 44
A = 104

P = ?
A = ?

2.

P = ?
A = ?

12

P = 24
A = 27

9

5.4 Exercises and Applications

1. The perimeters of two similar pentagons are 20 in. and 45 in. What is the ratio of lengths of corresponding sides? What is the ratio of the areas?

Extra Practice exercises on page 567

2. Two similar triangles have areas of 49 cm² and 36 cm². What is the ratio of lengths of corresponding sides? What is the ratio of the perimeters?

Find the missing values for each pair of similar figures.

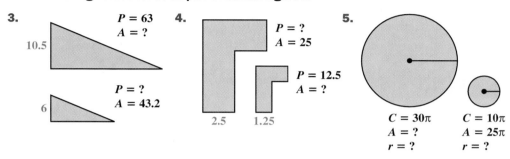

3. $P = 63$ $A = ?$ 10.5 $P = ?$ $A = 43.2$ 6

4. $P = ?$ $A = 25$ $P = 12.5$ $A = ?$ 2.5 1.25

5. $C = 30\pi$ $A = ?$ $r = ?$ $C = 10\pi$ $A = 25\pi$ $r = ?$

6. Caitlin wants to enclose a 20 ft by 24 ft rectangular area with a fence. She is told that the fence will cost $100. Since this is more than she can afford, she decides to enclose a smaller area, a 10 ft by 12 ft rectangle. What price should she expect to pay for the fence now? Explain.

7. Dee's Pizza has 10 in. and 14 in. diameter pizzas on the menu.

   a. What is the ratio of the areas of the two pizzas?

   b. The price of a 10 in. pizza is $5. What do you think the price of a 14 in. pizza should be? Explain your reasoning.

   c. **Writing** Dee is thinking of adding a 20 in. pizza to the menu. Should she charge twice as much as she does for the 10 in. pizza? Explain your thinking.

8. **CONSTRUCTION** Carpenters refer to a blueprint when building a home. The one shown has a scale of 1 in. = 15 ft, or 1 in. = 180 in. Use a ruler to measure the dimensions of the family room. Find the actual area of the family room in *square yards*. How did you get your answer?

9. **Challenge** Suppose the ratio of the side lengths of two cubes is 1:3.

   a. The surface area of a cube is $6x^2$ where x is the length of one side of the cube. What is the ratio of the surface areas of the cubes?

   b. What is the ratio of the volumes of the cubes? How do you know?

   c. Suppose you have two spheres with diameters of 3 m and 4 m. What do you think is the ratio of the volumes of the spheres? Explain.

10. **PHOTOGRAPHY** Suppose you have a 5 in. by 7 in. photo framed. A framing store charges you $7 for the wood you select, $3 for the glass to cover the photo, and $20 for the labor to assemble the parts. What should you expect to pay for having a 10 in. by 14 in. photo framed? Explain your reasoning.

Graphic designers are often hired to create interesting pictographs. A *pictograph* is a bar graph that uses pictures or symbols to represent quantities. In Exercises 11–14, you will explore the accuracy of pictographs.

Cost of Mailing a One-Ounce Letter

Cost of Mailing a One-Ounce Letter

11. Without measuring, choose a stamp on each pictograph that seems to be twice the size of the May 29, 1978, stamp. Explain your choice.

12. Use a ruler to measure the height of each May 29, 1978, postage stamp to the nearest millimeter. Find the stamp on each pictograph that is about twice the height of the May 29, 1978, stamp on the same pictograph. Does each stamp represent about twice as much postage?

13. Now find the area of each May 29, 1978, postage stamp. Find the stamp on each pictograph that has about twice the area of the May 29, 1978, stamp of the same pictograph. Does this stamp represent about twice as much postage? Are these the same stamps you found for Exercise 12?

14. **Open-ended Problem** Which pictograph do you think is more accurate? Why? Describe another way to use a stamp to illustrate the data accurately.

15. SAT/ACT Preview If the side of a square is increased by 20%, the area of the square is increased by:

   A. 20%   **B.** 40%   **C.** 37.5%   **D.** 44%

ONGOING ASSESSMENT

16. Writing Suppose the lengths of the sides of rectangle *ABCD* are *x* and *y*, and the lengths of the sides of rectangle *EFGH* are *kx* and *ky*. Show that the ratio of the perimeters is $k:1$ and the ratio of the areas is $k^2:1$.

SPIRAL REVIEW

Solve. If an equation is an *identity* or there is *no solution*, say so.
(Sections 4.2 and 4.3)

17. $\frac{2}{3}t - 7 = -\frac{1}{3}$ **18.** $-4(3 + 5p) = 48$ **19.** $-1 = \frac{5d}{7} + 1$

20. $\frac{3m - 7}{5} = -1.2$ **21.** $29 = \frac{3}{5}n + 2$ **22.** $13(3 - 4s) = 0$

ASSESS YOUR PROGRESS

VOCABULARY

similar (p. 206)

1. Writing A tabletop air hockey game (top right) is $18\frac{1}{2}$ in. by 40 in.

An ice hockey rink (bottom right) is 100 ft by 200 ft. Is the tabletop game an accurate representation of a real hockey rink? Explain your reasoning. *(Section 5.3)*

In the drawing, $\triangle ABC \sim \triangle DEF$. Use this drawing for Exercises 2–5.

2. Find the length of $\overline{AC}$. *(Section 5.3)*

3. Find the length of $\overline{EF}$. *(Section 5.3)*

4. Find the ratio of the perimeters of $\triangle ABC$ and $\triangle DEF$. *(Section 5.4)*

5. Find the ratio of the areas of $\triangle ABC$ and $\triangle DEF$. *(Section 5.4)*

6. Draw two squares of appropriate size to compare the areas of Alaska (591,004 mi²) and Texas (266,807 mi²). *(Section 5.4)*

7. Journal Describe an occupation that might require knowledge of similar figures and their perimeters and areas. How would someone with this occupation use the skills taught in Sections 5.3 and 5.4?

5.5

Learn how to...

- calculate theoretical and experimental probability

So you can...

- predict the outcome of events, such as whether a basketball player will make a free throw

Exploring Probability

There's a 70% chance of rain tomorrow. Melissa has a 50-50 chance of making a basket in basketball. You hear statements about chance, or *probability*, every day. Probability can help you to understand sports, politics, science, and other aspects of your life.

Probability is a ratio that measures how likely an **event** is to happen. When you roll a die, there are six possible results or **outcomes**. How often do you think you will roll a 2?

EXPLORATION
COOPERATIVE LEARNING

Rolling a Die

Work with a partner.
You will need:
- one die

1 How many times do you think each number will occur in 24 rolls? Explain your reasoning.

2 Roll the die 24 times. Record your data on a graph as shown. Were your predictions close to your results?

3 How does your graph compare to the graphs of the other groups in your class?

4 Find the ratio of the number of times each number occurs to the total number of rolls. For example, if 2 occurs 3 times in 24 rolls:

$$\frac{\text{Number of 2's rolled}}{\text{Total number of rolls}} = \frac{3}{24} = \frac{1}{8}$$

5 What do you think the ratios found in Step 4 will be if you combine the results from all the groups in your class? Explain.

6 Combine your results with the rest of the class. Create a class graph and calculate the ratios in Step 4 for each number. Are these numbers close to your prediction in Step 5?

Experimental Probability

In Step 4 of the Exploration, you found the *experimental probability* of rolling each number. **Experimental probability** is the ratio of the number of times an event actually occurs to the number of times the experiment is done.

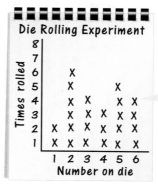

This student's experimental probability of rolling a 6 was $\frac{4}{24}$, or about 17%.

One way to think of experimental probability is as the following ratio:

$$\text{Experimental probability} = \frac{\text{Number of successes}}{\text{Number of tries}}$$

EXAMPLE 1 **Application: Sports**

Basketball player Jerry West of the Los Angeles Lakers attempted 1507 free throws during playoff games from 1961–1974. He made 1213 of these. What was the experimental probability of Jerry West making a free throw during a playoff game? Give your answer as a percent.

SOLUTION

$$\frac{\text{Number of free throws made}}{\text{Number of free throws attempted}} = \frac{1213}{1507} \approx 0.80$$

The probability that Jerry West would make a free throw was about 80%.

Jerry West holds the NBA playoff record for the most free throws made throughout his career. He also holds the record for the most free throws made during a single series, making 86 free throws in 6 games in 1965.

Probabilities are always between 0 and 1 and can be expressed as fractions, decimals, or percents.

The closer a probability is to 0, the less likely the event. An event with a probability of 0 can never happen. The closer a probability is to 1, the more likely the event. An event with a probability of 1 is certain to happen.

The probability of rolling a 7 is 0.

The probability of rolling a number less than 7 is 1.

Theoretical Probability

The more data you collect in a probability experiment, the less the results will vary. When students combined their data from the Exploration shown on page 218, the experimental probability came much closer to the *theoretical probability*.

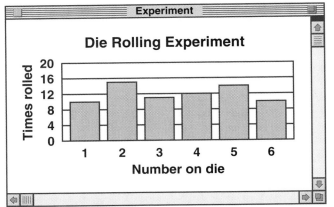

Theoretical probability is the ratio of the number of *favorable outcomes* to the total number of possible outcomes. An outcome is **favorable** if it makes the event happen.

You may have found a theoretical probability in Step 1 of the Exploration. The die has six numbers on it. Each number is equally likely to occur, so you should have expected a favorable outcome of 2 about $\frac{1}{6}$ of the time. The theoretical probability of rolling a 2 is $\frac{1}{6}$.

EXAMPLE 2

A die is rolled. What is the theoretical probability of rolling a number greater than 4?

SOLUTION

The two favorable outcomes are **5** and **6** because both of these numbers are greater than 4.

$$\begin{array}{l}\text{Probability of rolling a} \\ \text{number greater than 4}\end{array} = \begin{array}{l}\text{Probability} \\ \text{of rolling a 5}\end{array} + \begin{array}{l}\text{Probability} \\ \text{of rolling a 6}\end{array}$$

$$= \frac{1}{6} + \frac{1}{6}$$

$$= \frac{2}{6}$$

$$= \frac{1}{3}$$

The probability of rolling a number greater than 4 is $\frac{1}{3}$.

THINK AND COMMUNICATE

1. Explain why a probability cannot be greater than 1. (*Hint:* Think of the ratio of the number of successes to the number of tries.)

2. How do you know that the probability in Example 1 is experimental?

3. In Example 1, what does experimental probability take into account that theoretical probability does not?

☑ CHECKING KEY CONCEPTS

1. Suppose you roll a die like the one used in the Exploration on page 218. Find the theoretical probability of rolling:

   a. a number greater than 6.

   b. a number greater than 0.

   c. a 1 or a 2.

2. You roll a 3 twice in 17 rolls of a die. What is the experimental probability of rolling a 3?

3. There are two sides to a coin, so the probability of a coin landing heads up is 50%. Is this an experimental or theoretical probability?

5.5 | **Exercises and Applications**

Make a copy of this number line. On your number line mark your estimate of the probability of each event described in Exercises 1–5.

Extra Practice exercises on page 568

```
0                    1/2                    1
├────────────────────┼────────────────────┤
```

1. It will rain sometime during the next two days where you live.

2. You will have breakfast tomorrow morning.

3. You will finish your homework before 9 P.M. tonight.

4. You will receive a Nobel Prize next week.

5. You will talk to a friend today.

6. **Writing** Choose one of the events from Exercises 1–5 and explain your estimate.

7. **Open-ended Problem** Describe an event that has the given theoretical probability. Will the experimental probability of each event necessarily be the same as the theoretical probability? Explain your reasoning.

   a. 1   **b.** 0   **c.** $\frac{1}{3}$ **d.** $\frac{3}{4}$

BY THE WAY

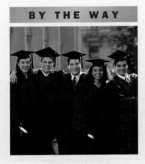

In 1992, there were over 14 million people enrolled in higher education in the United States. The average student spent $12,250 for school and received $745 in financial support.

According to the United States Census, these were the median yearly income levels for heads of households with different levels of education in 1992. Probability can help you understand statistics like these.

Education and Income of Heads of Households in the United States

Education level attained	Number of people	Median income
Less than 9th grade	9,060,000	$13,383
9th to 12th grade, no diploma	9,933,000	$17,375
High school graduate	30,103,000	$29,006
Some college, no degree	15,387,000	$35,327
Associate degree	5,502,000	$38,382
Bachelor's degree or more	21,382,000	$54,117

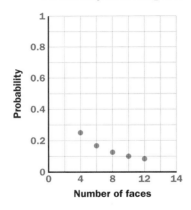

8. How many people were "heads of households" in 1992? How do you know?

9. What is the probability that a head of a household had less than a 9th grade education in 1992? How did you get your answer?

10. What is the probability that a head of a household had some college education in 1992, but no degree?

11. Challenge What is the probability that a head of a household earned more than the median income for his or her group?

12. a. A 3 occurs 23 times when a die is rolled 180 times. What is the experimental probability of rolling a 3?

b. What is the theoretical probability of rolling a 3?

Most dice come with six faces. You can also find dice with more than six faces. The graph shows the probability of rolling a 1 as a function of the number of faces on the die. Use the graph for Exercises 13–15.

13. Describe the domain and the range of the function. Write an equation describing the function. How did you get your answer?

14. Would you connect the points on the graph? Why or why not?

15. Suppose the graph showed the probability of rolling a 2 as a function of the number of faces on the die. Would the graph look different? Why or why not?

Probability of Rolling a 1

16. a. The probability that you will get an even number when you roll a die is 50%. What is the probability that you will get an odd number?

b. If there is a 70% chance that it will rain tomorrow, what is the probability that it will not rain?

c. If Kirby Puckett's *batting average*, the probability that he would get a hit, was .329 in 1992, what was the probability that he would not get a hit?

d. Suppose *x* represents the probability that something will happen. Write an expression for the probability that it will not happen. Explain your reasoning.

17. SAT/ACT Preview Which one of the following equals 7.5%?

A. 0.75   **B.** $\dfrac{3}{4}$ **C.** $\dfrac{75}{100}$ **D.** 0.075

18. Writing Look back at the Exploration on page 191 in Section 5.1. How does the capture-recapture method rely on probability? Explain.

19. Cooperative Learning Work with a partner to toss a pair of pennies 50 times. Record the number of times you get two heads, two tails, or a head and a tail.

a. Before you begin, discuss the possible outcomes and favorable outcomes for each event with your partner.

b. One of you should calculate the theoretical probability for each event. The other person should calculate the experimental probability for each event. (*Hint:* Think of the events as head and head, tail and tail, head and tail, tail and head.)

c. Are your predictions close to your results?

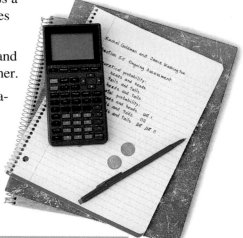

Graph each point in a coordinate plane. Name the quadrant (if any) in which the point lies. (*Section 2.4*)

20. $A(-1, 1)$ **21.** $B(4, 0)$ **22.** $C(0, -5)$

23. $D(5, -1)$ **24.** $E(-5, -4)$ **25.** $F\left(5, \dfrac{1}{2}\right)$

Solve each equation. (*Sections 4.3 and 4.4*)

26. $\dfrac{a}{8} = \dfrac{5}{56}$ **27.** $\dfrac{7}{3}b - 9 = -10\dfrac{2}{3}$ **28.** $0.3x - 1.33 = 6.5$

29. $d + \dfrac{1}{2}d + \dfrac{1}{3}d = 1$ **30.** $7 - 0.5x = 7.5x - 9$ **31.** $f - 1 = 2(f - 2) - f$

5.6

Geometric Probability

Did you know that over 2000 meteorites weighing 1 kg land on Earth every year? Scientists say most will land in water. But how do they know?

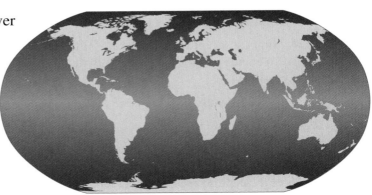

When a meteorite falls to Earth, it can hit either land or water. Because $\frac{7}{10}$ of Earth's surface is covered by water, the probability that an object falling to Earth will land in water is 70%. The chance that a meteorite will land in water is an example of **geometric probability**, a probability based on area.

EXAMPLE 1

Suppose a randomly thrown dart hits the target shown. (A random throw means that you do not aim for any particular part of the target, and each part has an equal chance of being hit.) What is the theoretical probability of the dart hitting the shaded area?

SOLUTION

The square is divided into 8 triangles having equal area, and 2 of them are shaded.

$$\text{Probability of hitting a shaded area} = \frac{\text{Shaded area}}{\text{Total area}}$$

$$= \frac{\text{Number of shaded triangles}}{\text{Total number of triangles}}$$

$$= \frac{2}{8}$$

$$= \frac{1}{4}$$

The probability of a randomly thrown dart hitting the shaded area is 25%.

EXAMPLE 2

Suppose you have the chance to win a prize by throwing a dart at either of the two targets shown and hitting a balloon. Which target gives you a better chance of winning a prize?

SOLUTION

Compare the probabilities of hitting a balloon on each of the two targets.

$$\text{Probability of hitting the balloon on Target } A = \frac{\text{Area covered by large balloon}}{\text{Area of Target } A}$$

$$= \frac{\pi(6)^2}{12^2}$$

$$= \frac{36\pi}{144}$$

$$\approx 0.785$$

$$\text{Probability of hitting a balloon on Target } B = \frac{4 \times (\text{Area covered by one small balloon})}{\text{Area of Target } B}$$

$$= \frac{4\pi(3)^2}{12^2}$$

$$= \frac{36\pi}{144}$$

$$\approx 0.785$$

Since the probability of hitting a balloon is the same for both targets, neither one gives you a better chance of winning a prize.

THINK AND COMMUNICATE

1. In Section 5.4, you learned that if the ratio of the lengths of corresponding sides of two similar figures is $a:b$, then the ratio of the areas is $a^2:b^2$. How can you use this fact to answer the question in Example 2?

2. Suppose a square board with side length 12 in. has nine balloons attached to it. If the radius of each balloon is 2 in., what is the probability that a randomly thrown dart that hits the target hits a balloon?

☑ CHECKING KEY CONCEPTS

The targets shown below are a square, an equilateral triangle, and a circle. Find the probability that a randomly thrown dart that hits each target hits the shaded area.

1.

2.

3.

5.6 Exercises and Applications

Extra Practice exercises on page 568

The targets shown below are squares, equilateral triangles, and circles. Find the probability that a randomly thrown dart that hits each target hits the shaded area.

1.

2.

3.

4.

5.

6.

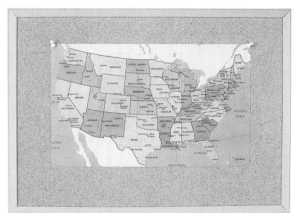

7. Suppose you and a friend want to take a vacation somewhere in the 48 states shown in the map. You can't decide where to go, so you agree to throw a dart randomly at the map and go where the dart lands.

   a. Which state do you have the highest probability of visiting? Why?

   b. **Writing** Explain why your method is *not* equivalent to writing the names of the 48 states on slips of paper and picking one slip of paper at random.

In Exercises 8–10, the targets and the shaded areas are all equilateral triangles. Find the probability that a randomly thrown dart that hits each target hits the shaded area.

8.

3

18

9.

3

4.5

10.

12

5

11. Challenge The map shows the location of sunken ships in a 28.5 mi by 34.5 mi section of the Atlantic Ocean just off the North Carolina coast. Suppose you want to search for a sunken ship in these waters, but you don't know the location of any of these ten ships. If you select a point in this area at random and search within a 0.5 mi radius of that point, what is the probability that you will find one of the ships? Show how you got your answer.

35°00′

F.W. Abrams   Keshena

76°00′   Dixie Arrow

British Splendor   E.M. Clark

Proteus   U.S.S. Tarpon

Nordal

Manuela

Malchace   75°30′

34°30′

Connection   SPORTS

At the Olympics, archers shoot at a target like the one shown. They earn the most points when they hit the center, or the bull's-eye, of the target. You can use geometric probability to see just how difficult it is to hit the bull's-eye.

12.2 cm   122 cm

12. The target has a diameter of 122 cm. The bull's-eye has a diameter of only 12.2 cm. What is the probability that a randomly shot arrow that hits the target hits the bull's-eye? Is this probability experimental or theoretical?

13. Open-ended Problem Do you think the probability you found in Exercise 12 takes into account the skill of the archer? Explain your reasoning.

14. Suppose you are playing a board game that comes with the spinner shown. Find the probability that you will:

    a. lose your turn.   **b.** spin a number greater than 2.

15. Suppose you create an elaborate target in stages, as shown below. For Stage 0, you have a square, blank target. For Stage 1, you divide the target into 9 squares and shade the center one. For Stage 2, you divide each of the unshaded squares into 9 squares and shade the center one.

        Stage 0   Stage 1   Stage 2

    a. At each stage, suppose you randomly throw a dart at the target. What is the probability of hitting the target's unshaded area? Answer this question by copying and completing the table. Express the probabilities as fractions.

Stage	Probability of hitting the target's unshaded area
0	?
1	?
2	?

    b. What do you think is the probability for a Stage 3 target?

16. The probability that a randomly thrown dart that hits the square target shown hits the shaded area is $\frac{5}{9}$.

The area of the inner square is 16 in.^{2}. Find the area of the target.

17. 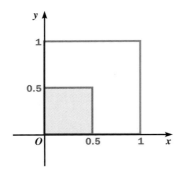 **Technology** Work with a partner. You need a scientific or graphing calculator with a random-number feature. This feature will give you randomly chosen decimals between 0 and 1. You will use this feature to simulate the random throwing of darts at a dart board. Repeat the following steps 40 times:

    • Get a pair of random numbers from your calculator. Treat these as the coordinates of a point in the 1-by-1 square shown.

    • Have your partner record whether the point lies in the shaded area of the square shown. Assume the point $(0.5, 0.5)$ lies in the shaded area.

    a. What must be true about x and y if the point (x, y) lies in the shaded area?

    b. What is the theoretical probability of hitting the shaded area?

    c. What experimental probability do you get as a result of this activity? Is this close to the theoretical probability?

18. Cooperative Learning Work in a group of 3–5 students. Each student needs notebook paper and a small dried bean.

   a. Design a target to cover a piece of notebook paper. Shade in part of it so that the probability that a bean hits the shaded area when dropped on the target is 75%. What kind of probability is this? Explain.

   b. Place another student's target on the floor and drop the bean on the target twenty times.

   c. Find the ratio of the number of times the bean lands clearly inside the shaded area to the number of times the bean hit the target. What kind of probability is this? Explain.

S P I R A L   R E V I E W

Evaluate each variable expression when $a = -2$, $b = 10$, and $c = \frac{3}{4}$.
(Section 1.5)

19. $a^2 + b^2$ **20.** $c^2 - b^2$ **21.** $9a^2 + 16b^2$

22. $a(bc + a)$ **23.** $\left(\dfrac{a + b}{2}\right) + 5c$ **24.** $b + c(a + b) - a$

ASSESS YOUR PROGRESS

VOCABULARY

probability (p. 218)   theoretical probability (p. 220)
event (p. 218)   favorable outcome (p. 220)
outcome (p. 218)   geometric probability (p. 224)
experimental probability (p. 219)

Find the theoretical probability of each event. Then choose one event and find its experimental probability. *(Section 5.5)*

   1. A die is rolled and lands showing a number less than 5.

   2. A calendar is opened randomly to a month that begins with "M."

For Exercises 3–5, find the theoretical probability that a randomly thrown dart that hits each target hits the shaded area. *(Section 5.6)*

3.

4.

5.

   6. Journal List five real-world situations that involve probability. One of them should involve geometric probability.

Using a Cartogram

Most maps of the United States show the size, shape, and location of each state. Some other maps are drawn so that each state's size is distorted to reflect a particular group of data, such as population. This type of map is called a cartogram.

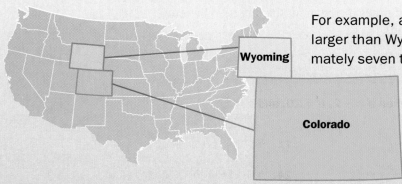

For example, although Colorado is only a little larger than Wyoming, in 1990 it had approximately seven times as many people. The cartogram shows how these states look when their sizes are distorted to represent population instead of area.

PROJECT GOAL Make a cartogram of a selected group of states. Then make a poster of your cartogram.

Gathering Data

Work in a group of 2–3 students. First choose a set of statistical data to illustrate with your cartogram. Some possible topics are:

- number of physicians
- amount of land suitable for farming
- amount of a natural resource produced each year, such as oil or coal

Select three to five neighboring states to show on your cartogram, and find the relevant data for each. Organize the information in a table. For example, this table shows the population densities of Nebraska, Iowa, and Illinois.

State	Population density, 1993
Nebraska	20.9
Iowa	50.4
Illinois	210.4

Constructing Your Cartogram

Use a computer drawing program or graph paper to form the shapes of the states in your cartogram with squares.

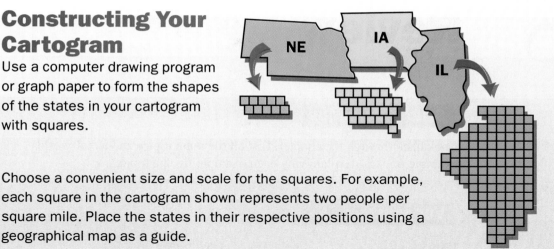

Choose a convenient size and scale for the squares. For example, each square in the cartogram shown represents two people per square mile. Place the states in their respective positions using a geographical map as a guide.

Making a Poster

Display your cartogram on a poster. Label each state and describe the scale you used. Include a labeled geographical map of the states and a table of the data.

You and your classmates can extend your study of cartograms by exploring some of the ideas below:

- Create a class cartogram of the entire United States.

- Make three or four cartograms that illustrate different data for the same group of states.

- Start a scrapbook of cartograms that you find in newspapers, magazines, or reference books.

Self-Assessment

Write a paragraph explaining your project. Describe how you chose the scale and how you created each state. Why is it important to use the same sized squares for all the states? How does it help to have a geographical map next to your cartogram? Describe any difficulties you had. What would you do differently if you were to do the project over again?

5

Review

Look back through the chapter. List all the main topics and a real-world example of each. Explain why each example fits the topic.

VOCABULARY

ratio (p. 192)
proportion (p. 193)
means (p. 193)
extremes (p. 193)
scale (p. 198)
scale factor (p. 199)
similar (p. 206)

probability (p. 218)
event (p. 218)
outcome (p. 218)
experimental probability (p. 219)
theoretical probability (p. 220)
favorable outcome (p. 220)
geometric probability (p. 224)

SECTIONS | 5.1 *and* 5.2

A **ratio** is a quotient that compares two numbers. The ratio of 1 to 2 can be written in three ways:

$$\frac{1}{2} \qquad 1:2 \qquad 1 \text{ to } 2$$

A **proportion** is an equation that shows two ratios to be equal. For any proportion, the product of the **extremes** is equal to the product of the **means**.

$$\frac{9}{12} = \frac{6}{8}$$
$$9 \cdot 8 = 12 \cdot 6$$
$$72 = 72$$

A **scale** is a ratio of two different measurements. The photograph of *Columbia* has a scale of about $\frac{1 \text{ cm}}{6 \text{ m}}$. Each centimeter in the picture represents 6 meters in real life.

A **scale factor** is the ratio of a new size to an old size. You can write the scale factor of the photograph as $\frac{1}{600}$, 0.0017, or 0.17%.

Two figures are **similar** when corresponding angles have equal measure and the ratios of the lengths of corresponding sides are equal. Similar figures are always the same shape.

$$ABCD \sim EFGH$$

Triangles are similar when two angles of one triangle have the same measures as two angles of the other triangle.

> Always name corresponding angles of similar figures in the same order.

If the ratio of corresponding sides of similar figures is $x : y$, then the ratio of the perimeters of these figures is $x : y$, and the ratio of the areas is $x^2 : y^2$.

For figure *ABCD* and figure *EFGH* above, the ratio of the lengths of corresponding sides is $2 : 3$. The ratio of the perimeters is $2 : 3$, and the ratio of the areas is $2^2 : 3^2$ or $4 : 9$.

Probability is a ratio that measures how likely an **event** is to happen.

- The probability that water will freeze at 100°F is 0.
- The probability that the sun will rise tomorrow is 1.

Experimental probability is the ratio of the number of times an event actually happens to the number of times the experiment is performed.

- June oversleeps an average of once every four days, so the probability that she will oversleep tomorrow is 25%.

Theoretical probability is the ratio of the number of **favorable outcomes** to the total number of outcomes.

- Three of the six numbers on a die are prime, so the probability of rolling a prime number is 0.5.

Geometric probability is a probability based on area.

- One ninth of the target is shaded, so the probability that a randomly thrown dart that hits the target will hit the shaded area is $\frac{1}{9}$.

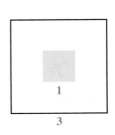

5 | Assessment

VOCABULARY QUESTIONS

For Questions 1–4, complete each statement.

1. A _?_ is an equation that shows two ratios are equal. In this equation, the product of the _?_ equals the product of the _?_ .

2. _?_ figures always have the same shape. The corresponding _?_ are equal and the corresponding _?_ have the same ratio of lengths.

3. The _?_ probability is the ratio of the number of favorable outcomes to the total number of possible outcomes. The _?_ probability is the ratio of the number of successes to the number of tries.

4. An event with a probability of _?_ will certainly occur. An event with a probability of _?_ will never occur.

SECTIONS 5.1 *and* 5.2

Write each ratio as a fraction in lowest terms.

5. 6 in. to 5.5 ft

6. $42.50 to $5.00

7. A 1.25 acre school sits on a 7 acre campus. One acre of the grounds is devoted to parking and 4 acres are devoted to playing fields. For parts (a)–(c), find the ratio of the first quantity to the second. Write each answer as a fraction in lowest terms.

   a. school area to campus area

   b. school area to playing field area

   c. unused area to campus area

Solve each proportion.

8. $\dfrac{15}{35} = \dfrac{82.5}{x}$

9. $\dfrac{3.14}{50.24} = \dfrac{n}{16}$

10. $\dfrac{3}{5} = \dfrac{n-12}{n}$

11. $4 : p = 6 : p + 9$

12. $a : 8 = 7 : 1$

13. $5 : c = 9 : 8$

14. **Writing** Explain how to solve the proportion $5 : 8 = 30 : n$ without using the means-extremes property.

15. A 12 ft by 20 ft room is being remodeled. What scale should be used in order to fit the largest possible drawing on an $8\frac{1}{2}$ in. by 11 in. paper? What is the scale factor of the drawing?

16. Writing The graph shows the number of school-age children in Hopedale. Suppose you want to use the graph to convince the town council that Hopedale needs a new school. Redraw the graph using a different scale on the vertical axis. Explain how the scale you chose changes the visual effect of the graph.

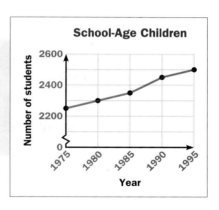

SECTIONS 5.3 *and* 5.4

17. Open-ended Problem Draw a pair of similar figures so that the ratio of lengths of corresponding sides is 7 : 2.

For Questions 18–22, use the fact that *ABCD* ~ *EFGH*.

18. Find *AD* and *FG*.

19. Find ∠ *B*, ∠ *C*, ∠ *E*, and ∠ *H*.

20. The perimeter of *ABCD* is 19.2. Find the perimeter of *EFGH*.

21. The area of *EFGH* is 31.5. Find the area of *ABCD*.

22. Tell whether each statement is *True* or *False*.

    a. *BCDA* ~ *FGHE*

    b. *CBAD* ~ *GHEF*

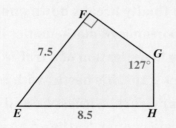

SECTIONS 5.5 *and* 5.6

Number rolled	1	2	3	4	5	6
Times rolled	8	12	5	9	10	6

23. The table shows results of 50 rolls of a die. Find the theoretical and experimental probabilities of rolling either a 2 or a 5 on one roll.

24. Darts are tossed at the square board pictured. What is the geometric probability that a dart will land inside the circle?

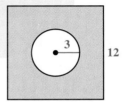

PERFORMANCE TASK

25. Choose an exercise or quiz question you solved incorrectly. Explain what you did wrong. Correct it. Explain your correction.

6 Working with Radicals

On the scene of the CRASH

Nathan Shigemura

Many of us think we know from television shows and movies what a police investigation is all about. The detective arrives at the scene of the crime, searching for clues and questioning witnesses until a suspect finally breaks down and confesses. But few people realize the importance of mathematics in an investigation, particularly an investigation of a traffic accident. "When I get to an accident scene, it's hectic, with rescue personnel, police officers, survivors, witnesses, and victims all intermingled," explains Trooper Nathan Shigemura, an accident reconstructionist with the Illinois State Police. "It is often dark and always chaotic. Out of that chaos, we try to figure out what happened and why."

> " **Math enables people to understand the world around them.** "

236

" All accidents obey the same laws of physics and math. "

Trooper Shigemura shown measuring the road surface's coefficient of friction.

Solving a Puzzle

An accident is like a puzzle. Shigemura arrives after the collision and tries to put the pieces together. Mathematics plays a big role in this process because "everything happens according to the laws of physics and many of these can be represented by mathematical equations." By examining the position of the vehicles, the nature and extent of the damage they have sustained, and other physical evidence, he works backwards to identify the factors leading up to the crash.

Making Cars and Roads Safer

Knowledge gained from these investigations can highlight safety problems that weren't recognized before. "Based on what we see, we might learn that a particular make of car has an engineering defect or that the roadway itself is dangerous," says Shigemura. He is pleased that his studies of tragic accidents can eventually pave the way for safer roads and cars. He also finds satisfaction in the task of taking a complex situation like an accident scene and gradually piecing together what happened. "Math enables people to understand the world around them. When you can understand how things work—rather than just looking around in wonder and scratching your head—it opens up a whole new world."

The Physical Evidence

Shigemura uses physics and math to reconstruct an accident, charting the precise path the vehicles must have taken before the crash to end up where they did. For example, he uses the formula $S = \sqrt{30df}$ to calculate a car's minimum speed S in miles per hour at the moment the car begins to skid. In the formula, d = the length in feet of the skid mark left by the car and f = the coefficient of friction. The coefficient of friction is a measure of the slipperiness or stickiness of the road. Shigemura finds d with a tape measure, and determines f with a special device. The matrix at the right shows speeds calculated for several different combinations of d and f.

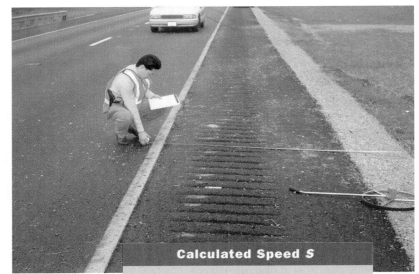

Calculated Speed S

		d	
	76	98	123
0.40	30.19	34.29	38.41
f **0.80**	42.70	48.49	54.33
1.20	52.30	59.39	66.54

$$\sqrt{30 \cdot 123 \cdot 0.40} \approx 38.41$$

If d = 123 and f = 0.40, the car was traveling about 38 mi/h when it began to skid.

EXPLORE AND CONNECT

Nathan Shigemura standing in the dent left in the front of a truck by its impact with a telephone pole.

1. Writing Use the formula in the article. If the value of f does not change, what happens to S as d increases? If d does not change, what happens to S as f increases? If a car is traveling at 55 mi/h, will it skid farther on a road with a high coefficient of friction or a low one?

2. Research Find out the requirements for becoming a police officer. What mathematics do officers need to know? What are the course requirements for a degree in law enforcement? Report your findings to your class.

3. Project Use a toy car and a ramp. Let the car roll down the ramp on to various surfaces: rough concrete, a smooth floor, and so on. Observe how far the car travels before it stops. What do you notice?

Mathematics
& Nathan Shigemura

In this chapter, you will learn more about how mathematics is related to police investigations.

Related Examples and Exercises

Section 6.1
• **Example 2**

Section 6.2
• **Example 1**
• **Exercises 28–30**

Section 6.3
• **Exercises 19 and 20**

6.1

Learn how to...

- identify right triangles using the Pythagorean theorem

So you can...

- solve real-world problems, such as checking whether a corner is square

The Pythagorean Theorem

Surveyors in ancient Egypt may have used stretched ropes like the one shown to make right angles when laying out land boundaries. The surveyors knew that a triangle with side lengths of 3, 4, and 5 units is a *right triangle*.

A **right triangle** has one right, or 90°, angle. As shown on the map below, people have been aware for centuries of a special relationship among the sides of a right triangle. The *Pythagorean theorem* describes this relationship. The theorem is named after Pythagoras, a Greek mathematician.

The Pythagorean Theorem

In a right triangle, the square of the length of the **hypotenuse** is equal to the sum of the squares of the lengths of the **legs**.

$$a^2 + b^2 = c^2 \qquad\qquad 3^2 + 4^2 = 5^2$$

| 2000 B.C. | 1500 B.C. | 1000 B.C. | 500 B.C. | 100 B.C. | 100 A.D. |

Clay tablet
Babylonia
1900-1600 B.C.

Sulbasutras
India
1500-800 B.C.

Pythagoras
Greece
560-480 B.C.

Zhōu bì suàn jīng China
100 B.C.-100 A.D.

BY THE WAY

A German book published in 1525 included the first known use of the symbol $\sqrt{}$ in print. The symbol eventually replaced R_x, an older notation. R_x came from the first and last letters of the word *radix*, a Latin translation of an Arabic word for *root*.

Working with the Pythagorean theorem involves using *squares* and *square roots*. The number 9 is called a **perfect square** because it is the square of the integer 3. The number 3 is called a **square root** of 9 because $3^2 = 9$. The symbol $\sqrt{}$ is called a **radical sign**. The radical sign indicates the positive square root of a number. All positive numbers have two square roots. Negative numbers do not have real square roots.

The positive square root is written as $\quad \sqrt{9} = 3.$

The negative square root is written as $\quad -\sqrt{9} = -3.$

Both square roots are written as $\quad \pm\sqrt{9} = 3 \text{ or } -3.$

The symbol $\pm$ is read as *plus or minus*.

EXAMPLE 1 **Application: Recreation**

Kevin is parasailing. He is wearing a special parachute and is connected to the boat by a 250 ft rope. As the speed of the boat increases, he is lifted into the air. Kevin is 200 ft behind the boat when the rope is taut. What is his height in feet above the water?

Not drawn to scale

h 250 ft

200 ft

SOLUTION

Use the Pythagorean theorem. Let h = Kevin's height in feet above the water.

$$h^2 + 200^2 = 250^2$$
$$h^2 + 40{,}000 = 62{,}500$$
$$h^2 = 22{,}500$$
$$h = \pm\sqrt{22{,}500}$$
$$h = \pm 150$$

Find the *square root* of both sides of the equation.

The value of h can be either the positive square root or the negative square root of 22,500. Only the positive solution makes sense in this situation. Kevin is 150 ft above the water.

THINK AND COMMUNICATE

Find each square root. You may be able to use mental math. You can also use a calculator or the Table of Squares and Square Roots on page 613.

1. $\sqrt{1}$ **2.** $-\sqrt{4}$ **3.** $\pm\sqrt{289}$ **4.** $\sqrt{0}$

As you will see in Example 2 below, Nathan Shigemura uses the *converse* of the Pythagorean theorem in his work as an accident reconstructionist.

Converse of the Pythagorean Theorem

If the lengths of the sides of a triangle satisfy the relationship $a^2 + b^2 = c^2$, then the triangle is a right triangle.

EXAMPLE 2   Interview: Nathan Shigemura

To measure the extent of damage to a car, Nathan Shigemura creates a rectangle around the damaged car to represent its original shape. By comparing the shape of the damaged car to its original shape, he can see how badly the car is bent.

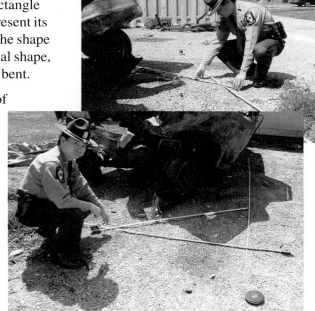

To make sure that the corners of his rectangle are right angles, Shigemura uses measuring tapes to form a triangle at each corner. He uses the converse of the Pythagorean theorem to check whether each triangle is a right triangle. Tell whether the given lengths can be the sides of a right triangle.

a. 5 in., 12 in., 13 in.

b. 16 in., 11 in., 19 in.

SOLUTION

Use the converse of the Pythagorean theorem.

a. $5^2 + 12^2 \overset{?}{=} 13^2$
$25 + 144 = 169$
These lengths *can* be the sides of a right triangle.

b. $16^2 + 11^2 \overset{?}{=} 19^2$
$256 + 121 \neq 361$
These lengths *cannot* be the sides of a right triangle.

THINK AND COMMUNICATE

5. Choose a corner of something. Use the converse of the Pythagorean theorem and a ruler to check whether the corner is a right angle.

6. To check the angles of his rectangle, Nathan Shigemura always uses a right triangle with side lengths of 3, 4, and 5 units. Why do you think he chooses to use this right triangle?

For Questions 1–3, refer to the right triangle below. Find the unknown length for each right triangle.

1. $a = 8, b = 15, c = \underline{\ ?\ }$

2. $a = \underline{\ ?\ }, b = 120, c = 169$

3. $a = 1.5, b = \underline{\ ?\ }, c = 2.5$

For Questions 4–7, tell whether the given lengths can be the sides of a right triangle.

4. 3 cm, 4 cm, 5 cm

5. 7 in., 9 in., 15 in.

6. 6 ft, 8 ft, 12 ft

7. 1 m, 2.4 m, 2.6 m

6.1 **Exercises and Applications**

Extra Practice exercises on page 568

Find each square root.

1. $\sqrt{25}$

2. $-\sqrt{64}$

3. $\pm\sqrt{100}$

4. $\sqrt{36}$

5. $\pm\sqrt{400}$

6. $-\sqrt{144}$

Find the unknown length for each right triangle.

7.

8.

9.

The legs of a right triangle have lengths a and b. The hypotenuse has length c. Find the unknown length for each right triangle.

10. $a = 60, b = 11$

11. $b = 18, c = 82$

12. $a = 2291, c = 3541$

13. $a = 12, c = 37$

For Exercises 14–19, tell whether the given lengths can be the sides of a right triangle.

14. 2.4 cm, 1.8 cm, 3.0 cm

15. 0.6 in., 1.0 in., 1.2 in.

16. 48 ft, 74 ft, 55 ft

17. 91 m, 84 m, 35 m

18. 33 mm, 19 mm, 27 mm

19. Challenge $2x, 3x, 4x$

20. An equilateral triangle has sides of equal length. Use algebra to show why an equilateral triangle cannot be a right triangle.

21. a. The size of a television screen is given by the length of the diagonal of the screen. What size is a television screen that is 21.6 in. wide and 16.2 in. high?

b. Open-ended Problem Why do you think that manufacturers might choose to measure the size of television screens this way? Use algebra to support your answer.

Toolbox p. 595
Angles

Connection ▸ GEOMETRY

The Pythagorean theorem is important in both algebra and geometry. Exercises 22 and 23 will help you see why the theorem applies only to right triangles.

Acute triangle
All angles < 90°

Right triangle
1 angle = 90°

Obtuse triangle
1 angle > 90°

The **area of a triangle** is $\frac{1}{2} \times$ base $\times$ height.

22. Cooperative Learning Work in a group of 3–4 students. You will need at least six pieces of graph paper and a ruler.

Work together to repeat the steps below at least six times. Use two right triangles, two acute triangles, and two obtuse triangles.

Step 1 Draw a triangle on each piece of graph paper so that each corner is on a grid point. Label the sides of each triangle *a*, *b*, and *c*, with *c* being the longest side (if there is a longest side). Label each triangle *right*, *acute*, or *obtuse*.

Step 2 Create a square on each side of a triangle.

Step 3 Find the area of each square. If a square is not aligned with the grid, divide it into right triangles and a small square, as shown. Add the areas of the pieces to find the area of the square.

Step 4 Compare the sum of the areas of the two smaller squares to the area of the largest square.

Area of this triangle
$\frac{1}{2} \cdot 5 \cdot 3 = \frac{15}{2}$

Area of large blue square =
$4 \cdot \frac{15}{2} + 4 = 34$

Area = 25
b

c

a

right triangle

Area = 9

$25 + 9 = 34$

23. Writing Based on your observations, use > or < to complete each statement. Explain how your observations show that the Pythagorean theorem applies only to right triangles.

a. In an obtuse triangle, $a^2 + b^2 \underline{\ ?\ } c^2$.

b. In an acute triangle, $a^2 + b^2 \underline{\ ?\ } c^2$.

24. URBAN PLANNING Choose two locations on the map below. Copy part of the map to show two different routes between the points. One route should use the diagonal streets. The other route should not. Estimate the length of each route. Use a city block as one unit. Round your answer to the nearest tenth. Explain why city planners would choose to include the diagonal streets.

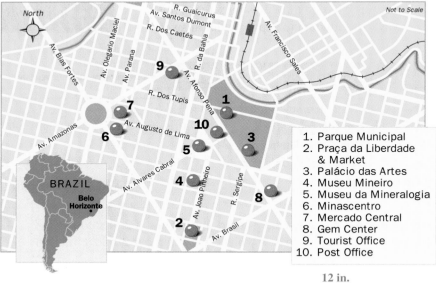

1. Parque Municipal
2. Praça da Liberdade & Market
3. Palácio das Artes
4. Museu Mineiro
5. Museu da Mineralogia
6. Minascentro
7. Mercado Central
8. Gem Center
9. Tourist Office
10. Post Office

25. Challenge Find the length d of the diagonal of the rectangular box.

ONGOING ASSESSMENT

26. a. A *Pythagorean triple* is a group of three numbers that could be the lengths of the sides of a right triangle. Copy and complete the table. Are the three numbers in each row a Pythagorean triple?

n	$2n$	$n^2 - 1$	$n^2 + 1$
3	6	8	10
4	?	?	?
5	?	?	?
6	?	?	?

b. Do you think this pattern will always work? Try some other values of n to test your theory.

SPIRAL REVIEW

27. In 1990, the average price for gasoline was $1.22 per gallon. Write and graph a direct variation equation based on this fact. *(Section 3.2)*

Find the reciprocal of each number. *(Section 4.2)*

28. $\dfrac{23}{5}$ **29.** $-\dfrac{13}{36}$ **30.** 15   **31.** -11

6.2

Learn how to...

- **find noninteger square roots**

So you can...

- **estimate the speed of a car before it started to skid**

Irrational Numbers

Nathan Shigemura can estimate the minimum speed a car was traveling before a crash. He measures the distance the car skidded and finds the coefficient of friction, or slipperiness, for the road surface. He uses a formula that involves finding a square root.

EXAMPLE 1   Interview: Nathan Shigemura

An accident reconstructionist measures the distance a car skidded before coming to a stop and finds the coefficient of friction for the road. Was the car exceeding the 25 mi/h speed limit when the driver applied the brakes?

length of the skid mark is 83 ft

coefficient of friction is 0.5

$S = \sqrt{30\,df}$

SOLUTION

Use the formula from the article on pages 236–238. Let S = the speed of the car in miles per hour when it started to skid. Let d = the length of the skid mark in feet. Let f = the coefficient of friction for the road.

$$S = \sqrt{30\,df}$$

Substitute the measured values for d and f into the formula.

$$S = \sqrt{(30)(83)(0.5)}$$

The radical sign is a grouping symbol. First do the operations under the radical sign.

$$= \sqrt{1245}$$

$$\approx 35.28$$

A calculator gives an approximate value for $\sqrt{1245}$.

The car was traveling about 35 mi/h. The car was exceeding the 25 mi/h speed limit.

Rational and Irrational Numbers

Any number that can be written as a ratio of two integers $\frac{a}{b}$, when $b \neq 0$, is a **rational number**. For example 7.5 and $0.\overline{3}$ are rational because $7.5 = \frac{15}{2}$ and $0.\overline{3} = \frac{1}{3}$. When rational numbers are written in decimal form, they either terminate or repeat.

The number $\sqrt{1245}$ cannot be written as a ratio of two integers. When written as a decimal, it never terminates or repeats. It is an **irrational number**. The square root of any integer that is not a perfect square is irrational.

a terminating decimal — a repeating decimal

REAL NUMBERS

Rational Numbers

$0.\overline{3}$ $1\frac{1}{2}$ 2.25   3.555...

−1   0   1   2   3

$-\sqrt{2}$ $-0.121121112...$ $\sqrt{2}$ $\sqrt{5}$ π

Irrational Numbers

a non-repeating decimal

All **real numbers** are either *rational* or *irrational* and can be located on the real number line. The real numbers are **dense**. This means you can always find another real number between any two real numbers.

THINK AND COMMUNICATE

Tell whether each number is *rational* or *irrational*. Explain how you know.

1. $\frac{1}{23}$ **2.** 0.37   **3.** $\sqrt{3}$ **4.** $\sqrt{16}$

5. Use a calculator to find a decimal approximation for $\frac{1}{23}$.

Does the decimal appear to terminate or repeat? Why can't you use a calculator to tell the difference between a rational number and an irrational number?

6. Find a real number between 0.3 and 0.4. Now find another real number between 0.4 and the number you found. How many times can you repeat this process?

7. Explain how to write 0.37 as a ratio of two integers.

EXAMPLE 2

Write the repeating decimal 0.272727... as a fraction in lowest terms.

Toolbox p. 587
Fractions, Decimals, and Percents

SOLUTION

Write 0.272727... as $0.\overline{27}$. Let $b = 0.\overline{27}$. Because there are 2 digits in the repeating block, multiply both sides of the equation $b = 0.\overline{27}$ by 100.

$$100b = 100(0.\overline{27})$$
$$100b = 27.\overline{27}$$

Subtract the original equation from the new equation to eliminate the decimal part of $27.\overline{27}$.

$$\begin{aligned} 100b &= 27.\overline{27} \\ -b &= -0.\overline{27} \\ \hline 99b &= 27 \end{aligned}$$

$$b = \frac{27}{99}$$

$$b = \frac{3}{11}$$

The decimal 0.272727... can be written as $\frac{3}{11}$.

THINK AND COMMUNICATE

8. Describe the steps you would use to write $0.\overline{2}$ as a fraction in lowest terms. What steps would you use to write $0.\overline{2397}$ as a fraction in lowest terms?

☑ CHECKING KEY CONCEPTS

For Questions 1–4, decide which numbers below belong in each category.

$$-5,\ -4.24,\ -\sqrt{29},\ -\sqrt{16},\ 0.0,\ 3,\ 4.1,\ 0.\overline{72},\ \sqrt{11}$$

1. integer   **2.** rational   **3.** irrational   **4.** real

Write each decimal as a fraction in lowest terms.

5. 0.4   **6.** 0.77777...   **7.** 1.26   **8.** $0.\overline{37}$

9. At the scene of a car accident, an accident reconstructionist found a 112 ft skid mark. The coefficient of friction for the road was 0.6. Use the formula $S = \sqrt{30df}$ to find the speed of the car when the driver applied the brakes.

6.2 Exercises and Applications

Extra Practice
exercises on
page 569

For Exercises 1–4, decide which numbers below belong in each category.

$$-6, -4.3, -\sqrt{25}, -\sqrt{29}, -\frac{3}{5}, 0, 0.42, 0.45, 8, \sqrt{961}, \sqrt{19}, 0.\overline{32}$$

1. integer   **2.** rational   **3.** irrational   **4.** real

Write each decimal as a fraction in lowest terms.

5. 0.35   **6.** 0.768   **7.** $0.\overline{63}$

8. $0.\overline{423}$ **9.** 0.4545…   **10.** $0.5\overline{04}$

AVIATION   **For Exercises 11–13, use the article below from an airline magazine.**

You can figure out how far you can see to the horizon on a clear day if you know how high your plane is flying. Your viewing distance v in miles is related to your altitude a in feet by the formula below:

$$v = 1.22\sqrt{a}$$

You can use this formula to calculate how far away the horizon is no matter where you are. For example, suppose the plane you are in is flying above Hawaii at an altitude of ⸺⸺ feet. A⸺

Honolulu, Hawaii, from the air. Using the formula at the left and the height at which this photo was taken, you can determine the distance to the horizon.

11. Suppose you are in a plane that is cruising at an altitude of 30,000 ft. How far away is a lake that you see on the horizon?

12. As the plane starts to descend, a city comes into view on the horizon. If the plane is at a height of 20,000 ft when you see the city, how far is the plane from the city?

13. Suppose you visit the observation tower at the top of a 1127 ft high skyscraper. How far can you see from the top of the building?

For Exercises 14–18, name a real number, if possible, that fits each description.

14. not rational   **15.** not irrational

16. rational but not an integer   **17.** not rational and not irrational

18. an integer that is not positive and not negative

19. Arrange the numbers below in order from smallest to largest.

$$0.26, 0.\overline{265}, 0.265, 0.\overline{26}, 0.2\overline{6}, 0.2, 0.\overline{2}$$

The legs of a right triangle have lengths _a_ and _b_. The hypotenuse has length _c_. Estimate the unknown length for each right triangle.

20. $a = 5, b = 8$ **21.** $b = 13, c = 17$ **22.** $b = 18, c = 27$

23. $a = 7, c = 19$ **24.** $a = 53, b = 28$ **25.** $a = 53, c = 91$

26. Writing Find three rational numbers between $\frac{2}{3}$ and $\frac{3}{4}$. How many rational numbers are there between $\frac{2}{3}$ and $\frac{3}{4}$? Explain your reasoning.

27. LANGUAGE ARTS Look up the word _dense_ in a dictionary. Which meanings help you understand the mathematical meaning of the word?

INTERVIEW Nathan Shigemura

Look back at the interview on pages 236–238.

A driver applies the brakes of a car and the car comes to a stop, leaving a skid mark on the pavement. The greater the initial speed of the car, the longer the skid mark, or stopping distance. Road conditions also affect stopping distance.

Nathan Shigemura finds the coefficient of friction of a road.

Stopping Distance		
C2		=SQRT(30*A2*B2)
	A	**B**	**C**
	_d_	_f_	_S_
1			
2	70	0.4	28.98
3	70	0.8	
4	70	1.2	
5	100	0.4	
6	100	0.8	
7	100	1.2	
8	130	0.4	
9	130	0.8	
10	130	1.2	

28. Spreadsheet Look back at the formula $S = \sqrt{30df}$ in Example 1 on page 245. The value of the coefficient of friction _f_ depends on road conditions. Copy the table and use the formula for car speed to complete it. Use a spreadsheet if you have one.

29. When the skid distance remains constant, how does the value of _f_ affect the calculated speed of the car? Why is it important to take road conditions into account when estimating the minimum speed of a car before an accident?

30. Open-ended Problem What types of road conditions would you expect to have a high coefficient of friction? a low coefficient of friction? Explain. What other factors might affect stopping distance?

31. BIOLOGY The maximum walking speed S in centimeters per second of any animal is related to the animal's leg length l in centimeters and the force of gravity g of the planet. You can use the formula $S = \sqrt{gl}$ to estimate an animal's maximum walking speed.

    a. Measure your own leg length. Use the formula to estimate your maximum walking speed. Does this speed seem reasonable? Explain how you decided. (*Note:* On Earth, $g \approx 981$ cm/s^2.)

    b. Use the formula to estimate your maximum walking speed on the moon. (*Note:* On the moon, $g \approx 162$ cm/s^2.)

32. Challenge Write $0.4\overline{5}$ as a fraction in lowest terms. Explain how you got your answer.

ONGOING ASSESSMENT

33. Writing Sue Boswell purchased a king-size mattress and box spring. Use the Pythagorean theorem to explain how she was able to fit the 76 in. by 80 in. mattress through her bedroom door. Why do you think king-size box springs are sold as two smaller pieces, as shown in the picture at the right?

SPIRAL REVIEW

For each pair of similar figures, find the ratio of the perimeters. Find the ratio of the areas. (*Section 5.4*)

34.

35.

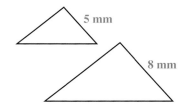

5 mm

8 mm

The legs of a right triangle have lengths a and b. The hypotenuse has length c. Find the unknown length for each right triangle. (*Section 6.1*)

36. $a = 1.5$ in., $b = 2$ in.   **37.** $a = 5$ cm, $b = 12$ cm

38. $a = 15$ km, $c = 39$ km   **39.** $a = 5$ ft, $c = 13$ ft

6.3

Calculating with Radicals

Learn how to...
- **multiply and divide radicals**
- **simplify radicals**

So you can...
- **find your reaction time or the escape velocity for a planet**

There are many ways to write the square root of eight. You can use a calculator to find a decimal approximation. You can write the number in radical form as $\sqrt{8}$. In the Exploration, you will find other ways to write the square root of 8.

EXPLORATION
COOPERATIVE LEARNING

Comparing Radicals

Work with a partner.
You will need:
- graph paper

1 Draw a right triangle with legs 1 unit long, as shown. Use the Pythagorean theorem to show that the hypotenuse is $\sqrt{2}$ units long.

2 Draw a right triangle with legs 2 units long. Show that the hypotenuse is $\sqrt{8}$ units long. Divide the triangle into smaller triangles like the one in Step 1 to show that the hypotenuse is also $2\sqrt{2}$ units long.

3 Draw a right triangle with legs 3 units long, one with legs 4 units long, and one with legs 5 units long. For each triangle, write the length of the hypotenuse two different ways. Describe any patterns you see.

4 The number 8 can also be written as the product 4 · 2. You can write $\sqrt{8}$ as $\sqrt{4 \cdot 2}$. How does writing $\sqrt{8}$ this way help explain what you saw in Step 2? How does it help to explain what you saw in Step 3?

Properties of Square Roots

For all nonnegative numbers a and b:

$\sqrt{a^2} = a$

$\sqrt{ab} = \sqrt{a} \cdot \sqrt{b}$

$\sqrt{\dfrac{a}{b}} = \dfrac{\sqrt{a}}{\sqrt{b}},\ b \ne 0$

For example:

$\sqrt{3^2} = 3$

$\sqrt{4 \cdot 3} = \sqrt{4} \cdot \sqrt{3} = 2\sqrt{3}$

$\sqrt{\dfrac{3}{4}} = \dfrac{\sqrt{3}}{\sqrt{4}} = \dfrac{\sqrt{3}}{2}$

An expression that is written using the symbol $\sqrt{\ }$ is in **radical form**. You can use the properties of square roots to simplify an expression in radical form. An expression in radical form is in *simplest form* when:

1. there is no integer under the radical sign with a perfect square factor,

2. there are no fractions under the radical sign, and

3. there are no radicals in the denominator.

For example, $\sqrt{8}$ can be simplified because it has a perfect square factor.

$$\sqrt{8} = \sqrt{4 \cdot 2} = \sqrt{4} \cdot \sqrt{2} = 2\sqrt{2}$$

EXAMPLE 1

Simplify each expression.

a. $\sqrt{\dfrac{15}{81}}$

b. $\sqrt{24} + 5\sqrt{6}$

SOLUTION

a. Use the property $\sqrt{\dfrac{a}{b}} = \dfrac{\sqrt{a}}{\sqrt{b}}$.

$$\sqrt{\dfrac{15}{81}} = \dfrac{\sqrt{15}}{\sqrt{81}}$$

$$= \dfrac{\sqrt{15}}{9}$$

b. Simplify each term in the expression.

> Look for a factor of 24 that is a perfect square.

$$\sqrt{24} + 5\sqrt{6} = \sqrt{4 \cdot 6} + 5\sqrt{6}$$

$$= \sqrt{4} \cdot \sqrt{6} + 5\sqrt{6}$$

$$= 2\sqrt{6} + 5\sqrt{6}$$

> You can combine terms that have the same radical part.
> $2 \cdot \sqrt{6} + 5 \cdot \sqrt{6} = (2 + 5)\sqrt{6}$

$$= 7\sqrt{6}$$

Estimating Square Roots

Sometimes you may want to estimate the square root of an integer. To estimate $\sqrt{59}$, find the two perfect squares closest to 59 on either side of it. They are 49 and 64.

$$\sqrt{49} \ < \ \sqrt{59} \ < \sqrt{64}$$

$$7 \ < \ \sqrt{59} \ < \ 8$$

The square root of 59 is between 7 and 8.

EXAMPLE 2   **Application: Physics**

Elisa and Zack are testing their reaction time in physics lab. Elisa holds a yardstick above Zack's hand. He tries to catch it as quickly as he can after Elisa drops it. The yardstick falls 9 in. before Zack catches it. Use the formula in the lab notebook to find his reaction time.

Reaction Time Lab

t = reaction time in seconds

d = distance the yardstick falls in inches

$$t = \sqrt{\frac{d}{192}}$$

SOLUTION

Method 1 Use estimation.
Substitute 9 for d in the formula.

$$t = \sqrt{\frac{d}{192}}$$

$$= \sqrt{\frac{9}{192}}$$

$13^2 < 192 < 14^2$
Because 192 is very close to 14^2, $\sqrt{192} \approx 14$.

$$= \frac{\sqrt{9}}{\sqrt{192}}$$

$$\approx \frac{3}{14}$$

$$\approx 0.21$$

Zack's reaction time is about 0.2 s.

Method 2 Use a calculator.
Substitute 9 for d in the formula.

$$t = \sqrt{\frac{d}{192}}$$

$$= \sqrt{\frac{9}{192}}$$

$$= \sqrt{0.046875}$$

$$\approx 0.22$$

> Divide before you find the square root.

Zack's reaction time is about 0.2 s.

THINK AND COMMUNICATE

1. Can $\sqrt{2} + \sqrt{3}$ be simplified? Can $11 + 2\sqrt{3}$? Explain why or why not.

2. The property $\sqrt{ab} = \sqrt{a} \cdot \sqrt{b}$ also works in reverse. Use this fact to simplify each expression.

   a. $\sqrt{2} \cdot \sqrt{3}$　　　b. $\sqrt{5} \cdot \sqrt{7}$　　　c. $3\sqrt{2} \cdot \sqrt{8}$

3. Explain how to use the prime factors of 720 to simplify $\sqrt{720}$.

4. Explain why the properties of square roots are true only for nonnegative numbers.

☑ CHECKING KEY CONCEPTS

Simplify each expression.

1. $\sqrt{12}$　　　　　2. $\sqrt{18}$　　　　　3. $\sqrt{\dfrac{13}{36}}$

4. $5\sqrt{6} \cdot \sqrt{3}$　　　5. $\sqrt{20} + \sqrt{5}$　　　6. $\sqrt{14} + 2\sqrt{6}$

Estimate each square root within a range of two integers.

7. $\sqrt{5}$　　　　　8. $\sqrt{33}$　　　　　9. $\sqrt{84}$

6.3 Exercises and Applications

Extra Practice
exercises on
page 569

Simplify each expression.

1. $\sqrt{40}$　　　　　2. $\sqrt{75}$　　　　　3. $4 + 2\sqrt{10}$

4. $\sqrt{128}$　　　　5. $3\sqrt{20}$　　　　6. $8\sqrt{2} + 5\sqrt{2}$

7. $\sqrt{28} + \sqrt{7}$　　8. $\sqrt{\dfrac{27}{49}}$　　9. $\left(5\sqrt{3}\right)^2$

10. $5\sqrt{3} - \sqrt{75}$　　11. $\sqrt{\dfrac{5}{16}}$　　12. $4\sqrt{3} \cdot \sqrt{18}$

13. Nina says that $11 + 6\sqrt{2}$ cannot be simplified. Ellen thinks it can be simplified to $17\sqrt{2}$. Who is correct? Explain how you know.

14. **Technology** Graph the equations $y = \left(2\sqrt{2}\right)x$ and $y = \left(\sqrt{8}\right)x$ on a graphing calculator. What do you observe about the two graphs? Explain what you see.

GEOMETRY Find the perimeter and area of each rectangle.

15.

$3\sqrt{2}$　　　10

16.

$2\sqrt{3}$　　　$5\sqrt{2}$

---

17. **MEDICINE** For making medical decisions, such as how much medicine to prescribe, a doctor may need to know a person's body surface area. You can use a formula to find a person's body surface area A in square meters. Let h = height in inches. Let w = weight in pounds.

$$A = \sqrt{\frac{h \cdot w}{3131}}$$

Find the body surface area of a person who is 5 ft 4 in. tall and weighs 160 lb. Round your answer to the nearest hundredth.

18. **Challenge** A square and a rectangle have the same area. The side lengths of the rectangle are integers, and the perimeter is 28 cm. Which of the areas below cannot be the area of the square? Explain.

    A. 33 cm^2 **B.** 40 cm^2 **C.** 42 cm^2 **D.** 45 cm^2

INTERVIEW **Nathan Shigemura**

Look back at the interview on pages 236–238.

In addition to investigating accidents that have already happened, Nathan Shigemura is interested in preventing future accidents. One way drivers can avoid getting into accidents is by considering braking distance and reaction time.

zero mark

19. **Cooperative Learning** Work with a partner. Try the experiment described in Example 2 on page 253. You will need a yardstick.

    a. Hold the yardstick so that the zero mark is just between your partner's fingers. Without warning, drop the yardstick. Record your partner's "reaction distance" in inches. Then have your partner measure your reaction distance.

    b. Let d = your reaction distance from part (a). Let t = your reaction time in seconds. Use the formula $t = \sqrt{\frac{d}{192}}$ to estimate your reaction time to the nearest tenth of a second.

20. a. A study found that a typical driver's reaction time to brake lights is about 0.66 s. Suppose a driver is traveling at 55 mi/h. Suddenly, the car directly ahead slows down. About how far will the driver travel during the time it takes to react and apply the brake?

    b. **Open-ended Problem** Based on your answer to part (a), what advice would you give a new driver about preventing accidents?

Driver sees brake lights. | Driver applies brakes. | Driver stops car.

Reaction distance | Stopping distance

21. GEOMETRY Make a spiral by following these steps.

a. Start close to the center of a clean sheet of paper. Draw a right triangle with legs 1 unit long. The hypotenuse has length $\sqrt{2}$. Use the hypotenuse as one leg of a second right triangle. Make the other leg 1 unit long. Use algebra to find the length of the hypotenuse of the second triangle. Write your answer as a radical in simplest form.

b. Use the hypotenuse of the second triangle as one leg of a third right triangle. Make the other leg 1 unit long. Find the length of the hypotenuse of the third triangle. Write your answer as a radical in simplest form.

c. Continue to make triangles until you run out of room on the paper. What pattern do you see in the hypotenuse lengths?

22. **Spreadsheets** A spacecraft must reach *escape velocity* to escape from the gravitational pull of a planet or moon. Let v = the escape velocity, g = the force of gravity on the planet or moon, and r = the radius of the planet or moon. Use the formula $v = \sqrt{2gr}$ to find each escape velocity in the chart below. Use a spreadsheet if you have one.

Escape Velocity			
D2		=SQRT(2*B2*C2)	
	A	**B**	**C**	**D**
1		g (km/s^2)	r (km)	v (km/s)
2	Mercury	0.00358	2,436	4.18
3	Venus	0.00887	6,059	
4	Earth	0.00981	6,378	
5	Moon	0.00162	1,740	
6	Mars	0.00374	3,384	
7	Jupiter	0.02601	69,833	
8	Saturn	0.01117	58,280	
9	Uranus	0.01049	23,495	
10	Neptune	0.01325	22,740	
11	Pluto	0.00221	5,706	

23. SAT/ACT Preview $\dfrac{900}{10} + \dfrac{90}{100} + \dfrac{9}{1000} = \underline{}$.

    A. 99.09   **B.** 90.99   **C.** 90.909   **D.** 99.009   **E.** 90.09

ONGOING ASSESSMENT

24. Writing Is the equation $\sqrt{a + b} = \sqrt{a} + \sqrt{b}$ true for all values of a and b? Explain how you reached your conclusion. Describe a quick check you can use if you forget your conclusion.

SPIRAL REVIEW

Solve each equation. *(Section 4.3)*

25. $7x = 35 + 2x$ **26.** $4 + 4x = 20x$ **27.** $3x = 24 - 3x$

28. $4x - 15 = 16 - 8x$ **29.** $3(x - 2) = 9$ **30.** $4(x - 8) = 2x - 12$

31. $\triangle ABC \sim \triangle PQR$. Find each missing side length and angle measure.
(Section 5.3)

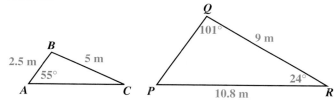

ASSESS YOUR PROGRESS

VOCABULARY

right triangle (p. 239)
Pythagorean theorem (p. 239)
hypotenuse (p. 239)
leg (p. 239)
perfect square (p. 240)
square root (p. 240)
radical sign (p. 240)

converse of the Pythagorean theorem (p. 241)
real number (p. 246)
rational number (p. 246)
irrational number (p. 246)
dense (p. 246)
radical form (p. 252)

The legs of a right triangle have lengths a and b. The hypotenuse has length c. Find the unknown length for each right triangle. *(Sections 6.1 and 6.2)*

  1. $a = 9, c = 15$ **2.** $a = 6, b = 15$ **3.** $a = 15, c = 22$

Write each decimal as a fraction in lowest terms. *(Section 6.2)*

  4. 3.65   **5.** 3.5757…   **6.** $2.\overline{5}$

Simplify each expression. *(Section 6.3)*

  7. $\sqrt{45}$ **8.** $\sqrt{5} + 2\sqrt{20}$ **9.** $\sqrt{\dfrac{28}{121}}$

10. Journal What do you find hard about working with radical numbers? What do you find easy? Why?

6.4 Multiplying Monomials and Binomials

Learn how to...

- find products of monomials and binomials

So you can...

- find the area of a complex figure, such as the sidewalk around a fountain

You can visualize the product $20 \cdot 40$ by thinking of the surface area of a 20 ft by 40 ft pool. To help visualize a product that involves a variable, you can use algebra tiles.

	1-tile	**x-tile**	**x^2-tile**
Dimensions	1 by 1	1 by x	x by x
Area	$1 \cdot 1 = 1$	$1 \cdot x = x$	$x \cdot x = x^2$

EXAMPLE 1

Write an equation modeled by the algebra tiles.

SOLUTION

The length of one side of the rectangle is $3x$. The length of the other side is $2x$. Therefore this model shows the product $(3x)(2x)$.

There are 6 x^2-tiles.

The tiles model the equation $(3x)(2x) = 6x^2$.

Finding Products with Algebra Tiles

Work with a partner.

You will need:

- algebra tiles
- ruler
- paper

SET UP Use tiles to find the product $2x(x + 3)$.

1 Draw two perpendicular lines on your paper.

2 Using tiles, mark off the dimensions $2x$ and $(x + 3)$.

3 Create and completely fill in a rectangle using x^2-tiles, x-tiles, and 1-tiles.

4 Count the tiles to see that $2x(x + 3) = 2x^2 + 6x$.

Use algebra tiles to find each product. Make a drawing of each tile model and show its dimensions.

1. $3x(x + 2)$ **2.** $(x + 2)(x + 3)$ **3.** $(2x + 1)(2x + 3)$

4. Choose your own product to model. Make a drawing of your tile model and show its dimensions.

Not all products can be modeled conveniently with algebra tiles. You can use the distributive property to find many products, as shown in Example 2.

EXAMPLE 2

Find each product.

a. $x(x - 5)$
b. $(4 - \sqrt{3})(2 + \sqrt{3})$

SOLUTION

a. $x(x - 5) = x \cdot x + x(-5)$ Use the distributive property.

$= x^2 - 5x$

b. Use the distributive property twice.

$(4 - \sqrt{3})(2 + \sqrt{3}) = (4 - \sqrt{3})2 + (4 - \sqrt{3})\sqrt{3}$

$= (4 - \sqrt{3})2 + (4 - \sqrt{3})\sqrt{3}$

$= 8 - 2\sqrt{3} + 4\sqrt{3} - (\sqrt{3})(\sqrt{3})$

$= 8 + 2\sqrt{3} - 3$

$= 5 + 2\sqrt{3}$

Notice that $(\sqrt{3})(\sqrt{3}) = \sqrt{9} = 3$. If $a \geq 0$, $\sqrt{a} \cdot \sqrt{a} = \sqrt{a^2} = a$.

You can name a variable expression by the number of terms in the expression. Each term can be a constant, a variable, or the product of a constant and one or more variables.

A **monomial** is an expression with *one* term.

-2 x $6x^2$ $-3xy$

A **binomial** is an expression with *two* unlike terms.

$3x + 1$ $2x^2 - 6x$ $x + y$ $y - y^2$

A **trinomial** is an expression with *three* unlike terms.

$x^2 + 5x - 6$ $3x + 2y + 1$ $4a - 3a^2 - 2$ $x - xy + y$

THINK AND COMMUNICATE

1. When do you think it is easier to model a product using algebra tiles? When do you think it is easier to use the distributive property?

2. What are some other words with the prefixes *mono-*, *bi-*, and *tri-*? Talk to someone who knows another language. Are these prefixes used in other languages?

☑ CHECKING KEY CONCEPTS

For Exercises 1–4, write an equation for each tile model.

1.

2.

3.

4.

Use algebra tiles to find each product. Make a drawing of your model.

5. $x(5x)$ **6.** $3(x + 2)$ **7.** $(x + 1)(x + 1)$

8. Look back at the Exploration on page 259. Use the distributive property to find each of the products you found using algebra tiles.

Tell whether each expression is a *monomial*, a *binomial*, a *trinomial*, or *none of these*.

9. $3p + 4$ **10.** $x^2 + xy + y^2$ **11.** 0 **12.** $a + b + c + d$

6.4 | Exercises and Applications

For Exercises 1–4, write an equation for each tile model.

Extra Practice exercises on page 569

1.

2.

3.

4.

Use algebra tiles to find each product. Make a drawing of your model.

5. $x(3x)$ **6.** $2(4x)$ **7.** $3(x + 4)$

8. $3x(x + 3)$ **9.** $(x + 1)(2x + 5)$ **10.** $(2x + 1)(2x + 1)$

Find each product.

11. $4(x + 1)$ **12.** $3(5 - 2x)$ **13.** $4x(x + 3)$

14. $(x + 2)(x + 4)$ **15.** $(x + 4)(2x - 1)$ **16.** $(5x + 1)(3x + 2)$

17. $3(5 + \sqrt{2})$ **18.** $(6 - \sqrt{5})(4 - \sqrt{3})$ **19.** $(3\sqrt{3} + 2)(\sqrt{2} + 1)$

20. URBAN PLANNING A town council is considering a proposal to create a rectangular fountain as part of a new park. The fountain will be surrounded by a concrete walkway that is the same width on each side.

a. Write an expression for the area of the fountain.

b. Write a variable expression for the combined area of the fountain and walkway. Simplify the expression.

c. Use your answers to parts (a) and (b) to write a variable expression for the area of the walkway.

d. **Open-ended Problem** The walkway must be at least 5 ft wide to allow two wheelchairs to pass. Compare the area of a 5 ft wide walkway and a 6 ft wide walkway. What are some reasons for choosing one width over the other?

Connection   HISTORY

To prove a theorem means to show that it is always true. Historians believe that around 275 A.D., the commentator Zhao Shuang added the diagram at the right to the *Zhōu bì suàn jīng* (100 B.C. – 100 A.D.). This diagram can be used to prove the Pythagorean theorem.

21. a. The length of each side of the large square can be represented as $a + b$. Write a variable expression for the area of the large square. Simplify the expression.

b. The large square can be divided into four identical right triangles shown in blue and a smaller red square. Write a variable expression for the area of the large square as the sum of the areas of the blue triangles and the red square.

22. Write an equation that states that the variable expressions you wrote in parts (a) and (b) of Exercise 21 are equal. Simplify both sides of the equation. Explain how your results prove the Pythagorean theorem.

23. Writing A student tried to find the product $2x(x + 5)$, as shown. Explain what is wrong with the student's work. Find the correct product.

$$2x(x + 5) = 2(x + 5) + x(x + 5)$$
$$= 2x + 10 + x^2 + 5x$$
$$= x^2 + 7x + 10$$

24. Challenge Suppose you own a business and are looking for new office space to rent. Each of the 100 employees will get a cubicle that is 10 ft by 10 ft. The aisles between the cubicles need to be at least 3.5 ft wide. Draw a sketch like the one shown and write an expression for the area of the cubicles and aisles. What minimum amount of office space should you look for to arrange the 100 cubicles? Explain your reasoning.

One way to arrange office cubicles is in groups of four with aisles around the groups. A square arrangement of cubicle groups will need the least amount of space.

BY THE WAY

Office aisle space requirements vary from city to city. Most cities require that each aisle be at least 3.5 ft wide and that an aisle be left between most windows and cubicles.

ONGOING ASSESSMENT

25. Cooperative Learning Work in a group of 3–4 students. Make a list of examples of the product of a monomial and a binomial and of the product of two binomials. Discuss how you would find each product. Would you use algebra tiles or the distributive property? Why?

SPIRAL REVIEW

Exercises 26–28 refer to a six-sided die. *(Section 5.5)*

26. What is the theoretical probability of rolling a number less than 3?

27. What is the theoretical probability of rolling a number greater than 3?

28. Add your values from Exercises 26 and 27. Is the sum equal to 1? Explain why or why not.

Use an equation to model each set of data. *(Section 3.6)*

29.

30.

6.5

Finding Special Products

The previous section showed how to multiply binomals. Sometimes it is helpful to know about special cases, such as when a binomial is multiplied by itself.

Learn how to...

- recognize some special products of binomials

So you can...

- find products of some variable expressions more quickly

The square of a binomial follows a special pattern. You may see a pattern in the diagrams below that will help you remember the product.

$$(x + 1)^2 = x^2 + 2x + 1$$

Read this as "the square of the quantity x plus 1," or as "x plus 1, quantity squared."

$$(x + 2)^2 = x^2 + 4x + 4$$

THINK AND COMMUNICATE

1. Use the diagram below to find the product $(x + 3)^2$.

2. What pattern do you see in the second term of all three products shown on this page?

3. What pattern do you see in the third term?

4. Use the patterns you saw in Questions 2 and 3 to find the product $(x + 4)^2$. Draw a diagram or use the distributive property to check your answer.

5. Find the product $(x + 10)^2$ using only mental math.

Squaring a Binomial

The diagrams on page 264 show you a pattern that can help you find the square of a binomial. Remembering this pattern can help you square binomials quickly. For any numbers a and b:

$$(a + b)^2 = a^2 + 2ab + b^2$$
$$(a - b)^2 = a^2 - 2ab + b^2$$

The square of a binomial is called a **perfect square trinomial**.

EXAMPLE 1

Find each product.

a. $(x + 9)^2$

b. $(x - 15)^2$

SOLUTION

a. $(x + 9)^2 = x^2 + 2(9)x + 9^2$

$$= x^2 + 18x + 81$$

b. $(x - 15)^2 = x^2 - 2(15)x + 15^2$

$$= x^2 - 30x + 225$$

Check

Use the distributive property.

$(x + 9)(x + 9) = (x + 9)x + (x + 9)9$

$$= x^2 + 9x + 9x + 81$$

$$= x^2 + 18x + 81 ✔$$

EXAMPLE 2

Find each product.

a. $\left(3 - \sqrt{5}\right)^2$

b. $(2t + 3)^2$

SOLUTION

a. $\left(3 - \sqrt{5}\right)^2 = 3^2 - 2(3)\left(\sqrt{5}\right) + \left(\sqrt{5}\right)^2$

$$= 9 - 6\sqrt{5} + 5$$
$$= 14 - 6\sqrt{5}$$

b. $(2t + 3)^2 = (2t)^2 + 2(2t)(3) + 3^2$

$$= 4t^2 + 12t + 9$$

The Difference of Two Squares

Joelle Rozene designs office spaces. In estimating the cost of doing a job, she often uses mental math for multiplying numbers. For example, to find the area of a room that is 42 ft by 38 ft, she does this calculation in her head:

$$42 \cdot 38 = (40 + 2)(40 - 2) = 40^2 - 2^2$$

Joelle finds the product of the sum of the numbers 40 and 2 and the difference of the numbers 40 and 2.

The product of the sum and difference of two terms is called the **difference of two squares**.

$$(a + b)(a - b) = a^2 - b^2$$

EXAMPLE 3

Find the product $(x + 3)(x - 3)$.

SOLUTION

Notice that you are finding the product of the sum of two terms (x and 3) and the difference of the same two terms.

$$(a + b)(a - b) = a^2 - b^2$$
$$(x + 3)(x - 3) = x^2 - 3^2$$
$$= x^2 - 9$$

THINK AND COMMUNICATE

6. Use the distributive property to show that the solution found in Example 3 is correct. Explain why the pattern $(a + b)(a - b) = a^2 - b^2$ always works. (*Hint:* What happens to two of the terms when you use the distributive property?)

7. Explain how to use the difference of two squares to find the product $14 \cdot 26$.

8. Explain how you could find 42^2 by using mental math and the pattern for squaring binomials.

9. Why is $\left(2 + \sqrt{3}\right)\left(2 - \sqrt{3}\right)$ the product of the sum and difference of two terms? Use mental math to find the product.

Find each product.

1. $(x + 8)^2$ **2.** $(x - 5)^2$ **3.** $(x + 9)(x - 9)$

4. $\left(3 + \sqrt{5}\right)^2$ **5.** $\left(7 - \sqrt{3}\right)^2$ **6.** $\left(2 + \sqrt{6}\right)\left(2 - \sqrt{6}\right)$

7. Find 54^2 by writing it as the square of a binomial.

8. Look back at page 266. Use Joelle Rozene's method to find the product $53 \cdot 47$.

6.5 Exercises and Applications

Find each product.

Extra Practice exercises on page 570

1. $(w - 2)^2$ **2.** $(x + 5)^2$ **3.** $(x - 3)^2$

4. $(x + 8)(x - 8)$ **5.** $(y - 4)(y + 4)$ **6.** $(c + 0.2)(c - 0.2)$

7. $(2r + 6)^2$ **8.** $(4x - 2)^2$ **9.** $(10t + 6)(10t - 6)$

10. Investigation Repeat the following steps several times using different groups of integers:

> **Step 1** Write down three consecutive integers, such as 3, 4, and 5.
> **Step 2** Square the middle number.
> **Step 3** Multiply the first and third numbers together.

a. Describe the relationship between the results of Steps 2 and 3.

b. Let n = an integer. Then $n - 1$, n, and $n + 1$ are three consecutive integers. Use these variable expressions to do Steps 2 and 3. How do your results explain the relationship you saw in part (a)?

c. Open-ended Problem Design and carry out an investigation of other groups of integers, such as four consecutive integers or three consecutive even integers. Can you identify any patterns? Try to find an algebraic explanation for your observations.

11. CONSTRUCTION A contractor needs to estimate the cost of building a stone path around a square garden, as shown. Explain why the cost of building the path depends on the width of the path. Use a variable expression to support your answer.

20 ft

20 ft

Find each product using mental math.

12. $18 \cdot 22$

13. 39^2

14. $25 \cdot 35$

15. 2.1^2

16. $\left(6\frac{1}{2}\right)^2$

17. $\left(9\frac{1}{2}\right)\left(10\frac{1}{2}\right)$

Find each product.

18. $(x + y)^2$

19. $(x - y)^2$

20. $(3x - 5)^2$

21. $(x + y)(x - y)$

22. $(3x + 6y)(3x - 6y)$

23. $(4x - 7)(4x + 7)$

24. $\left(1 - \sqrt{6}\right)^2$

25. $\left(4 + 2\sqrt{3}\right)^2$

26. $\left(6 + \sqrt{13}\right)^2$

27. SAT/ACT Preview For which value of x does a rectangle with side lengths $x + 4$ and $x - 4$ have an area of 84?

   A. 5   **B.** 10   **C.** 15   **D.** 20   **E.** 25

28. Challenge Rewrite each polynomial as the product of two binomials.

   a. $y^2 + 6y + 9$ **b.** $25x^2 - 10x + 1$ **c.** $36 - y^2$

Connection   GENETICS

Chicken breeders use *genetics* to breed chickens for desired characteristics. For example, the shape of a chicken's comb is determined by a combination of two genes, one inherited from each of its parents. Let G_R represent the gene for a "rose comb." Let G_S represent the gene for a "single comb." For Exercises 29–31, suppose a hen and rooster each have a mixed pair of comb-shape genes.

This geneticist is inspecting chicken genes that have been separated so that they can be identified.

29. The table at the right is called an *inheritance table*. The table is used to show possible gene combinations for the offspring of the hen and rooster. Copy and complete the table to find which gene combinations are possible.

single-comb   rose-comb

		Rooster	
		G_R	G_S
Hen	G_R	$G_R G_R$	?
	G_S	?	?

30. Suppose the chick has an equal chance of inheriting either G_R or G_S from each of its parents. The table shows that the probability of the chick inheriting a $G_R G_R$ pair is 0.25. What is the probability of inheriting a $G_S G_S$ pair? a mixed pair?

31. Writing Square the binomial $0.5G_R + 0.5G_S$. Explain how the product is related to your answer to Exercise 30.

32. Writing A magician asks a member of the audience for a number less than 100. The magician then gives a second number. The audience is amazed that the magician can find the product of the two numbers mentally before the audience member can find it with a calculator. Use what you have learned in this section to write instructions for performing this trick.

Find an equation of the line through the given points. *(Section 3.5)*

33. $(1, 4)$ and $(3, 12)$ **34.** $(0, 3)$ and $(-2, 7)$ **35.** $(4, -2)$ and $(-8, -8)$

Find an equation of the line with the given slope and through the given point. *(Section 3.5)*

36. slope $= 3$; $(0, 5)$ **37.** slope $= \dfrac{2}{3}$; $(3, 6)$ **38.** slope $= -1.4$; $(-5, 4)$

Find each product. *(Section 6.4)*

39. $3m(2m - 5)$ **40.** $(x + 1)(2x + 3)$ **41.** $(5x - 1)(1 - 5x)$

42. $(y - 1)(y + 3)$ **43.** $5s(3s - 4)$ **44.** $(3x + 5)(5x - 6)$

ASSESS YOUR PROGRESS

VOCABULARY

monomial (p. 260)   **perfect square trinomial** (p. 265)
binomial (p. 260)   **difference of two squares** (p. 266)
trinomial (p. 260)

Find each product. *(Sections 6.4 and 6.5)*

1. $5x(x + 3)$ **2.** $5(2 + \sqrt{3})$ **3.** $(x - 5)(x + 5)$

4. $(6 + \sqrt{10})^2$ **5.** $3x(2x + 3)$ **6.** $(x - 4)^2$

7. $5x(3x)$ **8.** $(2x + 1)(x + 5)$ **9.** $(x + \sqrt{8})^2$

10. Write a variable expression for the area of a square with sides of length $(n + 8)$ units. Write your expression as a trinomial. *(Section 6.5)*

11. Write variable expressions for the perimeter and the area of the rectangle shown.
(Section 6.5)

$x + 7$

$x - 7$

12. Journal How do you keep track of terms when you use the distributive property to multiply binomials? Do you think your method works well?

Investigating Pendulum Length

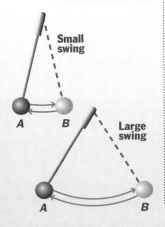

Small swing

A   B

Large swing

A   B

The pendulum of a clock swings back and forth, causing the clock to tick off the seconds. A pendulum clock runs down over time. With each swing, the pendulum travels a slightly smaller distance.

Even as the clock starts to run down, it can still keep good time. This is because the **period** of a pendulum, or the amount of time it takes the pendulum to swing from point *A* to point *B* and back to point *A*, is the same for large swings and small swings.

PROJECT GOAL Investigate how the length of a pendulum affects the period.

Conducting an Experiment

1. MAKE your pendulum by taping several coins to a piece of string. Use a pencil to hold the pendulum away from the table.

Work with a partner to design and carry out an experiment in which you measure the period of a pendulum for at least 10 different lengths of string. Here are some hints for organizing your work:

2. DETERMINE how many seconds it takes your pendulum to make 10 full swings. Divide by 10 to find the period.

3. ORGANIZE your data in a table. Then make a scatter plot. Put string length on the horizontal axis and period on the vertical axis.

Use a range of string lengths between 10 cm and 30 cm.

Based on your observations, find a string length that produces a period of one second. Can you keep time with this pendulum?

Using a Formula

The Italian scientist Galileo Galilei (1564 – 1642) showed that the period of a pendulum is proportional to the square root of the length of the pendulum. You can use a formula to describe how the period **P** of a pendulum is related to its length **L**.

$$P = 0.2\sqrt{L} \qquad \textbf{\textit{P} is in seconds and \textit{L} is in centimeters.}$$

• Calculate **P** for the **L**-values you used in your experiment. (Make sure you express the **L**-values in centimeters.) Compare your calculated **P**-values to your measured values. What could cause differences?

• Graph the equation **P** = 0.2√**L** on a graphing calculator. How does the graph compare with your scatter plot?

• Use the formula to find **L** when the period is one second. Compare this **L**-value to the length you found from experimenting.

Writing A Report

Write a report about your experiment. Describe your procedure and tell what conclusions you made about the relationship between the length of a pendulum and its period. Include pictures, graphs, and tables to support your ideas.

You may want to extend your report to explore some of the ideas below:

• The first pendulum clock was invented in 1657. How does a pendulum clock work?

• Pendulum clocks have a weight that can be raised and lowered. Why would this come in handy?

• How does the weight of a pendulum affect the period? Design and carry out an experiment to find out.

Self-Assessment

In your report, be sure to describe any difficulties you had. Tell what went well in your experiment and why. Is there anything that you would like to remember for a future experiment, such as a useful way to organize data, or a way to use your data to set up a good viewing window on a graphing calculator?

6 | **Review**

Find the exercises in each lesson that you had the most difficulty answering. Do these exercises again without looking at your previous work. Check your new work against your previous work.

VOCABULARY

right triangle (p. 239)
Pythagorean theorem (p. 239)
hypotenuse (p. 239)
leg (p. 239)
perfect square (p. 240)
square root (p. 240)
radical sign (p. 240)
converse of the Pythagorean theorem (p. 241)
rational number (p. 246)

irrational number (p. 246)
real number (p. 246)
dense (p. 246)
radical form (p. 252)
monomial (p. 260)
binomial (p. 260)
trinomial (p. 260)
perfect square trinomial (p. 265)
difference of two squares (p. 266)

SECTION | 6.1

A **right triangle** has one 90° angle.

Pythagorean theorem: In a right triangle, the square of the length of the hypotenuse is equal to the sum of the squares of the lengths of the legs.

$$a^2 + b^2 = c^2$$

Converse of the Pythagorean theorem: If the lengths of the sides of a triangle satisfy the relationship $a^2 + b^2 = c^2$, then it is a right triangle.

A **perfect square** is a number whose **square root** is an integer. A positive number has two square roots. A negative number has no square roots.

$$\sqrt{25} = 5 \qquad -\sqrt{25} = -5$$

SECTIONS 6.2 *and* 6.3

Every **real number** is either a rational number or an irrational number.

A **rational number** can be written as the ratio of two integers. When a rational number is written as a decimal it either repeats or terminates.

An **irrational number** cannot be written as the ratio of two integers. When an irrational number is written as a decimal it never repeats or terminates.

Properties of Square Roots

For nonnegative numbers *a* and *b*:

$$\sqrt{a^2} = a$$
$$\sqrt{ab} = \sqrt{a} \cdot \sqrt{b}$$
$$\sqrt{\frac{a}{b}} = \frac{\sqrt{a}}{\sqrt{b}}, b \neq 0$$

For example:

$$\sqrt{4^2} = 4$$
$$\sqrt{4 \cdot 6} = \sqrt{4} \cdot \sqrt{6} = 2\sqrt{6}$$
$$\sqrt{\frac{64}{49}} = \frac{\sqrt{64}}{\sqrt{49}} = \frac{8}{7}$$

SECTIONS 6.4 *and* 6.5

A variable expression can be named by the number of terms in the expression.

monomial	binomial	trinomial
$2x$	$2r + 3$	$3xy - 5y + 7$
1 term	2 terms	3 terms

You can use the distibutive property to multiply binomials. For example, find the product $(x + 26)(2x - 5)$.

$$(x + 26)(2x - 5) = (x + 26)2x - (x + 26)5$$
$$= 2x^2 + 52x - 5x - 130$$
$$= 2x^2 + 47x - 130$$

The product $(2x + 8)^2$ is a **perfect square trinomial**.

$$(2x + 8)^2 = (2x)^2 + 2(2x)(8) + 8^2$$
$$= 4x^2 + 32x + 64$$

The product $\left(5 + \sqrt{7}\right)\left(5 - \sqrt{7}\right)$ is the **difference of two squares**.

$$\left(5 + \sqrt{7}\right)\left(5 - \sqrt{7}\right) = 5^2 - \left(\sqrt{7}\right)^2$$
$$= 25 - 7$$
$$= 18$$

Assessment

VOCABULARY REVIEW

1. Complete each sentence. In triangle ABC, $\angle C$ measures ? degrees. Side $\overline{AB}$ is called the ? and sides $\overline{AC}$ and $\overline{BC}$ are called ? . Describe what the Pythagorean theorem says about the side lengths of $\triangle ABC$.

2. What is a *perfect square*? Give an example.

3. Define *rational number*. Give three examples of rational numbers between 1 and 3. Define *irrational number*. Give three examples of irrational numbers between 1 and 3.

SECTIONS 6.1, 6.2, *and* 6.3

For Questions 4–6, find the unknown length for each right triangle.

4.

5.

6.

For Questions 7 and 8, tell whether the given lengths can be the sides of a right triangle.

7. 5 ft, 3.75 ft, 6.25 ft

8. 13 cm, 14 cm, 15 cm

9. **ART** Rosalie is making a frame for a drawing with the dimensions shown. To check whether the corners of her frame are right angles, she measures the lengths of both diagonals. If the corners are right angles, how long are the diagonals?

8.25 in.

11 in.

Write each decimal as a fraction in lowest terms.

10. 0.684

11. $0.777\ldots$

12. $0.\overline{95}$

Write each expression in simplest form.

13. $\sqrt{48}$

14. $3 + \sqrt{200}$

15. $\sqrt{50} + \sqrt{18}$

16. $\sqrt{\dfrac{36}{100}}$

17. $2\sqrt{3} \cdot \sqrt{240}$

18. $\sqrt{\dfrac{192}{121}}$

Estimate each number within a range of two integers.

19. $\sqrt{54}$ **20.** $\sqrt{107}$ **21.** $\sqrt{72}$

22. Find the perimeter and area of the rectangle. Find the length of a diagonal of the rectangle. Write each dimension in simplest form.

$3\sqrt{5}$

$5\sqrt{5}$

SECTIONS 6.4 *and* 6.5

Find each product.

23. $7(2x + 5)$ **24.** $3y(y - 9)$

25. $(2m - 7)(3m + 5)$ **26.** $\sqrt{3}(8 + \sqrt{3})$

27. $(5 + 2\sqrt{3})(5 - 2\sqrt{3})$ **28.** $(3 + 2\sqrt{20})(5 + 3\sqrt{45})$

29. a. Write a trinomial that represents the area of the rectangle.

$x - 5$

$2x - 6$

     b. What is the smallest integer value of x that will give positive values for the length and the width of the rectangle?

Find each product.

30. $(n - 8)^2$ **31.** $(2a + 3)^2$ **32.** $(7 + 2\sqrt{3})^2$

33. $(c - 9)(c + 9)$ **34.** $\left(6x - \dfrac{1}{5}\right)\left(6x + \dfrac{1}{5}\right)$ **35.** $(5 + \sqrt{5})(5 - \sqrt{5})$

Show how to use special products to find each product.

36. $(5.5)(4.5)$ **37.** $(23)(37)$ **38.** $(28)^2$

39. Writing Explain the error in the product below. Show how to find the correct product.

$$(w - 7)^2 = w^2 + 14w + 49$$

PERFORMANCE TASK

40. Use the Pythagorean theorem to approximate the distance between two places that cannot be connected directly by a straight line. Explain how you approximated the distance. Include a sketch. An example is shown.

Cumulative Assessment
CHAPTERS $4-6$

CHAPTER 4

Solve. If an equation is an *identity* or there is *no solution*, say so.

1. $-\dfrac{2}{5}z = 8$

2. $\dfrac{2}{3}c - \dfrac{1}{2}c = 1$

3. $7x = 15 - 3x$

4. $1.3t - 0.7 = 2.7t$

5. $6(2u - 3) = -8 + 8u$

6. $3(n + 4) - n = 2(n + 5)$

Solve each inequality. Graph each solution on a number line.

7. $5 - x > 1$

8. $3m - 6 \geq 3$

9. $8p + 5 \geq 23p$

PERSONAL FINANCE One car rental company charges customers **$225.00 per week with unlimited mileage to rent a mid-sized car. Another company charges $200.00 per week plus $.25 per mile. A person needs to rent a car for one week.**

10. Model this situation with a table and with a graph.

11. Let m = the number of miles. For what number of miles is it less expensive to rent from the company with unlimited mileage? Write your answer as an inequality.

12. Write and solve an equation to find when the costs are equal.

13. Writing Do you find a table, a graph, or an equation most helpful when you compare car-renting options? Explain.

14. Open-ended Problem What factors might you consider besides cost when deciding which company to rent from?

CHAPTER 5

For Questions 15–17, solve each proportion.

15. $13 : 65 = 3 : s$

16. $\dfrac{7}{8.4} = \dfrac{r}{1.8}$

17. $\dfrac{15}{p} = \dfrac{20}{p + 3}$

18. The blueprint of a 60 ft by 75 ft building has a scale of 1 in. = 8 ft. What are the dimensions of the blueprint?

19. An $8\dfrac{1}{2}$ in. by 11 in. drawing is reduced to $6\dfrac{3}{8}$ in. by $8\dfrac{1}{4}$ in. What scale factor was used to reduce the drawing?

For Questions 20–24, use the fact that △*ABC* ~ △*XYZ*. For Questions 20–22, complete each statement.

20. $AC = \underline{\ ?\ }$

21. $YZ = \underline{\ ?\ }$

22. $\angle A = \angle \underline{\ ?\ }$

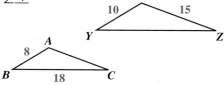

23. Find the ratio of the perimeter of △*ABC* to the perimeter of △*XYZ*.

24. Find the ratio of the area of △*ABC* to the area of △*XYZ*.

25. Writing Is every rectangle similar to every other rectangle? Explain.

26. What is the probability that someone born in September was born on or before September 12?

27. Open-ended Problem Give an example of experimental probability.

28. A dart lands on a square of a checkerboard at random. What is the probability that the dart lands on one of the four corner squares?

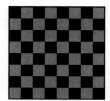

CHAPTER 6

Tell whether the given lengths can be sides of a right triangle. Explain your reasoning.

29. 15, 36, 39

30. 160, 340, 300

31. 8.2, 9.1, 11.9

32. Open-ended Problem Use a right triangle with lengths 3, 4, and 5.

    a. Multiply the length of each side of the triangle by 3. Is the new shape a right triangle?

    b. Add 3 to the length of each side. Is the new shape a right triangle?

    c. Choose a different right triangle. Repeat parts (a) and (b). Compare your answers for both triangles. Do you think your results would be the same for any right triangle you chose?

Write each repeating decimal as a fraction.

33. $0.030303\ldots$

34. $17.\overline{04}$

35. $0.\overline{961}$

Simplify each expression.

36. $\sqrt{63}$

37. $7\sqrt{2} + 12\sqrt{2}$

38. $\sqrt{\dfrac{144}{9}}$

39. $4\sqrt{5} \cdot \sqrt{15}$

40. $\sqrt{\dfrac{6}{49}}$

41. $10\sqrt{7} - \sqrt{28}$

Find each product.

42. $3q(2q - 5)$

43. $\left(5\sqrt{3} - 1\right)\left(6\sqrt{3} + 4\right)$

44. $\left(5 - 2\sqrt{3}\right)^2$

45. $\left(6x + \sqrt{5}\right)\left(6x - \sqrt{5}\right)$

46. Writing Explain the error in the product below. Show how to find the correct product.

$$(m + 4)(m - 4) = m^2 + 16$$

7 Systems of Equations and Inequalities

Feeding a need for athletes

Nancy Clark

Just about everyone knows that being a top athlete takes talent and hard work. But fewer people appreciate the importance of diet in helping athletes perform at their best. Sports nutritionist Nancy Clark, M.S., R.D., is trying hard to change this situation. "Many people train hard but neglect their training diet; this is a big mistake," Clark says. "They're shooting themselves in the foot." Clark meets with athletes of all ages, sizes, and shapes — from high school and college students to Olympic athletes and members of the Boston Celtics and Red Sox. Though each of her clients has different needs, Clark is guided by one general principle: finding the right balance of training, nutrition, and rest.

"You can't compete at your best if you don't train at your best."

Finding the Right Balance

For example, some of the high school students Clark works with tend to skip both breakfast and lunch. When they show up for after-school sports, they are running on empty. "They come to me looking for a quick energy fix, but no amount of fix can make up for a poor breakfast and lunch," she says. "Together we figure out some smart eating choices—a combination of foods that they like to eat which will provide them with energy they need throughout the day."

Fueling Up

"Food is really just fuel, not so different from the gasoline you put in a car," she says. "You need to fill up before strenuous exercise and refuel afterwards." Food provides energy from three sources. Proteins build muscles but get used for energy in emergency situations, carbohydrates fuel muscles, and fats contain energy reserves. Of the three,

carbohydrates are the most critical for athletes because they provide the body with glycogen, a form of sugar that is stored within muscles as a source of energy. Foods that are high in carbohydrates include bread, pasta, potatoes, and rice. Athletes should load up on carbohydrates while preparing for and recovering from demanding events such as triathlons, marathons, or long-distance ski races. Between 60 and 70 percent of the calories they consume daily should come from carbohydrates, according to Clark. She stresses that these carbohydrate guidelines make sense for everybody, not just athletes. "It's what we call a healthy diet," she says.

Nancy Clark serving food at the annual pasta fest on the eve of the Boston Marathon.

" You can't train at your best if you don't eat right on a daily basis. "

How Much is Enough?

The number of calories that a particular athlete should consume depends on many factors including age, height, and activity level. The graph shown below can help an athlete figure out how many calories from carbohydrates to include in a daily diet. Once the athlete's daily calorie needs are known, the number of calories from carbohydrates should be within the shaded region of the graph. As you will learn in this chapter, the graph represents a system of inequalities.

Nutrition labeling can help create daily meals with appropriate amounts of proteins, carbohydrates, and fats.

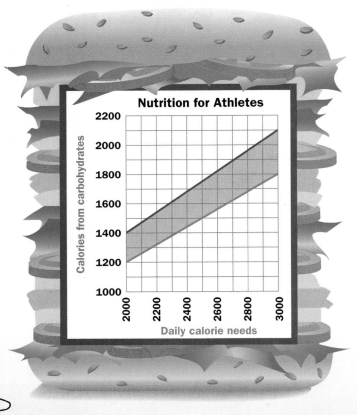

Nutrition for Athletes

Calories from carbohydrates (y-axis): 1000, 1200, 1400, 1600, 1800, 2000, 2200

Daily calorie needs (x-axis): 2000, 2200, 2400, 2600, 2800, 3000

foods
- proteins
  - build muscles
- carbohydrates
  - provides fuel
    - bread
    - pasta
    - potatoes
    - rice
- fats

EXPLORE AND CONNECT

Nancy Clark gathering data, to better advise on nutrition issues.

1. Writing According to the graph, if an athlete needs to consume 2200 Cal each day, about how many should come from carbohydrates? If another athlete needs to consume 1700 Cal from carbohydrates each day, about how many total calories does that athlete need?

2. Research In physics, a calorie is defined as a unit of heat. What is the relationship between calories used in physics and food calories? Find out how the numbers of calories in foods are determined.

3. Project The other 30–40% of an athlete's daily calorie intake should come from proteins and fats. What kinds of foods are high in proteins and fats? Create a poster showing how carbohydrates, fats, and proteins help the body to function.

Mathematics & Nancy Clark

In this chapter, you will learn more about how mathematics is related to sports nutrition.

Related Examples and Exercises

Section 7.1
- Example 1
- Exercises 34–36, 38

Section 7.5
- Exercises 29–31

Section 7.6
- Exercises 15–18

7.1

Using Linear Equations in Standard Form

Learn how to...

- write and graph equations in standard form

So you can...

- solve problems with two variables, such as nutritional planning or telecommunications problems

Nancy Clark recommends that runners replace their muscles' energy stores after a marathon. A runner should eat at least 0.5 g of foods containing carbohydrates for each pound of body weight within 2 h after the race. The athlete should eat at least this amount again 2 h later.

EXAMPLE 1 **Interview: Nancy Clark**

Leon should eat at least 70 g of carbohydrates after running a marathon. One cup of apple juice contains **30 g** of carbohydrates, and 1 oz of pretzels contains **23 g**. Write and graph an equation showing combinations of pretzels and juice that provide 70 g of carbohydrates.

SOLUTION

Step 1 Write an equation.
Let j = number of cups of juice. Let p = number of ounces of pretzels.

$$\begin{array}{ccc} \text{Carbohydrates} \\ \text{from juice} \end{array} + \begin{array}{c} \text{Carbohydrates} \\ \text{from pretzels} \end{array} = 70$$

$$30j \quad + \quad 23p \quad = 70$$

Step 2 Graph the equation.

First make a table of values.

Then graph each point and draw a line through the points.

j	p
0	3.0
1.0	1.7
2.3	0

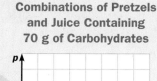

Combinations of Pretzels and Juice Containing 70 g of Carbohydrates

Juice (cups)

BY THE WAY

After the Boston Marathon, runners eat snacks such as pretzels and bread supplied by their sponsors. Ten small pretzels weigh about one ounce.

THINK AND COMMUNICATE

1. Why does the graph in Example 1 only make sense in the first quadrant?

2. How would the graph in Example 1 be different if you put cups of juice on the vertical axis?

As Example 1 shows, the graph of $30j + 23p = 70$ is a line. All linear equations can be written in the form **$ax + by = c$**, where a and b are both real numbers.

Linear equations written like this are in **standard form**, where a, b, and c are integers.

The **variable terms** are on one side of the equation.

There is an integer on the other side.

$$30j + 23p = 70$$

The coefficients are integers.

EXAMPLE 2

Graph the equation $2x + 5y = 1720$.

SOLUTION

Step 1 The graph is a line. Use the equation to find two points on the line. Choose two points that are easy to find.

Find y when $x = 0$.

$$2(0) + 5y = 1720$$
$$5y = 1720$$
$$y = 344$$

One point on the line is $(0, 344)$.

Find x when $y = 0$.

$$2x + 5(0) = 1720$$
$$2x = 1720$$
$$x = 860$$

Another point on the line is $(860, 0)$.

Step 2 Graph the two points and draw the line through them.

The **vertical intercept**, or **y-intercept**, is 344.

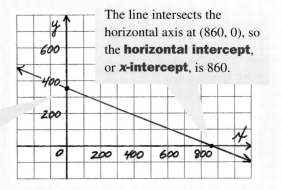

The line intersects the horizontal axis at $(860, 0)$, so the **horizontal intercept**, or **x-intercept**, is 860.

THINK AND COMMUNICATE

3. Explain how to find two points to graph when graphing an equation in standard form.

If you are given a linear equation in standard form, you can rewrite it in slope-intercept form. Many graphing calculators cannot graph equations in standard form. Rewriting the equation can help you to graph the equation on a graphing calculator.

EXAMPLE 3

Graph the equation $-5x + 2y = 4$ on a graphing calculator.

SOLUTION

First rewrite the equation in slope-intercept form.

Solve for y.

$$-5x + 2y = 4$$
$$-5x + 5x + 2y = 4 + 5x$$
$$2y = 4 + 5x$$
$$\frac{2y}{2} = \frac{4 + 5x}{2}$$
$$y = 2 + \frac{5}{2}x$$

Now you can graph the equation on your calculator.

$y = 2 + \frac{5}{2}x$

THINK AND COMMUNICATE

4. Example 3 shows an equation written in standard form and slope-intercept form. Describe an advantage and a disadvantage of writing an equation in slope-intercept form.

5. Explain how to rewrite $x + 2y = 4$ in slope-intercept form.

☑ CHECKING KEY CONCEPTS

1. Adult tickets to a museum cost $8 each and tickets for children cost $5 each.

    a. What is the cost of tickets for a adults and c children?

    b. Write an equation showing combinations of adults' tickets and children's tickets that cost $40.

Tell whether each equation is in standard form. If it is not, explain why not.

  2. $3x + 5y = 30$ **3.** $y = 2x - 5$ **4.** $8x - 9y = 36$

  5. Find the x-intercept and the y-intercept of the graph of $3x + 5y = 30$.

  6. Explain how to rewrite $3x + 5y = 30$ in slope-intercept form.

  7. Describe two ways to graph the equation $3x + 5y = 30$.

Exercises and Applications

Extra Practice
exercises on
page 570

SPORTS **Exercises 1–5 refer to a soccer league in which a team gets 2 points for each win and 1 point for each tie game. How many points does a team with each record have?**

1. 6 wins and 3 ties

2. 8 wins and 2 ties

3. no wins and 5 ties

4. w wins and t ties

5. a. Write an equation showing combinations of wins and ties that result in a total of 13 points.

   b. Graph your equation. What do the intercepts of your graph represent?

For Exercises 6–10, use the advertisement. How much does each combination of compact disks and tapes cost?

BARRYS
Music Store

Annual SALE !

All CDs $**15**·00

All Tapes $**9**·00

6. 3 compact disks and 2 tapes

7. 4 compact disks and 7 tapes

8. 5 compact disks and no tapes

9. c compact disks and t tapes

10. a. Write an equation that shows how many compact disks and tapes you can buy with exactly $60.

   b. Graph your equation. Should the graph be a line? Explain.

Graph each equation by using the method in Example 1 on page 281.

11. $3x + 2y = 24$ **12.** $4x + 3y = 18$ **13.** $8x - 3y = 12$

14. $-5x + 2y = 20$ **15.** $2x - 4y = -5$ **16.** $-3s + 2t = -1$

17. $1.5a - 0.5b = 3.0$ **18.** $\dfrac{x}{2} - \dfrac{y}{3} = 1$ **19.** $3x + 4y = 1.2$

 Technology **Rewrite each equation in slope-intercept form. Then graph it. Use a graphing calculator if you have one.**

20. $2x + y = 8$ **21.** $3x + 2y = 10$ **22.** $2x - 7y = 21$

23. $-4x + 5y = 12$ **24.** $2x + 2y = 7$ **25.** $8 - 3y = 4x$

26. $50 = 5x + 4y$ **27.** $0.3x + 0.2y = 1.2$ **28.** $\dfrac{2}{3}x - \dfrac{1}{2}y = 2$

29. Which equations in Exercises 11–28 are not in standard form? Explain your answers.

30. SAT/ACT Preview A line crosses the axes at (15, 0) and (0, 12). Which of the equations below could be an equation of the line?

   A. $y = 15 + x$ **B.** $y = 12 + x$ **C.** $x + y = 27$

   D. $y = 12 - \dfrac{4}{5}x$ **E.** $y = \dfrac{5}{4}x$

Cooperative Learning For Exercises 31–33, work with a partner. One of you should write an equation for the situation in Exercise 31. The other should write an equation for the situation in Exercise 32. Work together on Exercise 33.

31. The roller coaster at the state fair costs $1.50 to ride and the bumper cars cost $1.00. Ben has $10.00 left and decides to spend it all on x roller coaster rides and y bumper car rides.

32. At Macomber's Market, you can buy grapes for $1.50 per pound and mushrooms for $1.00 per pound. Suppose you spend $10.00 on x pounds of grapes and y pounds of mushrooms.

33. Writing Explain to your partner how your equation represents your situation. Which of the graphs describes the situation in Exercise 31? Which describes the situation in Exercise 32? Explain your answers.

A.

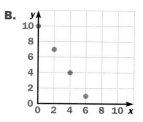
B.

INTERVIEW Nancy Clark

Look back at the article on pages 278–280.

In an article that appeared in the journal American Fitness, Nancy Clark discusses sources of carbohydrates.

NUTRITION

Counting Carbohydrates
Maintain the optimal athlete's diet to refuel active muscles.
By Nancy Clark, M.S., R.D.

34. a. How many grams of carbohydrates are in c cups of a wheat and barley cereal?

b. How many grams of carbohydrates are in m cups of milk?

reading the nutrition labels on packaged foods. For example, at breakfast, a label-reading athlete who needs 400 grams of carbohydrates per day can get 150 grams under the belt by eating one cup of a wheat and barley cereal (92 grams) with 1/4 cup raisins (25 grams), one cup milk (10 grams) plus one cup of orange juice (25 grams). This carbohydrate counting system is particularly educational and helpful

c. Write an equation showing combinations of cereal and milk that provide 100 g of carbohydrates.

d. List three possible solutions of your equation. Which solution describes the combination of cereal and milk that you like best?

Oatmeal Nutrition Information	
Serving size	$\frac{1}{2}$ cup dry (40 g)
Calories	150
Fat	3 g
Carbohydrates	27 g

35. How many cups of oatmeal and how many cups of raisins do you need to eat to get 100 g of carbohydrates? Write an equation and list three possible solutions.

36. Research What do you eat for breakfast? Look at nutrition labels and calculate the total amount of carbohydrates. If some foods are not labeled, you may want to look at a nutrition guidebook from the library.

37. **Open-ended Problem** Write an equation in standard form. Graph your equation and describe the method you used to graph it.

38. **FITNESS** The energy you expend while exercising can be measured in calories (Cal). Based on a table in *Nancy Clark's Sports Nutrition Guidebook,* John finds that he may use 6 Cal/min while doing calisthenics and 8.5 Cal/min while jogging. At football practice, John does calisthenics and jogs for a total of 30 min.

    a. Copy and complete the table.

    b. Write an equation relating x and y. Graph the equation.

    c. Give the x- and y-intercepts of the graph. What do they tell you about the situation?

Minutes of calisthenics (x)	Minutes of jogging (y)	Total calories used (c)
0	30	?
5	25	?
10	?	?
18	?	?
?	0	?

39. a. **Challenge** Graph the equations $3x - 2y = 12$, $3x - 2y = 6$, and $3x - 2y = 0$ in the same coordinate plane.

    b. Do any of these equations describe direct variation?

    c. How can you tell by looking at a linear equation in standard form whether it describes direct variation? Support your answer with examples.

ONGOING ASSESSMENT

40. **Open-ended Problem** Make up an equation in standard form that represents some situation in your life. Explain what your equation and the variables represent. Give at least three solutions of your equation.

SPIRAL REVIEW

Technology Graph each equation. Use a graphing calculator if you have one. Use the graph to find the missing value. *(Section 2.6)*

41. $y = 3.2x - 1.5$; find the value of x when $y = 0$.

42. $y = 10 + 1.75x$; find the value of x when $y = -1$.

43. $y = -3 - 4.7x$; find the value of x when $y = -2$.

44. $y = -0.1x - 0.6$; find the value of x when $y = 0.2$.

Solve each equation. *(Section 2.2)*

45. $4x - 3 = 7$	46. $2y - 10 = 1$	47. $1.5 + 2z = 2.5$
48. $9 = 14 - 3t$	49. $2r + 5 = \dfrac{3}{7}$	50. $4s - \dfrac{3}{2} = 1$

7.2

Solving Systems of Equations

Learn how to...
- **solve systems of equations**

So you can...
- **solve problems with two variables, such as planning flight training**

How would you like a career flying a plane? In addition to flying between cities, commercial pilots bring mail and food to remote towns and even stock lakes with fish.

An experienced private pilot must train for fifty hours to get a commercial license. Pilots can train in airplanes and on the ground in a flight simulator.

Plane Instruction = $98/h

Simulator = $46/h

EXAMPLE 1 Application: Aviation

Nancy Loesch is training for her commercial pilot's license. She can afford to spend only **$3500** on the required **50 hours** of instruction. Nancy would like to spend as much time as possible training in an airplane, but airplane instruction costs **$98** per hour. Training in the simulator costs only **$46** per hour.

Let a = number of hours spent training in an airplane.

Let s = number of hours spent training in a flight simulator.

Write two equations relating a and s.

SOLUTION

Hours spent training in an airplane	+	Hours spent training in a simulator	=	Hours that Nancy must train
a	+	s	=	50

Cost of a hours in an airplane	+	Cost of s hours in a simulator	=	Amount that Nancy can spend
$98a$	+	$46s$	=	3500

THINK AND COMMUNICATE

1. Give two solutions of the equation $a + s = 50$. Give two solutions of the equation $98a + 46s = 3500$. Do you think that there is an ordered pair that is a solution of both equations?

The two equations in Example 1 are a **system of linear equations**. Any pair of numbers that satisfies both equations is a **solution** of the system. If you graph the equations together, you can see that they have one common solution.

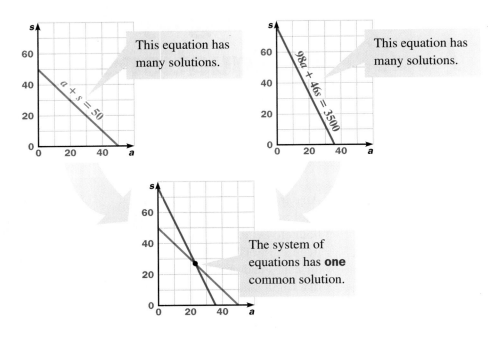

This equation has many solutions.

This equation has many solutions.

The system of equations has **one** common solution.

EXAMPLE 2   Method: Graphing

Use a graphing calculator to solve the system of equations.

$$a + s = 50$$
$$98a + 46s = 3500$$

SOLUTION

Step 1 Rewrite each equation using x and y as variables.

$$x + y = 50 \qquad 98x + 46y = 3500$$

Step 2 Solve each equation for y so you can enter it into the calculator.

$$y = 50 - x \qquad 46y = 3500 - 98x$$
$$y = \frac{3500 - 98x}{46}$$

Step 3 Graph each equation and find the point where the graphs intersect.

Intersection
X=23.076923 Y=26.923077

Write your answer in terms of the original variables. The solution (a, s) is about $(23.1, 26.9)$.

Solving Systems by Substitution

You can substitute variable expressions into equations just as you substitute numbers. Substitution gives you another method for solving systems of equations.

For example, to solve the system

$$y = 2x$$
$$3x - y = 5$$

you can substitute **2x** for y in the **second equation**.

$$3x - 2x = 5$$
$$x = 5$$

Now that you have a value for x, you can find a value for y.

$$y = 2x$$
$$y = 2 \cdot 5$$
$$y = 10$$

Substitute 5 for x in the **first equation**.

The ordered pair (5, 10) is the solution of the system of equations.

EXAMPLE 3 Method: Substitution

Use substitution to solve the system of equations from Example 1. Explain what your solution tells you about the situation.

$$a + s = 50$$
$$98a + 46s = 3500$$

SOLUTION

Step 1 Solve the **first equation** for a.

$$a + s = 50$$
$$a = 50 - s$$

Step 2 Substitute $50 - s$ for a in the **second equation** and solve for s.

$$98a + 46s = 3500$$
$$98(50 - s) + 46s = 3500$$
$$4900 - 98s + 46s = 3500$$
$$1400 = 52s$$
$$s \approx 26.9$$

Step 3 To find a value for a, substitute 26.9 for s in either of the original equations.

$$a + s = 50$$
$$a + 26.9 \approx 50$$
$$a \approx 23.1$$

Check
Substitute 26.9 for s and 23.1 for a in each equation.

$$a + s = 50$$
$$23.1 + 26.9 \overset{?}{=} 50$$
$$50 = 50 ✔$$
$$98a + 46s = 3500$$
$$98(23.1) + 46(26.9) \overset{?}{=} 3500$$
$$2263.8 + 1237.4 \overset{?}{=} 3500$$
$$3501.2 \approx 3500 ✔$$

The solution (a, s) is about (23.1, 26.9). Nancy should spend about 23 h training in an airplane and about 27 h training in a flight simulator.

THINK AND COMMUNICATE

Tell what you would do first to solve each system by substitution.

2. $3x + 5y = 18$
$2x + y = 4$

3. $x + 3y = 15$
$5x - 6y = 19$

4. $x + y = 1$
$4x - 3y = 18$

5. In step 3 of Example 3, what happens if you substitute $s = 26.9$ into $98a + 46s = 3500$ instead of into $a + s = 50$?

6. Which method do you prefer for solving the system of equations in Example 1? Why?

☑ CHECKING KEY CONCEPTS

1. Use the substitution method to solve each system in Think and Communicate Questions 2–4.

2. Match each of the systems in Think and Communicate Questions 2–4 with one of the graphs below.

A.   **B.**   **C.**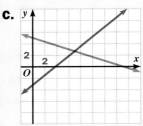

7.2 Exercises and Applications

Extra Practice exercises on page 570

 Technology For Exercises 1–4, solve each system of equations by graphing. Use a graphing calculator if you have one.

1. $2x - 3y = 24$
$2x + y = 8$

2. $3x - y = 9$
$x + 2y = 10$

3. $6x + 9y = -5$
$x + y = 1$

4. $7x - 14y = 2$
$y - \dfrac{3}{5}x = 0$

Use substitution to solve each system of equations.

5. $3x + 4y = 4$
$y = x - 6$

6. $2x + 7y = 3$
$x = 1 - 4y$

7. $x + 4y = 1$
$2x - 3y = -9$

8. $3x + 2y = 13$
$2x + y = 7$

9. $x - 3y = 8$
$4x + 5y = -2$

10. $2a + 3b = 12$
$a + b = 5$

11. $4c + 5d = 11$
$3c - d = 13$

12. $5r - s = 5$
$5s - 4r = 17$

13. $5x - 4y = -6$
$7x - y = 10$

14. **SAT/ACT Preview** Suppose a section of the test has 50 multiple-choice questions. Trevor earns one point for each right answer and loses $\frac{1}{4}$ point for each wrong answer. For example, if he gives 32 right answers, 8 wrong answers, and leaves 10 questions blank, his score is $32 - \frac{1}{4}(8) = 30$.

   a. Let r and w represent the number of questions Trevor got right and wrong. Write an equation stating that Trevor answered 41 questions.

   b. Write an equation stating that Trevor's score for answering 41 questions was $34\frac{3}{4}$.

   c. Solve your system of equations from parts (a) and (b).

15. a. **FINANCE** On March 1, 1995, one French franc was worth $.19 and one German mark was worth $.68. Write two equations expressing these facts.

   b. Write one equation relating the value of a German mark to a French franc. Explain how to use your equation to find the value in marks of 2000 francs.

16. **Technology** Use algebra or a graphing calculator to decide whether or not these three equations have a common solution.

   $$2x - y = -10$$
   $$-x + y = 7$$
   $$-3x - y = 6$$

17. **SPACE SCIENCE** Use the information in the article to write and solve a system of equations.

   Let $N = $ the heart's rate of pumping blood under normal conditions.

   Let $L = $ the heart's rate of pumping blood during launch.

   As a space shuttle or rocket is launched, the astronauts inside experience a great deal of stress. An astronaut's heart pumps twice as much blood per minute during rocket launch as under normal conditions—the heart pumps 10 pints more blood each minute during launch than when the astronaut is at rest on the ground.

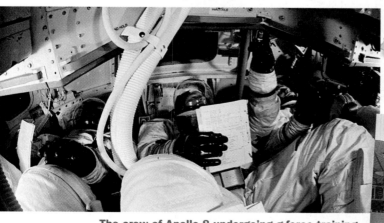

The crew of Apollo 8 undergoing g-force training. The Apollo program used a centrifuge, like the one shown at the right, in g-force training.

Many airlines provide meals or snacks on their flights. These meals are prepared in the huge kitchens of catering firms. The airlines must pay different amounts for the meals based on what they contain.

18. Mario Clementi is in charge of ordering meals for a flight on a jet plane. There will be 164 coach-class passengers and 24 first-class passengers. From past experience, Mario knows that first-class meals cost about twice as much as coach-class meals. How much can he spend on each type of meal if his budget is $1300? Write and solve a system of equations.

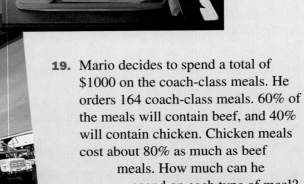

19. Mario decides to spend a total of $1000 on the coach-class meals. He orders 164 coach-class meals. 60% of the meals will contain beef, and 40% will contain chicken. Chicken meals cost about 80% as much as beef meals. How much can he spend on each type of meal? Write and solve a system of equations.

20. a. Challenge Verify that $(2, -4)$ is a solution of the first system of equations at the right.

$$3x - 5y = 26$$
$$6x + y = 8$$

b. Compare the two systems of equations. How are they alike? How are they different?

$$3\left(\frac{1}{x}\right) - 5\left(\frac{1}{y}\right) = 26$$

c. Use your answers to parts (a) and (b) to find a solution of the second system of equations. Explain your reasoning.

$$6\left(\frac{1}{x}\right) + \left(\frac{1}{y}\right) = 8$$

21. Open-ended Problem Find equations of at least three lines that intersect each other at the point $(6, 3)$. Explain why there are other possible answers.

22. a. **Open-ended Problem** Use the information in the advertisement to write a system of equations. Explain what each equation represents.

   b. Solve your system of equations. Explain what the solution represents.

23. **GEOMETRY** The Cosmoclock 21 in Yokohama City, Japan, is one of the largest Ferris wheels in the world. Its diameter is 328 ft. The 60 arms holding the gondolas serve as second hands for the clock.

   a. How many feet would you travel in one revolution of the Ferris wheel? *(Toolbox, page 593)*

   b. Find your speed in miles per hour if the wheel makes one revolution per minute. Use the fact that 1 mi = 5280 ft. *(Section 3.1)*

ASSESS YOUR PROGRESS

VOCABULARY

standard form of a linear equation (p. 282)

vertical intercept, y-intercept (p. 282)

horizontal intercept, x-intercept (p. 282)

system of linear equations (p. 288)

solution of a system (p. 288)

Give the x- and y-intercepts of each graph. Graph each equation. *(Section 7.1)*

1. $4x + 3y = -12$

2. $-3x + 8y = 12$

Rewrite each equation in slope-intercept form. Then graph the equation. *(Section 7.1)*

3. $5x - y = 5$

4. $2x + 3y = 9$

5. Shira gets four points for each correct answer on her biology quiz. She loses a point for each wrong answer. *(Section 7.2)*

   a. Write an equation showing that Shira earned 30 points on the quiz.

   b. Write an equation showing that Shira answered 20 questions.

   c. Solve the system of equations in parts (a) and (b) by graphing.

Use substitution to solve each system of equations. *(Section 7.2)*

6. $6x + 5y = 1$
    $y = 1 - 2x$

7. $2a + 7b = 24$
    $3a - b = 13$

8. Journal You can solve a system of equations by graphing or by using substitution. Describe some advantages and disadvantages of each method.

7.3 Solving Linear Systems by Adding or Subtracting

Learn how to...

- **solve systems of equations by adding or subtracting**

So you can...

- **solve problems in two variables, such as understanding a restaurant bill**

Many Chinese restaurants serve *dim sum*. Instead of ordering from a menu, you choose items from carts of appetizers and desserts that circulate through the restaurant. Some items are more expensive than others. To distinguish between them, some restaurants serve the more expensive items from larger plates.

The character 小 means *small* and 大 means *large*.

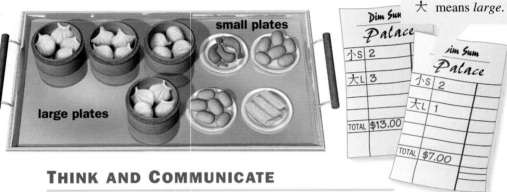

small plates

large plates

Dim Sum Palace

| 小 S | 2 |
| 大 L | 3 |

| TOTAL | $13.00 |

Dim Sum Palace

| 小 S | 2 |
| 大 L | 1 |

| TOTAL | $7.00 |

THINK AND COMMUNICATE

1. Can you figure out how much a large *dim sum* plate costs?

2. How much does a small *dim sum* plate cost? Explain your reasoning.

EXAMPLE 1 Method: Subtraction

Write and solve a system of equations to find the cost of a large *dim sum* plate.

SOLUTION

Let l = the cost of a large *dim sum* plate.

Let s = the cost of a small *dim sum* plate.

According to the bills above, $3l + 2s = \$13$ and $l + 2s = \$7$.

If you **subtract** each side of the second equation from the corresponding side of the first, you can solve for l.

$$
\begin{array}{rcr}
3l + 2s & = & 13 \\
- (l + 2s) & = & -(7) \\
\hline
2l & = & 6
\end{array}
$$

$$l = 3$$

Notice that the coefficients of s are the same.

A large *dim sum* plate costs $3.

3. To keep both sides of an equation balanced, you must always perform the same operation on both sides. In Example 1, was the same amount subtracted from both sides of the first equation? Explain.

4. In Example 1, how can you find how much an item on a small *dim sum* plate costs?

In Example 1, the coefficients of s in each equation are the same, so you can eliminate s from the system by subtracting. In Example 2, the coefficients of y are opposites. If you add corresponding sides of each equation, you can eliminate y.

EXAMPLE 2 **Method: Addition**

Solve the system of equations.

$$8x - 9y = 13$$
$$3x + 9y = 9$$

SOLUTION

Step 1 Add corresponding sides of each equation to eliminate one variable.

$$
\begin{array}{rcl}
8x - 9y &=& 13 \\
+ (3x + 9y) &=& + (9) \\
\hline
11x &=& 22 \\
x &=& 2
\end{array}
$$

Step 2 Now you can solve for x.

Step 3 Substitute 2 for x in either equation and solve for y.

$$8x - 9y = 13$$
$$8(2) - 9y = 13$$
$$-9y = -3$$
$$y = \frac{1}{3}$$

The solution (x, y) is $\left(2, \frac{1}{3}\right)$.

Check

Substitute 2 for x and $\frac{1}{3}$ for y in each equation.

$$8x - 9y = 13 \qquad\qquad 3x + 9y = 9$$
$$8(2) - 9\left(\frac{1}{3}\right) \stackrel{?}{=} 13 \qquad 3(2) + 9\left(\frac{1}{3}\right) \stackrel{?}{=} 9$$
$$16 - 3 \stackrel{?}{=} 13 \qquad\qquad 6 + 3 \stackrel{?}{=} 9$$
$$13 = 13 \checkmark \qquad\qquad\qquad 9 = 9 \checkmark$$

THINK AND COMMUNICATE

5. When can you solve a system by subtracting? When can you solve a system by adding?

Tell whether you would use addition or subtraction to solve each system of equations. Then solve the system.

1. $3x + y = 10$
$x - y = 2$

2. $4x + 3y = 14$
$-4x + 5y = 2$

3. $4a - 2b = -6$
$3a - 2b = 8$

4. $4x + 5y = 55$
$2x + 5y = 35$

5. $2a + 3b = -12$
$2a - 3b = 0$

6. $8x + 3y = 20$
$5x + 3y = 17$

7.3 Exercises and Applications

*Extra Practice
exercises on
page 571*

1. Writing At the Peking Sky restaurant, you can order small, medium, or large plates. The waiters keep track of how much you spend by stamping the appropriate section of your bill each time you ask for a plate. Susan Kim went to Peking Sky with a large group of friends for *dim sum*. They sat at two tables and each table got a separate bill.

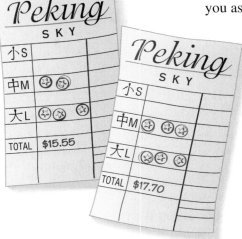

a. At Peking Sky, how much does a medium plate cost? How do you know?

b. How much does a large plate cost? How do you know?

Solve each system of equations by adding or subtracting.

2. $x + y = 8$
$x - y = 4$

3. $2a + b = 5$
$a - b = 1$

4. $6 = 5n + 2p$
$22 = 9n + 2p$

5. $c - 3d = 7$
$c + 2d = 2$

6. $3r - 5s = -35$
$-2r + 5s = 30$

7. $13a + 5b = -11$
$13a + 11b = 7$

8. $-4x = -7 + 5y$
$8x = 9 - 5y$

9. $-13 = 3x + 5y$
$-5 = 3x + y$

10. $a - 2b - 9 = 0$
$-2b + 3a = 9$

11. PERSONAL FINANCE John Haag knows that the Thacker Tree Company charges its customers a certain amount per cord of fire wood, plus a fixed price for delivery. The checks that John wrote for the last two deliveries were $260 for 1 cord of wood and $377.50 for $1\frac{1}{2}$ cords of wood. How much will it cost to have 2 cords of wood delivered?

Solve each system of equations by graphing, substituting, adding, or subtracting.

12. $y = 5x - 9$
$3x + 4y = 10$

13. $3a + 5b = -11$
$6a - 5b = -16$

14. $6c + 9d = 4$
$6c + 3d = 8$

15. $4y - 3z = 10$
$y + 4z = -7$

16. $3y = 21 - 5x$
$x + y = 9$

17. $2y + x = 17$
$3y = x + 8$

Technology Graph each system of equations. Add corresponding sides of the equations and graph the resulting equation with the original system. What do you notice about the graph of the third equation? Use a graphing calculator if you have one.

18. $2x + 3y = 6$
$-2x + 5y = 10$

19. $2x + 3y = 6$
$3x - 2y = 24$

20. Open-ended Problem How can you tell if these three equations have a common solution?

$3x - 5y = 1$ $\qquad$ $4x - y = 24$ $\qquad$ $2x - 3y = 2$

Connection   COOKING

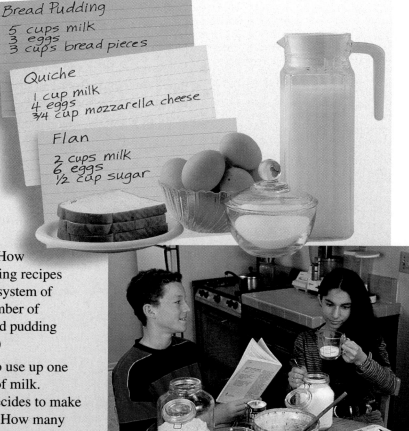

Some recipes, like the ones shown for bread pudding and quiche, are excellent for using leftover ingredients. If you don't have enough ingredients to follow a recipe exactly, you can make smaller batches by scaling down the amounts.

Bread Pudding
5 cups milk
3 eggs
3 cups bread pieces

Quiche
1 cup milk
4 eggs
3/4 cup mozzarella cheese

Flan
2 cups milk
6 eggs
1/2 cup sugar

21. If you want to make bread pudding but only have two eggs, how many cups of bread should you use?

22. Sandi Gubin wants to use up 14 eggs and 12 cups of milk. How many quiches and bread pudding recipes can she make? (*Hint:* Write a system of equations that involves the number of quiches, x, and number of bread pudding recipes, y, and the ingredients.)

23. Challenge Paul Gill wants to use up one dozen eggs and a half gallon of milk. (*Note:* 1 gal = 16 cups) He decides to make some flan and bread pudding. How many recipes of each can he make?

24. AVIATION On a typical day with mild winds, the 2459 mi flight from Los Angeles to New York City takes about $5\frac{1}{4}$ h. The return flight takes about $\frac{3}{4}$ h longer because the plane is flying into the wind.

New York City

Los Angeles

a. What is the plane's average speed on the way to New York City? on the return flight?

b. Let $s =$ the speed of the plane when there is no wind.
Let $w =$ the speed of the wind.
On the way to New York City, the plane's total speed is $s + w$. Write a variable expression for the plane's speed on the return trip.

c. Use your answers to parts (a) and (b) to write and solve a system of equations. How fast would the plane fly if there were no wind?

25. Cooperative Learning Work with a partner. You and your partner will explore two methods for finding an equation in the form $y = mx + b$ for the line through the points (3, 2) and (5, 10). One of you should do part (a). The other should do part (b). Work together on part (c).

a. Look back at Section 3.5. Use the method described in Example 2 on page 126.

b. Substitute 3 for x and 2 for y in the equation $y = mx + b$. Then substitute 5 for x and 10 for y in the equation $y = mx + b$. Solve this system of equations to find m and b. Then write an equation.

c. Explain the method you used to your partner. Make sure that you understand both methods. Which method do you prefer? Why?

ONGOING ASSESSMENT

26. Writing Describe four methods for solving systems of equations. When can you use each method?

SPIRAL REVIEW

Solve each equation. *(Section 4.4)*

27. $3.2x + 8.7 = 2.3$

28. $0.4 - 2.54n = 0.108$

29. $\frac{2}{3}x - \frac{1}{3} = 7$

30. $\frac{a}{3} + \frac{a}{4} = 5$

For Exercises 31–33, graph each equation. Then write an inequality to represent the x-values when $y \le 2$. *(Section 4.5)*

31. $y = 2x - 4$

32. $y = \frac{2}{3}x + 2$

33. $y = -2x + 3$

7.4 Solving Linear Systems Using Multiplication

Learn how to...

- write equivalent systems of equations

So you can...

- make purchasing decisions, such as deciding how much fabric to buy

You can solve any system of linear equations by using multiplication to rewrite the system so that addition or subtraction will eliminate a variable. Complete the Exploration to discover how solving by multiplication works.

EXPLORATION
COOPERATIVE LEARNING

Graphing Multiple Equations

Work in a group of 3 students.
You will need:
- graph paper

1 Choose one of the equations below and graph it. Each of you should choose a different equation.

- **a.** $2x + 2y = 5$
- **b.** $4x + 4y = 10$
- **c.** $6x + 6y = 15$

2 Compare your graphs. What do you notice?

3 Compare the three equations. How are they related?

4 Choose one of the systems of equations below and solve it by graphing. Each of you should choose a different system.

- **a.** $2x + 2y = 5$
  $5x - 6y = 7$
- **b.** $4x + 4y = 10$
  $5x - 6y = 7$
- **c.** $6x + 6y = 15$
  $5x - 6y = 7$

5 Compare your solutions. What do you notice? Explain your results.

6 Which of the systems in step 4 is easiest to solve algebraically? Why?

You can use multiplication to rewrite a system of equations as an equivalent system that can be solved by adding or subtracting.

EXAMPLE 1 **Method: Multiplication**

Use multiplication to solve the system of equations from Example 1 on page 287.

$$98a + 46s = 3500$$
$$a + s = 50$$

SOLUTION

Multiply one equation by **46**, so the coefficients of s in both equations are the same.

$$98a + 46s = 3500$$
$$a + s = 50 \quad \times \textbf{46}$$

$$98a + 46s = 3500$$
$$- (46a + 46s) = - (2300)$$
$$\overline{52a } = \overline{1200}$$
$$a \approx 23.1$$

Then use subtraction to solve the system.

$$a + s = 50$$
$$23.1 + s \approx 50$$
$$s \approx 26.9$$

The solution (a, s) is about $(23.1, 26.9)$.

EXAMPLE 2

Solve the system of equations.

$$5x + 3y = 2$$
$$3x - 2y = 5$$

SOLUTION

Transform both equations so the coefficients of y are opposites.

$$5x + 3y = 2 \quad \times \textbf{2}$$
$$3x - 2y = 5 \quad \times \textbf{3}$$

$$10x + 6y = 4$$
$$+ (9x - 6y) = + (15)$$
$$\overline{19x } = \overline{19}$$
$$x = 1$$

Solve by addition.

$$5x + 3y = 2$$
$$5(1) + 3y = 2$$
$$5 + 3y = 2$$
$$y = -1$$

WATCH OUT! ▶

Always check your solutions by substituting the values of x and y in both equations.

The solution (x, y) is $(1, -1)$.

A system of equations can have one solution, many solutions, or no solutions. When you solve a system with **many solutions**, you get a statement that is **always true**. When you solve a system with **no solutions**, you get a statement that is **never true**.

One solution	Many solutions	No solutions
The lines **intersect**.	The lines are the **same**.	The lines are **parallel**.
$x + 2y = 5$	$x + 2y = 5$	$x + 2y = 5$
$-\ (x + y = 2)$	$-\ (x + 2y = 5)$	$-\ (x + 2y = 0)$
$y = 3$	$0 = 0$	$0 = 5$
one solution	**always true**	**never true**

EXAMPLE 3   Application: Crafts

Ella Robinson needs 20 yd of fabric for a quilt. She has $80 to spend and wants to buy two different fabrics. She likes the four fabrics shown.

A.   **B.**   **C.**   **D.**

$10/yd $10/yd $4/yd $4/yd

a. Ella thinks A and B will look the best in her quilt. How many yards of each can she buy?

b. How many yards each of fabrics C and D can Ella buy?

SOLUTION

a. Problem Solving Strategy: Use a graph.

Let a = amount of fabric A (in yards). $a + b = 20$
Let b = amount of fabric B (in yards). $10a + 10b = 80$

> Ella wants 20 yd of fabric.

> Ella has $80 to spend.

Graph the system on a graphing calculator.

The lines are parallel, so they will never intersect. This means that the two equations do not have any common solutions.

Ella cannot buy 20 yd of fabrics A and B with $80.

Solution continued on next page.

Pyramid, pieced by Gussie Wells, who began quilting at age nine and continued to quilt into her 80's.

In *Links and Lineage* (1986), artist Paul Goodnight adapted African tribal designs from the Yoruba, the Dogon, the Ashanti, and other African tribes for the quilt design.

SOLUTION *continued*

b. Let c = the amount of fabric C (in yards). $c + d = 20$

   Let d = the amount of fabric D (in yards). $4c + 4d = 80$

Graph the system on a graphing calculator.

The equations have the same graph, so every solution of $c + d = 20$ is also a solution of $4c + 4d = 80$.

Ella can buy as much of each fabric as she likes, as long as she buys 20 yd total.

THINK AND COMMUNICATE

1. Solve each system in Example 3 by multiplying. Explain your results.

2. How does the system in part (b) of Example 3 change if Fabric C costs $8/yd?

Choosing Strategies

You have learned many methods for solving systems of equations. When should you use each strategy?

Method	Example	When to use
Graphing	 Intersection X=23.076923 Y=26.923077	When you want a visual understanding of the problem
Substitution	$y = 4 + 2x$ $2x + 7y = 3$	When one equation is already solved for x or y
Addition	$3x - 5y = 23$ $4x + 5y = 14$	When coefficients of one variable are opposites
Subtraction	$3x + 5y = 23$ $3x + 2y = 51$	When coefficients of one variable are the same
Multiplication	$5x + 3y = 2$ $3x + 2y = 5$	When no corresponding coefficients are the same

Describe at least two ways to solve each system of equations.

1. $2x + 3y = -2$
$2x - 3y = -16$

2. $x - 5y = 15$
$7x + 9y = -2$

3. $2x + 5y = 13$
$9x + 4y = 3$

Solve each system of equations.

4. $2x + y = 4$
$-4x - 2y = -8$

5. $3x + 4y = 11$
$4x - 3y = 23$

6. $6 = 21x - 15y$
$3 = 7x - 5y$

7.4 Exercises and Applications

For each system of equations, give the solution or state whether there are
no solutions or *many solutions*.

Extra Practice
*exercises on
page 571*

1. $3x + 8y = 1$
$7x + 4y = 17$

2. $-3x + 5y = -7$
$9x + 2y = 38$

3. $7r + 3s = 13$
$14r - 15s = -16$

4. $2x - 7y = 42$
$4x - 14y = 64$

5. $9x + 12y = 15$
$3x + 4y = 5$

6. $2 = 4s - 3t$
$2 = 12s - 9t$

Connection ▸ HISTORY

Mathematicians in ancient China used counting
rods to solve systems of equations. The illustration
at the right shows how these rods were used.

**Use the illustration at the right to solve the
system of equations in each rod diagram.**

The illustration shows this system of equations:
$4x - 3y = 11 \qquad -5x + 2y = -12$

A vertical rod
stands for 1.

A horizontal rod
stands for 10.

Black represents
negative.

Red represents
positive.

7.

8.

9.

10. Challenge The problem at the right is
from the *Jiǔzhāng suànshù*, a mathematical
text described on page 210.

  a. Draw a rod diagram to represent
the problem.

  b. What is the solution?

"Five large containers
and one small container
have a total capacity of
3 hu. One large container
and five small containers
have a capacity of 2 hu.
Find the capacities of
one large container and
one small container."

Solve each system of equations.

11. $2x + 5y = 4$
$3x - 7y = -10$

12. $2x - 3y = 6$
$3x + 4y = -2$

13. $3a - 6b = 18$
$4a + 8b = 24$

14. $4x + 8y = 5$
$5x + 10y = 13$

15. $13 = 8a - 8b$
$6 = 4a - 5b$

16. $160 = 2x - 10y$
$120 = 3x - 15y$

Connection LITERATURE

Christian Morgenstern first published *Songs from the Gallows* in Germany in March 1905. Some of his verses were set to music, and some inspired the artist Paul Klee. Morgenstern's artistic strength is not so surprising, considering that his father and both of his grandfathers had been artists.

Die zwei Parallelen
by Christian Morgenstern

Es gingen zwei Parallelen
ins Endlose hinaus,
zwei kerzengerade Seelen
und aus solidem Haus.

Sie wollten sich nicht schneiden
bis an ihr seliges Grab:
Das war nun einmal der beiden
geheimer Stolz und Stab.

Doch als sie zehn Lichtjahre
gewandert neben sich hin,
da ward's dem einsamen Paare
nicht irdish mehr zu Sinn.

War'n sie noch Parallelen?
See wussten's selber nicht,—
sie flossen nur wie zwei Seelen
zusammen durch ewiges Licht.

Das ewige Licht durchdrang sie,
da wurden sie eins in ihm;
die Ewigkeit verschlang sie
als wie zwei Seraphim.

The Two Parallels
translated by
B. Anderson

There run two parallels
out into endlessness,
two candle-straight souls
of the best families.

They won't let themselves be cut
until their quiet grave:
This was actually once
their secret pride and staff.

But when they had wandered
ten light years side by side,
then the lonesome pair was
no longer of earthly mind.

Were they still parallel?
They didn't know themselves,
they only flow like two souls
together through eternal light.

The eternal light transfused them,
they were made one with it;
eternity had twisted them,
as would two seraphim.

17. Open-ended Problem Give an example of a system of equations whose graph could be described by this poem.

19. a. Research Find someone who can read German and ask him or her to read the original poem to you. Compare the original and its translation. What are the advantages and disadvantages of each?

18. Writing Graph a system of equations and write a poem or story about the graph.

b. When you graph a system of equations, you can think of it as translating from the language of symbols to the language of pictures. What are the advantages and disadvantages of representing a linear system by its equations and by its graph?

20. An exam will have 20 questions worth a total of 100 points. There will be True-False questions worth 3 points each and short essay questions worth 11 points each. Several students ask how many essay questions there will be, but the teacher won't tell. How many essay questions will there be?

21. Sarah wants to buy 15 flower bulbs. Freesia bulbs cost $3 and dahlia bulbs cost $3. She has $45 to spend. How many of each type of bulb can she buy?

22. a. Visual Thinking Match lines (a), (b), and (c) with equations (1), (2), and (3). Explain your reasoning.

    (1) $x + y = 6$ **(2)** $x + 2y = 12$ **(3)** $x + 2y = 6$

  b. What are the coordinates of points P and Q?

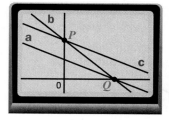

ONGOING ASSESSMENT

23. Writing Describe two ways to tell how many solutions a system of equations has.

SPIRAL REVIEW

Graph each equation and each point. Tell whether the ordered pair is a solution of the equation. *(Section 2.5)*

24. $y = 2x + 3$; $(0, 3)$ **25.** $y = 6 - \frac{2}{3}x$; $(1, 5)$ **26.** $y = 5x - 10$; $(2, 0)$

Graph each equation. *(Section 3.4)*

27. $y = 0$ **28.** $y = 3x + 3$ **29.** $x = 5$

ASSESS YOUR PROGRESS

Solve each system of equations. *(Sections 7.3 and 7.4)*

1. $7c + 9d = 13$ **2.** $-4x + 11y = 93$ **3.** $-4r + 5s = 27$
   $5c + 9d = 17$ $4x - 7y = -65$ $r - 6s = -2$

4. $5 = 2t - s$ **5.** $4t - 6s = 2$ **6.** $12a + 20b = 15$
   $20 = 8t - 4s$ $-6t + 9s = -3$ $9a + 15b = 8$

7. Emily Burnes's health plan has a monthly cost of $35 and a health benefit for which she pays $5 per visit to a fitness facility. One month she paid $75 total. Adam Katz has the same health plan and spent $135 one month. Between the two of them, they went to the health facility a total of 28 times. How many times did each person visit the health facility?

8. Journal What is the purpose of using multiplication as the first step when solving a system of equations?

7.5

Learn how to...
- graph linear inequalities

So you can...
- visualize problems involving inequalities

Linear Inequalities

You can tell that $(5, 3)$ and $(5, 4)$ are both solutions of the inequality $x - y \geq 1$ because they both make the inequality true. What other solutions does the inequality have? What does the graph of all of the solutions of the inequality look like?

EXPLORATION
COOPERATIVE LEARNING

Graphing $x - y \geq 1$

Work with a partner.
You will need:
- a ruler
- colored pencils

1 Make a large coordinate grid like the one in the picture. Both axes should extend from -4 to 4. Draw the grid lines.

2 At each point where two grid lines cross, write the value of $x - y$ for that point. For example, at the point $(3, -2)$, $x - y = 3 - (-2) = 5$.

3 With a colored pencil, circle each point that represents a solution of the inequality $x - y \geq 1$. What do you notice?

4 Find the point $(2.5, 0.5)$ on your grid. Do you think this is a solution of the inequality? Do you think $(2.5, 1.5)$ is a solution of the inequality? Check your answers.

5 Shade your grid to show all of the solutions of the inequality $x - y \geq 1$. How is the border between the shaded and the unshaded regions related to the inequality $x - y \geq 1$? Explain.

The inequality $x - y \geq 1$ is an example of a **linear inequality**. As you saw in the Exploration, the graph of a linear inequality is a region on a coordinate plane whose edge is a line.

EXAMPLE 1

Graph the inequality $y \leq 4 - \dfrac{x}{2}$.

SOLUTION

Step 1 Graph the equation $y = 4 - \dfrac{x}{2}$. The line shows all ordered pairs for which y **equals** $4 - \dfrac{x}{2}$.

Step 2 Now shade all points for which y **is less than** $4 - \dfrac{x}{2}$.

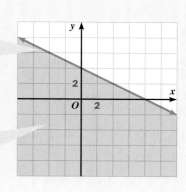

Check
Test any point in the shaded region, such as $(0, 0)$.

$$y \leq 4 - \frac{x}{2}$$

$$0 \overset{?}{\leq} 4 - \frac{0}{2}$$

$$0 \leq 4 \ \checkmark$$

THINK AND COMMUNICATE

1. In Example 1, why should you shade all points *below* the line? What does *below* the line mean?

EXAMPLE 2

Graph the inequality $3x - 2y < 12$.

SOLUTION

Step 1 Solve $3x - 2y < 12$ for y.

$$3x - 2y < 12$$

$$-2y < 12 - 3x$$

$$y > -6 + \frac{3}{2}x$$

Step 2 Graph the equation $y = -6 + \dfrac{3}{2}x$.

Make the line dashed, to show that the points on the line are not solutions of the inequality.

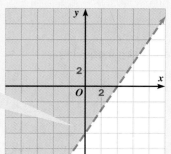

Step 3 Shade all points above the line.

◄ **WATCH OUT!**

When you multiply or divide both sides of an inequality by a negative number, remember to switch the direction of the inequality sign.

THINK AND COMMUNICATE

2. In Example 1, explain why the solutions of the equation $y = 4 - \dfrac{x}{2}$ are also solutions of the inequality $y \le 4 - \dfrac{x}{2}$. In Example 2, explain why the solutions of $y = -6 + \dfrac{3}{2}x$ are not solutions of $y > -6 + \dfrac{3}{2}x$.

3. When should you use a dashed line to graph an inequality? When should you use a solid line?

☑ CHECKING KEY CONCEPTS

Match each inequality with its graph.

1. $y > \dfrac{1}{2}x + 2$

2. $y < \dfrac{1}{2}x + 2$

3. $y \ge \dfrac{1}{2}x + 2$

A.   **B.**   **C.**

Graph each inequality.

4. $y \ge x + 1$ **5.** $x + y \le 3$ **6.** $2x - 5y < 20$

7.5 Exercises and Applications

Extra Practice exercises on page 571

Match each inequality with its graph.

1. $y < -x + 2$ **2.** $y > -x + 2$ **3.** $y \le 2$ **4.** $x \le 2$

A.   **B.**   **C.**   **D.**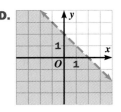

Graph each inequality.

5. $y > 2$ **6.** $x \ge -1$ **7.** $y \ge 2x$

8. $y < 3x$ **9.** $y \le 2x - 1$ **10.** $y > 5 - \dfrac{1}{2}x$

11. $y < -\dfrac{1}{3}x + 3$ **12.** $x + 2y \le 4$ **13.** $2x + 3y \ge 18$

14. $3x - y > 0$ **15.** $2x - y \ge 5$ **16.** $4x - 4y \le 9$

308   Chapter 7 *Systems of Equations and Inequalities*

Cooperative Learning Work with a partner. One of you should do Exercise 17. The other should do Exercise 18. Work together on Exercise 19.

17. Use the method described in Example 2 on page 307 to graph the inequality $4x - 3y \geq 12$.

18. a. Graph the equation $4x - 3y = 12$ by finding the x- and y-intercepts and drawing a line through them.

   b. Pick a point not on the line. Do the coordinates of the point make the inequality $4x - 3y \geq 12$ true? What does this tell you about the graph of the inequality $4x - 3y \geq 12$?

   c. Graph the inequality $4x - 3y \geq 12$.

19. Explain to your partner the method you used. Make sure that you both understand both methods. Which method do you prefer? Why?

Graph each inequality.

20. $2x + 4y > 5$

21. $x + 2y \leq 7$

22. $-3x + 2y < 9$

23. $5x - 3y \leq 10$

24. $3x - 8y > 10$

25. $-4x - 7y \geq 18$

26. $-\dfrac{2}{3}x + 4y \geq \dfrac{5}{6}$

27. $2.3x - 4.9y < 5.2$

28. $\dfrac{x}{2} - \dfrac{y}{3} < \dfrac{1}{6}$

INTERVIEW Nancy Clark

Look back at the article on pages 278–280.

In her Sports Nutrition Guidebook, Nancy Clark advises athletes to choose low-fat sources of calcium. Teenagers need at least 1200 mg of calcium each day to help build strong bones.

Nancy Clark uses calipers to help determine a healthy level of body fat for young adults.

29. One cup of low-fat milk supplies 300 mg of calcium. One cup of vanilla ice cream provides 180 mg of calcium.

   a. Write an inequality stating that x cups of milk and y cups of ice cream supply more than 1200 mg of calcium.

   b. Graph the inequality and list two different solutions.

30. One cup of 2% low-fat milk contains 5 g of fat. One cup of vanilla ice cream contains 14 g of fat. If you eat 2000 calories per day, you should eat between about 55 g and 65 g of fat.

   a. Write an inequality stating that x cups of milk and y cups of ice cream contain less than 65 g of fat.

   b. Graph your inequality. List two different solutions of the inequality.

31. Challenge Can you get 1200 mg of calcium from milk and ice cream without eating more than 65 g of fat? Explain.

32. When students move into a college dormitory, the elevators get a great deal of use. Suppose an elevator has a 3000 lb weight limit, that a typical student weighs 150 lb, and a typical box weighs 60 lb.

   a. Is the elevator safe for 8 students and 22 boxes?

   b. If there are 25 boxes in the elevator, how many students can be carried safely?

   c. Write and graph an inequality that describes the circumstances under which it is safe to load s students and b boxes into the elevator at the same time.

33. SAT/ACT Preview If $A = -y$ and $B = 2 - y$, then:

   A. $A > B$ **B.** $B > A$ **C.** $A = B$ **D.** relationship cannot be determined

34. Writing For his birthday, Jacob got a movie gift certificate worth $25. Matinees cost $4 and evening shows cost $7.

   a. Write an equation stating that Jacob saw m matinees and n evening shows for $25.

   b. Write an inequality stating that Jacob saw m matinees and n evening shows for at most $25.

   c. Compare the equation from part (a) with the inequality from part (b). Which do you think best represents the situation? Why?

35. Open-ended Problem Choose a problem from Sections 7.1–7.4 that can be represented by either an equation or an inequality. Write and graph an inequality that describes the problem.

ONGOING ASSESSMENT

36. Open-ended Problem Describe an everyday situation that this graph could represent.

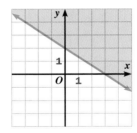

SPIRAL REVIEW

Find each product. *(Section 6.4)*

37. $2x(x + 7)$ **38.** $4x(5 - x)$ **39.** $(a + 7)(a - 5)$

40. $(3b + 5)(b + 3)$ **41.** $-2\left(3 - \sqrt{5}\right)$ **42.** $\left(4 + \sqrt{2}\right)\left(1 + \sqrt{2}\right)$

Use substitution to solve each system of equations. *(Section 7.2)*

43. $x = y$
$3x - 2y = 1$

44. $a = b - 2$
$3a + 2b = 29$

45. $3x - y = 1$
$2x + 5y = 63$

7.6

Learn how to...

- graph systems of inequalities

So you can...

- make decisions when there are many restrictions, such as when making food for a hiking trip

Systems of Inequalities

Because backpackers need large amounts of energy and must carry their food, they must plan meals carefully. Many hikers make a mixture of peanuts and chocolate chips. The mixture, sometimes called *gorp*, provides quick energy from carbohydrates.

One ounce of chocolate chips contains 20.9 g of carbohydrates.

One ounce of peanuts contains 7.8 g of carbohydrates.

EXAMPLE 1   Application: Nutrition

Andrea wants her gorp to contain at least 150 g of carbohydrates and weigh less than 16 oz. Write and graph two inequalities to show how much of each ingredient she should use.

SOLUTION

Let p = the number of ounces of peanuts.

Let c = the number of ounces of chocolate chips.

The gorp should weigh less than 16 oz.	Carbohydrates from peanuts	+	Carbohydrates from chocolate	≥ 150
$p + c < 16$	$7.8p$	+	$20.9c$	≥ 150

Graph the inequalities in the same coordinate plane.

The points that are in both the blue and the red regions satisfy both inequalities. For example, Andrea could use 7 oz of peanuts and 7 oz of chocolate chips.

$p + c < 16$

$7.8p + 20.9c \geq 150$

The inequalities in Example 1 are called a **system of inequalities**. The **solution of a system of inequalities** consists of all ordered pairs of numbers that satisfy both inequalities.

EXAMPLE 2 **Application: Nutrition**

The American Heart Association and the American Cancer Society recommend that you should eat at most 1 g of fat for every 5 g of carbohydrates you eat.

a. One ounce of peanuts contains **9.1 g of fat**. One ounce of chocolate chips contains **5.3 g of fat**. Write an inequality for the amount of each ingredient Andrea should use so that the gorp will contain **less than 30 g of fat**.

b. Graph your inequality along with the inequalities from Example 1. How much of each ingredient should Andrea use?

SOLUTION

a.

Fat from peanuts	+	Fat from chocolate	<	30
$9.1p$	+	$5.3c$	<	30

b.

The gorp contains **at least 150 g of carbohydrates** and weighs **less than 16 oz**.

The gorp contains less than 30 g of fat.
$9.1p + 5.3c < 30$

There are no points that satisfy all three inequalities.

It is impossible to make a gorp of peanuts and chocolate chips that contains more than 150 g of carbohydrates, contains less than 30 g of fat, and weighs less than 16 oz.

THINK AND COMMUNICATE

1. Is an inequality a function? Explain your answer.

2. The graph in Example 2 is the first quadrant of the graph of the system of inequalities $p + c < 16$

$$7.8p + 20.9c \geq 150$$
$$9.1p + 5.3c < 30.$$

Does this system have solutions in any other quadrants? Explain.

Graph each system of inequalities.

1. $x \geq 3$
$y \leq 5$

2. $y > x$
$y > 0$

3. $y < x + 4$
$y \geq -2x + 2$

4. $2x - y \geq 4$
$x + y \leq 3$

5. The graphs of the equations $y = x - 3$ and $y = 7 - x$ divide the coordinate plane shown at the right into four regions. Give a system of two inequalities that describes each region. Region A is done for you.

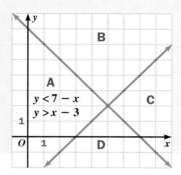

7.6 Exercises and Applications

Graph each system of inequalities.

1. $y \geq 0$
$x \geq -2$

2. $x < 0$
$y > \dfrac{1}{2}x$

3. $y > 2x$
$y < 12 - x$

Extra Practice exercises on page 572

4. $y \leq 4 - 2x$
$y \geq -2$

5. $y > 4x$
$y \leq 7 + x$

6. $x + y < 8$
$x + y > 2$

7. $2x + 3y < 6$
$3x + 2y > 6$

8. $5x - 2y \leq 12$
$-3x - y \leq 5$

9. $2 < x - 3y$
$x - 2y \leq 2$

Use the information about cat food for Exercises 10–12.

10. Write an inequality for each statement.

  a. Erik Mueller can spend at most $5 for cat food each week.

  b. Erik's cat eats at least 72 oz of cat food each week.

11. Graph the system of inequalities from Exercise 10. List some solutions of the system.

12. Erik prefers to give his cat more canned food than dry food. Express this preference as an inequality. Graph the inequality with your graph from Exercise 11. How much of each kind of food should Erik buy?

	Canned cat food	Dry cat food
Cost	$.50	$2.50
Ounces	6 oz	48 oz

Graph each pair of equations. The graphs will divide the coordinate plane into four regions. Write a system of inequalities that describes each region. Label the region with these inequalities.

13. $x = 6$
$y = -2$

14. $y = x - 3$
$y = 9 - x$

INTERVIEW Nancy Clark

Look back at the article on pages 278–280.

Nancy Clark stresses that a sports diet is healthy for all people, not just athletes. She says that, regardless of your sport or activity level, 60–70% of the calories you eat should come from carbohydrates.

15. **a.** Let c = calories from carbohydrates. Let t = total calories. Write an inequality that states that at least 60% of total calories should come from carbohydrates.

  b. Write an inequality that states that at most 70% of calories should come from carbohydrates.

16. Graph the system of inequalities from Exercise 15. Compare your graph to the graph on page 280. How are they alike? How are they different?

17. Writing How can you use your graph to find your ideal carbohydrate intake? What do you need to know?

18. Athletes in training require 70–80% of their calories to be from carbohydrates. Suppose a cyclist in the Tour DuPont, an international cycling stage race, uses about 963 Cal during one stage of the race. Write an inequality that represents the number of calories that should come from carbohydrates.

In 1995, the Tour DuPont followed a 1130 mile-long route from Wilmington, DE, to Piedmont Triad, NC, and took 12 days to complete. Cyclists often have to eat their meals while riding.

19. **a.** You can write $2 \le y \le 4$ as a short way of writing the system $2 \le y$ and $y \le 4$. Graph this system of inequalities.

  b. What system of inequalities does $0 \le x \le 5$ represent? Graph the system.

  c. Graph this system of four inequalities: $2 \le y \le 4$; $0 \le x \le 5$

20. **Technology** Find out how to graph a system of inequalities on your graphing calculator. You may need to use features designed for drawing and shading. Graph each system of inequalities.

  a. $y \ge 3$
     $y \le 2x$

  b. $y \ge 2x - 8$
     $y \le x + 1$

  c. Describe how to graph a system of inequalities on your graphing calculator. Be specific enough so that your directions can be followed accurately by someone who is familiar with your calculator but has never used it to graph inequalities before.

21. **SAT/ACT Preview** What is the area of the region described by the inequalities $x > 0$, $y > 0$, and $y < 8 - 2x$?

   A. 8   **B.** 16   **C.** 24   **D.** 32   **E.** 64

22. **Open-ended Problem** Create a system of inequalities whose solution region is a polygon.

23. **Challenge** How many pairs of positive integers x and y satisfy the inequality $2x + 5y \leq 10$?

ONGOING ASSESSMENT

24. **Writing** How is solving a system of inequalities like solving a system of equations? How is it different?

SPIRAL REVIEW

Solve each proportion. *(Section 5.1)*

25. $\dfrac{3}{7} = \dfrac{x}{2}$

26. $\dfrac{y}{3} = \dfrac{7}{8}$

27. $\dfrac{12}{n} = \dfrac{3}{5}$

28. $\dfrac{a}{2} = \dfrac{18}{a}$

29. $\dfrac{t-2}{2} = \dfrac{4}{3}$

30. $\dfrac{2n+1}{2} = \dfrac{3}{2}$

Find each product. *(Section 6.5)*

31. $(x + 2)^2$

32. $\left(4 + \sqrt{3}\right)^2$

33. $(x + 5)(x - 5)$

ASSESS YOUR PROGRESS

VOCABULARY

linear inequality (p. 307)   solution of a system
system of inequalities (p. 312)   of inequalities (p. 312)

Graph each inequality or system of inequalities. *(Sections 7.5 and 7.6)*

1. $y \leq 4 - \dfrac{1}{3}x$

2. $3x + 2y \leq -12$

3. $5x - 2y > 10$

4. $y > 2$
   $x < -1$

5. $4x + y \geq 0$
   $4x + y \leq 6$

6. $3y - x < 3$
   $2x + 7y \geq 7$

7. **Business** The members of the math team are selling T-shirts and sweatshirts. They need to sell a minimum of 10 T-shirts and 10 sweatshirts. They earn $5 for each T-shirt and $10 for each sweatshirt sold, and would like to raise at least $200. Write and graph a system of inequalities that shows how many T-shirts and sweatshirts the team needs to sell.

8. **Journal** How is graphing the inequality $3x - y < -6$ like graphing the equation $3x - y = -6$? How is it different?

Designing a Stairway

Have you ever found it difficult to climb a narrow or steep stairway? Stairs may appear to be all the same, but slight changes in the riser/tread ratio can greatly affect the safety and the cost of stairs.

Iowa City

PROJECT GOAL

Compare the dimensions of stairways in your community with some accepted rules for stairs, so that you can better design your own stairway.

Designing a Stairway

Work with a partner.

1. COLLECT riser and **tread** data for at least five different stairways. Take notes about how easy or hard it is to climb each stairway.

Number	Location	Tread Length(in)	Riser Length(in)
1	Near Gym	10	7

2. DECIDE on an ideal riser/tread ratio. Explain your decision.

Chicago

3. DESIGN your own stairway. Suppose the total distance from one floor to the next is 105 in. Determine lengths for the risers for a 9 in. tread and an 11 in. tread. Then make a scale drawing of your stairway.

riser   tread

Brooklyn

Cycle
by M.C. Escher
(1898 – 1972)

Analyzing Stairway Standards

Codes have been written to increase stair safety. Codes often concern the riser/tread ratio, lighting around stairs, and handrail height. Here are two generally accepted rules for stairway construction:

RULE 1: The sum of one riser and one tread should be from 17 in. to 18 in.

RULE 2: The sum of two risers and one tread should be from 24 in. to 25 in.

New Mexico

1. Make a graph showing all the riser/tread data that meet the conditions of these rules. (*Hint:* You will graph four inequalities.)

2. On the same coordinate grid, graph the data you collected about stairs as ordered pairs. Also graph the ordered pair for the stairway you designed.

3. Describe any patterns you see. Did all the stairways follow the generally accepted rules? Did any not follow the rules?

Presenting Your Results

Give a report that includes your graph, your table, and your scale drawing. Write a paragraph comparing your local stairways' dimensions with the accepted rules for stairs. Give your recommendations for riser/tread ratios.

You may want to extend your report by examining some of the ideas below:

- Learn more about constructing stairs, including how a carpenter's square is used.

- Collect photographs of stairways from other periods of time and from other countries.

- Find the slope for each stairway on your list. Then find the range of slopes for the stairways.

- Interview architects and/or carpenters in your area. Ask what they would consider the best riser/tread ratio for stairs.

Questions

1. What was the most interesting stairway you've designed?

2. What's the difference between designing a stairway for a home and for an office building?

3. What different types of stairways have you designed?

Self-Assessment

What method(s) did you use to create your graph? How did you and your partner decide on an ideal riser/tread ratio? Overall, was there enough variety in the stairs you studied to give you a sense of how the riser/tread ratio affects stairway use? What was the most difficult aspect of the project? the easiest? Why?

7 | Review

STUDY TECHNIQUE

For each vocabulary word or phrase listed below, give an example or draw a graph to illustrate.

VOCABULARY

standard form of a linear equation (p. 282)
vertical intercept, *y*-intercept (p. 282)
horizontal intercept, *x*-intercept (p. 282)
system of linear equations (p. 288)
solution of a system of linear equations (p. 288)

linear inequality (p. 307)
system of linear inequalities (p. 312)
solution of a system of inequalities (p. 312)

SECTIONS | 7.1 *and* 7.2

You can solve a system of equations by graphing or by substitution.

By graphing

Graph the intercepts, or solve for *y* and use a graphing calculator.

The **red line** contains all solutions of the equation $2x + 5y = 11$.

The **green line** contains all solutions of the equation $x - 4y = -1$.

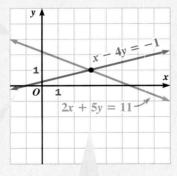

The solution
(x, y) is **(3, 1)**.

By substitution

$$2x + 5y = 11$$
$$x - 4y = -1$$
$$x = 4y - 1$$

$$2(4y - 1) + 5y = 11$$
$$13y - 2 = 11$$
$$y = 1$$

$$x - 4y = -1$$
$$x - 4(1) = -1$$
$$x = 3$$

The solution (x, y) is **(3, 1)**.

7.3 *and* 7.4

You can solve a system of equations by **adding**, **subtracting**, or **multiplying**.

$$
\begin{array}{rcl}
2x + 3y &=& 1 \\
+ (-2x + 4y) &=& + (-8) \\
\hline
0 + 7y &=& -7 \\
y &=& -1 \\
x &=& 2
\end{array}
$$

$$
\begin{array}{rcl}
3x - 4y &=& -6 \\
- (5x - 4y) &=& - (-2) \\
\hline
-2x + 0 &=& -4 \\
x &=& 2 \\
y &=& 3
\end{array}
$$

$$
\begin{array}{rcl}
4x + 2y = 3 & \times\ 3 \\
6x + 3y = 1 & \times\ 2
\end{array}
$$

$$
\begin{array}{rcl}
12x + 6y &=& 9 \\
- (12x + 6y) &=& -(2) \\
\hline
0 &=& 7
\end{array}
$$

no solution

A graph can show whether a system of equations has one solution, many solutions, or no solution.

The lines **intersect**.	The lines are the **same**.	The lines are **parallel**.
one solution	**many solutions**	**no solutions**

7.5 *and* 7.6

The graph of a linear inequality is a region on the coordinate plane whose edge is a line. The region shows all ordered pairs that satisfy the inequality.

The **blue region** contains all points that satisfy the inequality *y* > 2*x* − 3.

The **purple region** contains all points that satisfy **both** inequalities.

The **pink region** and the **red line** contain all points that satisfy the inequality 2*x* + 3*y* ≤ 6.

Assessment

VOCABULARY REVIEW

1. Give an example of a linear equation in standard form. Give an example of a linear equation in slope-intercept form.

2. Give an example of a linear inequality.

3. What is the difference between a horizontal intercept and a vertical intercept?

SECTIONS 7.1 *and* 7.2

Give the horizontal intercept and the vertical intercept of each equation. Then graph each equation.

4. $4x - 3y = 24$

5. $2x + 3y = 17$

 Technology **Solve each equation for *y*. Then use a graphing calculator to graph each equation.**

6. $5x - 2y = 11$

7. $2x - 3y = 12$

8. Anita makes a blend of teas for a party. Sunrise Tea costs $6/lb and Great Taste Tea costs $4/lb.

a. Write an equation showing that Anita spends $20 altogether for tea.

b. Write an equation showing that Anita buys 4 lb of tea.

c. Solve the system of equations to find how much of each kind of tea Anita purchases.

Solve each system of equations by graphing or by substitution.

9. $4x + 3y = 24$
   $x - 2y = -5$

10. $3x + 3y = 7$
   $4x - y = 6$

SECTIONS 7.3 *and* 7.4

Solve each system of equations.

11. $4x + 2y = 14$
   $x = 1 + 2y$

12. $4x + 3y = -2$
   $4x - 2y = 8$

13. $4x + 5y = 14$
   $10x + 5y = 20$

14. $4x + 3y = 10$
   $8x - 6y = -20$

15. $2x - 3y = 11$
   $6y = -22 + 4x$

16. $3y = 5 - 2x$
   $3x + 4y = 2$

17. At a county fair one day, 600 tickets were sold. Adult tickets cost $10 each and tickets for children cost $5 each. If the total money collected from selling tickets was $4750, how many children's tickets were sold?

18. Writing Use the graphs at the right. Tell what you know about the equation of each line and about the solution(s) of each system of equations.

a.

b.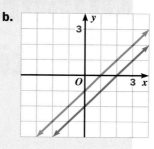

SECTIONS 7.5 *and* 7.6

Graph each inequality or system of inequalities.

19. $y < x + 6$

20. $y > \frac{2}{3}x - 2$

21. $2x + 4y \le 17$

22. $3y < 16 - 3x$

23. $y \ge 3$
$y \le -4x$

24. $2x + 3y > 5$
$6y < 1 - 4x$

25. $2x + 3y < 11$
$y \ge \frac{4}{5}x - 3$

26. $y < \frac{3}{4}x + 1$
$y \ge \frac{1}{3}x - 3$

27. $4x + 5y > 6$
$1 \ge x + y$

28. For the opening of an art gallery, Gina DiFazio and Eric Kornfeld need to buy at least 10 lb of nuts. Peanuts cost $2/lb and deluxe mixed nuts cost $5/lb. Gina and Eric can spend up to $25 on nuts.

  a. Write an inequality stating that x pounds of peanuts and y pounds of mixed nuts come to a total of at least 10 lb.

  b. Write an inequality stating that Gina and Eric can spend up to $25 on x pounds of peanuts and y pounds of mixed nuts.

  c. They cannot buy negative amounts of nuts, so $x \ge 0$ and $y \ge 0$. Graph these inequalities, as well as those you wrote in parts (a) and (b), in the same coordinate plane.

  d. What is the greatest amount of mixed nuts Gina and Eric can buy? Explain.

PERFORMANCE TASK

29. Draw three diagrams that show the possibilities for a solution of a system of equations. Give the equations for the lines and solve each system.

8 Quadratic Functions

Meeting the CHALLENGE

Bill Pinkney

"In sailing you have to use your mental abilities to navigate safely from one point unseen to another point unseen. It's completely mathematics."

On June 9, 1992, former businessman Bill Pinkney sailed into Boston Harbor—and into history books. After 16 months and 32,000 miles of battling weather, loneliness, and equipment failures, Pinkney became the first African American to sail solo around the world. His route made his achievement even more remarkable. Pinkney circled much of the globe 40 to 60 degrees south of the equator. Here the Pacific, Indian, and Atlantic Oceans merge around Antarctica, creating extreme conditions for sailors: strong winds, high waves, icebergs, and snowstorms, even in summer.

Using these winds required many calculations. "You want to match wind speed with the amount of surface area of sail you have up," Pinkney explains. "Too much wind hitting too much sail can knock you flat on your side, which happened to me twice."

Using the Wind

Strong west-to-east winds powered *Commitment*, Pinkney's 47 foot boat. "Cruising sailors like to sail in trade winds of 20 to 22 knots. But ideal sailing weather for me was just below true gale-force winds," Pinkney says. "That meant a 35 knot wind and moderate waves, anywhere from 15 to 25 feet." (The word knot means nautical miles per hour. One knot is about 1.15 miles per hour.)

Storms and Squalls

What did Pinkney do when he encountered storms and squalls during his voyage? "You evaluate a storm as bad, terrible, or life threatening," he says, "and if it's bad, you go on and do other things. I'd take care of my housekeeping, maintenance, and navigation duties, and I'd read, sleep, and eat."

Pinkney experienced the worst storm of the trip as he rounded Cape Horn, the southern tip of

South America. Cold, 65 to 70 knot winds forced Pinkney to pull down his sails. High waves crashed in from two directions. During this 40 hour ordeal, Pinkney called on all his sailing and navigating skill to clear Cape Horn's rocky coast.

Pacific Ocean

Atlantic Ocean

Cape Horn

Wind Speed and Wave Height

Like all sailors, Pinkney knows that strong winds mean big waves. The duration of high winds, the amount of open water, and the strength of water currents affect wave height, but wind speed is one of the most important factors. The graph shows that as the wind speed increases, the height of ocean waves increases dramatically. The dependence of wave height on wind speed is an example of a *quadratic function.*

Wind Speed and Wave Height

Wave height (ft) vs. Wind speed (mi/h)

Navigational compass

1. Writing Use the graph to determine what happens to wave height when wind speed doubles. Give examples to justify your conclusion.

2. Project The Egyptians first began using sails for boats about 5000 years ago. The Phoenicians, the Polynesians, and the Vikings also developed a high degree of sailing skill. Prepare a report on the seafaring accomplishments of one of these civilizations.

3. Research In addition to sailing around Cape Horn, Bill Pinkney sailed around four other "great capes." Locate these five great capes on a map or a globe. Roughly trace Bill Pinkney's route.

Mathematics & Bill Pinkney

In this chapter, you will learn more about how mathematics is related to sailing.

Related Exercises

Section 8.3
• **Exercises 14–16**

Section 8.4
• **Exercises 5–8**

8.1 Nonlinear Relationships

Graphs are not always lines. Often the relationship between two variables is *nonlinear,* as you will see in the Exploration.

Learn how to...

- analyze the shape of a graph

So you can...

- decide whether a relationship is linear

Exploring Nonlinear Graphs

Work in a group of 2–4 students.
You will need:
- graph paper
- a graphing calculator

You want to design a rectangular pen for a dog, using 30 ft of fencing. You decide that the length of each side must be an integer.

Perimeter = 30 ft
Area = 14 ft²

1 Draw all the pens you can build. Then copy and complete the table. Add rows until you have listed all possibilities.

Length of Side 1 (ft)	Length of Side 2 (ft)	Perimeter (ft)	Area (ft²)
1	14	30	14
14	1		

2 Which dimensions give the dog the most area? Which dimensions give the dog the longest run? Discuss which dimensions you would recommend. Explain your decision.

3 Make a scatter plot of the data. Put the length of Side 1 on the horizontal axis and the area of the pen on the vertical axis. Describe the shape of the graph.

4 Let x = the length of Side 1 of the pen. Write an expression in terms of x for the length of Side 2. Then write and graph an equation for the area of the pen as a function of the length of Side 1. Use a graphing calculator.

5 Discuss with your group how the graphs in Steps 3 and 4 compare.

If the graph of an equation is a line, then the equation shows constant change and the relationship that it models is linear. Otherwise, the equation and the relationship that it models are both **nonlinear**.

EXAMPLE 1

Tell whether each relationship is *linear* or *nonlinear*.

a. the relationship between the perimeter P of a square and the length s of one side

b. the relationship between the area A of a square and the length s of one side

SOLUTION

a. Write and graph an equation for P as a function of s.

$$P = 4s, s \geq 0$$

The graph is part of a line. The relationship between the perimeter of a square and the length of one side is linear.

b. Write and graph an equation for A as a function of s.

$$A = s^2, s \geq 0$$

The graph is not a line or part of a line. The relationship between the area of a square and the length of one side is nonlinear.

THINK AND COMMUNICATE

1. An equation for the volume V of a cube with edge length s is $V = s^3$. A graph of the equation is shown.

a. What is a reasonable domain for this function?

b. Is the relationship between volume and edge length *linear* or *nonlinear*? Explain.

2. In the Exploration, is the relationship between the area of the pen and the length of Side 1 *linear* or *nonlinear*? Explain.

EXAMPLE 2

Is the equation $y = |x|$ linear or nonlinear?

SOLUTION

Graph the equation.

X	Y₁
-2	2
-1	1
0	0
1	1
2	2

Y1 = abs X

A table of values can also help you to see what's happening.

The graph is not one continuous line. The equation $y = |x|$ is nonlinear.

THINK AND COMMUNICATE

3. In Example 2, how does the graph of the equation $y = |x|$ show that the absolute value of a number x is never negative?

4. How is the graph of the equation $y = |x|$ different from the graph of the equation $y = x$? Which equation is linear and which is nonlinear?

5. The relationship in Example 1(a) is linear, but the relationship in Example 2 is nonlinear. How are the graphs alike? How are they different?

☑ CHECKING KEY CONCEPTS

Tell whether each graph represents a *linear* or a *nonlinear* relationship.

1.

2.

3.
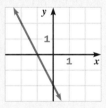

Graph each equation. Tell whether each equation is *linear* or *nonlinear*.

4. $y = 3x + 2$ **5.** $y = 3x^2$ **6.** $y = 3|x|$

7. A convenience store has a self-service copier. The rates for using the copier are shown in the table. Is the relationship between the cost of making copies and the number of copies *linear* or *nonlinear*? Explain.

Number of copies	Cost per copy (cents)
1–10	15
11–25	10
26 +	7

Exercises and Applications

Extra Practice
exercises on
page 572

Graph each equation. Tell whether each equation is *linear* or *nonlinear*.

1. $y = 5x$

2. $y = 5|x|$

3. $y = 5x^2$

4. $y = -2|x|$

5. $y = -3x + 1$

6. $y = |x| + 3$

7. $y = \frac{1}{2}x + 4$

8. $y = 4x^2 - 3$

9. $y = x^3 + 2$

10. Writing Based on the results of Exercises 1–9, generalize how you can tell from an equation whether its graph will be a line.

GEOMETRY **Tell whether the relationship between the radius *r* and each given variable is *linear* or *nonlinear*. Explain your reasoning.**

11. Circumference C

$$C = 2\pi r$$

12. Area A

$$A = \pi r^2$$

13. Volume V

$$V = \frac{4}{3}\pi r^3$$

14. Open-ended Problem Find two different examples of real-world relationships that can be modeled with a graph, one linear and one nonlinear. Model each relationship with a graph.

For Exercises 15–18, use the table.

The drama club needs to decide on the ticket price for a school play. They estimate that about 500 people will buy tickets if the ticket price is $3. Based on previous performances, they estimate that for every $.50 increase in the ticket price, they will sell 50 fewer tickets.

15. Copy and complete the table.

16. **Technology** Make a scatter plot of the data. Put ticket price on the horizontal axis and income on the vertical axis.

17. Do you think the relationship between ticket price and income is *linear* or *nonlinear*? Explain.

18. Assume that the estimates are accurate. What price would you charge for a ticket? Give a reason for your answer.

Expected Income from Ticket Sales		
Ticket price (dollars)	Number of tickets sold	Income (dollars)
3.00	500	1500
3.50	450	1575
4.00	?	?
4.50	?	?
5.00	?	?

A professional baker uses the table at the right as a guide. It shows the greatest number of servings that you can expect from a four-layer cake of a given diameter.

19. **Technology** Make a scatter plot of the data. Put diameter on the horizontal axis and number of servings on the vertical axis.

20. Do you think the relationship between the diameter and the number of servings is *linear* or *nonlinear*? Explain.

21. **Technology** Use the formula below for the volume of a cylinder. Assume that each layer is 1.5 in. high. Decide whether the relationship between number of servings and volume is *linear* or *nonlinear*. Use a spreadsheet or a graphing calculator to make a scatter plot to support your answer.

Volume $= \pi r^2 h$
$\quad\quad\quad = 6\pi r^2$

Servings from a Four-Layer Cake	
Diameter (in.)	Number of servings
6	20
8	30
10	45
12	65
14	95
16	125

$h = 6$ in.

22. The numbers in the sequence below are called *triangular numbers* because you can arrange each number of dots to form a triangle.

1st	2nd	3rd	…	*n*th
1	3	6	…	?

The triangular numbers follow a pattern.

$$1 = \frac{1 \times 2}{2} \quad\quad 3 = \frac{2 \times 3}{2} \quad\quad 6 = \frac{3 \times 4}{2} \quad\quad \ldots \quad\quad ?$$

a. What are the 4th, 5th, 6th, and 10th triangular numbers?

b. Write a formula for the *n*th triangular number.

c. **Technology** Graph your formula from part (b) on a graphing calculator. Is the formula *linear* or *nonlinear*?

Cooperative Learning For Exercises 23–25, work with a partner. You will each need a graphing calculator.

23. **Technology** Each of you should graph the equation $y = |x|$ on a graphing calculator. Then one of you should do part (a). The other should do part (b). Work together on part (c).

   a. Graph the equations $y = 2|x|$, $y = 3|x|$, and $y = 4|x|$ along with the graph of $y = |x|$.

   b. Graph the equations $y = \frac{1}{2}|x|$, $y = \frac{1}{3}|x|$, and $y = \frac{1}{4}|x|$ along with the graph of $y = |x|$.

   c. Describe how the coefficients affect your original graphs.

24. **Technology** Work together to generalize how the graph of an equation in the form $y = a|x|$ changes as the coefficient a changes in value.

25. **Technology** Work together to write a procedure for exploring how the graph of an equation in the form $y = a|x|$ changes as the coefficient a becomes negative. Carry out the exploration and generalize the results.

26. **Challenge** Predict whether the equation $y = \dfrac{x^2 - 4}{x + 2}$ is *linear* or *nonlinear*. Make a graph to test your prediction. Explain the results.

ONGOING ASSESSMENT

27. **Open-ended Problem** Make a concept map that summarizes the distinction between linear and nonlinear relationships. Be sure to include examples. (For an example of a concept map, see the study technique on page 184 of Chapter 4.)

SPIRAL REVIEW

Find each product. *(Section 6.4)*

28. $5t(15t - 2)$　　　29. $(p - 3)(p + 7)$　　　30. $(u + 4)(2u - 3)$

31. $(2r + 5)(3r + 3)$　　32. $\left(5 + \sqrt{3}\right)\left(4 - \sqrt{3}\right)$　　33. $\left(2 + \sqrt{5}\right)\left(1 + \sqrt{3}\right)$

Rewrite each equation in slope-intercept form. Then graph it.
Use a graphing calculator if you have one. *(Section 7.1)*

34. $3x - 5y = 7$　　　35. $3x + 5y = 7$　　　36. $5y - 3x = 7$

37. $7y + 5x = 3$　　　38. $3x + 7y = 5$　　　39. $3x - 7y = 5$

8.2

Learn how to...

- **recognize characteristics of parabolas**

So you can...

- **predict the shape of a parabola**

- **recognize side views of real objects, such as a radio telescope**

Exploring Parabolas

Some nonlinear graphs are U-shaped curves called **parabolas**. The graph of the equation $y = x^2$ shown below is one example of a parabola. Every other parabola is a variation of this basic curve.

EXPLORATION
COOPERATIVE LEARNING

Analyzing the Shape of a Parabola

Work with a partner.
You will need:
- a graphing calculator
- graph paper

1 One person should graph the equations below. Set a good viewing window, such as $-9 \le x \le 9$ and $-2 \le y \le 10$. What do you notice about the shapes of the parabolas?

a. $y = x^2$ **b.** $y = 2x^2$ **c.** $y = 3x^2$

2 On graph paper, the other person should sketch what he or she thinks the graph of $y = 4x^2$ will look like in relation to the graphs in Step 1. Use a graphing calculator to check the prediction.

3 Switch roles and repeat Step 1 using the equations below.

a. $y = x^2$ **b.** $y = \frac{1}{2}x^2$ **c.** $y = \frac{1}{3}x^2$

4 On graph paper, sketch what you think the graph of $y = \frac{1}{4}x^2$ will look like in relation to the graphs in Step 3. Use a graphing calculator to check your prediction.

5 Work together to generalize what happens to the graph of the equation $y = ax^2$ as the value of a increases from 0 to 1 and beyond.

6 Work together to generalize what happens to the graph of the equation $y = ax^2$ as the value of a decreases from 0 to -1 and beyond. Try some examples. Be sure to set a good viewing window.

Quadratic Functions

A function or an equation whose graph is a parabola is called **quadratic**. The word *quadratic* comes from a Latin word that means *to make square*. All quadratic equations have a squared term in them.

The **vertex of a parabola** is the point where the function reaches a minimum or a maximum value. The plural of *vertex* is *vertices*.

Some examples of quadratic functions are shown below.

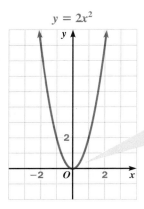

$y = 2x^2$

The function reaches a **maximum** value of $y = 0$ when $x = 0$.

The function reaches a **minimum** value of $y = 0$ when $x = 0$.

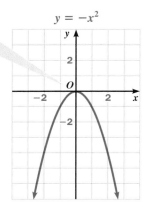

$y = -x^2$

Notice that the graph of the function with a positive coefficient opens up, while the graph of the function with a negative coefficient opens down.

EXAMPLE

Predict how the graph of the equation $y = -3x^2$ will compare with the graph of the equation $y = x^2$. Then sketch the graph.

SOLUTION

Here is one student's solution of the Example.

First look at the sign of the coefficient of x^2. The graph of $y = -3x^2$ will open down instead of up because the coefficient is negative. Now ignore the sign. The graph will be narrower and steeper than the graph of $y = x^2$, because 3 is greater than 1. The graph will go through the point $(1, -3)$ instead of $(1, 1)$.

THINK AND COMMUNICATE

1. Do you think there is a maximum value for the function graphed at the right? Explain.

2. How would the solution of the Example be different if the equation were $y = -\frac{1}{3}x^2$ instead of $y = -3x^2$?

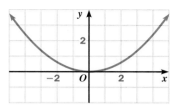

3. Which equation has the widest graph: $y = \frac{1}{4}x^2$, $y = -\frac{1}{2}x^2$, or $y = 4x^2$?

Symmetry

If you fold the graph of an equation in the form $y = ax^2$ along the y-axis, the two sides of the parabola line up. We say that the parabola is **symmetric** about the y-axis. The y-axis is the **line of symmetry**.

line of symmetry

As a rule, if you draw a vertical line through the vertex of any parabola, the parabola will be symmetric about that line.

☑ CHECKING KEY CONCEPTS

Is the graph of each equation *narrower* or *wider* than the graph of $y = x^2$? Does the graph open *up* or *down*? What does this tell you about the vertex?

1. $y = 5x^2$

2. $y = -5x^2$

3. $y = \frac{1}{5}x^2$

4. $y = -\frac{1}{5}x^2$

5. $y = 0.75x^2$

6. $y = -0.8x^2$

Sketch the graph of each equation.

7. $y = 5x^2$

8. $y = -\frac{1}{5}x^2$

9. $y = 0.75x^2$

8.2 Exercises and Applications

Extra Practice
exercises on
page 572

Match each equation with its graph.

1. $y = x^2$ **2.** $y = -4x^2$ **3.** $y = \frac{1}{4}x^2$ **4.** $y = 4x^2$

A.

B.

C.

D.
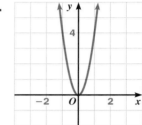

5. For each equation in Exercises 1–4, tell whether the function reaches a *maximum* or a *minimum* value at the vertex of the graph.

6. Writing Suppose someone shows you a function whose equation is in the form $y = ax^2$. Without graphing the equation, how can you tell whether the function will reach a minimum or a maximum value at the vertex of the graph?

 Technology For Exercises 7–15, use a graphing calculator to check your predictions.

a. Predict how the graph of each equation will compare with the graph of $y = x^2$. Explain how you made your prediction.

b. Sketch the graph of each equation.

7. $y = 6x^2$ **8.** $y = -2x^2$ **9.** $y = -10x^2$

10. $y = \frac{1}{6}x^2$ **11.** $y = -\frac{1}{3}x^2$ **12.** $y = \frac{2}{3}x^2$

13. $y = 0.5x^2$ **14.** $y = 1.25x^2$ **15.** $y = -2.5x^2$

16. Open-ended Problem Write an equation for a parabola that opens down and is wider than the graph of $y = x^2$. Then write an equation for a parabola that opens up and is narrower than the graph of $y = x^2$. Sketch all three equations in the same coordinate plane.

17. Challenge Is it possible for a quadratic equation in the form $y = ax^2$ to have a vertex at a point other than the origin? Why or why not?

Tell whether each graph is symmetric about the *y*-axis. Explain.

18.

19.

20.

Connection ▶ ASTRONOMY

Radio telescopes are designed to monitor radio waves. A common type of radio telescope has an antenna in the shape of a parabolic dish. This means that if you look at a cross section of the antenna, you will see a parabola.

The radio telescopes pictured below are part of a worldwide network. Data collected from these telescopes is pooled and then analyzed by computer to provide high-resolution images.

Each equation approximates the shape of the given telescope dish. Match each equation with its graph. Explain how you made each choice. (*Note:* Assume that the scale is the same on all three graphs.)

21. Parkes telescope, Australia: $y = 0.003x^2$

A.

22. Fortaleza telescope, Brazil: $y = 0.013x^2$

23. Hartebeesthoek telescope, South Africa: $y = 0.007x^2$

B.

C.

BY THE WAY

At the University of California at Berkeley, scientists use a network of radio telescopes to look for signs of extra-terrestrial civilizations.

Cooperative Learning For Exercises 24–27, work with a partner. You will need six sheets of graph paper and six objects of various sizes that you can use to trace circles.

24. Take turns tracing around each object on graph paper. Measure the radius of each circle drawn. Use the width of one square on the graph paper as your unit of measure.

25. Estimate the area of each circle drawn by counting the number of squares inside the circle. Include parts of squares to get as close an estimate as possible.

26. Make a scatter plot of your combined data. Put radius on the horizontal axis and area on the vertical axis.

27. Graph the equation for the area of a circle $A = \pi r^2$, $r \geq 0$, along with your scatter plot. Compare the graph and the scatter plot.

28. **SAT/ACT Preview** Which of the following equations has a graph that is a parabola?

   A. $y = 3x$ **B.** $y = 3|x|$ **C.** $y = 3x^2$ **D.** $y = 2x^3$

29. **Challenge** Suppose x represents both the radius of a circle and the length of a side of a square. As x increases, which area increases more quickly, the area of the circle or the area of the square? Explain.

ONGOING ASSESSMENT

30. **Writing** Suppose a friend has missed this lesson. Summarize how to predict what the graph of an equation in the form $y = ax^2$ will look like. Be sure to include examples.

SPIRAL REVIEW

Find each product. *(Section 6.5)*

31. $(y + 3)^2$ **32.** $(r - 4)^2$ **33.** $(2p + 1)^2$

34. $\left(3 + \sqrt{5}\right)^2$ **35.** $(x + 6)(x - 6)$ **36.** $(3n - 2)(3n + 2)$

Find the unknown length in each right triangle. *(Section 6.1)*

37.

38.

39.

8.3 Solving Quadratic Equations

Learn how to...

- **use square roots and graphs to solve simple quadratic equations**

So you can...

- **relate x-intercepts and solutions of quadratic equations**

- **solve simple motion problems involving acceleration**

Have you ever considered dropping an egg out of a window? At some schools this is an annual event! Art, physics, and engineering students test their skills by designing containers to prevent eggs from breaking when dropped from heights of 30 ft or more.

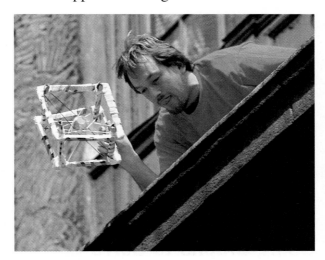

When an egg is dropped out of a window, the egg *accelerates*. That means it falls faster and faster over time. The formula $d = 16t^2$ can be used to model the distance d in feet that an egg falls in t seconds.

BY THE WAY

Students use all kinds of materials to cushion their eggs, including marshmallows, popcorn, gelatin, and peanut butter.

EXAMPLE 1 Application: Physics

Suppose an egg is dropped from a height of 80 ft. About how long will it take for the egg to reach the ground?

SOLUTION

Problem Solving Strategy: Use a formula.

Substitute 80 for d in the formula $d = 16t^2$ and solve for t.

$$80 = 16t^2$$
$$5 = t^2$$
$$\pm\sqrt{5} = t \qquad \text{Find the square root of both sides.}$$
$$\pm 2.24 \approx t$$

A negative time value does not make sense in this situation. Therefore, there is only one solution. It will take about 2.2 seconds for an egg to reach the ground when dropped from a height of 80 ft.

Quadratic equations can have as many as two solutions. You must decide which solutions apply to a given situation and ignore any solution that doesn't make sense, such as the negative time value in Example 1.

EXAMPLE 2

Solve the equation $x^2 - 9 = 0$.

SOLUTION

You can solve this quadratic equation algebraically. First get x^2 alone on one side of the equation.

$$x^2 - 9 = 0$$
$$x^2 = 9$$
$$x = \pm\sqrt{9} \quad \text{Find the square root of both sides.}$$
$$x = \pm 3$$

The solutions are 3 and -3.

THINK AND COMMUNICATE

1. Do Example 1 and Example 2 have the same number of solutions? Why or why not?

2. What is the first step you would take to solve the equation $x^2 - 25 = 0$?

3. What is the first step you would take to solve the equation $3x^2 = 48$?

Using Graphs to Solve Equations

A graph of the equation $y = x^2 - 9$ is shown below.

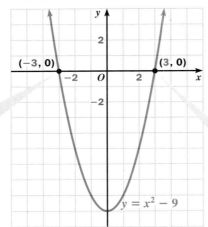

(−3, 0) (3, 0)

$y = x^2 - 9$

Notice that the *x*-intercepts **−3** and **3** are the solutions of the equation $x^2 - 9 = 0$ in Example 2. This makes sense, because the *x*-intercepts of $y = x^2 - 9$ occur when $y = 0$.

When $x = -3$, $y = 0$.
$$y = x^2 - 9$$
$$0 = (-3)^2 - 9$$

When $x = 3$, $y = 0$.
$$y = x^2 - 9$$
$$0 = 3^2 - 9$$

In general, the solutions of a quadratic equation in the form $ax^2 + c = 0$ are the *x*-intercepts of the graph of the related equation $y = ax^2 + c$. This is also true for quadratic equations in other forms.

EXAMPLE 3

Solve the equation $2x^2 - 4 = 24$.

SOLUTION

Method 1 Solve the equation algebraically.

Work to get x^2 alone on one side of the equation.

$$2x^2 - 4 = 24$$
$$2x^2 = 28$$
$$x^2 = 14$$
$$x = \pm\sqrt{14}$$
$$x \approx \pm 3.74$$

Find the square root of both sides.

The solutions are about -3.74 and about 3.74.

Method 2 Solve the equation using a graph.

Step 1 Rewrite the equation in the form $ax^2 + c = 0$.

$$2x^2 - 4 = 24$$
$$2x^2 - 28 = 0$$

Step 2 Graph the related equation $y = 2x^2 - 28$.

Step 3 Find the x-intercepts.

X=-3.740027   Y=-.0244021

X=3.7400266   Y=-.0244021

Use the trace and zoom features to get more accurate values.

The solutions are about -3.74 and about 3.74.

THINK AND COMMUNICATE

4. Contrast the very first steps of the two methods shown in Example 3.

5. What would you do first to solve the equation $x^2 + 25 = 75$ algebraically? What would you do first to solve it using a graph?

6. In Example 3, suppose you graphed the equation $y = 2x^2 - 4$. Would the x-intercepts be the solutions of $2x^2 - 4 = 24$? Explain.

7. What are the exact solutions of the equation $2x^2 - 4 = 24$ in Example 3? Which method will give you exact solutions? Explain.

Solve each equation using the given graph.

1. $2x^2 = 0$

2. $-x^2 + 4 = 0$

3. $\frac{1}{4}x^2 - 4 = 0$

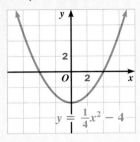

Solve each equation.

4. $x^2 = 100$

5. $2n^2 = 32$

6. $p^2 - 49 = 0$

7. $d^2 - 25 = 16$

8. $4x^2 + 18 = 27$

9. $70 = 5n^2 + 10$

8.3 | **Exercises and Applications**

*Extra Practice
exercises on
page 573*

Solve each equation algebraically.

1. $x^2 = 81$

2. $121 = n^2$

3. $3p^2 = 27$

4. $-10k^2 = -1000$

5. $\frac{1}{2}r^2 = 2$

6. $\frac{3}{4}x^2 = 12$

7. $k^2 - 144 = 0$

8. $0 = p^2 - 64$

9. $n^2 - 0.25 = 0$

10. a. COOKING A recipe for pizza dough says to use a 10 in. by 14 in. rectangular pan. Jake only has a circular pizza pan with a 12 in. diameter. If Jake makes the pizza in his circular pan, will the crust be thicker or thinner? Explain.

b. Suppose Jake wants to make a round pizza that is exactly the same thickness as the 10 in. by 14 in. pizza in the recipe. What will the diameter of the pan have to be? Is Jake likely to find such a pan? Why or why not?

Estimate the solutions of each equation using the given graph.

11. $-x^2 + 3 = 0$

12. $\frac{1}{2}x^2 - 3 = 0$

13. $-\frac{1}{4}x^2 + 5 = 0$

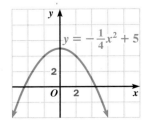

INTERVIEW Bill Pinkney

Look back at the article on pages 322–324.

Sailors like Bill Pinkney have to adjust their sails according to the speed of the wind in order to keep from being blown over. A mathematical formula that models the force F in lb/ft² on a sail when the wind is blowing perpendicular to the sail is $F = 0.00431v^2$, where v is the speed of the wind in knots (1 knot ≈ 1.15 mi/h).

14. According to the article, Pinkney's ideal wind speed was 35 knots. To the nearest tenth, what is the force on a sail of a 35-knot wind blowing perpendicular to the sail?

15. Estimate the wind speed that will produce a force of 4 lb/ft² when the wind is blowing perpendicular to the sail.

16. Challenge How could you use a graph of the equation $F = 0.00431v^2$ to answer Exercises 14 and 15? Are the intercepts of that graph the solutions of either exercise? Why or why not?

Cooperative Learning For Exercises 17–20, work with a partner. You will each need a graphing calculator.

17. **Technology** Each of you should graph the equation $y = x^2$ on a graphing calculator. Then one of you should do part (a). The other should do part (b). Work together on part (c).

    a. Graph the equations $y = x^2 + 1$, $y = x^2 + 2$, and $y = x^2 + 3$ along with the graph of $y = x^2$. Record the location of the vertex of each graph.

    b. Graph the equations $y = x^2 - 1$, $y = x^2 - 2$, and $y = x^2 - 3$ along with the graph of $y = x^2$. Record the location of the vertex of each graph.

    c. Describe how each graph compares with the original graph.

18. **Technology** Work together to generalize how the graph of an equation in the form $y = x^2 + c$ changes as the value of c changes.

19. **Technology** Work together to write a procedure for exploring how the graph of an equation in the form $y = |x| + c$ changes as the value of c changes. Carry out the exploration and generalize the results.

20. Predict how the graph of each equation will compare with the graph of $y = 2x^2$. Then sketch the graph of each equation. Check your predictions.

    a. $y = 2x^2 - 4$ **b.** $y = 2x^2 + 4$ **c.** $y = 2x^2 + 0.25$

Solve each equation algebraically or by using a graph.

21. $x^2 + 8 = 12$ **22.** $-n^2 + 10 = -15$ **23.** $4r^2 = 12$

24. $30 = 6p^2$ **25.** $0 = \frac{1}{4}x^2 - 1$ **26.** $4k^2 - 30 = 10$

27. $55 = 45n^2 - 25$ **28.** $-3r^2 + 10 = 4$ **29.** $\frac{2}{3}x^2 - 9 = 17$

30. Writing Solve the equations $3x - 12 = 0$ and $3x^2 - 12 = 0$. How is solving a quadratic equation in the form $ax^2 + c = 0$ like solving a linear equation in the form $ax + c = 0$? How is it different?

31. Open-ended Problem Write a quiz with five questions that could be used to check your understanding of the material in this section. Show how to answer all the questions on your quiz.

Connection   SPORTS

Skydivers refer to the period before they open their parachutes as *free fall*. If there were no air resistance, the formula $d = 16t^2$ could be used to model the distance d in feet a skydiver falls in t seconds of free fall. Because of air resistance, a skydiver eventually stops accelerating and reaches a terminal velocity.

32. The table shows the distances a skydiver falls in t seconds of free fall with air resistance. Make a scatter plot of the data in the table. Put time on the horizontal axis and distance on the vertical axis.

33. Graph the equation $d = 16t^2$ in the same coordinate plane as your scatter plot. Compare the graphs. How does comparing the graphs show you that the skydiver's acceleration is slowed down?

Time after jump (seconds)	Distance fallen (feet)
1	16
3	138
5	366
7	652
9	970
11	1,310
13	1,660
15	2,010
25	3,770
35	5,530
45	7,290
55	9,050
65	10,810
75	12,570

34. Challenge About how long does it take for the skydiver to reach terminal velocity? Estimate the terminal velocity. Give reasons for your answers.

35. Cooperative Learning Work in a group of three students. Take turns listing the key concepts of this section and examples that illustrate the concepts. Each of you should write a summary of the list.

A building is 150 ft long and 90 ft wide. Find each value. *(Section 5.2)*

36. A floor plan of the building is 20 in. long. How wide is the floor plan?

37. A cardboard model of the building is 8 in. wide. How long is the model?

38. An architect uses the scale 1 in. = 5 ft for the plans. What are the dimensions of the architect's plans?

Find the missing values for each pair of similar figures. *P* **represents perimeter and** *A* **represents area.** *(Section 5.4)*

39.

8

$P = 48$
$A = \underline{\ ?\ }$

6

$P = \underline{\ ?\ }$
$A = 72$

40.

10

$P = 24$
$A = \underline{\ ?\ }$

5

$P = \underline{\ ?\ }$
$A = 6$

ASSESS YOUR PROGRESS

VOCABULARY

nonlinear equation (p. 326)   **vertex of a parabola** (p. 332)
parabola (p. 331)   **symmetric** (p. 333)
quadratic (p. 332)   **line of symmetry** (p. 333)

Graph each equation. Tell whether each equation is *linear* or *nonlinear*.
(Section 8.1)

  1. $y = 2x + 1$ **2.** $y = 2x^2 + 1$ **3.** $y = 2|x| + 1$

Predict how the graph of each equation will compare with the graph of
$y = x^2$**. Explain your prediction, then sketch the graph.** *(Section 8.2)*

  4. $y = 7x^2$ **5.** $y = -7x^2$ **6.** $y = 0.7x^2$

Solve each equation algebraically or by using a graph. *(Section 8.3)*

  7. $x^2 = 144$ **8.** $\frac{1}{3}n^2 = 3$ **9.** $0 = k^2 - 4$

10. $-5p^2 + 20 = 0$ **11.** $2n^2 - 3 = 13$ **12.** $4x^2 + 25 = 49$

13. Journal Explain how you can solve the equation $3x^2 + 6 = 25$ algebraically. Then explain why the *x*-intercepts of the graph of the equation $y = 3x^2 - 19$ are the solutions of the equation $3x^2 + 6 = 25$.

8.4

Applying Quadratic Equations

Learn how to...

- **use graphs to solve more complicated quadratic equations**

So you can...

- **recognize quadratic equations with no solution**

- **solve any quadratic equation and find heights of thrown objects, for example**

Have you ever played or watched a game of table tennis? You may not have realized that the path the ball followed through the air was a parabola. There is a quadratic relationship between the height of the ball above the table and the ball's horizontal distance from where it bounced.

There is a *different* quadratic relationship between the height of the ball above the table and the *time* since the ball bounced. For the lower parabola in the photograph, the height-time relationship is modeled by this equation:

$$h = -4.9t^2 + 2.1t$$

In the equation, *h* stands for height in meters and *t* stands for time in seconds.

BY THE WAY

The celluloid ball used in table tennis was introduced in 1902. The diameter is about $1\frac{1}{2}$ in., and the ball weighs less than $\frac{1}{10}$ oz.

THINK AND COMMUNICATE

Use the graph of the equation $h = -4.9t^2 + 2.1t$ below.

1. Estimate the solutions of the equation $-4.9t^2 + 2.1t = 0$. What do the solutions tell you about the situation?

2. Estimate the maximum height the ball reached. About how much time had gone by when the ball reached this height?

Height of Ball over Time

All **quadratic equations** can be written in the form $ax^2 + bx + c = 0$, when $a \neq 0$. You can find the solutions of an equation in this form by using a graph. In fact, finding solutions of a quadratic equation is so important in mathematics that some graphing calculators have a *root* feature that will find the solutions for you. *Root* is another word for the solution of an equation.

EXAMPLE 1

Solve each equation.

a. $x^2 - 2x - 5 = 0$

b. $3x^2 + x = -2$

SOLUTION

a. Graph the related equation $y = x^2 - 2x - 5$. Then find the x-intercepts, which are where $y = 0$.

For more information about using the root feature, see the *Technology Handbook*, p. 607.

The solutions are about -1.45 and about 3.45.

b. Rewrite the equation in the form $ax^2 + bx + c = 0$.

$$3x^2 + x = -2$$
$$3x^2 + x + 2 = 0$$

Graph the related equation $y = 3x^2 + x + 2$. Notice that there are no x-intercepts. There are no values of x that make the equation $3x^2 + x + 2 = 0$ true.

The equation has no solution.

THINK AND COMMUNICATE

3. How can you use a graph to tell whether a quadratic equation has any solutions?

4. How could you use the trace and zoom features of a graphing calculator to find the solutions in Example 1(a)?

5. The vertex of the parabola $y = x^2 - 4x + 4$ lies on the x-axis. How many solutions does the related equation $x^2 - 4x + 4 = 0$ have?

6. Suppose a is negative. What are the possible numbers of solutions of the equation $ax^2 + bx + c = 0$?

EXAMPLE 2   **Application: Sports**

Janet threw a basketball. The equation $h = -16t^2 + 5t + 5$ describes the height h of the ball in feet t seconds after she threw it. About how many seconds went by before the ball hit the ground?

SOLUTION

Problem Solving Strategy: Use a graph.

To solve the equation using a graph, you must first substitute y for h and x for t in the equation $h = -16t^2 + 5t + 5$.

$$y = -16x^2 + 5x + 5$$

When the ball hit the ground, the height was 0. Solve the equation $0 = -16x^2 + 5x + 5$ by graphing the related equation $y = -16x^2 + 5x + 5$ and finding the x-intercepts.

Time cannot be negative, so ignore this intercept.

Root
X=.73669299 Y=0

Locate the positive x-intercept.

The ball hit the ground after about 0.7 seconds.

THINK AND COMMUNICATE

7. What does the vertex of the parabola in Example 2 tell you about the situation?

8. What equation would you have to solve to find out when the ball in Example 2 was 2 ft off the ground?

☑ CHECKING KEY CONCEPTS

Rewrite each equation in the form $ax^2 + bx + c = 0$.

1. $x^2 + 3x = 8$ **2.** $x^2 - 6x + 4 = 7$ **3.** $5x^2 + x - 8 = 6$

 Technology   Use a graphing calculator to solve each equation. If an equation has no solution, write *no solution*.

4. $x^2 - 2x - 3 = 0$ **5.** $6x^2 + 3x + 2 = 0$

6. $-x^2 + 4x = 0$ **7.** $3x^2 - x = 2$

8. $x^2 + 10x + 34 = 9$ **9.** $-4x^2 + 16x - 7 = 9$

8.4 Exercises and Applications

Estimate the solutions of each equation using the given graph. If an equation has no solution, write *no solution*.

Extra Practice exercises on page 573

1. $-x^2 + 4x - 3 = 0$

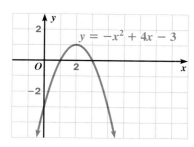

2. $x^2 + 2x + 1 = 0$

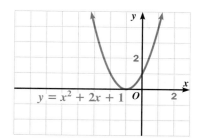

3. $\frac{1}{2}x^2 + 2x - \frac{1}{2} = 0$

4. $-x^2 + 3x - \frac{15}{4} = 0$

INTERVIEW Bill Pinkney

Look back at the article on pages 322–324.

According to the article, Pinkney's ideal wave height was "anywhere from 15 to 25 feet." An equation for the graph in the article is $h = 0.013s^2$, where h is wave height in feet and s is wind speed in miles per hour.

5. Use the graph on page 324 in the article to estimate the wind speeds that correspond to Pinkney's ideal wave heights.

6. **Technology** Use graphs of the equations $h = 0.013s^2 - 15$ and $h = 0.013s^2 - 25$ to estimate the wind speeds in Exercise 5.

7. Use equation solving to estimate the wind speeds in Exercise 5.

8. **Writing** Compare the methods used in Exercises 5–7. Which do you prefer? Why?

Technology Use a graphing calculator to solve each equation.
If an equation has no solution, write *no solution*.

9. $x^2 + 2x - 8 = 0$ **10.** $2x^2 + 7x - 4 = 0$ **11.** $-x^2 + 9 = 0$

12. $x^2 - 9x + 18 = 0$ **13.** $x^2 - 8x + 16 = 0$ **14.** $2x^2 - 3x + 5 = 0$

15. $-4x^2 - 12x = 9$ **16.** $x^2 - 4x + 9 = 4$ **17.** $2x^2 + 6x - 4 = 1$

18. $-3x^2 - 6x + 8 = 3$ **19.** $3x^2 - 15x = 24$ **20.** $-x^2 + 8 = 12x$

21. SAT/ACT Preview If S represents the greater of the two solutions of
the equation $x^2 - 4x - 3 = 5$, then:

 A. $S > 3$ **B.** $S < 3$ **C.** $S = 3$

 D. relationship cannot be determined

Connection HIGHWAY SAFETY

Speed limits vary from country to country. The speed limit
on highways in Mexico is 100 km/h. In France the limits
vary from 90 to 130 km/h. In Germany, some sections of
the highway have no speed limit. Other sections are
limited to either 180 km/h or 130 km/h.

The equation $d = 0.0056s^2 + 0.14s$
models the relationship between a
vehicle's stopping distance d in
meters and the vehicle's speed s in
kilometers per hour.

Germany

Mexico

France

 Technology For Exercises 22–24,
use a graphing calculator.

22. How many solutions does the equation $0.0056s^2 + 0.14s = 0$
have? How many of the solutions make sense in this situation?

23. About how fast should you be driving in order to be able to stop
within 100 m? In which of the above countries is this speed within
the speed limit?

24. Suppose a car is traveling at the speed of 130 km/h on a German
highway. What distance should the driver of the car leave between
the driver's car and the car in front of it?

25. Research Find out what is a reasonable length in meters for a car.
Then estimate the number of car lengths that corresponds to the
stopping distance you found in Exercise 24.

26. Open-ended Problem Pick one of the real-world equations in this section. Write your own set of problems about it. Solve each problem.

Challenge **For Exercises 27 and 28, use the three graphs below. Then do Exercises 29–31.**

a. $y = x^2 - 2x$

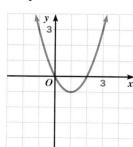

b. $y = x^2 - 4x$

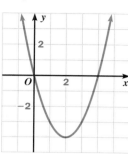

c. $y = x^2 + 2x - 3$

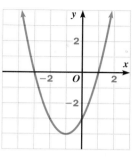

27. Identify the x-intercepts and the coordinates of the vertex of each graph.

28. Explain how the x-intercepts of each graph are related to the x-coordinate of the vertex.

29. The x-intercepts of the graph of the equation $y = x^2 - 12x + 32$ are 4 and 8. Explain how to find the x-coordinate of the vertex without graphing the equation.

30. For a graph of the equation $y = x^2 + bx + c$, the x-coordinate of the vertex is a function of b. Write an equation for the function.

31. Use the function you wrote in Exercise 30. For each equation, predict the x-coordinate of the vertex of the graph.

   a. $y = x^2 - 4x + 2$ **b.** $y = x^2 + 8x - 3$ **c.** $y = x^2 + 5x + 6$

ONGOING ASSESSMENT

32. Open-ended Problem A graph of the equation $y = -x^2 + 2x + 8$ is shown. Make a list of all the things you can think of to say about the graph based on what you have learned so far in this chapter.

SPIRAL REVIEW

Write each repeating decimal as a fraction in lowest terms. *(Section 6.2)*

33. $0.6666\ldots$ **34.** $0.7171\ldots$ **35.** $0.\overline{001}$

Write each expression in simplest form. *(Section 6.3)*

36. $\sqrt{50}$ **37.** $\sqrt{\dfrac{49}{9}}$ **38.** $\sqrt{\dfrac{7}{16}}$

39. $\sqrt{48}$ **40.** $7\sqrt{5} + 3\sqrt{5}$ **41.** $\sqrt{8} + 3\sqrt{2}$

8.5

Learn how to...

- **use the quadratic formula**

So you can...

- **solve quadratic equations without using a graph**

- **find more precise solutions to real-world problems**

The Quadratic Formula

Sometimes you want a more precise solution of a quadratic equation than you can get from a graph. You can use the *quadratic formula* to find exact solutions of quadratic equations in the form $ax^2 + bx + c = 0$, when $a \neq 0$:

$$x = \frac{-b + \sqrt{b^2 - 4ac}}{2a} \quad \text{or} \quad x = \frac{-b - \sqrt{b^2 - 4ac}}{2a}$$

The **quadratic formula** is usually written in this form:

$$x = \frac{-b \pm \sqrt{b^2 - 4ac}}{2a}$$

Example 1 shows you how to use the quadratic formula to find the values of x that make an equation in the form $ax^2 + bx + c = 0$ true.

EXAMPLE 1

Solve the equation $2x^2 + 6x - 8 = 0$.

SOLUTION

Find the values of a, b, and c: $2x^2 + 6x - 8 = 0$.

Substitute **2** for a, **6** for b, and **−8** for c in the quadratic formula.

$$x = \frac{-b \pm \sqrt{b^2 - 4ac}}{2a}$$

$$= \frac{-6 \pm \sqrt{6^2 - 4(2)(-8)}}{2(2)}$$

$$= \frac{-6 \pm \sqrt{100}}{4}$$

$$= \frac{-6 \pm 10}{4}$$

There are two exact solutions.

$$x = \frac{-6 + 10}{4} \quad \text{or} \quad x = \frac{-6 - 10}{4}$$

$$= \frac{4}{4} \qquad\qquad = \frac{-16}{4}$$

$$= 1 \qquad\qquad\quad = -4$$

The solutions are 1 and −4.

Check

Substitute each solution, 1 and −4, in the equation.

$$2x^2 + 6x - 8 = 0$$
$$2(1)^2 + 6(1) - 8 \stackrel{?}{=} 0$$
$$2 + 6 - 8 \stackrel{?}{=} 0$$
$$0 = 0 \checkmark$$

$$2(-4)^2 + 6(-4) - 8 \stackrel{?}{=} 0$$
$$32 - 24 - 8 \stackrel{?}{=} 0$$
$$0 = 0 \checkmark$$

THINK AND COMMUNICATE

1. What do you notice about the two solutions of the equation $2x^2 + 6x - 8 = 0$ from Example 1 and the graph of $y = 2x^2 + 6x - 8$ at the right?

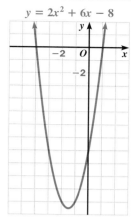
$y = 2x^2 + 6x - 8$

2. How would the answer to Example 1 be different if the equation were $2x^2 - 6x - 8 = 0$ instead of $2x^2 + 6x - 8 = 0$?

When you use the quadratic formula, you can write your answer with a radical sign or find a decimal approximation.

EXAMPLE 2   Application: Sports

A diver jumps from a platform which is 10 m above the water. Her height h in meters above the water t seconds after she jumps is modeled by the equation $h = -4.9t^2 + 3.5t + 10.8$. About how long will it take her to reach the water?

SOLUTION

The diver's height when she reaches the water is 0 m. Use the quadratic formula to solve the equation $0 = -4.9t^2 + 3.5t + 10.8$.

$$t = \frac{-b \pm \sqrt{b^2 - 4ac}}{2a}$$

$$= \frac{-3.5 \pm \sqrt{(3.5)^2 - 4(-4.9)(10.8)}}{2(-4.9)}$$

Substitute **−4.9** for *a*, **3.5** for *b*, and **10.8** for *c*.

$$= \frac{-3.5 \pm \sqrt{223.93}}{-9.8}$$

There are two solutions.

$$t = \frac{-3.5 + \sqrt{223.93}}{-9.8} \quad \text{or} \quad t = \frac{-3.5 - \sqrt{223.93}}{-9.8}$$

$$\approx -1.17 \qquad\qquad\qquad \approx 1.88$$

Since the time cannot be negative, −1.17 is not a realistic answer.

It will take the diver about 1.9 seconds to reach the water.

EXAMPLE 3

Solve the equation $x^2 - 8x + 18 = 2$.

SOLUTION

Rewrite the equation in the form $ax^2 + bx + c = 0$.

$$x^2 - 8x + 18 = 2$$
$$x^2 - 8x + 16 = 0$$

Substitute **1** for a, **−8** for b, and **16** for c in the quadratic formula.

$$x = \frac{-b \pm \sqrt{b^2 - 4ac}}{2a}$$

$$= \frac{-(-8) \pm \sqrt{(-8)^2 - 4(1)(16)}}{2(1)}$$

$$= \frac{8 \pm \sqrt{64 - 64}}{2}$$

$$= \frac{8 \pm \sqrt{0}}{2}$$

There is only one solution because $\sqrt{0} = 0$.

$$= \frac{8}{2}$$

$$= 4$$

The solution is 4.

☑ CHECKING KEY CONCEPTS

1. What does the graph of the equation $y = x^2 - 8x + 16$ at the right tell you about the solution of the equation $x^2 - 8x + 16 = 0$ from Example 3? Explain.

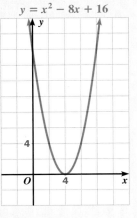

$y = x^2 - 8x + 16$

For each equation in the form $ax^2 + bx + c = 0$, identify the values of a, b, and c.

2. $x^2 + 10x + 25 = 0$

3. $x^2 - 3x - 10 = 0$

4. $5n^2 - n + 1 = 0$

5. $35t^2 + 3t = 0$

Rewrite each equation in the form $ax^2 + bx + c = 0$. Then solve.

6. $k^2 + 4k - 2 = 3$

7. $4x^2 - 2x = 1$

8. $n^2 - 6 = 4n$

9. $4.9t^2 = 7t$

8.5 Exercises and Applications

Solve each equation using the quadratic formula.

Extra Practice
*exercises on
page 573*

1. $x^2 - 3x - 18 = 0$ **2.** $y^2 - 2y + 1 = 0$ **3.** $3n^2 - 10n + 3 = 0$

4. $5p^2 + 4p - 1 = 0$ **5.** $-5x^2 + 9x + 2 = 0$ **6.** $2t^2 - 5t = 0$

7. Writing Jessie solved the equation $x^2 - 6x - 1 = 0$. When someone asked her how many solutions she found, she said there was only one: $\dfrac{6 \pm \sqrt{40}}{2}$. Write a letter to Jessie explaining why there are two solutions. Use a graph to make your explanation clearer.

Technology For Exercises 8–16, solve each equation using the quadratic formula. Then check each solution using a graphing calculator.

8. $y^2 + 12y + 10 = 0$ **9.** $4t^2 - 16t = 0$ **10.** $3n^2 - 9 = 0$

11. $-x^2 + 3x + 3 = 5$ **12.** $2n^2 + 9n + 4 = 1$ **13.** $-6y^2 + 2y = -3$

14. $3t^2 - 4t = 6$ **15.** $8y^2 + 2 = 10y$ **16.** $3x^2 = 4x + 5$

17. FIREFIGHTING A firefighter aims a hose at a window 25 ft above the ground. The equation $h = -0.01d^2 + 1.06d + 5$ models the path of the water, when h equals height in feet. Estimate, to the nearest foot, the horizontal distance d in feet between the firefighter and the building.

BY THE WAY

The first volunteer fire department in the American Colonies was founded in 1736 by Benjamin Franklin.

25 ft

d

18. Open-ended Problem You know how to solve quadratic equations using algebra, graphs, and the quadratic formula. Sometimes one method of solving is more convenient than another method. Describe how you would solve each equation. Give reasons for your answers.

a. $16x^2 = 1600$ **b.** $x^2 - 350 = 0$ **c.** $x^2 + 10x = 0$

19. Challenge Solve the equation $(x + 2)(x - 3) = 14$.

20. SAT/ACT Preview Which two numbers are solutions of the equation $x^2 - 6x + 8 = 0$?

A. 2 and -3 **B.** 2 and -4 **C.** -2 and 4   **D.** 2 and 4

In 1994, Brazil won the World Cup soccer tournament by defeating the Italian team 3 to 2. The championship game was decided by a round of penalty kicks, since neither team had scored during 90 min of regulation play and 30 min of overtime. The Italian team overshot the goal twice, sending the ball several feet over the goal's crossbar.

A player takes a penalty kick. The equation $h = -0.002d^2 + 0.36d$ models the height h in feet of the ball at distance d in feet along the ground from where the ball is kicked. Use this equation for Exercises 21–23.

21. Assume that the ball overshoots the goal and doesn't hit any obstacles before it lands on the ground. About how far will the ball go horizontally before it lands?

22. The penalty mark is 12 yd from the goal. The goal's crossbar is 8 ft high. By about how many feet does the ball go over the crossbar? (Remember that 3 ft = 1 yd.)

23. To the nearest foot, what is the ball's horizontal distance from the goal when it is 8 ft high?

ONGOING ASSESSMENT

24. **Writing** Explain why it is possible to solve the equation $3x^2 - 12 = 0$ algebraically, without using a graph or the quadratic formula, but not the equation $3x^2 - 14x + 12 = 0$.

SPIRAL REVIEW

Tell whether each number is *rational* or *irrational*. *(Section 6.2)*

25. $\sqrt{161}$

26. $\sqrt{121}$

27. $\sqrt{109}$

Use substitution to solve each system of equations. *(Section 7.2)*

28. $y = 2x - 1$
    $3x + 5y = 34$

29. $y = 5x - 6$
    $2x - 7y = 9$

30. $x = 3y - 6$
    $-7x - 5y = -10$

31. $a = \frac{4}{3}b$
    $15a - 4b = 4$

32. $p = 1$
    $7p - q = 6$

33. $y = 5x - 32$
    $3x - 4y = -59$

8.6

Learn how to...

- find the discriminant of a quadratic equation

So you can...

- find the number of solutions of a quadratic equation

Using the Discriminant

The expression $b^2 - 4ac$ in the quadratic formula is called the **discriminant**.

$$x = \frac{-b \pm \sqrt{b^2 - 4ac}}{2a}$$

The discriminant gives you information about the number of possible solutions when you solve a quadratic equation.

EXPLORATION
COOPERATIVE LEARNING

Solutions of Quadratic Equations

Work in a group of 2 or 3 students.
You will need:
- a graphing calculator

1 Take turns graphing the following equations. Use a table like the one shown to record the number of solutions of each equation when $y = 0$.

a. $y = -3x^2 + 5x + 5$

b. $y = 5x^2 + 3x + 2$

c. $y = 6x^2 + 12x + 6$

d. $y = x^2 + 8x - 3$

e. $y = x^2 - 2x + 4$

f. $y = x^2 + 4x + 4$

Equation	Number of solutions	$b^2 - 4ac$
$0 = -3x^2 + 5x + 5$	?	?
$0 = 5x^2 + 3x + 2$	?	?
$0 = 6x^2 + 12x + 6$	?	?
$0 = x^2 + 8x - 3$	?	?
$0 = x^2 - 2x + 4$	?	?
$0 = x^2 + 4x + 4$	?	?

2 Find the value of $b^2 - 4ac$ for each equation in Step 1. Record the results in your table.

3 Work together to make a generalization about the value of the discriminant and the number of solutions of an equation.

The Discriminant

The discriminant tells you how many real number solutions there are of an equation in the form $ax^2 + bx + c = 0$, when $a \neq 0$.

When $b^2 - 4ac = 0$, there is exactly **one** solution.

When $b^2 - 4ac > 0$, there are **two** solutions.

When $b^2 - 4ac < 0$, there are **no** solutions.

The graph of $y = ax^2 + bx + c$ has **one** x-intercept.

The graph of $y = ax^2 + bx + c$ has **two** x-intercepts.

The graph of $y = ax^2 + bx + c$ has **no** x-intercepts.

  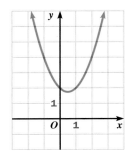

EXAMPLE

How many solutions are there of the equation $-3x^2 - 6x - 1 = 3$?

SOLUTION

In order to use the discriminant, the equation must be in the form $ax^2 + bx + c = 0$.

$$-3x^2 - 6x - 1 = 3$$
$$-3x^2 - 6x - 4 = 0$$

Subtract 3 from both sides of the equation.

Now you can identify a as -3, b as -6, and c as -4. To find the discriminant, substitute these values in the expression $b^2 - 4ac$.

$$b^2 - 4ac = (-6)^2 - 4(-3)(-4) = -12$$

The discriminant is negative, so there are no solutions.

When you solve the equation $-3x^2 - 6x - 4 = 0$ using the quadratic formula, this is what you get:

$$x = \frac{-(-6) \pm \sqrt{(-6)^2 - 4(-3)(-4)}}{2(-3)} = \frac{6 \pm \sqrt{-12}}{-6}$$

Because the square root of a negative number does not exist in the real numbers, there are no solutions.

THINK AND COMMUNICATE

1. **Technology** Graph the equation $y = -3x^2 - 6x - 4$. What does the graph tell you about the number of solutions when $y = 0$?

2. Describe how you would find the number of solutions of $-3x^2 - 6x + 4 = 7$. How could you check your answer?

☑ CHECKING KEY CONCEPTS

Find the discriminant for each equation. What does it tell you about the number of solutions?

1. $x^2 - 3x + 2 = 0$ 2. $5n^2 + 9n + 6 = 0$ 3. $3y^2 + 12 = 12y$

4. $-12k^2 + 5k - 9 = 0$ 5. $4x^2 - 4x + 1 = 0$ 6. $2t^2 - 7t - 4 = 0$

The density of water changes depending on its temperature. The equation $d = -0.0069T^2 + 0.0549T + 999.8266$ models the relationship between the density of water d in mg/cm^3 and its temperature T in °C.

7. For the equation $-0.0069T^2 + 0.0549T + 999.8266 = 0$, identify the values of a, b, and c that you would use to find the discriminant.

8. How many solutions are there when $d = 999.90$? when $d = 1000$?

9. Water freezes at a temperature of 0°C. What is the density of water at this temperature?

8.6 | Exercises and Applications

For each equation, find the number of solutions.

1. $x^2 - 9x + 14 = 0$ 2. $6y^2 - 8y + 3 = 0$ 3. $2p^2 - 13p - 3 = 0$

4. $0 = 6n^2 - n + 5$ 5. $-3x^2 - 5x + 1 = 0$ 6. $25k^2 - 10k = 0$

7. $2y^2 + 10 = 0$ 8. $n^2 + 49 = 14n$ 9. $x^2 = 64$

Extra Practice exercises on page 574

10. **GEOMETRY** Use the figures shown.

   a. Write an equation for the combined area of the two figures as a function of s.

   b. Find the discriminant of the equation you wrote in part (a) if the combined area is 48. What does this tell you about the number of possible solutions?

   c. Solve your equation using the quadratic formula. How many of the solutions make sense in this situation?

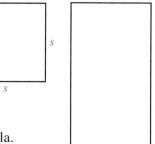

The Water Arc is a fountain that shoots recirculated water across the Chicago River from a water cannon. The path of the Water Arc can be modeled by the equation $y = -0.006x^2 + 1.2x + 10$, where x is the horizontal distance in feet and y is the vertical distance in feet.

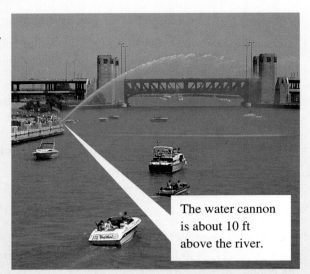

The water cannon is about 10 ft above the river.

11. How many solutions of the equation are there when $y = 0$? Find each solution using the quadratic formula.

12. How many solutions of the equation are there when $y = 70$? Find each solution using the quadratic formula.

13. Use the results of Exercises 11 and 12 to sketch the path of the Water Arc in a coordinate plane. Label your graph with key points, including the location of the water cannon.

14. Research The Water Arc was built to commemorate an engineering project which reversed the direction of the Chicago River. Find out about the history of this project.

For each equation, find the number of solutions. Then solve using the quadratic formula.

15. $x^2 + 7x + 6 = 0$ **16.** $y^2 - 5y + 6 = 0$ **17.** $x^2 - 2x + 1 = 0$

18. $n^2 - 8n + 5 = 15$ **19.** $x^2 = 2x + 2$ **20.** $3k^2 - 8k + 5 = 4$

21. $-x^2 - 16 = -8x$ **22.** $3n^2 + 11n = 4$ **23.** $-4x^2 = -12x - 9$

24. Writing Negative numbers do not have square roots in the real numbers. Use what you know about multiplying negative numbers to explain why.

25. Challenge What happens to the equation $0 = ax^2 + bx + c$ when $a = 0$, but $b \neq 0$ and $c \neq 0$? Can you use the discriminant to find the number of solutions? Can you use the quadratic formula to solve the equation? Give an example to support your answer.

26. Open-ended Problem Explain why some real-world problems involving quadratic equations have only one solution, even when the discriminant tells you that there are two. Include examples.

27. SAT/ACT Preview How many solutions does the equation $32 + 2x^2 = 16x$ have?

   A. none   **B.** one   **C.** two   **D.** three

28. SCIENCE An experiment shows that the relationship between a time t that it takes a person to react to a sound and the person's age a in years can be modeled by the equation $t = 0.0051a^2 - 0.3185a + 15.0008$.

(Each unit of time equals $\frac{1}{77}$ second.)

a. Do you think that the graph of this equation has a horizontal intercept? Why or why not? Check using the discriminant.

b. How many solutions are there to the equation when $t = 12$?

O N G O I N G A S S E S S M E N T

29. Cooperative Learning Work with a partner.

a. Each of you should make six flash cards with questions on the front and answers on the back. Your questions should cover the discriminant, the quadratic formula, and solutions of a quadratic equation.

b. Take turns holding up your flash cards and quizzing each other. Give each other helpful hints and go over any mistakes together.

S P I R A L R E V I E W

Solve each system of equations. *(Section 7.4)*

30. $2x - 3y = -5$
$\quad\ \ 3x - y = 3$

31. $5x - y = -1$
$\quad\ \ 3x - 2y = 5$

32. $4x + 9y = 8$
$\quad\ \ -10x - 3y = -7$

Solve each equation. *(Section 8.3)*

33. $x^2 - 4 = 0$

34. $4y^2 - 16 = 0$

35. $z^2 - 1 = 0$

36. $2 = r^2 - 7$

37. $z^2 - 5 = 0$

38. $2x^2 - 3 = 0$

ASSESS YOUR PROGRESS

VOCABULARY

quadratic equation (p. 345)   **discriminant** (p. 355)
quadratic formula (p. 350)

 Technology Solve each equation using a graphing calculator. *(Section 8.4)*

1. $x^2 - 5x - 4 = 0$ **2.** $-x^2 + 3x - 2 = 0$ **3.** $x^2 + x + 2 = 0$

Solve each equation using the quadratic formula. *(Section 8.5)*

4. $x^2 - 5x - 24 = 0$ **5.** $2x^2 + x - 15 = 0$ **6.** $2x^2 - x = 2$

Without graphing or solving each equation, find the number of x-intercepts of the graph of each equation. *(Section 8.6)*

7. $y = 3x^2 - 4x + 1$ **8.** $y = 3x^2 - 4x + 2$ **9.** $y = 9x^2 - 12x + 4$

10. Journal Summarize the ways you know to solve a quadratic equation.

Investigating Gravity

Downhill skiers take advantage of steep slopes to ski faster. Gravity enables them to accelerate as they go down the slope. Objects are pulled toward the ground because of gravity. The bigger the object, the greater its pull on other objects. Since Earth is the biggest thing around, it pulls objects (like skiers) toward it.

PROJECT GOAL Explore how gravity affects the time it takes a ball to roll down a ramp.

Collecting Data

Work with a partner to recreate an experiment first performed by Galileo Galilei over four hundred years ago. Measure the time it takes a solid heavy ball or marble to roll down a ramp for at least ten different distances. Here are some key steps you should follow:

1. MAKE a ramp by taping two meter sticks together side by side in a "V" shape.

2. PLACE one end of the ramp on a stack of books so that the other end forms a 10° angle with the floor. Check the angle with a protractor.

3. USE a stopwatch to measure the time it takes a ball to reach the end of the ramp. Record the time and the distance for ten starting points.

Based on your observations, how long do you think it would take the ball to roll 55 cm? How far would the ball roll in 0.75 s?

Comparing Graphs

Galileo showed that a ball rolling down an inclined plane accelerates at a constant rate. The equation below shows the relationship between the amount of time t in seconds that a solid ball rolls and the distance d it travels along a ramp inclined at a 10° angle:

$$d = 0.61t^2$$

• Calculate how long it would take a ball to roll 55 cm. Compare this value to the value you estimated using your scatter plot. What might explain any differences?

• Sketch a graph of the equation $d = 0.61t^2$ in the same coordinate plane as your scatter plot. How do the graph and the scatter plot compare?

Writing a Report

Write a report about your experiment. Include a brief description of your procedure. Describe the relationship between the distance a ball travels and the time it takes to travel that far.

4. MAKE a scatter plot of the data. Put time in seconds on the horizontal axis and distance in meters on the vertical axis.

You may want to extend your report to explore some of the ideas below:

• What effect does doubling the time a ball rolls have on the distance it travels?

• Will the results of the experiment change if you change the angle of the ramp? If so, how?

• Will the results of the experiment change if you use a lighter or a heavier solid ball? Why or why not?

Self-Assessment

Identify the mathematical skills you used to analyze the relationship you explored in this experiment. What are some possible sources of error in the way you collected your data? Describe how you and your partner worked together or could work together in the future to minimize errors.

8 Review

STUDY TECHNIQUE

Draw a concept map, as on page 184, to show what you learned in this chapter.

VOCABULARY

nonlinear equation (p. 326)
parabola (p. 331)
quadratic (p. 332)
vertex of a parabola (p. 332)
symmetric (p. 333)

line of symmetry (p. 333)
quadratic equation (p. 345)
quadratic formula (p. 350)
discriminant (p. 355)

SECTIONS | 8.1 *and* 8.2

When $a > 0$, the parabola opens up.

The vertex can be a maximum or a minimum.

When $a < 0$, the parabola opens down.

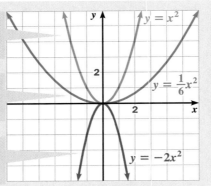

If the graph of an equation is not a line, then the equation and the relationship it models are both **nonlinear**. *Quadratic* functions and absolute value functions are nonlinear.

The graph of a **quadratic** equation is a **parabola**. A parabola that has an equation in the form $y = ax^2$ is **symmetric** about the y-axis. The shape of the parabola depends on the value of $|a|$.

SECTIONS | 8.3 *and* 8.4

To solve an equation like $2x^2 - 2 = 6$ algebraically, get the x^2 term alone on one side of the equation. Then find the square root of both sides.

$$2x^2 - 2 = 6$$
$$2x^2 = 8$$
$$x^2 = 4$$
$$x = \pm\sqrt{4} = \pm 2$$

The solutions are 2 and -2.

A quadratic equation can be written in the form $ax^2 + bx + c = 0$, when $a \neq 0$. The solutions are the x-intercepts of the graph of $y = ax^2 + bx + c$.

For example, you can find the solutions of the equation $3x^2 - 2x = 5$ by using a graph. First rewrite the equation in the form $3x^2 - 2x - 5 = 0$. Then graph the related equation $y = 3x^2 - 2x - 5$. The x-intercepts are the solutions.

You can use the root feature of a graphing calculator to find the solutions. They are -1 and $1\frac{2}{3}$.

SECTIONS 8.5 *and* 8.6

You can use the **quadratic formula** to solve a quadratic equation in the form $ax^2 + bx + c = 0$, when $a \neq 0$.

$$x = \frac{-b \pm \sqrt{b^2 - 4ac}}{2a}$$

For example, in the equation $8x^2 - 9x + 1 = 0$, $a = 8$, $b = -9$, and $c = 1$.

$$x = \frac{-(-9) \pm \sqrt{(-9)^2 - 4(8)(1)}}{2(8)}$$

$$= \frac{9 \pm \sqrt{49}}{16}$$

$$= \frac{9 \pm 7}{16}$$

There are two solutions.

$$x = \frac{9 + 7}{16} \quad \text{or} \quad x = \frac{9 - 7}{16}$$

$$= 1 \qquad\qquad = \frac{1}{8}$$

The solutions are 1 and $\frac{1}{8}$.

The value of the expression $b^2 - 4ac$ in the quadratic formula is called the **discriminant**. You can use the discriminant to find the number of solutions of a quadratic equation.

When $b^2 - 4ac = 0$, the quadratic equation has **one** solution.

When $b^2 - 4ac > 0$, the quadratic equation has **two** solutions.

When $b^2 - 4ac < 0$, the quadratic equation has **no** solutions.

8 Assessment

VOCABULARY REVIEW

1. Give an example of an equation whose graph is linear. Give an example of an equation whose graph is nonlinear.

2. Sketch a graph that is symmetric about the *y*-axis. What does the word *symmetry* mean?

3. What do all quadratic equations have in common? What do the graphs of all quadratic equations have in common?

4. What is the discriminant? What information does it give you about a quadratic equation?

SECTIONS 8.1 *and* 8.2

Tell whether each relationship is *linear* or *nonlinear*.

5. $y = 2|x| - 6$ **6.** $y = 5x + 1$ **7.** $y = 2x^2 + 10$

8. the relationship between *s* and the volume *V* of the box, where $V = 2s^3$

9. the relationship between *x* and the perimeter *P* of the equilateral triangle, where $P = 3x$

Describe how the graph of each equation compares to the graph of $y = x^2$.

10. $y = 2x^2$ **11.** $y = -6x^2$ **12.** $y = -\frac{1}{2}x^2$

Match each equation with its graph.

13. $y = -x^2$

14. $y = \frac{1}{3}x^2$

15. $y = 1.5x^2$

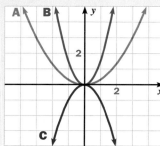

SECTIONS 8.3 *and* 8.4

Solve each equation algebraically.

16. $x^2 + 4 = 20$ **17.** $2x^2 = 50$ **18.** $3x^2 - 11 = 16$

19. $\frac{1}{2}x^2 + 5 = 13$ **20.** $\frac{2}{3}x^2 - 7 = 17$ **21.** $5x^2 - 43 = 87$

Estimate the solutions of each equation using the related graph.

22. $3x^2 - 4 = 0$

23. $-2x^2 + 5 = 0$

 Technology
Solve each equation using a graphing calculator.

24. $2x^2 - 5x = 3$ **25.** $3x^2 + 9x + 2 = 0$ **26.** $6x^2 + 5x + 7 = 0$

SECTIONS 8.5 *and* 8.6

Solve each equation using the quadratic formula.

27. $x^2 + 12x + 11 = 0$ **28.** $16x^2 - 8x + 1 = 0$ **29.** $5x^2 - 3x = 1$

For each equation, find the number of solutions.

30. $2x^2 + 9x - 2 = 0$ **31.** $3x^2 - x + 3 = 0$ **32.** $4x^2 + 4x + 1 = 0$

33. a. The square and the rectangle are equal in area. Write an equation that models this relationship.

b. How many solutions does your equation have?

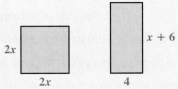

c. Writing How many of the solutions make sense? Explain.

PERFORMANCE TASK

34. Write a quadratic equation that has each combination of solutions. Use what you have learned in this chapter to support each answer.

a. no solutions   **b.** exactly one positive solution
c. exactly one negative solution   **d.** two positive solutions
e. two negative solutions   **f.** one solution of each sign

9 | Exponential Functions

Hands that heal

" **The patient and I have to work together toward the same goal: his wellness.** "

As a young girl growing up on the Navajo reservation in Crownpoint, New Mexico, Dr. Lori Arviso Alvord loved working with her hands. She learned to do beadwork from a Sioux friend and studied classical piano for six years. Both skills required fine work with her hands. "I always wanted to do something where I could use my hands," she says. "Surgery was the most appropriate field in medicine to be doing that." According to the Association of American Indian Physicians, Alvord is the first Navajo woman to become a surgeon. After completing her residency, she returned to New Mexico to work among the Navajo. Now, as a general surgeon at the Gallup Indian Medical Center, Alvord uses her hands to heal.

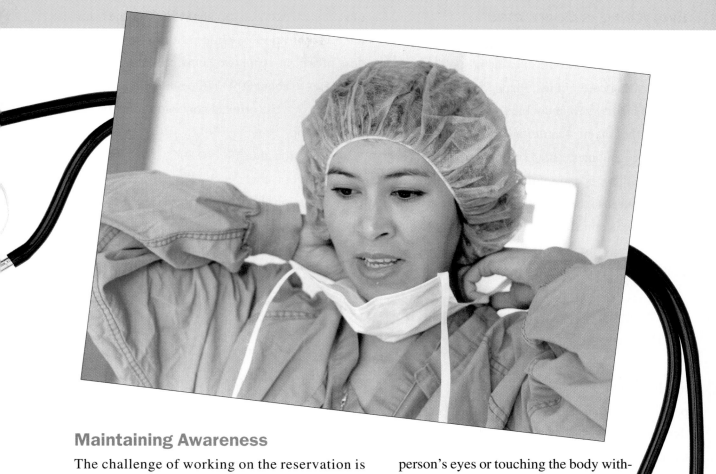

Maintaining Awareness

The challenge of working on the reservation is unique. Navajo people continue to turn to healers from within their own community when there is a need for medical care. But when so-called modern medicine is the only option, an awareness of Navajo customs is a must. For this reason, Alvord applies what she knows of her people's culture to her work.

For example, looking directly into another person's eyes or touching the body without asking permission is considered an insult. And although she does speak some Navajo, Alvord sometimes uses an interpreter to communicate with her patients. "If you don't relate well with Navajo people, they won't allow you to operate on them," she says. "They might just not show up."

Everything is Connected

At Alvord's recommendation, medicine men have been encouraged to come in to the hospital to perform healing ceremonies. Plans for an improved medical facility will undoubtedly include her suggestion that the new building have a round room for such ceremonies, called sings. Alvord says that, for her, "there is a great deal of spirituality" in surgery, because the Navajo philosophy holds that everything is connected. "The patient and I have to work together toward the same goal: his wellness."

Mathematics and Medicine

Part of Alvord's medical training involves understanding how long a particular medicine will be effective. In other words, what happens to the concentration of medicine in a person's bloodstream over time? Consider the case of salicylate, an active ingredient in aspirin. The graph shows that the concentration

of salicylate decreases to about 80% of the original concentration one hour after taking a small dose of aspirin. After two hours, the concentration decreases to about 80% of the remaining amount. If you let C = the original concentration, you can see a pattern:

Concentration after 1 h = $0.80C$

Concentration after 2 h = $0.80(0.80C)$

Concentration after 3 h = $0.80[0.80(0.80C)]$

In this chapter, you will learn that this pattern can be modeled by an *exponential function*.

Concentration of Salicylate in Bloodstream

EXPLORE AND CONNECT

Lori Arviso Alvord examines the x-rays of one of her patients.

1. Writing Based on what you read in the interview, describe how to find the concentration of salicylate in the body after four hours. After how many hours do you think the concentration of salicylate will be close to zero? Explain.

2. Research Find out some of the requirements for becoming a medical doctor. What are some of the reasons that doctors might need to know mathematics?

3. Project Many medicines used today are based on chemicals found in plants. Some examples are aspirin, ephedrine, digitalis, quinine, and reserpine. Find out about one of these medicines or some other medicine. Make a poster or write a report on its discovery and use.

Mathematics
& Lori Arviso Alvord

In this chapter, you will learn more about how mathematics is related to medicine.

Related Exercises

Section 9.2
• Exercises 19–21

Section 9.3
• Exercises 22–26

Section 9.5
• Exercises 18–21

9.1

Exponents and Powers

According to an ancient folktale from India, a man named Sissa Ben Dahir invented the game of chess. The king of India liked the game so much that he wanted to reward Sissa with 64 gold pieces, one for each square on a chessboard. Instead, Sissa asked for grains of wheat arranged on the chessboard in the pattern shown below.

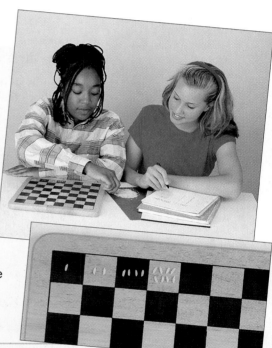

EXPLORATION
COOPERATIVE LEARNING

Exploring Exponents

Work with a partner.
You will need:
- a chessboard
- a handful of rice

SET UP Use rice to model Sissa Ben Dahir's pattern. Place the grains of rice on a chessboard as shown. Each square should contain twice as many grains as the square before it. Continue this pattern until you run out of rice. Then answer the questions below.

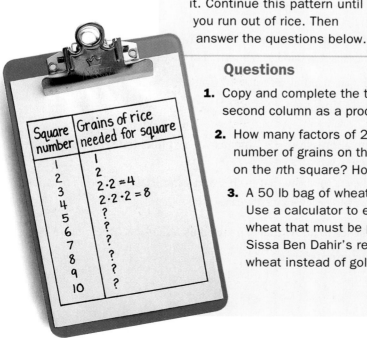

Square number	Grains of rice needed for square
1	1
2	2
3	2·2 = 4
4	2·2·2 = 8
5	?
6	?
7	?
8	?
9	?
10	?

Questions

1. Copy and complete the table. Express each number in the second column as a product of 2's.

2. How many factors of 2 would be in the product for the number of grains on the 30th square? on the 64th square? on the nth square? How do you know?

3. A 50 lb bag of wheat contains about 2,500,000 grains. Use a calculator to estimate the number of 50 lb bags of wheat that must be placed on the 30th square to satisfy Sissa Ben Dahir's request. Why do you think he asked for wheat instead of gold?

Using Powers

In the Exploration on page 369, the fifth square of the chessboard had $2 \cdot 2 \cdot 2 \cdot 2$ grains of rice. You can use a *power* to represent repeated multiplication by the same number.

$$\underbrace{2 \cdot 2 \cdot 2 \cdot 2}_{4 \text{ factors}} = 2^{\overset{\displaystyle 4 \; \text{exponent}}{}}$$

base

The **base** is the number being used as a factor. The **exponent** is the number of times the base is used as a factor.

The whole expression is called a **power**. You read 2^4 as "**two** to the **fourth** power" or "the **fourth** power of **two**." It means you use **2** as a factor **4** times.

EXAMPLE 1

Rewrite each product using exponents.

a. $-(9 \cdot 9 \cdot 9 \cdot 9 \cdot 9 \cdot 9 \cdot 9 \cdot 9 \cdot 9 \cdot 9)$

b. $xy \cdot xy \cdot xy \cdot xy \cdot xy$

c. $3\left(\dfrac{1}{4} \cdot \dfrac{1}{4} \cdot \dfrac{1}{4} \cdot \dfrac{1}{4} \cdot \dfrac{1}{4} \cdot \dfrac{1}{4} \cdot \dfrac{1}{4} \cdot \dfrac{1}{4}\right)$

SOLUTION

Find the factor repeated in the expression. This is the base. Then count the number of times it is used as a factor. This is the exponent.

WATCH OUT! ▶

Notice that $-9^{10} = -(9^{10})$. The expression -9^{10} is *not* equal to $(-9)^{10}$.

a. The number 9 is used as a factor 10 times.
$$-(9 \cdot 9 \cdot 9 \cdot 9 \cdot 9 \cdot 9 \cdot 9 \cdot 9 \cdot 9 \cdot 9) = -\left(9^{10}\right)$$
$$= -9^{10}$$

Note that the expression -9^{10} means you exponentiate the 9 before you apply the negative.

b. Use parentheses to show that both x and y are used as factors 5 times.
$$xy \cdot xy \cdot xy \cdot xy \cdot xy = (xy)^5$$

c. Use parentheses to show that $\dfrac{1}{4}$ is used as a factor 8 times, but 3 is used as a factor only once.
$$3\left(\frac{1}{4} \cdot \frac{1}{4} \cdot \frac{1}{4} \cdot \frac{1}{4} \cdot \frac{1}{4} \cdot \frac{1}{4} \cdot \frac{1}{4} \cdot \frac{1}{4}\right) = 3\left(\frac{1}{4}\right)^8$$

EXAMPLE 2

Evaluate each power.

a. 2^5 b. $\left(\dfrac{1}{2}\right)^3$ c. $(-3)^4$

SOLUTION

a. $2^5 = 2 \cdot 2 \cdot 2 \cdot 2 \cdot 2$
$\quad = 32$

b. $\left(\dfrac{1}{2}\right)^3 = \dfrac{1}{2} \cdot \dfrac{1}{2} \cdot \dfrac{1}{2}$
$\qquad = \dfrac{1}{8}$

c. $(-3)^4 = (-3)(-3)(-3)(-3)$
$\qquad = 9 \cdot 9$
$\qquad = 81$

The parentheses show that the base is -3, not 3.

THINK AND COMMUNICATE

1. Look back at the Exploration on page 369. Write a power representing the number of grains that must be placed on the nth square ($1 < n \le 64$) to satisfy Sissa Ben Dahir's request. Explain your reasoning.

2. Let $b =$ any real number and let $n =$ any positive integer. Complete each equation. Explain your answers.

   a. $b^1 = \underline{\ ?\ }$ b. $1^n = \underline{\ ?\ }$ c. $0^n = \underline{\ ?\ }$

3. **a.** Evaluate each of the powers below when $b = -2$.

   $b^1 \qquad\qquad b^2 \qquad\qquad b^3 \qquad\qquad b^4$

   b. Let $b =$ any negative number. When is b^n positive? When is b^n negative? Explain your reasoning.

Powers and Order of Operations

The list below tells you where evaluating powers fits into the order of operations you learned about in Section 1.2.

> **Order of Operations**
>
> 1. Simplify expressions inside parentheses.
> 2. Evaluate powers.
> 3. Do multiplication and division from left to right.
> 4. Do addition and subtraction from left to right.

EXAMPLE 3 **Application: Architecture**

Linda Savas is designing a building with a cube-shaped entrance, as shown by the model. The building is to be 50 ft tall and cover a square piece of land, but she is not sure about the other dimensions.

a. Write a variable expression for the volume of the building.

b. Find the volume of the building if the area the building covers is 60 ft by 60 ft, and the entrance is 25 ft by 25 ft by 25 ft.

SOLUTION

a. Problem Solving Strategy: Use a formula.

Let x = the length of one side of the area that the building covers.

Let y = the length of one side of the entrance.

You can think of the building as a box with a cube cut out of it.

$$\text{Volume} = \text{Length} \times \text{Width} \times \text{Height}$$

$$\text{Volume of box} = x \cdot x \cdot 50$$
$$= 50x^2$$

$$\text{Volume of cube} = y \cdot y \cdot y$$
$$= y^3$$

The building's volume is the volume of the box minus the volume of the cube.

$$\text{Volume of building} = \text{Volume of box} - \text{Volume of cube}$$
$$= 50x^2 - y^3$$

b. Evaluate the expression when $x = 60$ and $y = 25$.

$$50x^2 - y^3 = 50(60)^2 - (25)^3$$

Evaluate the powers first.

$$= 50 \cdot 3600 - 15{,}625$$
$$= 180{,}000 - 15{,}625$$
$$= 164{,}375$$

The volume of the building is 164,375 ft^3.

Rewrite each product using exponents.

1. $5 \cdot 5 \cdot 5 \cdot 5$ **2.** $4 \cdot 17 \cdot 17$ **3.** $2(y \cdot y \cdot y)$

Evaluate each power.

4. 4^6 **5.** $\left(\frac{1}{7}\right)^3$ **6.** $(-5)^3$

Evaluate each expression when $x = 3$.

7. $6x^2$ **8.** $4x^3 - 5$ **9.** $2(x + 1)^4$

9.1 **Exercises and Applications**

Rewrite each product using exponents.

Extra Practice
exercises on
page 574

1. $4.1 \cdot 4.1 \cdot 4.1 \cdot 4.1 \cdot 4.1$ **2.** $(-1.5)(-1.5)(-1.5)$

3. $\left(-\frac{3}{8}\right)\left(-\frac{3}{8}\right)\left(-\frac{3}{8}\right)$ **4.** $-a \cdot a \cdot a \cdot a \cdot a \cdot a \cdot a$

5. $pq(4r \cdot 4r \cdot 4r \cdot 4r)$ **6.** $(8 \cdot 8 \cdot 8 \cdot 8 \cdot 8 \cdot 8)(c - d)$

Evaluate each power.

7. 4^3 **8.** 3^4 **9.** 11^3 **10.** 6^4

11. $\left(\frac{1}{2}\right)^5$ **12.** $\left(-\frac{1}{2}\right)^5$ **13.** $(-2.8)^2$ **14.** $(1729.38)^1$

Write an algebraic expression for each phrase.

15. six to the seventh power   **16.** the sixth power of seven

17. the tenth power of y **18.** one fourth to the nth power

19. **Technology** Most calculators have a key for evaluating powers. This key is usually labeled , or .
For example, to evaluate 3^5 you might use this key sequence:

Use a calculator to evaluate each power. Round decimal answers to the nearest hundredth.

a. 5^8 **b.** $(1.65)^{15}$ **c.** $(-7)^9$

20. Michelle showed her classmates how to evaluate expressions like -3^2 and $(-3)^2$. Use Michelle's work shown here to help you evaluate the expressions in parts (a)–(d).

a. -2^5 and $(-2)^5$ **b.** -7^3 and $(-7)^3$

c. $-\left(\frac{1}{6}\right)^2$ and $\left(-\frac{1}{6}\right)^2$ **d.** $-x^9$ and $(-x)^9$

$$-3^2 = -(3 \cdot 3) = -9$$
$$(-3)^2 = (-3)(-3) = 9$$

A *byte* is the amount of computer memory needed to store a single character, such as "a" or "1." A computer's total memory is usually measured in *kilobytes* or *megabytes*. A kilobyte (K) is 2^{10} bytes. A megabyte (MB) is 2^{20} bytes.

Note: Your answers to Exercises 21–23 should not involve powers.

When computers were first invented, they filled entire rooms. Now some computers are so small that you can hold them on your lap.

21. a. Find the number of bytes in a kilobyte and in a megabyte.

   b. Research Look up the prefixes "kilo-" and "mega-" in a dictionary. Explain why it makes sense to use the words "kilobyte" and "megabyte" for 2^{10} and 2^{20} bytes, respectively.

22. One graphing program requires 250 K of computer memory to run. How many bytes are needed to run the program?

23. A certain brand of computer has 8 MB of memory. Express the computer's memory in bytes.

24. Challenge How many times larger than 2^n is 2^{n+1}? Explain.

Evaluate each expression when $x = 2$ and $y = 5$.

25. $3x^5 + 10$ **26.** $6y^2 - x^4$ **27.** $4(2 - y)^3$ **28.** $x^4 y$

29. xy^3 **30.** $(xy)^3$ **31.** $x^4 + y^4$ **32.** $(x + y)^4$

33. y^x **34.** x^y **35.** $10(x - y)^3$ **36.** $(x - y)^2 y^3$

GEOMETRY Formulas for the surface area S.A. and volume *V* of a sphere with radius *r* are given below. Use the formulas for Exercises 37–39.

$$\text{S.A.} = 4\pi r^2 \qquad V = \frac{4}{3}\pi r^3$$

37. Copy and complete the table. Leave your answers in terms of π.

38. How do the surface area and volume of a sphere change when the radius of the sphere is doubled? Use the table to support your answer.

39. Open-ended Problem Suppose a toy store sells rubber balls in two different sizes. The radius of the larger balls is twice the radius of the smaller balls. How do you think the store owner should price the two different kinds of rubber balls? Explain your reasoning.

40. SAT/ACT Preview If $A = 2^n$ and $B = 2n$ for a positive integer n, then:

   A. $A < B$ **B.** $A > B$ **C.** $A = B$

   D. relationship cannot be determined

Radius	Surface Area	Volume
1	?	?
2	?	?
4	?	?
8	?	?
n	?	?
2*n*	?	?

41. **CARPENTRY** Chris Taylor is building a storage unit like the one shown. The small cabinet portion is a cube with length, width, and height x. The tall shelf portion has length and width x and height y.

a. Write a variable expression for the volume of the storage unit.

b. Find the volume of the storage unit when $x = 2$ ft and $y = 6$ ft.

c. **Open-ended Problem** Substitute another pair of values for x and y and find the new volume of the storage unit. Be sure to choose reasonable values for each dimension.

42. **NUMBER THEORY** The largest known primes are found among a set of numbers named for the 17th-century French mathematician Marin Mersenne. A Mersenne number has the form $2^p - 1$, where p is a prime number. A Mersenne prime is a Mersenne number that is also prime.

a. Copy the table at the right. List the five smallest primes in the first column of your table. Then complete the rest of the table.

p	$2^p - 1$	Is $2^p - 1$ prime?
2	3	yes

b. **Writing** Is every Mersenne number a Mersenne prime? If so, explain how you can tell. If not, give an example of a Mersenne number that is not prime.

BY THE WAY

In 1978, two California students, Laura Nickel and Curt Noll, showed that the Mersenne number $2^{21,701} - 1$ is prime. This number has 6533 digits and was the largest known prime number at the time.

ONGOING ASSESSMENT

43. **Writing** When is the expression b^n negative? When is it zero? Explain your reasoning.

SPIRAL REVIEW

Describe a situation that can be represented by each expression. Then evaluate the expression when $m = 7$ and $p = 2$. *(Section 1.6)*

44. $32m$

45. $4.5p$

46. $\dfrac{2p}{3}$

Find each product. Then check the product by evaluating both sides of your equation when $r = -2$ and $s = 3$. *(Section 6.5)*

47. $(r + 5)^2$

48. $(2s + 3)(2s - 3)$

49. $(7 - 3r)^2$

9.2

Exponential Growth

Learn how to...

- interpret and evaluate exponential functions
- model real-world situations with exponential functions

So you can...

- make predictions about things that grow, such as population

Toolbox p. 587
Fractions, Decimals, and Percents

In 1980, there were only 75 peregrine falcons in Alaska. People decided to try to save the endangered bird. Because of their efforts, the population grew by about 12% per year from 1980 to 1992. You can use these facts to estimate the Alaskan peregrine population after one year.

$$\binom{\text{Population}}{\text{after } \mathbf{1} \text{ year}} = \binom{\text{Initial}}{\text{population}} + \binom{\text{Growth in}}{\text{population}}$$

$$\binom{\text{Population}}{\text{after } \mathbf{1} \text{ year}} = 75 + (12\% \text{ of } 75)$$

$$= 75 + 0.12(75) \qquad \text{0.12 is the same as 12\%.}$$

$$= 75(1 + 0.12)$$

$$= 75(1.12)^{\mathbf{1}}$$

$$= 84 \qquad \text{population in 1981}$$

The number 1.12 in the above calculations is called the *growth factor*. You can use the growth factor to find the peregrine population in other years.

EXAMPLE 1   **Application: Ecology**

Estimate the Alaskan population of peregrine falcons in 1983.

SOLUTION

Use the growth factor 1.12 to find the population in 1982 and in 1983.

$$\begin{aligned}
\text{Population after } \mathbf{2} \text{ years} &= \binom{\text{Population}}{\text{after } \mathbf{1} \text{ year}}(1.12) \\
&= [75(1.12)](1.12) \\
&= 75(1.12)^{\mathbf{2}} \qquad \text{population in 1982}
\end{aligned}$$

$$\begin{aligned}
\text{Population after } \mathbf{3} \text{ years} &= \binom{\text{Population}}{\text{after } \mathbf{2} \text{ years}}(1.12) \\
&= [75(1.12)(1.12)](1.12) \\
&= 75(1.12)^{\mathbf{3}} \qquad \text{population in 1983} \\
&\approx 105
\end{aligned}$$

In 1983, there were about 105 peregrine falcons in Alaska.

Exponential Equations

You can see from Example 1 that

Population after x years = 75(1.12)^x.

The number of peregrine falcons increased by a fixed percent each year. This kind of increase is called **exponential growth**. Any quantity y that grows by a fixed percent at regular intervals can be modeled by an equation in the form below. The x in the equation represents the number of intervals.

$$y = a(1 + r)^x$$

The constant a represents the initial amount before any growth occurs.

The constant r represents the **growth rate**. The quantity $1 + r$ represents the **growth factor**.

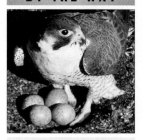

EXAMPLE 2   Application: Ecology

Use the information on page 376 to estimate the Alaskan population of peregrine falcons in 1991.

SOLUTION

Let y = the peregrine population x years after 1980.

$$y = a(1 + r)^x$$

The initial population in 1980 was 75 peregrine falcons, so **a = 75**.

The population grows by 12% each year after 1980, so **r = 0.12**.

Substitute these values into the equation.

$$y = 75(1 + 0.12)^x$$
$$= 75(1.12)^{11}$$
$$\approx 261$$

> 1991 is 11 years after 1980, so **x = 11**.

In 1991, there were about 261 peregrine falcons in Alaska.

THINK AND COMMUNICATE

The equation $y = 84(1.12)^x$ models the number of peregrine falcons in Alaska x years after 1981.

1. What does the number 84 represent in this equation?

2. What does the variable x represent in this equation?

3. What value of x would you substitute into this equation to estimate the number of peregrines in 1991? Explain your reasoning.

Graphing Exponential Equations

In Example 2 you saw that the Alaskan population of the peregrine falcon more than tripled in just 11 years. Another way to show the population increase over time is to graph the equation $y = 75(1.12)^x$.

As x becomes larger, y grows faster and the graph becomes steeper.

Seven years after 1980, the population was about 166.

For any graph that represents exponential growth, the value of y increases as x becomes larger.

THINK AND COMMUNICATE

4. Ming says that since the population increases by 12% each year, the population should increase by the same number of peregrines each year. Use the shape of the graph to show Ming his mistake.

5. The population y of New Zealand in thousands x years after 1980 can be modeled by the equation $y = 3113(1.006)^x$.

   a. What was the population of New Zealand in 1980? Explain.

   b. Predict where the graph of $y = 3113(1.006)^x$ crosses the vertical axis. Check your prediction by graphing the equation on a graphing calculator. Set a good viewing window, such as $0 \le x \le 50$ and $2500 \le y \le 4500$.

☑ CHECKING KEY CONCEPTS

Find the value of *y* when *x* = 8.

1. $y = 450(1.07)^x$ 2. $y = 12(1.5)^x$ 3. $y = 57(2)^x$

The exponential equation $y = 91.6(1 + 0.77)^x$ models the number of cellular telephone subscribers *y* in thousands in the United States *x* years after 1983.

4. What does the number 91.6 represent?

5. What does the number 0.77 represent?

6. Find the number of cellular telephone subscribers in the United States in 1992. Explain how you got your answer.

9.2 Exercises and Applications

Find the value of y when x = 5.

Extra Practice exercises on page 574

1. $y = 10(3)^x$ **2.** $y = 22(2.4)^x$ **3.** $y = 4.2(6)^x$

4. $y = (3.2)^x$ **5.** $y = 23.1(1.05)^x$ **6.** $y = 540(1.17)^x$

7. $y = 19(2.07)^x$ **8.** $y = 38(8.2)^x$ **9.** $y = 71(10)^x$

Each equation in Exercises 10–15 models the population y in thousands of the given country x years after 1980. For each country:

a. Give the population in 1980.

b. Give the annual growth rate.

c. Estimate the population in 1989 to the nearest thousand.

Explain how you found your answers.

10. Greece: $y = 9643(1.005)^x$ **11.** Austria: $y = 7549(1.002)^x$

12. Canada: $y = 24{,}070(1.01)^x$ **13.** Finland: $y = 4780(1.004)^x$

14. Japan: $y = 116{,}807(1.006)^x$ **15.** Italy: $y = 56{,}451(1.002)^x$

16. BIOLOGY A newly hatched channel catfish typically weighs about 0.3 g. This weight increases by an average of 10% per day for the first six weeks of a catfish's life.

    a. Write an equation that models the weight y of a catfish in grams x days after hatching. How did you find your equation?

    b. Estimate a catfish's weight 12 days after hatching.

PERSONAL FINANCE When you deposit money into a savings account, the money earns interest. For example, suppose you open a savings account that pays 2% annual interest. The money in your account will increase by 2% each year. For Exercises 17 and 18, assume no other deposits or withdrawals are made.

17. Suppose you use $200 to open an account paying 3.5% annual interest.

    a. Write an equation that gives the value y of the account in dollars x years from now.

    b. How much money will you have in the account after five years?

18. The value of an investment t years after you make an initial deposit of $500 depends on the interest rate I. Find the value of each account.

    a. $t = 10, I = 3\%$ **b.** $t = 20, I = 3\%$

    c. $t = 5, I = 6\%$ **d.** $t = 5, I = 12\%$

Lori Arviso Alvord

Surgeons like Lori Arviso Alvord need to protect patients from bacteria that can enter a surgical wound and reproduce. Most single-cell bacteria reproduce by splitting into two new cells. The pictures below show the increase in a population of bacteria. The amount of time that elapses between each split is called the generation time.

Look back at the article on pages 366–368.

19. a. By what percent does the population increase during each generation?

b. Suppose one cell of bacteria begins to reproduce by splitting into two. Write an equation that models the bacteria population y after x generations.

c. **Writing** Look back at your answer to part (b). How does your equation show that the population doubles in one generation?

20. One species of bacteria often found in infected surgical wounds is the *Pseudomonas aeruginosa*. It has a generation time of about 33 min.

a. Suppose one *P. aeruginosa* cell starts to reproduce. How many cells are there after 33 min? after 66 min? after 99 min?

b. Write an equation that models the number y of *P. aeruginosa* cells after $33x$ minutes have passed. Explain your reasoning.

BY THE WAY

Most of the bacteria that infect wounds actually live on normal healthy skin. They become harmful only when they enter the body through a wound.

21. **Challenge** The most common species of bacteria that infects surgical wounds is the *Staphylococcus aureus*. It has a generation time of about 30 min. For parts (a) and (b), let x = the number of generations.

a. Suppose two *S. aureus* cells have just split into four cells. Write and solve an equation to find the number y of *S. aureus* cells $15\frac{1}{2}$ h later.

b. Explain how you chose a value for x in part(a).

Technology How is the graph of $y = ab^x$ affected when you change the value of a? the value of b? For Exercises 22 and 23, use a graphing calculator. Set a good viewing window, such as $0 \le x \le 5$ and $0 \le y \le 25$.

22. Graph each equation on the same screen. How are the graphs similar? How are they different?

　　a. $y = 5(1.25)^x$　　　　**b.** $y = 5(1.5)^x$　　　　**c.** $y = 5(2)^x$

23. Graph each equation on the same screen. How are the graphs similar? How are they different?

　　a. $y = 5(1.2)^x$　　　　**b.** $y = 10(1.2)^x$　　　　**c.** $y = 15(1.2)^x$

For Exercises 24–26, match each function with its graph. Explain how you found your answers.

24. $y = 2(1.5)^x$ 　　　　**25.** $y = 2(1.2)^x$ 　　　　**26.** $y = 5(1.2)^x$

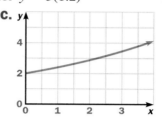

27. Open-ended Problem Describe a real-world situation that can be modeled by the equation $y = 150(1.06)^x$.

28. **Spreadsheets** In this exercise, you will compare the growth of three functions: the linear function $y = 2x$, the quadratic function $y = x^2$, and the exponential function $y = 2^x$.

　a. Use a spreadsheet to find the values of the three functions when $x = 1, 2, \ldots, 20$.

　b. Which function grows most quickly when $x > 4$? Use your spreadsheet to justify your answers.

　c. Graph the three functions in the same coordinate plane. Do the graphs support your answer to part (b)? If so, how?

In many spreadsheet programs, the symbol ^ is used for exponentiation.

Comparing Growth

B2 =2*A2

	A	B	C	D
1	x	2x	x^2	2^x
2	1	2	1	2
3	2	4	4	4
4	3			
5	4			
6	5			
7	6			
8	7			
9	8			
10	9			

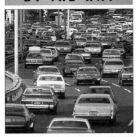
29. TRAFFIC In 1983, the traffic level in the United States was about 1653 billion vehicle-miles. (That means that all the vehicles in the country drove a total of 1653 billion miles that year.) The traffic level continued growing by about 4% per year for over a decade after that.

    a. Write an equation that models the traffic level y in billions of vehicle-miles x years after 1983.

    b. Estimate to the nearest billion the number of vehicle-miles traveled in 1989. What value did you substitute for x? Explain your reasoning.

30. SOCIAL STUDIES The British economist Thomas R. Malthus (1766–1834) observed that a human population tends to increase exponentially. A population model that assumes exponential growth is called a *Malthusian model*.

    a. The population of Portugal was about 9,778,000 in 1980 and about 10,365,000 in 1990. By what percent did the population increase from 1980 to 1990? Give your answer to the nearest percent.

$$\left(\textit{Hint: } \text{Percent increase} = \frac{1990 \text{ population} - 1980 \text{ population}}{1980 \text{ population}} \cdot 100\right)$$

    b. Create a Malthusian model by writing an exponential function that predicts Portugal's population y in millions x decades after 1980.

    c. Predict Portugal's population in 2020 assuming the growth rate remains the same. What value did you substitute for x? Explain.

ONGOING ASSESSMENT

31. Writing How do you know whether a quantity that grows can be modeled by an exponential equation? What information do you need in order to write the equation?

SPIRAL REVIEW

Solve each system of equations by adding or subtracting. *(Section 7.3)*

32. $x - y = 5$
$x + y = 1$

33. $15p - 7q = 11$
$3p + 7q = -11$

34. $m - 5n = 10$
$m + 7n = 4$

35. $3a - 5b = -12$

$3a + b = 6$

36. $\frac{3}{4}x + \frac{2}{3}y = 25$

$x - \frac{2}{3}y = -10$

37. $5x - 2y = 6$

$5x + 6y = 2$

Solve each equation by using the quadratic formula. *(Section 8.5)*

38. $x^2 + 2x - 2 = 0$
39. $y^2 - y - 1 = 0$
40. $z^2 - 5z + 3 = 0$

41. $2x^2 - 5x - 7 = 0$
42. $3x^2 + 4x = 1$
43. $1 + y^2 = 5y^2 - 2y$

9.3

Learn how to...

- **use equations to model real-world decay situations**

So you can...

- **make predictions about quantities that decrease, such as the height of a bouncing ball**

Number of times cup is emptied	Number of pennies left
0	100
1	?
2	?
3	?
4	?
5	?
6	?

Exponential Decay

If you have 100 pennies and take away half, how many do you have left? What if you take away half of those? What if you keep taking away half? This kind of decrease is called *exponential decay*.

EXPLORATION
COOPERATIVE LEARNING

Modeling Exponential Decay

Work with a partner.
You will need:
- 100 pennies
- a cup
- a graphing calculator

Follow Steps 1–3. Then answer the questions below.

1 Put the pennies in the cup. Then empty them onto a desk.

2 Remove all pennies that land heads up. Record the number of pennies remaining.

3 Repeat Steps 1 and 2 until no pennies remain.

Questions

1. Make a scatter plot of the data. Put the number of times the cup is emptied on the horizontal axis. Put the number of pennies remaining on the vertical axis.

2. You can expect to remove about half of the pennies each time. Use this fact to copy and complete the table at the right.

3. Write an equation that models the expected number *y* of pennies remaining after the cup is emptied *x* times. Graph the equation in the same coordinate plane as your scatter plot. What do you notice?

x = number of times cup is emptied	y = expected number of pennies remaining
0	100
1	$100\left(\frac{1}{2}\right)$
2	$\left(100 \cdot \frac{1}{2}\right)\left(\frac{1}{2}\right) = 100\left(\frac{1}{2}\right)^2$
3	?
4	?
5	?
6	?

In the Exploration, you could expect there to be about half as many pennies left after you cleared away the heads as there were before you emptied the cup.

$$\text{Pennies left after the cup} \atop \text{was emptied } \mathbf{1} \text{ time} = 100(0.5)^{\mathbf{1}}$$

$$\text{Pennies left after the cup} \atop \text{was emptied } \mathbf{2} \text{ times} = 100(0.5)^{\mathbf{2}}$$

$$\vdots$$

$$\text{Pennies left after the cup} \atop \text{was emptied } \mathbf{n} \text{ times} = 100(0.5)^{\mathbf{n}}$$

Patterns like the one above appear in many situations, as you will see in Example 1.

EXAMPLE 1   **Application: Physics**

Each time a certain rubber ball bounces, it reaches about 60% of the height from which it fell. Suppose this ball is dropped from a height of 300 cm. What height does the ball reach after the third bounce?

SOLUTION

$$\text{Height after the} \atop \textbf{first} \text{ bounce} = (300)(0.6)$$

$$\text{Height after the} \atop \text{second bounce} = [(300)(0.6)]0.6$$

The bounce height of the first bounce becomes the drop height of the second bounce.

$$\text{Height after the} \atop \text{third bounce} = [(300)(0.6)(0.6)]0.6$$

$$= 300(0.6)^3$$

$$= 64.8$$

After the third bounce, the ball reaches a height of about 64.8 cm.

The height reached by the ball in Example 1 decreases by 40% after each bounce. Any quantity y that decreases by a fixed percent at regular intervals is an example of **exponential decay**. Exponential decay can be modeled by an equation in the form below.

$$y = a(1 - r)^x$$

The constant a represents the initial amount before any decay occurs.

The constant r represents the **rate of decrease**. The quantity $1 - r$ is called the **decay factor**.

The x in the equation represents the number of intervals that have passed.

EXAMPLE 2 **Application: Meteorology**

At sea level, the barometric pressure is 29.92 in. of mercury. For each mile above sea level, the barometric pressure decreases by 18%.

a. Write an equation for the barometric pressure y in inches of mercury x miles above sea level.

b. Find the barometric pressure five miles above sea level.

SOLUTION

a. The initial pressure is 29.92, so $a = 29.92$.

The pressure decreases by 18% for each mile, so $r = 0.18$.

Substitute these values in the equation $y = a(1 - r)^x$.

$$y = 29.92(1 - 0.18)^x$$
$$y = 29.92(0.82)^x$$

b. Substitute 5 for x in the equation from part (a).

$$y = 29.92(0.82)^5$$
$$\approx 11.09$$

The barometric pressure five miles above sea level is about 11.09 in. of mercury.

THINK AND COMMUNICATE

1. How can you tell whether an exponential equation models decay?

2. a. Write an equation for the height y of the ball in Example 1 after x bounces. Graph the equation on a graphing calculator.

b. Use the graph from part (a) to estimate the number of times the ball bounces before it stops. How did you get your answer?

☑ CHECKING KEY CONCEPTS

Find the value of y when $x = 3$.

1. $y = 150(0.82)^x$ **2.** $y = 24(0.9)^x$ **3.** $y = \dfrac{9}{2}\left(\dfrac{2}{3}\right)^x$

The percent y of people living on farms in the United States can be modeled by the equation $y = 18(0.95)^x$, where $x =$ the number of years after 1946. Use this information for Questions 4–6.

4. What percent of people lived on farms in 1946? How do you know?

5. By how much did the percent decrease each year? Explain.

6. Estimate the percent of people who lived on farms in 1973.

Extra Practice
exercises on
page 575

Find the value of *y* when *x* = 6. Round decimal answers to the nearest hundredth.

1. $y = 120(0.25)^x$
2. $y = 278(0.93)^x$
3. $y = 7168(0.97)^x$

4. $y = 402(0.78)^x$
5. $y = 3012(0.875)^x$
6. $y = 0.8(0.6)^x$

7. $y = 48\left(\frac{1}{2}\right)^x$
8. $y = 117\left(\frac{1}{3}\right)^x$
9. $y = 1000\left(\frac{4}{5}\right)^x$

For Exercises 10 and 11, write an equation that models the situation described. Tell what each variable in the equation represents.

10. In 1990, the population of a city was 850,000. Each year since, the population has been about 97% what it was the previous year.

11. The amount *y* of a substance in grams decreases by 17% each hour after 3:00 P.M. At 3:00 P.M., *y* = 100.

12. **SPORTS** The women's United States Open tennis tournament is held each year. There are 128 players at the start of the tournament. During each round of play, half of the remaining players are eliminated.

   a. Write an equation that gives the number *y* of players remaining after *x* rounds.

   b. Find the number of players remaining after 4 rounds.

   c. A player competes in one match during each round. How many matches must a player win in order to win the tournament?

13. **Open-ended Problem** Write a word problem whose solution is the equation $y = 34{,}000(0.94)^x$.

For Exercises 14 and 15, tell whether each exponential graph represents *growth* or *decay*. Explain your answer.

14.

15.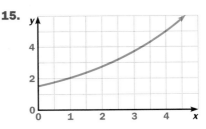

The equations in Exercises 16–21 describe quantities that grow or decay exponentially. For each equation:

a. Tell whether the quantity described *grows* or *decays*.

b. Give the initial amount of the quantity.

c. Give the growth or decay factor for the quantity.

16. $y = 300(1.15)^x$
17. $y = 14(0.23)^x$
18. $y = 20{,}000(0.72)^x$

19. $y = 0.25(1.88)^x$
20. $y = 1.01(0.99)^x$
21. $y = 0.99(1.01)^x$

Look back at
the article on
pages 366–368.

Doctors like Lori Arviso Alvord need to know how long certain medications are present in a patient's bloodstream. For example, when a person takes a 325 mg aspirin tablet, the amount of aspirin in the person's bloodstream decreases by about 20% per hour.

22. Write an equation that gives the amount of aspirin *y* in milligrams in a person's bloodstream *x* hours after taking a 325 mg aspirin tablet.

23. The directions on many aspirin bottles say to wait 4 hours between doses. How much aspirin is left in the bloodstream 4 hours after taking a 325 mg tablet?

Cooperative Learning In this exploration, you will perform an experiment that models exponential decay. Work in a group of 2–3 students. You will need a clear drinking glass, water, a measuring cup, and food coloring.

24. Pour 10 fluid ounces of water into the drinking glass. Add five drops of food coloring to the water. Dilute the mixture by pouring 2 oz of colored water out of the glass and adding 2 oz of clear water. How does the mixture change?

25. Continue diluting the mixture as in Exercise 24 until the water appears to be clear. How many times did you need to repeat the steps?

26. Writing Explain how this exploration is similar to the decrease of aspirin in a person's bloodstream over time.

OCCUPATIONAL SAFETY The equation $y = 2{,}046{,}322(0.87)^x$ models the maximum number of hours per day *y* that a worker is allowed to be exposed to a noise of *x* decibels (dB) according to federal standards. Use this information for Exercises 27 and 28.

27. Find the maximum number of hours per day that a person can work in each location at the given noise level.

a. room with amplified rock music: 110 dB

b. machine shop: 100 dB

c. subway train station: 90 dB

28. **Technology** Graph the equation $y = 2{,}046{,}322(0.87)^x$. Set a good viewing window, such as $80 \le x \le 120$ and $0 \le y \le 24$.

a. Based on the graph, what noise levels do you think are unhealthy during any amount of time? How did you get your answer?

b. Based on the graph, what noise levels do you think have no effect on an individual's hearing? Explain your reasoning.

c. Explain why it makes sense to set the window so that $0 \le y \le 24$.

To make the Chinese dish "dragon's beard noodles," fold a 5 ft strand of dough in half. Stretch the dough back to its original length so that two thinner strands are formed. Repeat this folding and stretching process over and over to produce increasing numbers of progressively thinner noodles.

5 ft

1 in. beginning radius

Fold in half.

Stretch to 5 ft in length.

? radius after dough is folded and stretched once

Fold in half.

?

Stretch to 5 ft in length.

Continue folding and stretching.

29. Write an equation that gives the number of noodles n formed by folding and stretching the dough x times.

30. Write an equation that gives the combined length l in feet of all the noodles after folding and stretching the dough x times. How did you get your answer?

BY THE WAY

The world's best chefs are able to complete 12 or 13 doublings of dragons's beard noodles. These noodles are only two or three times as thick as human hair.

31. **a.** The volume V of a noodle in cubic inches is given by the formula $V = 60\pi r^2$, where r is the radius of the noodle in inches. Find the volume of a noodle with a radius of 1 in. Leave your answer in terms of π.

b. Folding and stretching the dough does not change the total volume of all the noodles. Use this fact to find the value of x in the equation below. Explain your answer.

$$x \cdot \left(\begin{array}{c} \text{Volume of a noodle} \\ \text{after 1 fold and stretch} \end{array} \right) = \left(\begin{array}{c} \text{Volume of the} \\ \text{beginning noodle} \end{array} \right)$$

c. Challenge Use your completed equation from part (b) to complete the equation below.

$$\left(\begin{array}{c} \text{Radius of a noodle} \\ \text{after 1 fold and stretch} \end{array} \right) = \left(\begin{array}{c} \text{Radius of the} \\ \text{beginning noodle} \end{array} \right) \cdot \underline{\ ?\ }$$

d. Use the relationship in part (c) to write an equation giving the radius r of a noodle after x folds and stretches. By about what percent does the radius decrease after each fold and stretch?

32. Writing Explain why you need to know about percent to understand exponential growth and decay. Support your answer with an example.

Simplify each expression. *(Section 1.5)*

33. $\dfrac{-14 + 3 + 2 \cdot 7}{5}$

34. $-10 \cdot 11 \cdot \dfrac{3}{5}$

35. $\left| -6 + \dfrac{1}{2} \cdot \dfrac{1}{2} \right|$

Identify the reciprocal you would use to solve each equation. Then solve. *(Section 4.2)*

36. $\dfrac{4}{5}x = 7$

37. $-\dfrac{2}{3}a = 4$

38. $\dfrac{1}{2}r = 3$

39. $-\dfrac{3}{7}m = -21$

40. $\dfrac{5}{9}c = \dfrac{-1}{5}$

41. $\dfrac{1}{8}p = 8$

ASSESS YOUR PROGRESS

VOCABULARY

base (p. 370)
exponent (p. 370)
power (p. 370)
exponential growth (p. 377)
growth rate (p. 377)

growth factor (p. 377)
exponential decay (p. 384)
rate of decrease (p. 384)
decay factor (p. 384)

Evaluate each power. *(Section 9.1)*

1. 4^3

2. $(-2)^6$

3. $\left(\dfrac{1}{5}\right)^4$

Evaluate each expression when $x = 5$ and $y = -2$. *(Section 9.1)*

4. xy^4

5. $(x - y)^3$

6. $2x^2 + y^3$

7. $5y - x^2$

8. $y^3 + (x^2y - 1)$

9. $y^5 + (6x + 2)$

10. PERSONAL FINANCE Eric Janning has $250 in a savings account that pays 5% annual interest. *(Section 9.2)*

   a. Write an equation giving the amount y in dollars in Eric's account x years from now, assuming he makes no deposits or withdrawals.

   b. How much money will Eric have in his account 3 years from now?

11. FUEL ECONOMY The average annual fuel consumption per car in the United States has decreased by about 2% per year since 1980, when it was about 591 gallons per car. *(Section 9.3)*

   a. Write an exponential function to model this situation.

   b. Sketch the graph of the equation you wrote for part (a).

12. Journal Describe some real-world examples of exponential growth and decay. Sketch the graph that models one of these situations.

9.4 Zero and Negative Exponents

Learn how to...

- **evaluate powers with negative and zero exponents**
- **use negative exponents to express quotients**

So you can...

- **find past values, such as the population of Morocco years ago**
- **use formulas to find information, such as the amount of a car loan payment**

You have already seen how to use exponents to predict future values, such as population figures. You can use exponents to look into the past as well as into the future.

Look at the diagram below. Each bar is half as long as the previous bar. The diagram is a visual representation of several powers of 2.

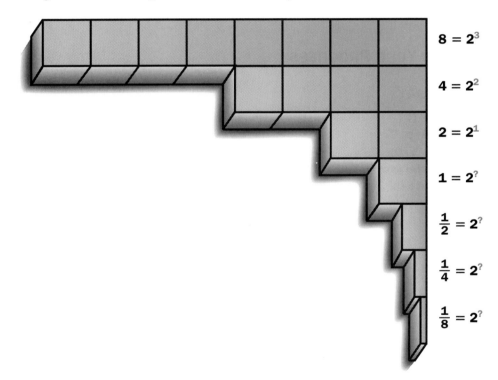

$$8 = 2^3$$

$$4 = 2^2$$

$$2 = 2^1$$

$$1 = 2^?$$

$$\frac{1}{2} = 2^?$$

$$\frac{1}{4} = 2^?$$

$$\frac{1}{8} = 2^?$$

THINK AND COMMUNICATE

1. As the bars in the diagram become shorter, how do the exponents change?

2. What do you think the last four exponents in the diagram should be? Explain your reasoning. Then use a calculator to check your answers.

3. Copy and complete the equations below. How did you get your answers?

$$9 = 3^? \qquad 3 = 3^? \qquad 1 = 3^? \qquad \frac{1}{3} = 3^? \qquad \frac{1}{9} = 3^?$$

Zero and Negative Exponents

For any nonzero number a and any integer n:

$$a^0 = 1 \qquad \text{and} \qquad a^{-n} = \frac{1}{a^n}$$

EXAMPLE 1

Evaluate each power.

a. $(7.2)^0$

b. 6^{-2}

SOLUTION

a. $(7.2)^0 = 1$

b. $6^{-2} = \dfrac{1}{6^2}$

$= \dfrac{1}{36}$

EXAMPLE 2

Rewrite each expression using only positive exponents.

a. $a^0 b^{-4}$

b. $\dfrac{ac^{-8}}{b^5}$

SOLUTION

a. $a^0 b^{-4} = 1 \cdot b^{-4}$

$= \dfrac{1}{b^4}$

b. $\dfrac{ac^{-8}}{b^5} = \dfrac{a}{b^5} \cdot c^{-8}$

$= \dfrac{a}{b^5} \cdot \dfrac{1}{c^8}$

$= \dfrac{a}{b^5 c^8}$

THINK AND COMMUNICATE

4. Chayla evaluated the power $\left(\dfrac{1}{4}\right)^{-2}$ as shown. Use her

method to evaluate each of the following powers.

a. $\left(\dfrac{1}{3}\right)^{-2}$ **b.** $\left(\dfrac{1}{2}\right)^{-5}$ **c.** $\left(\dfrac{1}{5}\right)^{-3}$

5. Based on your answers to Question 4, complete the
equation below. Explain your reasoning.

$$\left(\dfrac{1}{a}\right)^{-n} = \underline{\ ?\ }$$

$$\left(\tfrac{1}{4}\right)^{-2} = \frac{1}{\left(\tfrac{1}{4}\right)^2}$$

$$= \frac{1}{\frac{1}{16}}$$

$$= 1 \div \tfrac{1}{16}$$

$$= 1 \times 16$$

$$= 16$$

Applying Negative Exponents

The population y of Morocco in millions x years after 1980 can be modeled by this equation:

$$y = 20.5(1.025)^x$$

You already know how to use this equation to estimate the population of Morocco in years after 1980. You can also use it to estimate the population in years *before* 1980.

When $x = 1$, $y =$ the population **1 year after 1980**, in 1981.

When $x = 0$, $y =$ the population **in 1980**.

When $x = -1$, $y =$ the population **1 year before 1980**, in 1979.

EXAMPLE 3 **Application: Population**

Use the equation above to estimate the population of Morocco in 1975.

SOLUTION

Since 1975 is 5 years *before* 1980, substitute -5 for x.

$$y = 20.5(1.025)^{-5}$$
$$\approx 18.1$$

The population of Morocco was about 18.1 million in 1975.

THINK AND COMMUNICATE

6. a. How can you find the population of Morocco in 1980 just by looking at the equation $y = 20.5(1.025)^x$?

   b. Explain how to calculate the population of Morocco in 1980 by substituting a value for x in the equation. Tell what x value to use, and explain why you would choose this value.

   c. Describe how you can use the graph of the equation shown at the right to find the population of Morocco in 1980.

   d. How do the methods you described in parts (a), (b), and (c) compare? Does each method give you the same results?

7. In Example 3, what is assumed about the annual growth rate of Morocco's population from 1975 to 1980? How can you find out whether this assumption is reasonable?

Population of Morocco

CHECKING KEY CONCEPTS

Simplify each expression.

1. 11^{-2} **2.** 0.747966^0 **3.** $(9^0)(7^{-2})$ **4.** $\left(\frac{1}{5}\right)^{-2}$

Rewrite each expression using only positive exponents.

5. $7m^{-3}$ **6.** $t^{-6}r^0$ **7.** $\dfrac{a^{-4}}{c^6}$ **8.** $\dfrac{x^0}{y^{-2}}$

Evaluate each expression when $a = 3$, $b = 5$, and $c = -2$.

9. $(ac)^0$ **10.** b^2a^{-3} **11.** c^{-3} **12.** 3^{-b}

9.4 | **Exercises and Applications**

Simplify each expression.

Extra Practice exercises on page 575

1. 13^0 **2.** 5^{-6} **3.** 2^{-5} **4.** 8^{-2}

5. $\left(\frac{1}{5}\right)^0$ **6.** $\left(\frac{3}{5}\right)^{-3}$ **7.** $\left(\frac{2}{3}\right)^{-4}$ **8.** $\dfrac{1}{4^0}$

Rewrite each expression using only positive exponents.

9. a^{-2} **10.** $x^{-4}y^{-9}$ **11.** $4b^{-6}$

12. ab^2c^{-1} **13.** $\left(\frac{a}{b}\right)^{-3}$ **14.** $\left(\dfrac{k^{-7}}{12}\right)$

15. $2w^{-8}$ **16.** $\left(\dfrac{q^{-12}}{r^3}\right)$ **17.** $(yz)^{-2}$

Evaluate each expression when $x = 2$, $y = 3$, and $z = 7$.

18. $(x + z)^{-3}$ **19.** $x^{-2}z^0$ **20.** yz^{-2}

21. $(5x^0 - z)^{-4}$ **22.** $x^{-1} + y^{-2}$ **23.** $xz^0 - y^{-3}$

24. $\left(\dfrac{z}{x + y}\right)^{-2}$ **25.** $9^{x - y}$ **26.** yz^{-x}

27. Writing Tony was asked to evaluate the expression $3x^{-4}$ when $x = 2$. His work is shown at the right. Write a sentence or two explaining Tony's mistake. Then evaluate the expression correctly.

28. **Technology** Use a calculator to evaluate each power. Round your answers to the nearest ten thousandth.

     a. 25^{-1} **b.** 17^{-2} **c.** 1.42^{-3}

29. **HISTORY** The first census of the United States population was conducted in 1790. The population was found to be about 3.929 million and continued to increase by roughly 34% per decade until 1860.

   a. Write an equation that models the population y of the United States in millions x decades after 1790.

   b. Use your equation from part (a) to estimate the United States population in 1780, before the first census was taken.

   c. **Research** Use your equation from part (a) to estimate the United States population in 1990. Find out the actual 1990 population of the United States and compare it to your answer. What do you conclude about the growth rate of the population from 1860 to 1990?

Connection | **CONSUMER ECONOMICS**

When you take out a loan to buy a car, you must repay a portion of the loan with interest each month. The monthly payment P required to repay a loan of A dollars in n monthly installments is given by the formula

$$P = \frac{Am}{1 - (1 + m)^{-n}}$$

where m is the monthly interest rate expressed as a decimal. You can find the monthly interest rate by using the formula

$$\text{Monthly interest rate} = \frac{\text{Annual rate as a decimal}}{12}$$

For Exercises 30–32, suppose you take out a loan for the entire cost of the car in the advertisement.

30. What monthly interest rate is charged for a loan from the car dealership?

31. What is your monthly payment if you repay the loan over each time period?

   a. 24 months   b. 36 months

   c. 48 months   d. 60 months

32. **Challenge** What is the total amount of interest charged if you repay the loan over each time period?

   a. 24 months   b. 36 months

   c. 48 months   d. 60 months

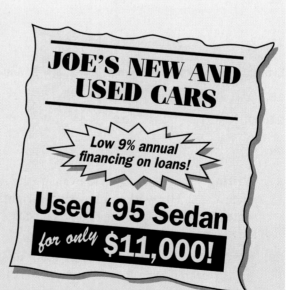

JOE'S NEW AND USED CARS

Low 9% annual financing on loans!

Used '95 Sedan for only $11,000!

33. **Open-ended Problem** Create your own ad that gives the price of a car and annual interest rate on a loan. Suppose you take out a loan with the dealership for the entire cost of the car. Calculate the monthly payment and total interest charged if you repay the loan over three years.

34. BIOLOGY The diagrams below show a growing bacteria colony at three different times. The mass of the colony is increasing by 30% per hour.

Time: 3:00 P.M.
Mass: 2.10 g

Time: 5:00 P.M.
Mass: 3.55 g

Time: 9:00 P.M.
Mass: 10.14 g

a. Write an equation that models the mass y of the bacteria colony in grams x hours after 3:00 P.M. Does your equation give the correct mass at 5:00 P.M.? at 9:00 P.M.?

b. Estimate the mass of the colony at 1:00 P.M., two hours before the diagram at the left above.

c. Visual Thinking Draw a diagram showing the approximate size of the colony at 1:00 P.M., two hours before the diagram at the left above.

35. SAT/ACT Preview If $A = 3^{-n}$ and $B = 3^n$ for a positive integer n, then:

A. $A < B$ **B.** $A > B$ **C.** $A = B$

D. relationship cannot be determined

36. Any rational number can be written as the sum of multiples of powers of 10. For example, you can write 268.43 like this:

$$268.43 = 2 \cdot 10^2 + 6 \cdot 10^1 + 8 \cdot 10^0 + 4 \cdot 10^{-1} + 3 \cdot 10^{-2}$$

Write each number as the sum of multiples of powers of 10.

a. 139.58   **b.** 4234.686   **c.** 73,009.5   **d.** 0.12345

ONGOING ASSESSMENT

37. Writing How would you use division to teach someone about negative and zero exponents? Use examples to support your answer.

SPIRAL REVIEW

Solve each inequality. *(Section 4.6)*

38. $2 + x \leq 7$ **39.** $3m \geq m - 6$ **40.** $p + 25 < 7 - p$

41. $3q + 5 > -9$ **42.** $4 \leq 3x - 2$ **43.** $2 - 2y \leq 10y$

9.5

Working with Scientific Notation

Learn how to...
- **write numbers in scientific notation**

So you can...
- **conveniently express numbers such as the weight of Earth or the diameter of a cell**

Have you ever had trouble entering a number into a calculator because it had too many digits? Mathematicians and scientists use a form of shorthand based on powers of 10 to write numbers with many digits.

The weights of the sun and the nine planets in tons are written in *scientific notation* in the diagram below. A number is expressed in **scientific notation** if it is in the form $a \times 10^n$ where $1 \leq a < 10$ and n is an integer.

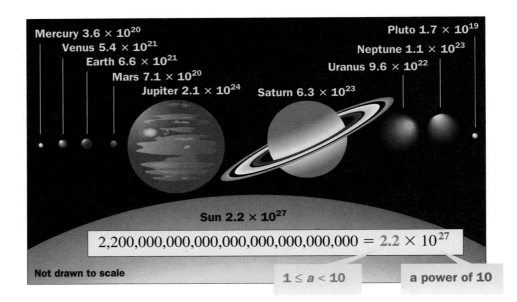

Mercury 3.6×10^{20}

Venus 5.4×10^{21}

Earth 6.6×10^{21}

Mars 7.1×10^{20}

Jupiter 2.1×10^{24} Saturn 6.3×10^{23}

Pluto 1.7×10^{19}

Neptune 1.1×10^{23}

Uranus 9.6×10^{22}

Sun 2.2×10^{27}

$$2{,}200{,}000{,}000{,}000{,}000{,}000{,}000{,}000{,}000{,}000 = 2.2 \times 10^{27}$$

Not drawn to scale

$1 \leq a < 10$ a power of 10

EXAMPLE 1

Write the weight of Earth in decimal notation.

SOLUTION

Follow the order of operations. Simplify the power before you multiply.

Notice that the power 10^{21} has **21** zeros.

$$6.6 \times 10^{21} = 6.6 \times 1{,}000{,}000{,}000{,}000{,}000{,}000{,}000$$

$$= 6{,}600{,}000{,}000{,}000{,}000{,}000{,}000$$

Earth weighs $6{,}600{,}000{,}000{,}000{,}000{,}000{,}000$ tons.

You can also write numbers less than 1 in scientific notation.

$$0.000002 = \frac{2}{1,000,000}$$

$$= \frac{2}{10^6}$$

$$= 2 \cdot \frac{1}{10^6}$$

$$= 2 \times 10^{-6}$$

Multiplying by 10^{-6} is the same as dividing by 10^6.

When $n > 0$, factoring $\mathbf{10}^n$ out of a number moves the decimal point n places to the **left**. Factoring out $\mathbf{10}^{-n}$ moves the decimal point n places to the **right**.

EXAMPLE 2

Write each number in scientific notation.

a. 72,000

b. 0.0000006093

SOLUTION

a. $72,000 = 7.2 \times 10^4$

b. $0.0000006093 = 6.093 \times 10^{-7}$

Factoring out 10^4 moves the decimal point **4** places to the **left**.

Factoring out 10^{-7} moves the decimal point **7** places to the **right**.

THINK AND COMMUNICATE

1. Look back at the diagram on page 396. How does the power of 10 show that Saturn weighs more than Earth?

2. Write each number in scientific notation.

   a. 0.1   **b.** 1   **c.** 10

3. Describe how to write a number a in scientific notation when $1 \le a < 10$.

EXAMPLE 3

Write each number in decimal notation.

a. 8.959×10^{-6}

b. 3.24×10^{-5}

SOLUTION

a. $8.959 \times 10^{-6} = 8.959 \times 0.000001$

$$= 0.000008959$$

Notice that the power 10^{-6} has **6** decimal places.

b. $3.24 \times 10^{-5} = 3.24 \times 0.00001$

$$= 0.0000324$$

THINK AND COMMUNICATE

4. a. Write each number in decimal notation and in scientific notation.

7 thousand   4.3 billion   6 millionths

b. How do the phrases in part (a) resemble numbers written in scientific notation?

5. Describe how to write a number in scientific notation. Illustrate your method with a number less than 1 and another number greater than 10.

☑ CHECKING KEY CONCEPTS

Write each number in decimal notation.

1. 6×10^5 **2.** 1.2×10^7 **3.** 5.03×10^6

4. 3×10^{-2} **5.** 7.72×10^{-4} **6.** 8.0430×10^{-3}

Write each number in scientific notation.

7. 8,000,000   **8.** 419,500   **9.** 34,005,600,000

10. 0.00001   **11.** 0.00074   **12.** 0.0000208

9.5 Exercises and Applications

Extra Practice exercises on page 575

Write each number in decimal notation.

1. 2×10^3 **2.** 7×10^{-4} **3.** 4.9×10^6

4. 5.12×10^{-5} **5.** 6.761×10^8 **6.** 8.35×10^0

Write each number in scientific notation.

7. 780   **8.** 0.000000046   **9.** 21,000

10. 0.0000003803   **11.** 9,700,000,000,000   **12.** 1.5

13. 3.2 million   **14.** 4 thousandths   **15.** 8.023 hundredths

16. ASTRONOMY The table gives each planet's approximate distance from the sun in scientific notation.

Earth: 9.3×10^7 mi	Mars: 1.4×10^8 mi	Jupiter: 4.8×10^8 mi
Uranus: 1.8×10^9 mi	Venus: 6.7×10^7 mi	Mercury: 3.6×10^7 mi
Pluto: 3.7×10^9 mi	Saturn: 8.9×10^8 mi	Neptune: 2.8×10^9 mi

a. Which planet is farthest from the sun? Which is closest to the sun?

b. Write each distance in decimal notation.

17. Research Create a diagram like the one on page 396 that gives a set of data in scientific notation.

Lori Arviso Alvord

Look back at the article on pages 366-368.

Doctors like Lori Arviso Alvord need to be aware of blood clotting when performing surgery. Blood clotting involves a series of chemical reactions that lead to the formation of a mesh of fibers that are too large for blood cells to pass through.

A micron is a unit of measurement 10^{-6} m long. Extremely small objects, such as blood cells, are measured in microns. For Exercises 18–20, give each measurement in microns. Explain how you got your answers.

18. A typical red blood cell is 0.0000072 m wide.

19. A white blood cell is 0.000007 m to 0.000012 m wide.

20. A platelet is about 0.000003 m wide.

21. Visual Thinking Use the measurements given in Exercises 18–20 to make a scale drawing of the three kinds of blood cells.

For Exercises 22 and 23, write the number in red in scientific notation.

22. A certain brand of "supercomputer" can perform about 2,500,000,000 arithmetic operations per second.

23. The probability of flipping a coin 20 times and getting all heads is about 0.000000954.

24. Human hair grows at a rate of 6 in./year. How fast does hair grow in miles per hour? Write your answer in scientific notation.

25. EARTH SCIENCE The Amazon River in South America discharges about 4,200,000 ft^3 of water into the Atlantic Ocean each second. Find the volume of water discharged from the Amazon River each day. Write your answer in scientific notation.

26. **Technology** In this exercise, you will explore how your calculator uses scientific notation.

a. Use a calculator to multiply a positive one-digit integer by 10. Then multiply the result by 10. Repeat this until the calculator gives the answer in scientific notation. What is the exponent of the power of 10?

b. Repeat part (a), but start with a two-digit integer. Is the exponent of the power of 10 the same as in part (a)?

c. Repeat parts (a) and (b), but divide by 10 instead of multiplying. How do your answers compare to your answers to parts (a) and (b)?

27. Writing Raj entered the following key sequence on his calculator:

2.5 [EE] 17 [−] 3.9 [EE] 5 [ENTER]

Explain why he got the answer shown.

```
2.5E17-3.9E5
          2.5E17
```

28. Open-ended Problem Describe three situations in which scientific notation might be more useful than decimal notation. Explain.

29. SAT/ACT Preview Which of the following numbers is (are) written in scientific notation?

   I. 21.8×10^7 **II.** 9.9005×10^{-2} **III.** 0.16×10^{-8}

   A. I only   **B.** II only   **C.** III only   **D.** I and II   **E.** I and III

30. Challenge Write each product in scientific notation. Explain how you found your answers.

   a. $\left(1 \times 10^2\right)\left(1 \times 10^2\right)$ **b.** $\left(4 \times 10^2\right)\left(1 \times 10^3\right)$

   c. $\left(1 \times 10^2\right)\left(2 \times 10^4\right)$ **d.** $\left(2 \times 10^3\right)\left(2 \times 10^4\right)$

ONGOING ASSESSMENT

31. Writing Describe a method for determining which of two numbers written in scientific notation is larger than the other. Show that your method works for two very large numbers and two very small numbers.

SPIRAL REVIEW

Write each expression in simplest form. *(Section 9.1)*

32. $2^4 \cdot 3^4$ **33.** $(2 \cdot 3)^4$ **34.** $\dfrac{6^5}{3^5}$

Simplify each power. *(Section 9.1)*

35. $(-5)^3$ **36.** $3\left(\dfrac{2}{3}\right)^4$ **37.** $\left(3 \cdot \dfrac{2}{3}\right)^4$

ASSESS YOUR PROGRESS

VOCABULARY

scientific notation (p. 396)

Evaluate each power. *(Section 9.4)*

1. 3.125^0 **2.** 4^{-5} **3.** $\left(\dfrac{4}{3}\right)^{-3}$

Evaluate each expression when $x = 4$, $y = -3$, and $z = 2$. *(Section 9.4)*

4. $z^{-5}x^0$ **5.** $x^{-2}y^2$ **6.** $\left(\dfrac{z}{x}\right)^{-2}$

7. The population of Australia in millions x years after 1980 can be modeled by the function $y = 14.6(1.014)^x$. Estimate the population of Australia for each year. *(Section 9.4)*

   a. 1976   **b.** 1980   **c.** 1972

Write each number in decimal notation. *(Section 9.5)*

8. 7.003×10^9 **9.** 4.21×10^{-5} **10.** 8×10^{-6}

Write each number in scientific notation. *(Section 9.5)*

11. 54,000,000,000   **12.** 3.28   **13.** 0.00000709

14. Journal How can you tell whether a number written in scientific notation is greater than or less than 1? Explain your reasoning.

9.6

Exploring Powers

A *light-year*, the distance that light travels in a year, is about 5.88×10^{12} mi. Astronomers use light-years to express large distances in space. To convert light-years to miles, you must often multiply two powers of 10.

EXPLORATION
COOPERATIVE LEARNING

Multiplying and Dividing Powers of 10

Work with a partner.
You will need:
- a calculator

Product of powers	Expanded product	Product as a single power
$10^2 \cdot 10^1$	$(10 \cdot 10)(10)$	10^3
$10^4 \cdot 10^2$	?	?
$10^7 \cdot 10^1$	?	?
$10^5 \cdot 10^6$	?	?
$10^0 \cdot 10^3$	?	?

Quotient of powers	Expanded quotient	Quotient as a single power
$\dfrac{10^3}{10^2}$	$\dfrac{10 \cdot 10 \cdot 10}{10 \cdot 10}$	10^1
$\dfrac{10^6}{10^4}$	?	?
$\dfrac{10^{12}}{10^3}$	?	?
$\dfrac{10^1}{10^3}$	?	?
$\dfrac{10^0}{10^2}$	?	?

Copy and complete the tables. Then answer the questions below.

Questions

1. Use any patterns you see in the tables to complete these rules for working with powers of 10:

   a. $10^m \cdot 10^n = 10^?$ **b.** $\dfrac{10^m}{10^n} = 10^?$

2. Use the rules you wrote in Question 1 to simplify each expression. Use a calculator to check your answers.

   a. $10^4 \cdot 10^7$ **b.** $10^{10} \cdot 10^3$ **c.** $\dfrac{10^9}{10^2}$ **d.** $\dfrac{10^8}{10^{14}}$

Rules for Products and Quotients of Powers

These rules are true for any nonzero number a and any integers m and n:

Product of Powers Rule

$$a^m \cdot a^n = a^{m+n}$$

$$2^4 \cdot 2^5 = 2^{4+5} = 2^9$$

Quotient of Powers Rule

$$\frac{a^m}{a^n} = a^{m-n}$$

$$\frac{5^8}{5^2} = 5^{8-2} = 5^6$$

EXAMPLE 1

Simplify each expression. Write your answers using positive exponents.

a. $a^3 \cdot a^7$

b. $\left(a^6 b^{-9}\right)\left(a^8 b^4\right)$

SOLUTION

a. $a^3 \cdot a^7 = a^{3+7}$

$ = a^{10}$

b. $\left(a^6 b^{-9}\right)\left(a^8 b^4\right) = \left(a^6 a^8\right)\left(b^{-9} b^4\right)$

$\phantom{\left(a^6 b^{-9}\right)\left(a^8 b^4\right)} = a^{6+8} b^{-9+4}$

$\phantom{\left(a^6 b^{-9}\right)\left(a^8 b^4\right)} = a^{14} b^{-5}$

$\phantom{\left(a^6 b^{-9}\right)\left(a^8 b^4\right)} = a^{14} \cdot \dfrac{1}{b^5}$

$\phantom{\left(a^6 b^{-9}\right)\left(a^8 b^4\right)} = \dfrac{a^{14}}{b^5}$

> Use the associative and commutative properties to group powers with the same base.

EXAMPLE 2

Simplify each expression. Write your answers using positive exponents.

a. $\dfrac{c^{11}}{c^5}$

b. $\dfrac{c^5 d^3}{c^6 d}$

SOLUTION

a. $\dfrac{c^{11}}{c^5} = c^{11-5}$

$\phantom{\dfrac{c^{11}}{c^5}} = c^6$

b. $\dfrac{c^5 d^3}{c^6 d} = \dfrac{c^5}{c^6} \cdot \dfrac{d^3}{d}$

$\phantom{\dfrac{c^5 d^3}{c^6 d}} = c^{5-6} d^{3-1}$

$\phantom{\dfrac{c^5 d^3}{c^6 d}} = c^{-1} d^2$

$\phantom{\dfrac{c^5 d^3}{c^6 d}} = \dfrac{1}{c} \cdot d^2$

$\phantom{\dfrac{c^5 d^3}{c^6 d}} = \dfrac{d^2}{c}$

> Group powers with the same base.

EXAMPLE 3   **Application: Astronomy**

The Milky Way galaxy has a diameter of about 5.88×10^{18} mi. The diameter of our solar system is about 7.29×10^9 mi. How many times as big as the diameter of our solar system is the diameter of the Milky Way?

SOLUTION

$$\frac{\text{Diameter of Milky Way}}{\text{Diameter of solar system}} = \frac{5.88 \times 10^{18} \text{ mi}}{7.29 \times 10^9 \text{ mi}}$$

$$= \frac{5.88}{7.29} \times \frac{10^{18}}{10^9}$$

$$\approx 0.8 \times 10^{18 - 9}$$

$$\approx 0.8 \times 10^9$$

$$\approx (8 \times 10^{-1}) \times 10^9$$

$$\approx 8 \times 10^{-1 + 9}$$

$$\approx 8 \times 10^8$$

Rewrite **0.8** in scientific notation.

The diameter of the Milky Way is about 8×10^8 times as big as the diameter of our solar system.

THINK AND COMMUNICATE

1. The diameter of the circle at the right is 1 in. Suppose this circle represents the size of our solar system. Based on the solution of Example 3, what would be the diameter of a circle that represents the size of the Milky Way? Give your answer in miles.

Solar System

2. Compare the diameter you found in Question 1 with the diameter of Earth, which is about 8000 mi.

☑ CHECKING KEY CONCEPTS

Write each expression as a single power.

1. $4^5 \cdot 4^3$

2. $\dfrac{7^{12}}{7^6}$

3. $\dfrac{2^3 \cdot 2}{2^4}$

Simplify each expression. Write your answers using positive exponents.

4. $m^4 m^2$

5. $t^6 t^{-2}$

6. $\dfrac{k^5}{k^9}$

7. $\dfrac{4a^2}{6a^2}$

8. $\dfrac{c^9 d^7}{c^3 d^2}$

9. $(3p^5 q^4)(5pq^2)$

10. **GEOGRAPHY** Lake Superior contains about 2.94×10^3 mi^3 of water. Find the number of gallons of water in Lake Superior. Use the fact that 1 mi$^3 \approx 1.10 \times 10^{12}$ gal.

9.6 Exercises and Applications

Extra Practice
exercises on
page 576

Write each expression as a single power.

1. $2^3 \cdot 2^4$

2. $\dfrac{10^6}{10^9}$

3. $3^5 \cdot 3^1$

4. $5^7 \cdot 5^{-4}$

5. $\dfrac{9^3}{9^{-10}}$

6. $\dfrac{4^{11}}{4^8 \cdot 4^{-2}}$

Simplify each expression. Write your answers using positive exponents.

7. $a^2 \cdot a^3$

8. $\dfrac{b^{-9}}{b^4}$

9. $c^{-5} \cdot c^{-2}$

10. $\dfrac{a^6}{4a^2}$

11. $\left(5d^4\right)\left(8d^6\right)$

12. $\dfrac{7x^5}{x^{-8}}$

13. $\dfrac{p^8 q^3}{p^6 q}$

14. $\left(5pq^3\right)\left(10p^{-3}q^2\right)$

15. $\dfrac{3a^{-2}b^5}{9a^4 b^4}$

Connection ▶ LITERATURE

The passage below is from *The Rolling Stones* by Robert A. Heinlein. In the story, the Stone family buys a spaceship and leaves their home on the moon to travel around the solar system.

Rock City

The Asteroid Belt is a flattened torus ring or doughnut in space encompassing thirteen thousand five hundred thousand million trillion cubic miles....

13,500,000,000,000,000,000,000,000 cubic miles of space.

Yet the entire human race could be tucked into one corner of a single cubic mile; the average human body is about two cubic feet in bulk....

Write the figure as 1.35×10^{25} cubic miles; that makes it easier to see if no easier to grasp. At the time the *Rolling Stone* arrived among the rolling stones of Rock City the Belt had a population density of one human soul for every two billion trillion cubic miles—read 2×10^{21}....

16. Challenge When *The Rolling Stones* was written in 1952, the world population was about 2.5 billion. According to the passage, what fraction of a cubic mile would the human race have filled?

17. Use the formula for population density given below to find the population of the Asteroid Belt.

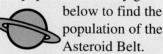

$$\dfrac{\text{Population}}{\text{density}} = \dfrac{\text{Population}}{\text{Volume}}$$

18. Open-ended Problem Write a short science fiction passage that includes calculations with numbers in scientific notation.

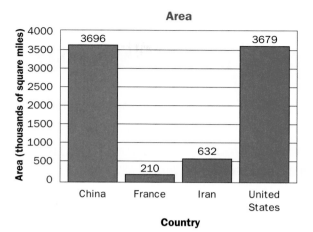

19. Make a table of the data above. Write all numbers in scientific notation.

20. a. Use the formula below to find each country's population density in persons per square mile.

$$\text{Population density} = \frac{\text{Population}}{\text{Area}}$$

b. **Writing** Explain why the population density of China is not $\frac{1192}{3696}$.

21. a. Recall that a light-year is the distance light travels in a year. Use the fact that light travels at 186,282 mi/s to show that a light-year is about 5.88×10^{12} mi. (Use 365.25 for the number of days in one year.)

b. **HISTORY** In 1054 A.D., Chinese and Japanese astronomers recorded a *supernova*. A supernova is the explosion of a star. The supernova produced a huge cloud of gas and dust known as the Crab Nebula about 4×10^3 light-years from Earth. Express this distance in miles.

ONGOING ASSESSMENT

22. Writing Explain the mistake that each student made. Then simplify the expression correctly.

$$2^3 \cdot 3^4 = (2 \cdot 3)^{3+4}$$
$$= 6^7$$
$$= 279,936$$

$$\frac{x^{-9}}{x^{-3}} = x^{-9-3}$$
$$= x^{-12}$$
$$= \frac{1}{x^{-12}}$$

SPIRAL REVIEW

Simplify each expression. *(Section 9.4)*

23. $\left(-\dfrac{8}{5}\right)^0$

24. $\left(-5\dfrac{12}{17}\right)^0$

25. $\left(\dfrac{4}{7}\right)^{-3}$

26. $\left(-\dfrac{4}{7}\right)^{-3}$

27. $-\left(\dfrac{2}{3}\right)^{11}$

28. $\dfrac{17.3^0}{6^{-2}}$

9.7

Working with Powers

Learn how to...

- **find the power of products and quotients**
- **find the power of a power**

So you can...

- **use numbers written in scientific notation in formulas**

The globe shown has a radius of 8 in. The radius of Earth is about 2×10^7 ft. To find their volumes, you may want to learn a few more rules of exponents. For example, to find Earth's volume, you can use the formula $V = \frac{4}{3}\pi r^3$, where r is the radius of Earth. If you enter $(2 \times 10^7)^3$ into a calculator, you get 8×10^{21}. Why?

EXAMPLE 1 **Application: Geology**

Show that the volume of Earth is $\frac{4}{3}\pi(8 \times 10^{21})$, or about 3.35×10^{22} ft³.

SOLUTION

Problem Solving Strategy: Use a formula.

Use the formula $V = \frac{4}{3}\pi r^3$. Substitute 2×10^7 for r.

$$\frac{4}{3}\pi r^3 = \frac{4}{3}\pi(2 \times 10^7)^3$$

$$= \frac{4}{3}\pi(2 \times 10^7)(2 \times 10^7)(2 \times 10^7)$$

$$= \frac{4}{3}\pi(2 \cdot 2 \cdot 2)(10^7 \cdot 10^7 \cdot 10^7) \qquad \text{Group the factors.}$$

$$= \frac{4}{3}\pi(8)(10^{7 + 7 + 7}) \qquad \text{Use the product of powers rule.}$$

$$= \frac{4}{3}\pi(8 \times 10^{21})$$

$$\approx 3.35 \times 10^{22}$$

In Example 1, you saw that $(2 \times 10^7)^3 = 2^3 \times (10^7)^3$. The exponent **3** tells you to multiply each factor inside the parentheses **3** times. This result illustrates the *power of a product rule*.

To find how the ratio of the radii of two spheres relates to the ratio of the volumes of the spheres, you may need to find the power of a quotient, as in the Example below.

EXAMPLE 2 Application: Geology

The center region of Earth is an extremely hot sphere called the *inner core*. The radius of the inner core is one fifth the radius R of Earth. Write the volume of the inner core as a multiple of the volume of Earth.

SOLUTION

Let R = the radius of Earth. Then $\frac{R}{5}$ = the radius of the inner core.

$$\text{Volume of inner core} = \frac{4}{3}\pi\,(\text{Radius of inner core})^3$$

$$= \frac{4}{3}\pi\left(\frac{R}{5}\right)^3$$

$$= \frac{4}{3}\pi\left(\frac{R}{5}\right)\left(\frac{R}{5}\right)\left(\frac{R}{5}\right)$$

Rewrite the numerator and the denominator as powers.

$$= \frac{4}{3}\pi\left(\frac{R\cdot R\cdot R}{5\cdot 5\cdot 5}\right)$$

$$= \frac{4}{3}\pi\left(\frac{R^3}{5^3}\right)$$

$$= \frac{4}{3}\pi\left(R^3\cdot\frac{1}{5^3}\right)$$

Group the factors. Since R = the radius of Earth, $\frac{4}{3}\pi R^3$ = the volume of Earth.

$$= \frac{1}{5^3}\cdot\left(\frac{4}{3}\pi R^3\right)$$

$$= \frac{1}{125}\cdot(\text{Volume of Earth})$$

The volume of the inner core is $\frac{1}{125}$ the volume of Earth.

In Example 2, you saw that $\left(\frac{R}{5}\right)^3 = \left(\frac{R}{5}\right)\left(\frac{R}{5}\right)\left(\frac{R}{5}\right) = \frac{R^3}{5^3}$. The exponent **3** tells you to use both the numerator and the denominator as factors **3** times. This result illustrates the *power of a quotient rule*.

These rules are true for any nonzero numbers a and b and any integer m:

Power of a Product Rule

$$(ab)^m = a^m b^m$$

$$(4 \cdot 3)^2 = 4^2 \cdot 3^2$$

$$(12)^2 = 16 \cdot 9$$

$$144 = 144$$

Power of a Quotient Rule

$$\left(\frac{a}{b}\right)^m = \frac{a^m}{b^m}$$

$$\left(\frac{4}{3}\right)^2 = \frac{4^2}{3^2}$$

$$\frac{16}{9} = \frac{16}{9}$$

In Example 1, you saw that $\left(10^7\right)^3 = 10^7 \cdot 10^7 \cdot 10^7 = 10^{7+7+7} = 10^{21}$. This result illustrates the *power of a power rule*.

Power of a Power Rule

For any nonzero number a and any integers m and n:

$$\left(a^m\right)^n = a^{m \cdot n}$$

For example: $\left(2^2\right)^3 = 2^{2 \cdot 3}$

$$(4)^3 = 2^6$$

$$64 = 64$$

THINK AND COMMUNICATE

1. This globe from page 406 has radius 8 in., or $\frac{2}{3}$ ft. Use the power of a quotient rule to find the volume of the globe in cubic feet.

2. The radius of Earth is about 3.75×10^8 times the radius of the globe. How many times greater than the globe's volume is Earth's volume?

☑ CHECKING KEY CONCEPTS

Evaluate each power.

1. $(3 \cdot 2)^4$

2. $(3 \cdot 5)^3$

3. $\left(2^3\right)^2$

4. $\left(10^5\right)^2$

5. $\left(\frac{7}{8}\right)^2$

6. $\left(\frac{5}{4}\right)^3$

Simplify each expression.

7. $(ab)^9$

8. $\left(x^5\right)^6$

9. $\left(z^4\right)^4$

10. $\left(\frac{2}{d}\right)^3$

11. $(5c)^2$

12. $\left(\frac{m}{n}\right)^7$

9.7 Exercises and Applications

Evaluate each power. Write your answers in scientific notation.

1. $\left(3 \times 10^4\right)^2$

2. $\left(2 \times 10^9\right)^3$

3. $\left(5 \times 10^7\right)^4$

Extra Practice exercises on page 576

Evaluate each power.

4. $\left(\dfrac{1}{2}\right)^4$

5. $\left(\dfrac{4}{7}\right)^3$

6. $\left(\dfrac{5}{3}\right)^5$

Simplify each expression.

7. $(ab)^5$

8. $\left(a^4\right)^8$

9. $\left(\dfrac{a}{b}\right)^6 \left(2c^3 d^2\right)^4$

10. $(2a)^3$

11. $(3ab)^4$

12. $\left(5c^7 d^8\right)^2$

13. $\left(p^{-2} q^5\right)^7$

14. $\left(\dfrac{a^2}{b^3}\right)^9$

15. $\left(\dfrac{b^6 c}{d^3}\right)^5$

16. Open-ended Problem Leon and Ed were asked to find the product $(x + 3)^2$. Identify any mistakes in their solutions and assign each student a score from 1 to 10, where 10 represents a completely correct solution.

Ed
$(x + 3)^2 = x^2 + 3^2$
$\qquad\ \ = x^2 + 9$

Leon
$(x + 3)^2 = (x + 3)(x + 3)$
$\qquad\qquad = x^2 + 3x + 3x + 9$
$\qquad\qquad = x^2 + 6x + 9$

17. PHYSICS If you stand in a river flowing at s mi/h, the pressure p that the water exerts on your legs is given by the equation $p = 4.125s^2$, where p is measured in pounds per square inch.

a. Suppose you stand in a river flowing at $\dfrac{4}{3}$ mi/h. Find the pressure that the water exerts on your legs.

b. Suppose you stand in a river flowing at 4 mi/h. Find the pressure that the water exerts on your legs.

c. Refer to the diagrams at the right. Complete this equation:

$$\text{Pressure exerted by River B} = \underline{\ ?\ } \times \left(\text{Pressure exerted by River A}\right)$$

Current A: $s = a$ mi/h

Current B: $s = 3a$ mi/h

The plant shown is called the papyrus plant. The ancient Egyptians used this plant to make the papyruses they wrote on. One papyrus written in 1850 B.C. gives evidence that the ancient Egyptians knew the formula for the volume of a pyramid. This papyrus can now be seen at the Museum of Fine Arts in Moscow.

$$V = \frac{1}{3}s^2h$$

18. For the Great Pyramid in Egypt, $h = 481$ ft and $s = 755$ ft. Find the volume of the Great Pyramid.

19. Show that the volume of Pyramid A is twice the volume of Pyramid B.

Pyramid A

Pyramid B

20. The Moscow papyrus also gives evidence that the ancient Egyptians knew how to find the volume of a partially built pyramid, like the one in the diagram. Let f = the fraction of the pyramid that has been built. Then f is given by this equation:

$$f = 1 - \left(\frac{x}{s}\right)^3$$

a. Suppose $s = 90$ ft and $x = 60$ ft. What fraction of the pyramid has been built?

b. **Writing** Suppose $x = \frac{1}{2}s$. Is it true that the pyramid is half built? Explain your reasoning.

21. **BIOLOGY** The formula $w = 446l^3$ gives the approximate weight w in grams of a hognose snake whose length is l meters.

   a. Find the weight of a hognose snake whose length is 0.4 m.

   b. One snake is twice as long as another. Show that the weight of the longer snake is eight times the weight of the shorter snake.

22. **Challenge** Use the power of a product rule and the power of a power rule to prove the power of a quotient rule. (*Hint:* Rewrite the quotient as a product involving a negative exponent.)

23. SAT/ACT Preview Which of these expressions equals $\left(\dfrac{x^3 y^2}{z^4}\right)^3$?

A. $\dfrac{x^6 y^5}{z^7}$

B. $\dfrac{x^9 y^6}{z^{12}}$

C. $\dfrac{x^{15} y^6}{z^{12}}$

D. $\dfrac{x^9 y^6}{z^7}$

E. none of these

ONGOING ASSESSMENT

24. Open-ended Problem Use the rules you learned in this section to write the expression $\dfrac{a^2 b^6}{c^4}$ in two other ways. Explain how you used each rule.

SPIRAL REVIEW

Simplify each variable expression. *(Section 1.7)*

25. $-3(5u + 3v) + u$

26. $6 + b^2 - \left(3 - b^2\right)$

27. $2(5 - m) - (4 - m)$

28. $8(4 - p) - (p + 1)$

29. $\dfrac{1}{2}(6q + 10)$

30. $(d + 5) + 2(3 - d)$

Solve each proportion. *(Section 5.1)*

31. $\dfrac{f}{12} = \dfrac{3}{1}$

32. $\dfrac{7}{2} = \dfrac{5}{s}$

33. $\dfrac{15}{35} = \dfrac{p}{100}$

34. $\dfrac{m}{m + 6} = \dfrac{2}{5}$

35. $\dfrac{10}{x} = \dfrac{5}{x - 1}$

36. $\dfrac{3c}{c + 2} = \dfrac{9}{4}$

ASSESS YOUR PROGRESS

Simplify each expression. *(Sections 9.6 and 9.7)*

1. $a^9 \cdot a^7$

2. $\dfrac{b^{12}}{b^8}$

3. $\dfrac{a^{10} b^{-1}}{a^3 b^3}$

4. $(ab)^4$

5. $\left(c^7\right)^8$

6. $b\left(\dfrac{c}{d}\right)^{10}$

7. $\left(\dfrac{p^2 q^6}{r^4}\right)^3$

8. $\left(\dfrac{x^2 y}{y^3 z^0}\right)^4$

9. TRANSPORTATION In 1990, there were about 1.44×10^8 cars in the United States, and the average car was driven about 1.06×10^4 mi. Estimate the total number of miles traveled by cars in the United States during 1990. Write your answer in scientific notation. *(Section 9.6)*

10. PHYSICS The equation $d = 16t^2$ gives the distance d in feet traveled by a falling object in t seconds. Suppose Rock B has been falling for twice as long as Rock A. Complete the following equation. Be sure to justify your answer algebraically. *(Section 9.7)*

Distance traveled by Rock B = $\underline{\ ?\ }$ × Distance traveled by Rock A

11. Journal List the five rules of exponents you learned in the last two sections. Tell whether each was easy or difficult for you to learn.

Models for Spirals

Many patterns found in nature can be modeled by mathematical equations. One pattern that often appears in flowers, spider webs, and sea shells is the spiral. In this project, you will use exponential equations to create a spiral.

PROJECT GOAL Find an exponential formula that models a spiral. Then create your own spiral.

Measuring the Nautilus Shell

Work with a partner. Find a photograph of a nautilus shell, or use the one shown here. Measure the distance in millimeters from the center of the spiral to the outside corner of the first chamber. Label this distance d_1 as shown. Measure the corresponding distances for the next seven chambers. Label these distances d_2, d_3, and so on.

Calculating the Ratios

As you move inward along the spiral, the distances you measure become shorter. They are decreasing exponentially. Organize the distances you measured in a table or spreadsheet. Find the ratio of the distance for each chamber to the distance for the previous chamber.

Spiral Project		
	A	**B**	**C**
1	n	dn	dn+1/dn
2	1	100	0.95
3	2	95	
4	3	86	
5	4	77	
6	5	72	

Modeling the Spiral with an Equation

Let r = the mean of the ratios you found. Substitute your values for d_1 and r into the formula below to create an equation for d_n, the distance for the nth chamber.

$$d_n = d_1 \cdot r^n$$

Calculate values for d_n by substituting values of n into your equation. How close do the calculated values come to the distances you measured? Make a scatter plot of your data and graph your equation in the same coordinate plane. Describe your results.

Creating Your Own Spiral

1. **CHOOSE** values for r and d_1. Make sure your r-value is between 0 and 1.

2. **DRAW** a line segment of length d_1.

3. **MULTIPLY** the length of the line segment you drew by r. Draw a line segment of this new length at a 25° angle to the previous line.

4. **REPEAT** Step 3 until the lengths become very close to zero.

5. **CONNECT** the endpoints of the line segments with a smooth curve.

Find an equation to model your partner's spiral. How does the spiral change when you increase the value of r? when you increase the value of d_1? Write a paragraph answering these questions. Include sketches of your work.

Self-Assessment

Tell what you liked and didn't like about this project. How did you and your partner divide up the work in the project? Describe any difficulties you had and discuss how you solved problems. What would you do differently if you were to do the project over again? List some ways you could extend your study of spirals.

9 | Review

Write at least six questions about this chapter. Focus on concepts you had the most trouble learning. Three should be short answer questions about specific situations or details. Three should be more involved questions, concentrating on more general ideas. Then answer the questions.

VOCABULARY

base (p. 370)

exponent (p. 370)

power (p. 370)

exponential growth (p. 377)

growth rate (p. 377)

growth factor (p. 377)

exponential decay (p. 384)

rate of decrease (p. 384)

decay factor (p. 384)

scientific notation (p. 396)

SECTIONS | 9.1, 9.2, *and* 9.3

The expression $(-3)^6$ is called a **power**. The number -3 is the **base**, and the number **6** is the **exponent**. The power can be evaluated by using the number -3 as a factor **6** times.

$$(-3)^6 = \underbrace{(-3) \cdot (-3) \cdot (-3) \cdot (-3) \cdot (-3) \cdot (-3)}_{\text{6 factors}} = 729$$

The order of operations for evaluating expressions that include powers is as follows:

1. Simplify expressions inside parentheses.

2. Evaluate powers.

3. Do multiplication and division from left to right.

4. Do addition and subtraction from left to right.

For example:

$$8 - (3 + 4) + 2 \cdot 6^2 = 8 - (7) + 2 \cdot 6^2$$
$$= 8 - 7 + 2 \cdot 36$$
$$= 8 - 7 + 72$$
$$= 73$$

Evaluate the powers.

You can use equations and graphs to model **exponential growth** and **exponential decay**.

The graph shows that money earning interest *grows* exponentially. The value y of \$500 earning 9.5% interest per year for x years is modeled by the equation $y = 500(1 + 0.095)^x$. The constant 0.095 is called the **growth rate**. The quantity $(1 + 0.095)$ is called the **growth factor**.

The number of teams left in a single-elimination tournament *decreases* exponentially by 50% each round of the tournament. If there are 64 teams at the start of the tournament, the number of teams y left at the end of x rounds is given by $y = 64(1 - 0.50)^x$. The constant 0.50 is called the **rate of decrease**. The quantity $(1 - 0.50)$ is called the **decay factor.**

SECTIONS   9.4 *and* 9.5

For any nonzero number a and any integer n, $a^0 = 1$ and $a^{-n} = \dfrac{1}{a^n}$. For example, $4^{-3} = \dfrac{1}{4^3}$ and $4^0 = 1$.

$$(-2)^0 = 1 \qquad\qquad 6^{-2} = \frac{1}{6^2} = \frac{1}{36} \qquad\qquad a^5 b^{-6} = a^5 \cdot \frac{1}{b^6} = \frac{a^5}{b^6}$$

A number is expressed in **scientific notation** if it is in the form $a \times 10^n$, where $1 \le a < 10$ and n is an integer.

$$6{,}270{,}000{,}000{,}000 = 6.27 \times 10^{12} \qquad 0.00000000013 = 1.3 \times 10^{-10}$$

Move the decimal point
12 places to the left.

Move the decimal point
10 places to the right.

SECTIONS   9.6 *and* 9.7

Five basic rules for working with powers are:

Product of Powers: $a^m \cdot a^n = a^{m+n}$ $\qquad\qquad$ $3^2 \cdot 3^3 = 3^{2+3} = 3^5 = 243$

Quotient of Powers: $\dfrac{a^m}{a^n} = a^{m-n}, a \ne 0$ $\qquad$ $\dfrac{3^5}{3^3} = 3^{5-3} = 3^2 = 9$

Power of a Product: $(ab)^m = a^m b^m$ $\qquad\qquad$ $(-2x)^5 = (-2)^5 x^5 = -32x^5$

Power of a Quotient: $\left(\dfrac{a}{b}\right)^m = \dfrac{a^m}{b^m}, b \ne 0$ $\qquad$ $\left(\dfrac{5}{2}\right)^3 = \dfrac{5^3}{2^3} = \dfrac{125}{8}$

Power of a Power: $(a^m)^n = a^{m \cdot n}$ $\qquad\qquad$ $(4^2)^3 = 4^{2 \cdot 3} = 4^6 = 4096$

Assessment

VOCABULARY REVIEW

For Exercises 1–3, complete each sentence.

1. In the expression 7^6, 7 is called the _?_ , 6 is called the _?_ , and the whole expression is called a _?_ .

2. The population of Sweden increased by 0.3% per year throughout the 1980s. This situation is an example of _?_ . The constant 0.3% is the _?_ , and the quantity 1.003 is the _?_ .

3. The amount of aspirin in a person's blood stream decreases by 20% per hour. This situation is an example of _?_ . The quantity 0.80 is the _?_ .

4. Writing Explain what is meant by *scientific notation*. Give an example of a number written in scientific notation.

SECTIONS 9.1, 9.2, *and* 9.3

Evaluate each power.

5. 9^3 **6.** $(-3)^4$ **7.** $\left(\frac{1}{4}\right)^5$ **8.** 0.5^3

Write each expression as a power. Then evaluate the power.

9. the third power of two   **10.** three to the second power

Evaluate each expression when $x = 4$ and $y = -2$.

11. $x^2y^3 + (5 - y)^2$ **12.** $(y - x)^3 + xy^2$ **13.** $3y^3 - 2x^2 + (xy)^2$

14. $2000 is placed in an Individual Retirement Account that earns 7.5% interest annually. No other deposits or withdrawals are made.

   a. Write an equation giving the amount of money y in the account x years from now.

   b. How much money will be in the account 40 years from now?

15. A new car was valued at $13,000 in 1995. Each year, the car's value decreases by 12%.

   a. Write an equation giving the value y of the car x years after 1995.

   b. How much will the car be worth in the year 2005?

   c. Graph the equation you wrote in part (a). Where does the graph cross the vertical axis? What does this point represent?

Evaluate each power.

16. 4.37^0 **17.** 2^{-5} **18.** $(-3)^{-6}$ **19.** $\left(\frac{2}{5}\right)^{-3}$

Evaluate each expression when $x = -5$, $y = 2$, and $z = 3$.

20. $x^{-3}z^0$ **21.** x^2y^{-3} **22.** $\left(\frac{yz}{x}\right)^{-1}$ **23.** $y^{-2}z + x^2$

24. BIOLOGY The mass y in grams of a colony of bacteria x hours after 3:00 P.M., May 2, is given by the equation $y = 2.1(1.35)^x$. Estimate the mass at each of the times below.

  a. 1:00 P.M., May 2   **b.** 3:00 P.M., May 2   **c.** 3:00 A.M., May 2

Write each number in decimal notation.

25. 4.57×10^7 **26.** 3.8×10^{-6} **27.** 7.023×10^4

Write each number in scientific notation.

28. 0.00000000678   **29.** 23,000,000,000   **30.** 5.0004

31. Open-ended Problem Give a value of x that satisfies the given inequality. Write your answers in scientific notation.

  a. $x < 1$ **b.** $1 \le x < 10$ **c.** $x > 10$

Simplify each expression.

32. $\dfrac{d^4m^5}{d^7m^3}$ **33.** $\left(2b^2\right)^4$ **34.** $\left(3c^5d^7\right)\left(-2c^2d^8\right)$

35. $\dfrac{x^{-3}y^4}{y^3z^{-1}}$ **36.** $\left(\dfrac{-2a^2b^{-1}}{3ab^0}\right)^3$ **37.** $\left(\dfrac{s^3t}{t^2s^0}\right)^4$

38. HISTORY According to the 1990 Census, the population of the United States was 2.49×10^8. The cost of taking the census in 1990 was 2.6×10^9 dollars. How much did the census cost per person in the United States?

PERFORMANCE TASK

39. Open-ended Problem Find an ad for a car you would like to buy. Call a local bank and find out what its interest rate is for a car loan. Set up a plan for buying the car, assuming you make $100 a week at a part-time job. Use the formula given on page 392 to find your monthly payment. How long would it take you to pay for the car? How would your payment plan change if you made $75 a week? $150 a week?

Cumulative Assessment
C H A P T E R S $7-9$

CHAPTER 7

1. Give the horizontal intercept and the vertical intercept of the graph of $5x - y = -10$. Then solve the equation for y.

For each system of equations, give the solution or state whether there are no solutions or many solutions.

2. $x + 3y = 15$
   $2x - 3y = 3$

3. $y = 5x + 60$
   $10x - 2y = -120$

4. $5x - 2y = 4$
   $3x + 7y = 27$

5. **Writing** Tell which method you used to solve each equation in Questions 2–4. Explain why you chose each method.

RECREATION Sharon Page wants to plant a border of impatiens and marigolds along a walkway. A tray of impatiens costs $2.50. A tray of marigolds costs $2.00. Each tray holds six plants.

6. Write an equation showing that Sharon buys a total of 54 plants.

7. Write an equation showing that Sharon spends $20 altogether for plants.

8. Solve the system of equations from Questions 6 and 7 to find how many trays of each type of plant Sharon buys.

Graph each inequality or system of inequalities.

9. $y \le \frac{4}{3}x - 3$

10. $x < 2$
    $y > -x + 2$

11. $x - y \le 2$
    $x - y > -1$

12. **Open-ended Problem** In a coordinate plane, draw a simple four-sided figure and shade the interior. Using the endpoints of each side of the figure, write a system of four inequalities that describes your shaded figure.

CHAPTER 8

13. Graph the equation $y = 3|x| - 1$. Tell whether the relationship between x and y is *linear* or *nonlinear*.

Solve each equation algebraically.

14. $4x^2 = 25$

15. $n^2 - 100 = 300$

16. $3a^2 - 8 = 19$

Match each equation with its graph.

17. $y = 3x^2$ **18.** $y = -3x^2$ **19.** $y = -\frac{1}{3}x^2$

A. **B.** **C.**

Technology Estimate the solution(s) of each equation using a graphing calculator. Then find the exact solution(s) using the quadratic formula.

20. $x^2 - 3x - 7 = 0$ **21.** $-3x^2 + 5x + 6 = 3$

22. Writing Find the discriminant of the equation $4x^2 - 8x + 4 = 0$. What does it tell you about the number of solutions of the equation? In general, what does the discriminant tell you about the number of solutions of a quadratic equation?

23. Open-ended Problem Write a quadratic equation that has no solution. Use a graph to show why the equation has no solution.

CHAPTER 9

Evaluate each expression when $a = -3$ and $b = 2$.

24. $(ab)^3 - 4b^5$ **25.** $(2a + b)^2 + ab^3$ **26.** $a^0 + b^{-4}$

Simplify each expression. Write each answer using positive exponents.

27. $(ab)^2(a^3b)^4$ **28.** $(x^0y^5)^3$ **29.** $(4m)^2(2m^3)^3$

30. $\dfrac{(2c)^3}{c^2d}$ **31.** $\left(\dfrac{2r}{3s}\right)^5$ **32.** $\left(\dfrac{5k^6}{k^8}\right)^3$

POPULATION The population of the Netherlands in 1990 was about 14,952,000. The annual growth rate for the period from 1980 through the year 2000 is about 0.6%. Use this information for Questions 33–35.

33. Write an equation that models the size y of the population x years after 1990.

34. Estimate the population in 1980 to the nearest thousand.

35. Predict the population in the year 2000 to the nearest thousand.

36. Writing Write *thirty-five billion, thirty million* in decimal notation and in scientific notation. Using your answer as an example, explain how to write a number in scientific notation.

37. Open-ended Problem Give an example of an equation that models exponential decay. Graph the equation. Does the graph ever cross the horizontal axis? Explain.

10 | Polynomial Functions

INVESTING in Equality

INTERVIEW ## Linda Pei

What are your dreams for the future? Going to college? Traveling the world? Making dreams come true often takes money, sometimes lots of it. How can you save money and make it work for you? You could open a savings account at a bank and make regular deposits. The bank would pay you a small amount of interest on the money in your account. Or you might ask financial adviser Linda Pei about investing in the stock market. She will tell you that some stocks earn 20 percent or more each year, but they can lose that much, too. You can avoid losing money by investing in a variety of carefully chosen companies. "You need to spread your risk around—if one investment does not perform well, hopefully another one will do better," Pei says. "In the end you won't lose everything because you've invested in many different companies."

> " **When I was a child growing up in Japan, we got red envelopes at New Year's. ...** "

Investing in Change

Pei founded a *mutual fund* to help people invest wisely and support companies that promote equality at the same time. A mutual fund is a company that employs experts to invest money in stocks they think will rise in value. The money comes from thousands of people who share the profits and losses on the stocks owned by the fund.

Pei's experience in banking and consulting gave her a chance to see how different companies operate. "That view made me think about how people manage other people and how senior manage-

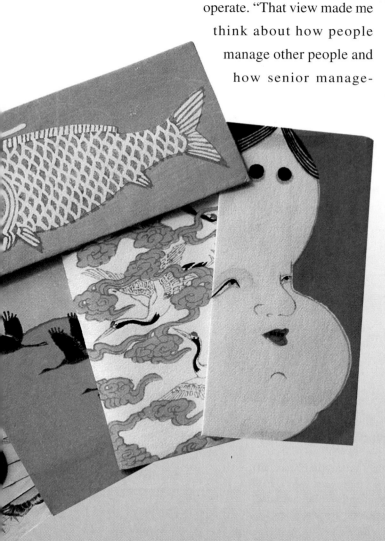

> **"...Whatever money you got in them, you put some away for the future."**

ment directs and treats employees." All too often, Pei noticed women being treated in an unfair way.

Determined to fight for equal treatment of all employees, Pei believed she could become a force for change by supporting companies that promote equality. She worked with others to design a mutual fund to do just that. They developed a set of standards to identify companies actively working to promote equality. They invest only in companies that meet these standards.

421

Calculating Your Returns

Financial advisers like Linda Pei encourage people to save and invest on a regular basis. One consequence of this is that the amount of money earning interest grows. For instance, suppose you invest $1000 at the start of each year over a period of three years in a mutual fund that earns a consistent return of 10% annually. You can calculate the total value of your investment after three years by writing an expression like this one:

Total value of investment after three years $= 1000(1 + 0.1)^3 + 1000(1 + 0.1)^2 + 1000(1 + 0.1)$

value of investment made the first year

value of investment made the second year

value of investment made the third year

If the annual growth rate is r, written as a decimal, the expression above becomes:

$$1000(1 + r)^3 + 1000(1 + r)^2 + 1000(1 + r)$$

This is an example of a *polynomial*, the topic of this chapter.

Many newspapers publish information about the performance of mutual funds.

1. **Fund** Names of fund groups.
2. **NAV** Net asset value.
3. **N** Signifies no load. The fund does not charge any sales fees.
4. **Dly. Chg.** Change in **NAV** from previous session.
5. **12 mo. % ret.** Percent return of fund for last 12 month period.

	1		2	4	5
Mutual funds					
Fund	NAV		Dly. Chg.	12 mo. % ret.	
abc XYZ Fund	17.65N		+.03	+8.1	
def	17.65N		+.03	+8.1	
hij	17.65N		+.03	+8.1	

3

Much of Pei's business is done over the phone and via computer.

1. **Writing** Calculate the total value of the investment represented by the first expression above. What will be the value after four years? Explain.

2. **Research** What are stocks? How do they make money for investors? How are they different from savings accounts at banks? Learn about an aspect of stocks or the stock market. Report your findings to your class.

3. **Project** Linda Pei hopes that her mutual fund will encourage businesses to treat all employees fairly. Find out about another person or organization that supports the equal treatment of people at work. Report your findings to your class.

Mathematics & Linda Pei

In this chapter, you will learn more about how mathematics is related to investing.

Related Exercises

Section 10.1
• **Exercises 22–24**

Section 10.4
• **Exercises 30–33**

10.1 Adding and Subtracting Polynomials

Learn how to...
- recognize and classify polynomials
- add and subtract polynomials

So you can...
- use polynomials to model real-world situations, such as the total value of an investment

Financial advisers like Linda Pei sometimes write formulas like the one on page 422 in terms of an annual growth factor g. If you substitute g for $1 + r$, you get the *polynomial* shown below.

A **polynomial** is an expression that can be written as a monomial or as a sum of monomials with exponents that are whole numbers. The polynomial below is in **standard form**, which means it is simplified and the exponents of the terms are in order from largest to smallest.

The **degree of a term** of a polynomial is the exponent of the variable. The degree of this term is 2.

The **degree of a polynomial** in standard form is the largest degree of all the terms. The degree of this polynomial is 3.

$$1000g^3 + 1000g^2 + 1000g$$

For a term with more than one variable, the degree is the *sum* of the exponents of the variables. For example, the degree of $5xy$, or $5x^1y^1$, is 2.

Classifying Polynomials

You can classify a polynomial by its degree.

Degree	Example	Classification
1	$3x + 2y$	linear polynomial
2	$-2x^2 + 3x + 1$	quadratic polynomial
3	$5x^3 - x^2 + 6$	cubic polynomial

THINK AND COMMUNICATE

1. Use what you know about exponents to explain why $\dfrac{4}{x}$ is not a polynomial.

2. The expression $8x^2 + \dfrac{1}{2}x^3 - 4$ is a polynomial. Write it in standard form and identify the degree of each term. Then identify the degree of the polynomial and its classification.

3. Explain why the polynomial $4x - 2 - 4x$ has degree 0.

Algebra tiles can help you understand how to add polynomials. You can represent positive and negative terms with algebra tiles as shown below.

Positive and negative tiles of the same size cancel each other out.

$$x^2 + (-x^2) = 0$$

EXAMPLE 1

Add $3x^2 + 2x + 3$ and $2x^2 - x - 2$.

SOLUTION

Method 1 Use algebra tiles.

Model each polynomial with algebra tiles.

$3x^2 + 2x + 3$

$2x^2 - x - 2$

Group tiles of the same size.

$x + (-x) = 0$

$2 + (-2) = 0$

Combine the remaining tiles. The sum is $5x^2 + x + 1$.

Method 2 Add vertically.

Line up the like terms. Then add the coefficients.

$$3x^2 + 2x + 3$$
$$\underline{2x^2 - 1x - 2} \qquad x = 1x$$
$$5x^2 + \;\; x + 1$$

Method 3 Add horizontally.

$$(3x^2 + 2x + 3) + (2x^2 - x - 2) = (3x^2 + 2x^2) + (2x - 1x) + (3 - 2) \qquad \text{Group like terms.}$$
$$= 5x^2 + x + 1$$

You can subtract polynomials by using a method similar to the one you used in Chapter 1 to simplify expressions like $5x - (3x + 2)$.

EXAMPLE 2

Subtract $(a^3 - a + 7) - (4a^2 - 3a + 1)$.

SOLUTION

Method 1 Subtract vertically.

Line up the like terms. You may need to insert placeholders, like $0a^3$ and $0a^2$.

Change each term being subtracted to its opposite. Then *add* like terms.

$$1a^3 + 0a^2 - 1a + 7 \qquad\qquad 1a^3 + 0a^2 - 1a + 7$$
$$\underline{0a^3 + 4a^2 - 3a + 1} \qquad\qquad \underline{-0a^3 - 4a^2 + 3a - 1}$$
$$\qquad\qquad\qquad\qquad\qquad\qquad a^3 - 4a^2 + 2a + 6$$

Method 2 Subtract horizontally.

Change each term being subtracted to its opposite and then add.

$$(a^3 - a + 7) - (4a^2 - 3a + 1) = (a^3 - a + 7) + (-4a^2 + 3a - 1)$$
$$= a^3 + (-4a^2) + (-a + 3a) + (7 - 1)$$
$$= a^3 - 4a^2 + 2a + 6$$

THINK AND COMMUNICATE

4. Suppose the polynomials being added in Example 1 were $3x^2 + 2xy + 3$ and $2x^2 - xy - 2$. How would the answer be different?

5. To simplify $(-n^3 + 7n^2 - 3n) - (8n^2 - 5n - 2)$, you must find the *sum* of $-n^3 + 7n^2 - 3n$ and what polynomial?

Tell whether each expression is a polynomial. Write *Yes* or *No*. If not, explain why not.

1. $3a + 4b + 7c$

2. $2x^4 - x^6 + x^3$

3. $6n^4 + n^{-3}$

4. -8

5. $0.7xyz$

6. $\dfrac{3}{x^2} + \dfrac{4}{x} + \dfrac{1}{2}$

Write each polynomial in standard form. Then classify the polynomial as *linear, quadratic, cubic,* or *none of these*.

7. $1 - 2x + 5x^2$

8. $0.9t + 0.1t^2$

9. $1 - x^4$

10. $x - 7x^3 + 4x^2 - 3$

11. $14 - \dfrac{3}{4}h$

12. $k^3 + 3k$

Add or subtract each pair of polynomials as indicated. You may want to use algebra tiles.

13. $(7x^2 + 8x - 6) + (-x^2 - 8x - 3)$

14. $(9t^2 - 3t + 1) + (12t^2 + 5)$

15. $(9x + 3xy + 6y) - (x - 8xy + 4y)$

16. $(2x^3 + 5x^2 - 7x) - (6x^2 - 7x)$

10.1 **Exercises and Applications**

Extra Practice exercises on page 576

Match each phrase with an algebraic expression.

1. linear binomial

    A. $3a^2 - 4a + 2$

2. quadratic monomial

    B. $\dfrac{2}{m} + 3$

3. quadratic trinomial

    C. $z^3 - z^2 + z$

4. cubic trinomial

    D. $\dfrac{1}{2}y^2$

5. not a polynomial

    E. $2x + 3y$

6. Writing Explain why $\dfrac{x^4}{3}$ is a polynomial, but $\dfrac{3}{x^4}$ is not.

Add. Give the degree of each sum.

7. $\begin{aligned} 5x^2 - 3x - 7 \\ x^2 + 2x + 3 \end{aligned}$

8. $\begin{aligned} -y^2 + 2y - 8 \\ y^2 + 3y - 2 \end{aligned}$

9. $\begin{aligned} 2r^3 - 7r^2 - r + 2 \\ 4r^2 - r + 1 \end{aligned}$

10. Open-ended Problem Write the polynomial modeled by the algebra tiles below. Find two polynomials whose sum is that polynomial.

Add.

11. $(9n^2 + 3n - 5) + (8n^2 - n + 2)$ **12.** $(4x^2 - x - 5) + (6x^2 + x)$

13. $(y^2 + 3y + 2) + (-7y - 9)$ **14.** $(3k - 5) + (k^2 - 8k)$

15. $(a - 3b + 7) + (-2a - b)$ **16.** $(9x^2 + 6xy + y^2) + (x^2 - y^2)$

17. $(b^3 + b^2 - 8) + (b^3 - 6b + 9)$ **18.** $(a^3 + 4a - 1) + (6a^2 - 7a + 2)$

GEOMETRY **Express the perimeter of each triangle as a polynomial in standard form.**

19.

$4x - 2$

$x^2 - x - 6$

$5x$

20.

$7x - 10$

$2x^2 + 4$

$x^2 + 8x$

21. Suppose the perimeter of the triangle in Exercise 19 is 120 cm. Find the length of each side of the triangle.

INTERVIEW **Linda Pei**

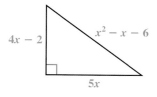

Financial advisers like Linda Pei sometimes handle one-time investments. The growth of a one-time investment can be modeled by an exponential function of the annual growth factor g.

22. Does the polynomial in the article on page 422 represent a one-time investment of $1000? Why or why not?

23. a. Write a polynomial in terms of g to represent the value of a one-time investment of $2000 at the end of four years.

b. Suppose you invest $2000 at the beginning of every year for four years. Write a polynomial in terms of g to represent the total value of the investment at the end of four years.

c. Find the sum of the polynomials in parts (a) and (b). What is the degree of the new polynomial?

24. Writing One aspect of a company that Pei might consider when making investments is the number of women in management positions. Use the graph to describe trends in management from 1983 to 1991.

People in Management Jobs

Men

Women

Percent

100

80

60

40

20

0

1983   1985   1987   1989   1991

Year

Subtract.

25. $3c^2 + c + 8$
$\underline{2c^2 - 2c + 4}$

26. $4x^2 - 5x - 7$
$\underline{-x^2 - 5x + 1}$

27. $4z^3 - 6z^2 + z + 4$
$\underline{ 2z^2 + 9z - 4}$

28. $(3a^2 - 5a + 3) - (a^2 + a - 2)$ **29.** $(-6m^2 - 5m + 1) - (2m^2 - 3)$

30. $(5n^2 - 4n + 8) - (-4n + 2)$ **31.** $(x^2 + 8) - (3x^2 - x + 1)$

32. $(8r + s + 5) - (r + 8s - 5)$ **33.** $(2y^3 - 2y + 9) - (5y^3 + y^2 - 2y)$

34. $(n^3 - n^2 + 5n - 8) - (9n^2 - 2)$ **35.** $(4a^2 - 9b^2) - (5a^2 - 12ab + 9b^2)$

36. Challenge Luis claimed that the sum of any four consecutive integers is an even integer. Use algebra to show that Luis is correct. Then use a logical argument to show that Luis is correct.

ONGOING ASSESSMENT

37. Cooperative Learning Work in a group of three students. Check each other's work at every stage.

a. First each of you should write a polynomial on the same piece of paper using the variable x.

b. Then you should take turns adding your teammates' polynomials.

c. Finally you should take turns subtracting your original polynomial from the sum of your teammates' polynomials.

SPIRAL REVIEW

Find each product. *(Section 6.4)*

38. $2t(t + 2)$ **39.** $(x + 1)(x - 1)$ **40.** $(2n + 1)(n - 4)$

Estimate the solutions of each equation using the given graph. *(Section 8.3)*

41. $x^2 - 2 = 0$ **42.** $-2x^2 + \dfrac{5}{2} = 0$ **43.** $\dfrac{2}{3}x^2 - 3 = 0$

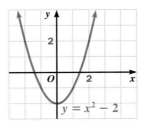

Distribution of Federal Funds			
	A	**B**	**C**	**D**
1	Region	1988	1993	Change
2	Northwest	175,419	250,215	
3	Midwest	182,666	257,590	
4	South	303,828	448,269	
5	West	187,577	268,615	

44. 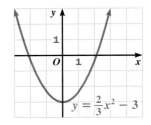 **Spreadsheets** Use a spreadsheet. Find the change in the distribution of federal funds for each region from 1988 to 1993. (The numbers shown are in millions of dollars.) Then find the average change for the four regions. What formulas did you use?
(Section 1.8)

10.2

Learn how to...

- **use the FOIL method to multiply binomials**
- **multiply polynomials**

So you can...

- **use polynomials to model lengths, areas, and volumes, such as the volume of an origami box**

Multiplying Polynomials

People from many different cultures enjoy the art of paper folding, often referred to as *origami*. The word comes from the Japanese *ori*, "to fold," and *kami*, "paper." In traditional origami, you fold a square piece of paper into an object without cutting or pasting.

EXPLORATION
COOPERATIVE LEARNING

Modeling Dimensions with Polynomials

Work in a group of 4 students.
You will need:

- rectangular pieces of paper that are not square
- a ruler

Each of you should build a box and write expressions for its dimensions as outlined in Steps 1–8.

1 Fold a rectangular piece of paper into 16 equal parts as shown. Always fold toward the front of the piece of paper.

Fold in half...

then in quarters...

then turn and repeat.

Exploration continued on next page.

Origami flourished in South America as well as in Japan. This model of a parrot perched on a glass was designed by Ligia Montoya of Argentina.

2 Hold your paper with the longer side facing you and fold the outer flaps back in as shown.

3 Fold in the four corners so the edges line up with the horizontal fold lines as shown.

4 Fold back two vertical strips in the middle, one to the left and the other to the right, to "lock" the corners.

5 Lift up gently as shown to form your box.

6 Unfold your box as shown. Measure the length and the width of the original piece of paper. Also measure the width of the vertical strips that you folded back.

7 What do *x* and the region outlined in red represent in terms of the box?

8 Use your measurements from Step 6 to write expressions for the length, width, and height of the box in terms of *x*.

9 Compare expressions with those of the members of your group. How are the expressions alike? How are they different?

EXAMPLE 1   Application: Crafts

Manuel used a 6 in. by 4 in. piece of paper to do the Exploration on pages 429 and 430. His group expressed the dimensions of the folded box as shown.

x

$5 - 2x$ $4 - 2x$

a. Write a polynomial to represent the area of the base of the box.

b. Write a polynomial to represent the volume of the box.

SOLUTION

a. Area of base = Length × Width = $(5 - 2x)(4 - 2x)$

You can use the *FOIL* method to multiply binomials. *FOIL* stands for "**First terms** in the binomials," "**Outer terms** in the expression," "**Inner terms** in the expression," and "**Last terms** in the binomials." You will get the same result as if you used the distributive property.

First
Last

$$(5 - 2x)(4 - 2x) = 5 \cdot 4 + 5(-2x) + (-2x)4 + (-2x)(-2x)$$

Inner
Outer

$$= 20 - 10x - 8x + 4x^2$$
$$= 4x^2 - 18x + 20 \qquad \text{standard form}$$

b. Volume = Area of base × Height

$$= (4x^2 - 18x + 20)x \qquad \text{Use the area of the base from part (a).}$$
$$= 4x^2 \cdot x - 18x \cdot x + 20 \cdot x$$
$$= 4x^3 - 18x^2 + 20x$$

Use the distributive property.

You saw on page 259 that area models can be used to represent the product of two binomials. Volume models can be used to represent the product of three binomials, as shown below. The length of each side of the cube is represented by the expression $x + 1$.

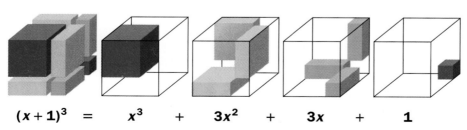

$(x + 1)^3$ = x^3 + $3x^2$ + $3x$ + 1

EXAMPLE 2

Multiply $(3x - 2)(2x^2 + 3x - 6)$.

SOLUTION

Use the distributive property.

Method 1 Multiply horizontally.

$$(3x - 2)(2x^2 + 3x - 6) = 3x(2x^2 + 3x - 6) - 2(2x^2 + 3x - 6)$$
$$= 6x^3 + 9x^2 - 18x - 4x^2 - 6x + 12$$
$$= 6x^3 + 5x^2 - 24x + 12$$

Method 2 Multiply vertically.

Multiply each term by
$3x$: $3x \cdot 2x^2 = 6x^3$.

$$
\begin{array}{r}
2x^2 + 3x - 6 \\
3x - 2 \\
\hline
6x^3 + 9x^2 - 18x \\
-4x^2 - 6x + 12 \\
\hline
6x^3 + 5x^2 - 24x + 12
\end{array}
$$

Add.

Multiply each term by -2. Line up like terms.

THINK AND COMMUNICATE

1. Show how you can use the distributive property to do Example 1(a) on page 431.

2. In Example 2, what expression do you get when you distribute $3x$? when you distribute -2?

3. Why do you think a polynomial with degree three is called a *cubic* polynomial? (*Hint:* Look at the diagram at the bottom of page 431.)

☑ CHECKING KEY CONCEPTS

Identify the *First*, *Outer*, *Inner*, and *Last* terms of each expression.

1. $(x + 4)(x + 8)$ 2. $(a + 7)(a - 9)$ 3. $(2n + 4)(n - 8)$

Replace each ? with the correct expression.

4. $8a(4a^2 + 5a - 6) = 8a(\underline{?}) + 8a(\underline{?}) - 8a(\underline{?})$

5. $(a + 4b)(a + 3b) = a^2 + 3ab + \underline{?} + 12b^2$

6. $(2x^2 - 5x + 1)(4x + 5) = (\underline{?})4x + (\underline{?})5$

Multiply.

7. $(m + 4)(m + 5)$ 8. $(2z - 1)(z - 7)$

9. $k(k^2 + 2k - 24)$ 10. $2x(3x^2 - 5x + 9)$

11. $(y - 1)(3y^2 - 8y - 4)$ 12. $(2a + 6)(2a^2 - 3a + 9)$

10.2 | Exercises and Applications

Multiply.

1. $(c + 5)(c + 7)$
2. $(r + 5)(r - 8)$
3. $(s - 4)(s - 2)$
4. $(3t + 2)(t - 4)$
5. $(6a + 1)(2a + 6)$
6. $(11m - 6)(2m + 5)$
7. $n(n^2 + 10)$
8. $p(4p^2 - 9)$
9. $5b(9b^2 + 6)$
10. $3x(2x^2 - x + 6)$
11. $-z(z^2 - 5z + 4)$
12. $\frac{1}{2}y(8y^2 + 4y - 2)$

Extra Practice exercises on page 577

13. **Writing** Write a note to a friend explaining how to do exercises like Exercises 1–12. Include some worked-out examples.

14. Write each product in standard form. Then compare parts (a) and (b). What do you notice? Does the same apply to parts (c) and (d)? Explain.

    a. $(x + 2)(x + 3)$
    b. $(x + 2y)(x + 3y)$
    c. $(4a + 5)(a - 7)$
    d. $(4a + 5b)(a - 7b)$

Connection ▸ CRAFTS

When quilters design a quilt, they have to decide how wide to make the squares, the lattice, and the border so that the quilt fits nicely on a bed of a given size.

For the quilt pattern shown, suppose the width of each square is x, the width of the lattice is y, and the width of the border is 10 in. all around.

Lattice y in. **Squares** x in. by x in. **Border** 10 in.

15. a. Write expressions in terms of x and y for the dimensions of the quilt.

    b. Write an expression in terms of x and y for the perimeter of the quilt. Do the same for the area of the quilt.

16. Suppose you want your quilt to be 98 in. wide by 113 in. long (king-size). Write and solve a system of equations to find the widths of the squares and the lattice used in the quilt pattern.

17. Use your answer from Exercise 16 to check the polynomials you wrote in Exercise 15(b).

18. **Open-ended Problem** Suppose you want your quilt to be 83 in. by 107 in. (queen-size). How might you adjust the pattern to fit?

GEOMETRY Write a polynomial in standard form to represent the area of each figure.

19.

$3x + 2$
$6x + 1$

20.
$n + 3$
$8n - 3$

21.

$y + 3$
$3y + 1$

Multiply.

22. $(x + 4)(x^2 + 2x - 5)$

23. $(n - 3)(6n^2 + 8n + 9)$

24. $(3x + 2)(x^2 + 7x - 6)$

25. $(5x - 1)(x^2 - 3x - 10)$

26. $(2x^2 - 5x + 10)(x + 4)$

27. $(7j^2 - j + 4)(6j - 1)$

28. $(3t^2 - 2t - 4)(5t + 9)$

29. $(2x^2 + 9x - 4)(3x - 2)$

30. Writing Explain why the product of a quadratic polynomial and a linear polynomial must be a cubic polynomial.

31. Challenge Show how to multiply $(2y - 3)^3$.

Cooperative Learning Work with a partner. In Exercises 32–35, you will each make a box with an attached lid and explore its volume.

32. Each of you will need a 6 in. by 4 in. piece of paper. Follow the steps below to form two boxes, using two different values for x.

Step 1 Fold your paper in half.

Step 2 Draw and cut along the solid lines to form flaps.

$2x$
x x
x x

Step 4 Form a box. The double flap will hold the two short flaps in place.

Step 3 Fold inward along each dotted line.

33. Write polynomials in terms of x to represent the length, width, and height of your box. Compare results with your partner.

34. Write a function in terms of x to represent the volume of your box. Write the function as a polynomial in standard form. What is a reasonable domain for the function? Compare results with your partner.

35. [icon] **Technology** Graph your volume function from Exercise 34 on a graphing calculator. Find the value of x that gives the maximum volume.

36. SAT/ACT Preview If $A = (x + 1)^2$ and $B = (x + 3)(x - 1)$, then:

    A. $A > B$ **B.** $B > A$ **C.** $A = B$

    D. relationship cannot be determined

ONGOING ASSESSMENT

37. Writing Use a vertical multiplication method to find the products $(4x + 1)(2x^2 + 3x + 5)$ and $41 \cdot 235$. How is multiplying polynomials like multiplying numbers? How is it different?

SPIRAL REVIEW

For each system of equations, give the solution or state whether there are no solutions or many solutions. *(Section 7.4)*

38. $3x + 2y = 3$
    $4x - y = -7$

39. $-2 = 5m - 4n$
    $5 = 7m - 3n$

40. $2x - 3y = -18$
    $6x - 9y = 18$

Predict how the graph of each equation will compare with the graph of $y = x^2$. Explain how you made your prediction. *(Section 8.2)*

41. $y = -4x^2$

42. $y = \frac{3}{4}x^2$

43. $y = -1.2x^2$

ASSESS YOUR PROGRESS

VOCABULARY

polynomial (p. 423)

standard form of a

  polynomial (p. 423)

degree of a term (p. 423)

degree of a polynomial (p. 423)

linear polynomial (p. 423)

quadratic polynomial (p. 423)

cubic polynomial (p. 423)

1. Is $x - \dfrac{1}{x}$ a polynomial? Explain your answer. *(Section 10.1)*

Add or subtract each pair of polynomials as indicated. Write each answer in standard form and give its degree.

2. $(3x^2 - 4x + 2) + (2x^2 - x - 5)$ **3.** $(5y^2 + 7y - 9) + (-8y^2 + 6)$

4. $(6a^2 + a - 1) - (6a^2 - a - 1)$ **5.** $(z^3 + 2z - 1) - (2z^2 - 5z - 4)$

Multiply. *(Section 10.2)*

6. $(4y - 7)(2y - 5)$ **7.** $7n(n^2 + 3n - 8)$ **8.** $(5j + 3)(j^2 - 4j + 6)$

9. Show two different ways to find a polynomial that represents the area of the border of the square textile at the right. Then write the polynomial in standard form.

10. Journal What do you find most tricky about multiplying two polynomials such as $7k - 4$ and $3k^2 - k + 5$? Can you think of a way to avoid making mistakes with the part you find tricky?

Hmong textile

10.3 Exploring Polynomial Equations

Learn how to...
- **factor out a linear factor from a polynomial**
- **use the zero-product property**

So you can...
- **begin to use factoring to solve equations**

A **polynomial equation** is an equation in which both sides are polynomials. In Chapter 8 you learned how to use a graph to visualize the solutions of a polynomial equation. Another way that you can easily identify the solutions involves *factoring*.

EXPLORATION
COOPERATIVE LEARNING

Factors, Intercepts, and Solutions

Work with a partner.
You will need:
- graphing calculators

1 One of you should graph equations 1–4 while the other graphs equations A–D. Match the equations that have the same graph.

STANDARD FORM

1. $y = x^2 + 2x - 3$

2. $y = x^3 - 12x^2 + 44x - 48$

3. $y = x^2 - 4x$

4. $y = x^3 + 4x^2 + 3x$

FACTORED FORM

A. $y = x(x - 4)$

B. $y = (x + 3)(x - 1)$

C. $y = x(x + 3)(x + 1)$

D. $y = (x - 2)(x - 4)(x - 6)$

2 Find the *x*-intercepts of each graph in Step 1. Then look at the factored form of the equation for each graph. How are the *x*-intercepts and the factored form related?

3 Suppose you want to solve the polynomial equation $x^2 - 6x = 0$. How can the graph of $y = x^2 - 6x$ help you solve the equation? How can the factored form $x(x - 6) = 0$ help you solve the equation?

Rewriting a polynomial as a product of polynomial factors is called **factoring a polynomial**. For example:

standard form $\qquad$ $x^3 + 2x^2 - 3x = x(x + 3)(x - 1)$ $\qquad$ factored form

EXAMPLE 1

Factor $x^2 + 2x$.

SOLUTION

Method 1 Use algebra tiles.

Arrange one x^2-tile and two x-tiles to form a rectangle. The total area of the rectangle is $x^2 + 2x$. The dimensions of the rectangle will be the factors.

$x + 2$

x

$$\text{Area} = \text{Length} \times \text{Width}$$
$$x^2 + 2x = (x + 2) \cdot x$$
$$= x(x + 2)$$

Method 2 Factor algebraically.

Identify the greatest common factor (GCF) and write each term as a product.

Use the distributive property.

$$x^2 + 2x = x \cdot x + 2 \cdot x \qquad \text{The GCF is } x.$$
$$= x(x + 2)$$

THINK AND COMMUNICATE

1. a. Use the graph of $y = x^2 + 2x$ at the right. Explain the relationship between the factors of $x^2 + 2x$, found in Example 1, and the x-intercepts of the graph.

  b. How can you use the graph to solve the equation $x^2 + 2x = 0$?

2. Replace each $\underline{?}$ with a monomial to factor each polynomial.

  a. $2x^3 - 6x^2 = 2x^2(\underline{?} - \underline{?})$ **b.** $6n^3 + 3n^2 + 9n = \underline{?}(2n^2 + n + 3)$

The property below can help you solve equations in factored form. It is based on the fact that zero times any number is zero. For example, $0 \cdot 3 = 0$ and $3 \cdot 0 = 0$.

Zero-Product Property

If a product of factors is equal to zero, then one or more of the factors is equal to zero.

EXAMPLE 2

Solve $(x + 3)(4x - 5) = 0$.

SOLUTION

Use the zero-product property. One of the factors must equal zero.

$$(x + 3)(4x - 5) = 0$$

$$x + 3 = 0 \qquad \text{or} \qquad 4x - 5 = 0$$
$$x = -3 \qquad\qquad\qquad x = \frac{5}{4} = 1.25$$

The solutions are -3 and 1.25.

BY THE WAY

The dimples on golf balls have a lot to do with their performance. Other factors that affect the flight of a golf ball are air conditions, ball mass, ball diameter, spin, velocity, and launch angle.

EXAMPLE 3 **Application: Physics**

Many forces affect the height of a golf ball in flight. Suppose you ignore all forces except gravity. Then under certain initial conditions the equation $h = -16t^2 + 40t$ can be used to model the height h in feet of a golf ball t seconds after the ball has been hit. Using this model, about how long does it take for the golf ball to hit the ground if it is hit on level ground?

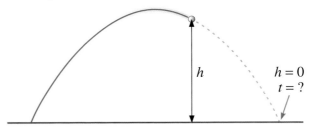

SOLUTION

The ball hits the ground when the height h is zero.

$$h = -16t^2 + 40t$$

Substitute **0** for **h**.

$$0 = -16t^2 + 40t$$

$$-1 \cdot 0 = -1(-16t^2 + 40t)$$

Multiply both sides by **−1**.

$$0 = 16t^2 - 40t$$

$$0 = 8t(2t - 5)$$

Factor.

Use the zero-product property.

$$8t = 0 \qquad \text{or} \qquad 2t - 5 = 0$$

$t = 0$ when the ball is hit.

$$t = 0 \qquad\qquad\qquad t = \frac{5}{2} = 2.5$$

It takes the golf ball about 2.5 s to hit the ground.

3. Can you solve the equation $2x(x - 3) = 4$ using the zero-product property without rewriting the equation first? Why or why not?

4. Is zero a solution of the equation $2(x - 3) = 0$? Why or why not?

5. What would you do to solve the equation $(x + 3)(9x - 45) = 0$?

☑ CHECKING KEY CONCEPTS

Find the greatest common factor of each group of monomials.

1. $5x^2, 3x$

2. $8n^2, 12n$

3. $4y^3, 6y^2, 2y$

Factor each polynomial.

4. $m^2 + 5m$

5. $14x^2 + 7x$

6. $6y^2 - 9y$

7. $30n^2 - 45n$

8. $4z^2 + 20z + 8$

9. $3t^3 - 6t^2 + 9t$

Solve each equation.

10. $(x + 6)(x - 4) = 0$

11. $(5n - 2)(2n + 5) = 0$

12. $y^2 - 2y = 0$

13. $16k^2 + 12k = 0$

10.3 Exercises and Applications

Factor each polynomial.

Extra Practice exercises on page 577

1. $x^2 - 4x$

2. $2m^2 + 11m$

3. $4y^2 + 4$

4. $8j + 10$

5. $2x^2 + 6x$

6. $21k^2 - 3k$

7. $15n^2 + 10n$

8. $36v^2 - 30v$

9. $a^3 - 4a^2 + a$

10. $9t^3 + 6t^2 + 9t + 3$

11. $54x^3 + 18x^2 + 2x$

12. $12r^3 - 4r^2 - 28r$

13. Visual Thinking Which polynomial from Exercises 1–12 is modeled by the diagram at the right? How does the diagram show how to factor that polynomial?

14. Writing What do you find the most difficult about factoring polynomials? Why? How might you overcome this difficulty?

Solve each equation.

15. $x(x - 2) = 0$

16. $2t(t + 7) = 0$

17. $0 = 3n(n - 4)$

18. $k(5k + 6) = 0$

19. $0 = 7m(3m - 1)$

20. $(k + 3)(k - 3) = 0$

21. $(a + 4)(a - 7) = 0$

22. $(8y - 1)(y - 4) = 0$

23. $0 = (4z + 3)(2z + 1)$

A team of economists must find the most profitable retail price for their company's new laptop computer. The equation $n = -100p^2 + 300{,}000p$ models the expected relationship between net sales n and the retail price p of the laptop computer, in dollars.

24. Show how to solve the equation $-100p^2 + 300{,}000p = 0$ by factoring. What information does the solution give you about the expected sales?

Expected Sales

Price
(thousands of dollars)

25. At the left is a graph of the equation $n = -100p^2 + 300{,}000p$. What does the graph tell you about the relationship between price and net sales?

26. What price should the team propose for the laptop computer? Explain your reasoning.

BY THE WAY

The Altair 8800 was the first personal computer widely available. The kit cost $397 when it was introduced in 1975, but that price did not include a terminal, floppy disks, or a printer.

27. Writing Jasmine and Patty solved the equation $5x^2 = 15x$. Who got the correct answer? What happens when you use the other student's method?

Jasmine

$5x^2 = 15x$
$5x^2 - 15x = 0$
$5x(x-3) = 0$

The solutions are 0 and 3.

Patty

$5x^2 = 15x$

$\dfrac{5x^2}{5x} = \dfrac{15x}{5x}$

$x = 3$

The solution is 3.

Solve each equation.

28. $r^2 + 10r = 0$ **29.** $3s^2 + 7s = 0$ **30.** $2x^2 - 18x = 0$

31. $0 = 3y^2 + 3y$ **32.** $4b^2 - 6b = 0$ **33.** $n^2 = 5n$

34. $14x^2 = -x$ **35.** $12z^2 = 24z$ **36.** $18p = -15p^2$

37. Open-ended Problem Find at least two different possible quadratic equations for the graph shown at the right. Explain why each is possible.

38. Use the zero-product property to solve the equation $(x + 1)(x - 2)(2x - 5) = 0$. What must be true about the graph of $y = (x + 1)(x - 2)(2x - 5)$? Explain your reasoning.

Cooperative Learning **For Exercises 39–43, work with a partner.**

39. Take turns using the zero-product property to solve each equation.

    a. $(x - 2)^2 = 0$ **b.** $(x + 4)^2 = 0$

    c. $(x + 5)^2 = 0$ **d.** $(x - 3)^2 = 0$

40. Use the results of Exercise 39 and what you know about factors and x-intercepts to make a conjecture about the graph of each equation.

    a. $y = (x - 2)^2$ **b.** $y = (x + 4)^2$

    c. $y = (x + 5)^2$ **d.** $y = (x - 3)^2$

41.   **Technology** Test your conjectures by graphing each equation in Exercise 40. Generalize your results by describing the graph of an equation in the form $y = (x + k)^2$ in relation to the graph of $y = x^2$.

42. Challenge Use what you learned in Chapter 8 and in Exercises 39–41 to describe how to transform the graph of $y = x^2$ to obtain the graph of $y = 2(x + 3)^2 - 4$.

43.   **Technology** Make a conjecture about how the graphs of $y = |x - 2|$ and $y = |x + 2|$ are related to the graph of $y = |x|$. Test your conjectures using a graphing calculator. Generalize your results by describing the graph of an equation in the form $y = |x + k|$ in relation to the graph of $y = |x|$.

ONGOING ASSESSMENT

44. Writing Suppose one of your friends missed this lesson. Write a description of at least two ways you can solve an equation in the form $0 = ax^2 + bx$. Be sure to include examples.

SPIRAL REVIEW

Simplify each expression. *(Section 9.4)*

45. $(0.17)^0$ **46.** 2^{-3} **47.** 165^0

48. $\left(\dfrac{1}{3}\right)^{-4}$ **49.** $\left(\dfrac{5}{11}\right)^{-2}$ **50.** $\dfrac{1}{5^{-2}}$

Multiply. *(Section 10.2)*

51. $(3x + 1)(x + 2)$ **52.** $(2y + 5)(y + 3)$ **53.** $(z - 7)(2z - 3)$

54. $(2a + 1)(2a + 1)$ **55.** $(2a + 1)(2a - 1)$ **56.** $(3b - 8)^2$

10.4

Exploring Factoring

In Section 10.2 you learned the FOIL method for multiplying binomials. Can you reverse that process? The Exploration will show you how.

Learn how to...

- **factor trinomials with positive quadratic terms and positive constant terms**

So you can...

- **use factoring to explore the value of an investment, for example**

EXPLORATION
COOPERATIVE LEARNING

Using Models to Factor Trinomials

Work with a partner.
You will need:
- algebra tiles

1 One of you should take one x^2-tile, five x-tiles, and four 1-tiles and arrange them to form a rectangle. The other should write the quadratic trinomial that represents the total area of the tiles.

2 Work together to decide what binomials represent the length and the width of your rectangle. Then use the formula *Area = Length × Width* to rewrite the trinomial you wrote in Step 1 in factored form.

3 Switch roles and repeat Steps 1 and 2 using one x^2-tile, six x-tiles, and eight 1-tiles.

4 Without overlapping tiles, each of you try to build a rectangle using one x^2-tile, two x-tiles, and three 1-tiles. What do you notice? What does this suggest about the trinomial $x^2 + 2x + 3$?

5 You may recall that a tile represents $-x$. The tiles at the right show how you can factor a trinomial with a negative linear term:

$$x^2 - 4x + 3 = (x - 3)(x - 1)$$

One of you should use tiles to factor $x^2 - 2x + 1$. The other should use tiles to factor $x^2 - 3x + 2$. Use the FOIL method to check each other's results.

$(-3)(-1) = +3$

EXAMPLE 1

Factor $x^2 + 5x + 6$.

SOLUTION

Method 1 Use algebra tiles.

Step 1 Try to build a rectangle with one x^2-tile, five x-tiles, and six 1-tiles. Begin by finding the possible arrangements of the x^2-tile and the six 1-tiles.

Step 2 Try to complete each rectangle with five x-tiles.

Five x-tiles are not enough to complete this rectangle.

Step 3 Find the dimensions of the completed rectangle: $x + 3$ and $x + 2$.

$$x^2 + 5x + 6 = (x + 3)(x + 2)$$

Method 2 Guess, check, and revise.

Your answer will have this form:

$$(x \pm \underline{\ ?\ })(x \pm \underline{\ ?\ })$$

The orange terms will be factors of 6:

$$6, 1 \qquad -6, -1 \qquad 3, 2 \qquad -3, -2$$

Try different combinations. Focus on getting the correct linear term.

$(x + 6)(x + 1)$	$(x - 6)(x - 1)$	$(x + 3)(x + 2)$
$6x$	$-6x$	$3x$
x	$-x$	$2x$
$7x$	$-7x$	$5x$

Stop when you find the correct linear term.

$$x^2 + 5x + 6 = (x + 3)(x + 2)$$

EXAMPLE 2

Factor $2x^2 - 11x + 15$.

SOLUTION

Toolbox p. 600
Problem Solving

Problem Solving Strategy: Guess, check, and revise.

Your answer will have this form:

$$(\underline{\:?\:}\,x \pm \underline{\:?\:})(\underline{\:?\:}\,x \pm \underline{\:?\:})$$

The blue terms will be factors of $2x^2$: $2x, x$

The orange terms will be factors of 15:

$$3, 5 \qquad -3, -5 \qquad 1, 15 \qquad -1, -15$$

Try different combinations. Focus on getting the correct linear term.

$(2x + 3)(x + 5)$	$(2x + 5)(x + 3)$	$(2x - 5)(x - 3)$
$3x$	$5x$	$-5x$
$\underline{10x}$	$\underline{6x}$	$\underline{-6x}$
$13x$	$11x$	$-11x$

$$2x^2 - 11x + 15 = (2x - 5)(x - 3)$$

the correct linear term

Examples 1 and 2 demonstrate the key things to look for when you factor a trinomial in the form $ax^2 + bx + c$ when c is positive.

When c is positive and b is positive...

$$x^2 + 5x + 6 = (x + 3)(x + 2)$$

...then these terms are both **positive**.

When c is positive and b is negative...

$$2x^2 - 11x + 15 = (2x - 5)(x - 3)$$

...then these terms are both **negative**.

THINK AND COMMUNICATE

1. Show how to use multiplication to check the result of Example 1.

2. Look back at Example 1 to figure out how to factor each trinomial.

    a. $x^2 + 7x + 6$ **b.** $x^2 - 7x + 6$ **c.** $x^2 - 5x + 6$

3. Look back at Example 2 to figure out how to factor each trinomial.

    a. $2x^2 + 11x + 15$ **b.** $2x^2 + 13x + 15$ **c.** $2x^2 - 13x + 15$

4. Factor $9x^2 - 6x + 1$ and $9x^2 - 10x + 1$. Which is a perfect square trinomial?

List all the pairs of factors of each number. Then tell which pair of factors has a sum of 7.

1. 10 **2.** 6 **3.** 12

Factor each trinomial. You may want to use algebra tiles.

4. $x^2 + 8x + 7$ **5.** $y^2 - 4y + 4$ **6.** $t^2 - 7t + 10$

7. $a^2 + 11a + 18$ **8.** $2y^2 - 5y + 3$ **9.** $3x^2 - 14x + 8$

10. $4z^2 + 5z + 1$ **11.** $25p^2 + 30p + 9$ **12.** $25b^2 - 50b + 9$

13. Show that $k^2 - 3k + 4$ is not factorable.

10.4 Exercises and Applications

Factor, if possible. If not, write *not factorable*.

1. $x^2 + 4x + 3$ **2.** $n^2 - 6n + 5$ **3.** $k^2 - 6k + 8$

4. $x^2 + 8x + 15$ **5.** $x^2 + 12x + 20$ **6.** $y^2 - 12y + 36$

7. $y^2 - 13y + 36$ **8.** $r^2 - 7r + 49$ **9.** $s^2 - 9s + 20$

10. $n^2 + 10n + 24$ **11.** $t^2 - 11t + 24$ **12.** $z^2 + 12z + 24$

Extra Practice exercises on page 578

13. Writing From Exercises 1–12, choose some trinomials that are factorable and find the value of the discriminant for each one. Then do the same for some trinomials that are not factorable. Explain how you can use the discriminant to decide whether a trinomial is factorable.

14. SAT/ACT Preview Which binomial is a factor of $x^2 - 10x + 9$?

A. $(x - 1)$ **B.** $(x - 2)$ **C.** $(x - 3)$ **D.** $(x - 5)$

15. Express the total area of the tiles shown at the right as a trinomial. Then express the total area as the product of two binomials.

16. Use algebra tiles or draw a diagram to show how to factor $4x^2 + 12x + 5$.

Factor each trinomial.

17. $5k^2 + 6k + 1$ **18.** $2d^2 - 7d + 6$ **19.** $3n^2 + 8n + 4$

20. $4a^2 + 12a + 9$ **21.** $4a^2 - 13a + 9$ **22.** $2m^2 + 15m + 18$

23. $5x^2 - 13x + 8$ **24.** $6n^2 + 17n + 7$ **25.** $10g^2 - 11g + 3$

26. $6q^2 - 11q + 4$ **27.** $9p^2 - 18p + 8$ **28.** $15x^2 + 19x + 6$

29. Open-ended Problem Find five different ways to complete the following trinomial so that it can be factored: $\underline{?}\; x^2 + 5x + \underline{?}$. Show how to factor each of your five trinomials.

Linda Pei

Look back at the article on pages 420–422.

Suppose a financial adviser like Linda Pei invests $1000 for someone at an annual growth rate r, where r is a decimal. The value of the investment at the end of two years can be represented by the polynomial $1000r^2 + 2000r + 1000$.

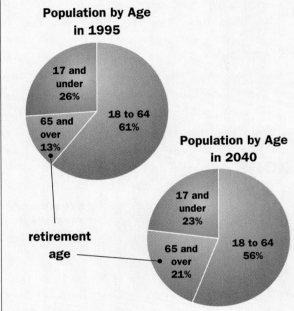

Population by Age in 1995

17 and under 26%

18 to 64 61%

65 and over 13%

retirement age

Population by Age in 2040

17 and under 23%

18 to 64 56%

65 and over 21%

30. Write the polynomial in factored form. (*Hint:* First factor out a common monomial factor.)

31. Explain why the polynomial represents the value of the investment at the end of two years.

32. a. Write a trinomial in terms of r to represent the value of a $5000 investment at the end of two years. Write the trinomial in standard form and in factored form.

b. Evaluate each trinomial from part (a) when $r = 0.07$. You should get the same result both times. If not, check your work.

33. a. What projected trend in the retirement-age population is shown by the graphs?

b. Suppose you invest $5000 at an annual growth rate of 7%. What will your investment be worth when you retire?

Challenge Factor completely. (*Hint:* First look for a common monomial factor.)

34. $2x^2 + 8x + 6$ **35.** $3a^2 + 9a + 6$ **36.** $16k^2 - 80k + 100$

37. $n^3 + 12n^2 + 27n$ **38.** $4y^3 - 5y^2 + y$ **39.** $9x^3 - 51x^2 + 30x$

ONGOING ASSESSMENT

40. Open-ended Problem Make a set of ten flash cards to help you practice your factoring skills. Include five exercises that you have already done in this section, and make up five exercises of your own.

SPIRAL REVIEW

Solve each equation algebraically. (*Section 8.3*)

41. $3y^2 = 147$ **42.** $\frac{2}{3}a^2 = 24$ **43.** $0 = k^2 - 225$

44. $91 = z^2 + 10$ **45.** $25x^2 - 9 = 0$ **46.** $-2n^2 + 63 = 11$

Solve each equation using the quadratic formula. (*Section 8.5*)

47. $n^2 + 7n + 12 = 0$ **48.** $2x^2 + 3x - 1 = 0$ **49.** $-2a^2 + 5a + 1 = 0$

50. $3x^2 - 5x = 0$ **51.** $p^2 - 2p - 1 = 2$ **52.** $3m^2 - 7 = 2m$

10.5 Applying Factoring

The pattern you use to factor a quadratic trinomial with a negative constant term is different than the pattern when the constant term is positive.

Learn how to...

- **factor quadratic polynomials with a negative constant term**

So you can...

- **solve quadratic equations by factoring**

THINK AND COMMUNICATE

1. Use the FOIL method to multiply each pair of binomials.

    a. $(x + 6)(x + 2)$ **b.** $(x - 6)(x - 2)$

    c. $(x + 6)(x - 2)$ **d.** $(x - 6)(x + 2)$

2. Use the results of Question 1 to help you answer these questions.

    a. When the signs in the binomials are the same, what is the sign of the constant term of the product?

    b. When the signs in the binomials are different, what is the sign of the constant term of the product?

EXAMPLE 1

Factor $x^2 - 2x - 8$.

SOLUTION

Problem Solving Strategy: Guess, check, and revise.

Your answer will have this form:

$$(x \pm \underline{\ ?\ })(x \pm \underline{\ ?\ })$$

The orange terms will be factors of -8.

    $1, -8$ $-1, 8$ $2, -4$ $-2, 4$

Try different combinations. Focus on getting the correct linear term.

$$
\begin{array}{ccc}
(x + 1)(x - 8) & (x - 1)(x + 8) & (x + 2)(x - 4) \\
x & -x & 2x \\
\underline{-8x} & \underline{8x} & \underline{-4x} \\
-7x & 7x & -2x
\end{array}
$$

the correct linear term

$$x^2 - 2x - 8 = (x + 2)(x - 4)$$

Check

$$(x + 2)(x - 4) \stackrel{?}{=} x^2 - 2x - 8$$
$$x^2 - 4x + 2x - 8 \stackrel{?}{=} x^2 - 2x - 8$$
$$x^2 - 2x - 8 = x^2 - 2x - 8 \checkmark$$

EXAMPLE 2

Factor $3x^2 + 5x - 8$.

SOLUTION

Problem Solving Strategy: Guess, check, and revise.

Your answer will have this form:

$$(\underline{\,?\,}\,x \pm \underline{\,?\,})(\underline{\,?\,}\,x \pm \underline{\,?\,})$$

The blue terms will be
factors of $3x^2$: $3x, x$

The orange terms will be factors of -8.

$1, -8 \qquad -1, 8 \qquad 2, -4 \qquad -2, 4$

Try different combinations. Focus on getting the correct linear term.

$$(3x + 1)(x - 8) \qquad\qquad (3x - 1)(x + 8)$$

$$\begin{array}{c} x \\ -24x \\ \hline -23x \end{array} \qquad\qquad \begin{array}{c} -x \\ 24x \\ \hline 23x \end{array}$$

$$(3x - 8)(x + 1) \qquad\qquad (3x + 8)(x - 1)$$

$$\begin{array}{c} -8x \\ 3x \\ \hline -5x \end{array} \qquad\qquad \begin{array}{c} 8x \\ -3x \\ \hline 5x \end{array}$$

the correct linear term

$$3x^2 + 5x - 8 = (3x + 8)(x - 1)$$

You may have noticed that the key to factoring trinomials of the form $ax^2 + bx + c$ is to position the signs and the factors of the constant term correctly. In Section 10.4 you saw how to do this when c is positive. Example 2 demonstrates the key things to do when c is negative.

When c is negative... $\qquad \mathbf{3x^2 + 5x - 8 = (3x + 8)(x - 1)} \qquad$ ...one factor of c is **positive** and the other is **negative**.

The factored form is either $(\underline{\,?\,}\,x + \underline{\,?\,})(\underline{\,?\,}\,x - \underline{\,?\,})$ or $(\underline{\,?\,}\,x - \underline{\,?\,})(\underline{\,?\,}\,x + \underline{\,?\,})$.

THINK AND COMMUNICATE

Decide whether each trinomial is factored correctly. If it is not, what changes are needed to make the factors correct?

3. $x^2 + 6x + 5 = (x - 1)(x + 5)$ **4.** $x^2 - 4x + 3 = (x + 3)(x + 1)$

5. $x^2 + 2x - 8 = (x - 2)(x + 4)$ **6.** $5x^2 - 13x + 6 = (5x - 2)(x - 3)$

EXAMPLE 3   **Application: Sports**

A pitcher throws a warm-up pitch in a softball game. The equation $h = -16t^2 + 24t + 1$ gives the height h in feet of the ball t seconds after the pitcher releases it. How long does it take for the ball to reach a height of 6 ft?

SOLUTION

Substitute **6** for h, and rewrite the equation in the form $ax^2 + bx + c = 0$, with $a > 0$.

$$h = -16t^2 + 24t + 1$$
$$6 = -16t^2 + 24t + 1$$
$$0 = -16t^2 + 24t - 5$$
$$-1 \cdot 0 = -1(-16t^2 + 24t - 5) \qquad \text{Multiply both}$$
$$\text{sides by } -1.$$
$$0 = 16t^2 - 24t + 5$$

Factor and use the zero-product property.

$$0 = (4t - 1)(4t - 5)$$

$$4t - 1 = 0 \qquad \text{or} \qquad 4t - 5 = 0$$
$$t = \frac{1}{4} \qquad\qquad\qquad t = \frac{5}{4}$$

The ball reaches a height of 6 ft after 0.25 s and after 1.25 s.

The diagram below illustrates how it is possible for the ball discussed in Example 3 to be 6 ft off the ground at two different times.

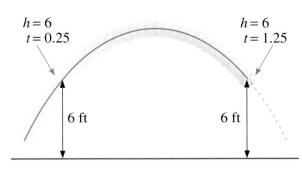

$h = 6$
$t = 0.25$

$h = 6$
$t = 1.25$

6 ft

6 ft

THINK AND COMMUNICATE

7. The notebook at the right shows how Kyle factored the expression $x^2 - 25$. When he was done, he recognized a pattern. What is it?

8. How could you use the pattern you identified in Question 7 to factor $x^2 - 4$? to factor $49x^2 - 16$?

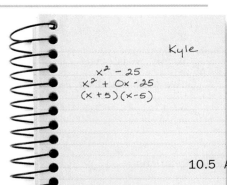

Kyle

$x^2 - 25$
$x^2 + 0x - 25$
$(x + 5)(x - 5)$

You now know three methods for solving quadratic equations. You can use a graph or the quadratic formula to solve any quadratic equation. You can use factoring to solve any quadratic equation whose discriminant is a perfect square, which tells you that the equation can be written in factored form.

Methods to Solve Quadratic Equations

$$x^2 - x - 6 = 0$$

Factor and use the zero-product property.

Use the quadratic formula.

$$x^2 - x - 6 = 0$$
$$(x + 2)(x - 3) = 0$$
$$x + 2 = 0 \quad \text{or} \quad x - 3 = 0$$
$$x = -2 \qquad x = 3$$

The solutions are -2 and 3.

Graph the related equation $y = x^2 - x - 6$. **The x-intercepts are the solutions.**

$$x = \frac{-b \pm \sqrt{b^2 - 4ac}}{2a}$$
$$= \frac{1 \pm \sqrt{1 - 4(1)(-6)}}{2(1)}$$
$$x = \frac{1 + 5}{2} \quad \text{or} \quad x = \frac{1 - 5}{2}$$
$$x = 3 \qquad x = -2$$

The solutions are -2 and 3.

One x-intercept is -2.

One x-intercept is 3.

The solutions are -2 and 3.

☑ CHECKING KEY CONCEPTS

1. Is it possible to solve $2x^2 + x + 1 = 0$ by factoring? Explain.

Choose the letter of the factored form of each polynomial. Then factor the polynomial.

A. $(\underline{?} + \underline{?})(\underline{?} + \underline{?})$ **B.** $(\underline{?} - \underline{?})(\underline{?} - \underline{?})$ **C.** $(\underline{?} + \underline{?})(\underline{?} - \underline{?})$

2. $x^2 + 7x - 8$ **3.** $3y^2 - 7y + 2$ **4.** $5t^2 + 6t + 1$

5. $4r^2 + 12r + 9$ **6.** $7z^2 + 2z - 5$ **7.** $p^2 - 100$

Solve each equation. Explain why you chose the method you used.

8. $x^2 - 4x - 5 = 0$ **9.** $x^2 - 4x + 1 = 0$ **10.** $4x^2 - 9 = 0$

10.5 Exercises and Applications

Extra Practice exercises on page 578

Replace each ? with the correct symbol or number.

1. $3m^2 - 20m - 7 = (3m \ \underline{?} \ 1)(m \ \underline{?} \ 7)$

2. $2x^2 + 3x - 5 = (2x + \underline{?})(x - \underline{?})$

3. $6r^2 - 17r + 12 = (3r \ \underline{?} \ 4)(2r \ \underline{?} \ 3)$

Factor each polynomial.

4. $r^2 - 2r - 3$ 5. $t^2 - 7t - 8$ 6. $y^2 - 7y + 12$

7. $z^2 - 5z - 36$ 8. $x^2 + 2x - 24$ 9. $y^2 - 36$

10. $3k^2 - 4k - 7$ 11. $2d^2 + 9d - 11$ 12. $3v^2 + 4v - 4$

13. $2x^2 - x - 6$ 14. $5w^2 + 9w + 4$ 15. $6y^2 - 11y + 4$

16. **GEOMETRY** If $x^2 + 6x - 40$ represents the area of a rectangle, find possible expressions for its length and width.

17. **SAT/ACT Preview** Which is the correct factorization of $12x^2 - 5x - 2$?

     A. $(6x + 2)(2x - 1)$ **B.** $(6x - 2)(2x + 1)$

     C. $(3x - 2)(4x + 1)$ **D.** $(3x + 2)(4x - 1)$

Find all integer values of *n* for which each trinomial can be factored.

18. $c^2 + nc - 6$ 19. $3s^2 + ns - 14$ 20. $4p^2 + np - 15$

21. **Writing** Describe how the diagram represents a factoring pattern.

Solve each equation by factoring.

22. $x^2 - 8x + 12 = 0$ 23. $x^2 - x - 2 = 0$ 24. $m^2 - 4m + 1 = -3$

25. $2r^2 - 5r - 12 = 0$ 26. $2y^2 + 15y + 10 = 3$ 27. $3q^2 + 7q - 4 = 2$

28. $6k^2 - k = 5$ 29. $9t^2 - 4 = 0$ 30. $16x^2 = 49$

31. **SPORTS** The equation $h = -16t^2 + 24t + 1$ models the height h in feet of a pitched softball t seconds after it is released.

     a. When is the ball 10 ft high?

     b. Is the ball ever 12 ft high? Explain.

Solve each equation using whichever method you prefer.

32. $(3x + 1)(x + 1) = 0$ 33. $2d^2 - 6d + 1 = 0$

34. $4r^2 + r = 0$ 35. $2y^2 + 5y = 3$

36. $k^2 - 4k = 4$ 37. $3x^2 + 2 = 7x + 8$

The table at the right gives data about the outfield walls of Boston's Fenway Park. Suppose the equation $h = -16t^2 + 64t + 4$ models the height h in feet of a ball t seconds after being hit. The equation $d = 77t$ gives the horizontal distance that the ball has traveled in feet from home plate after t seconds.

Outfield wall	Height (ft)	Distance from home plate (ft)
left	37	315
center	17	390
right	3–6	302

Position of Ball Over Time

	A	B	C
1	time	height	horizontal distance
2			
3			
4			
5			
6			

38. 📈 **Spreadsheets** Use a spreadsheet to track the height of the ball as a function of time.

a. Find the time interval when the ball is at least as high as the left field wall of Fenway Park. Write your answer as an inequality.

b. Add a column to your spreadsheet to calculate the horizontal distance at time t. Compare the height values with the distance values. If a batter hits the ball down the left field line, will it go over the wall?

39. Another hit is modeled by the equations $h = -16t^2 + 74t + 4$ and $d = 88t$. Use the distance equation to find t when the ball is 390 ft from home plate. How high is the ball at this time? Could it go over one of the outfield walls?

Challenge Factor completely. (*Hint:* First look for a common monomial factor.)

40. $3x^3 - 6x^2 - 24x$

41. $20x^3 + 38x^2 + 12x$

42. $10t^3 - 45t^2 + 45t$

43. $3v^2 - vw - 2w^2$

44. $4x^2 - 100y^2$

45. $6n^2 - nq - 2q^2$

O N G O I N G   A S S E S S M E N T

46. Open-ended Problem Make a concept map about factoring. Be sure to include examples.

Graph each point in a coordinate plane. Label each point with its letter. Name the quadrant (if any) in which the point lies. *(Section 2.4)*

47. $A(5, 4)$

48. $B(-3, 5)$

49. $C(4, -5)$

50. $D(0, -5)$

51. $E\left(-\frac{3}{4}, -\frac{1}{2}\right)$

52. $F(-4.5, 0)$

Tell whether the data show direct variation. Give reasons for your answer. *(Section 3.2)*

53.

Air Mail Rates to Asia in 1961

54.

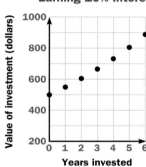

Growth of an Investment Earning 10% Interest

ASSESS YOUR PROGRESS

VOCABULARY

polynomial equation (p. 436)   **factoring a polynomial** (p. 437)

Solve each equation. *(Section 10.3)*

1. $(2x - 3)(4x + 5) = 0$

2. $x^2 - 3x = 0$

3. $2y^2 = -8y$

4. The equation $d = 0.05s^2 + s$ models the distance d in feet traveled by a car moving s miles per hour once the driver has decided to stop the car. Find the values of s for which $d = 0$. Interpret the solutions.

Factor each polynomial. *(Sections 10.4 and 10.5)*

5. $x^2 + 7x + 10$

6. $k^2 + 3k - 28$

7. $k^2 - 12k - 28$

8. $2a^2 - 11a + 5$

9. $25z^2 + 10z + 1$

10. $4m^2 - 100m + 49$

11. $9c^2 - 21c + 10$

12. $81t^2 - 64$

13. $9d^2 - d - 10$

Solve each equation. *(Section 10.5)*

14. $n^2 + 3n - 28 = 0$

15. $4x^2 + 15x + 14 = 0$

16. $24p^2 - 5p = 1$

17. Use the stopping-distance formula from Exercise 4. How fast can a car travel if it needs at least 15 ft to come to a complete stop?

18. **Journal** Discuss how you can use factoring to solve certain polynomial equations, and how you know when this method is appropriate.

Volume and Weight

Dolls and action figures are scale models of human beings, but not all such models are shaped like real people. You can learn something about the accuracy of a doll's proportions by measuring its volume.

PROJECT GOAL Find out how much a doll or an action figure would weigh if it were a real person.

Taking Measurements

Work with a partner to measure the volume of a model (either a doll or an action figure) by measuring the amount of water it displaces. You may want to do the experiment several times to be sure you are getting reliable results.

1. MEASURE the height of your model. Place a pitcher large enough to contain your model into a large pot. Fill the pitcher to the very top with water, being careful not to spill any into the pot.

2. TAPE over any places where water might leak in if your model is hollow. Then completely submerge your model in the water without letting your fingers go beneath the surface.

3. REMOVE your model and pitcher from the pot. Then use measuring cups and spoons to measure the volume of the water that was displaced, or spilled over, into the pot. This is the volume of your model.

4. CONVERT the volume of your model to liters using the conversion chart at the left.

U.S. Customary Units	Metric Units
1 cup	0.24 liters
1 tablespoon	0.015 liters
1 teaspoon	0.005 liters

Calculating Weight

The polynomial equation below shows the relationship between the volume of a real person V_p and the volume of a scale model V_m, in liters.

$$V_p = V_m s^3$$

The variable s is a scale factor.

$$s = \frac{\text{Person's height (in.)}}{\text{Model's height (in.)}}$$

Suppose the doll or action figure that you measured were a real person.

- About how tall do you think the person would be? What is the scale factor s for your model?

- What would be the person's volume?

- How much would the person weigh? Use the formula below to help you find out. The average weight density for males is 2.34 pounds per liter. The average weight density for females is 2.31 pounds per liter.

Weight = Volume × Weight density

Writing a Report

Summarize your experiment. Be sure to include:

- a brief description of the procedure you used

- the data you collected

- your calculations

- a discussion of some possible sources of error

Analyze your results. Is the weight you calculated realistic? Why or why not? If not, what could the designers of the model do to make the model more realistic? Explain your answer in terms of the volume relationship at the top of this page.

Self-Assessment

Conclude your report with a discussion of what went well for you during this experiment. Also discuss the aspect of the experiment that you and your partner found the most challenging and how you dealt with it. Describe what you have learned about the relationship between the volume of a scale model and the weight of the object it represents.

10 Review

Without looking at the book, outline this chapter's content. Use the book to check the outline. Fill in any important ideas that you missed.

VOCABULARY

polynomial (p. 423)

standard form of a polynomial (p. 423)

degree of a term (p. 423)

degree of a polynomial (p. 423)

linear polynomial (p. 423)

quadratic polynomial (p. 423)

cubic polynomial (p. 423)

polynomial equation (p. 436)

factoring a polynomial (p. 437)

SECTIONS 10.1 *and* 10.2

Polynomials can be added or subtracted by combining like terms.

$$2x^2 - 5x + 14$$
$$7x^2 - 9x + 24$$

To subtract, add the **opposite**.

$$2x^2 - 5x + 14$$
$$-7x^2 + 9x - 24$$
$$\overline{-5x^2 + 4x - 10}$$

You can use one of several methods to multiply polynomials.

Horizontal method

$$-2x(x^2 + 3x - 1) = -2x(x^2) + (-2x)(3x) - (-2x)(1)$$
$$= -2x^3 - 6x^2 + 2x$$

Vertical method

$$x^2 + 5x + 6$$
$$\underline{2x - 3}$$
$$2x^3 + 10x^2 + 12x$$
$$\underline{-3x^2 - 15x - 18}$$
$$2x^3 + 7x^2 - 3x - 18$$

FOIL method

$$(x + 2)(3x - 4) = x(3x) + x(-4) + 2(3x) + 2(-4)$$
$$= 3x^2 - 4x + 6x - 8$$
$$= 3x^2 + 2x - 8$$

Some polynomials can be **factored** by finding the greatest common factor (GCF).

$$4x^2 + 10x = 2x \cdot 2x + 2x \cdot 5$$

The GCF of $4x^2$ and $10x$ is **2x**.

$$= 2x(2x + 5)$$

Polynomials can also be factored using a guess and check method.

To factor $2k^2 - 13k + 21$, guess, check, and revise.

$$(\underline{?}\,k \pm \underline{?}\,)(\underline{?}\,k \pm \underline{?}\,)$$

The blue terms will be factors of **2k²**: $2k, k$.

The orange terms will be factors of 21:
$3, 7 \quad -3, -7 \quad 1, 21 \quad -1, -21$

Try different combinations. Focus on getting the correct linear term.

$(2k + 3)(k + 7)$	$(2k + 7)(k + 3)$	$(2k - 7)(k - 3)$
$3k$	$7k$	$-7k$
$\underline{14k}$	$\underline{6k}$	$\underline{-6k}$
$17k$	$13k$	$-13k$

So, $2k^2 - 13k + 21 = (2k - 7)(k - 3)$.

the correct linear term

You can use the FOIL method to check your results.

To factor a quadratic polynomial with a negative constant term, such as $x^2 + x - 6$, remember that the signs of the binomials will be different.

$$(x \pm \underline{?}\,)(x \pm \underline{?}\,)$$

The orange terms will be factors of -6:
$1, -6 \quad -1, 6 \quad 3, -2 \quad -3, 2$

$(x + 1)(x - 6)$	$(x - 1)(x + 6)$	$(x + 3)(x - 2)$
x	$-x$	$3x$
$\underline{-6x}$	$\underline{6x}$	$\underline{-2x}$
$-5x$	$5x$	x

the correct linear term

So, $x^2 + x - 6 = (x + 3)(x - 2)$.

You have seen three methods for solving quadratic equations.

1. Factor and use the zero-product property.

2. Use the quadratic formula: $x = \dfrac{-b \pm \sqrt{b^2 - 4ac}}{2a}$

3. Graph the related equation. The x-intercepts are the solutions.

10 Assessment

VOCABULARY REVIEW

1. Give an example of a polynomial with degree 3. Write the polynomial in standard form.

2. Give examples of a linear polynomial, a quadratic polynomial, and a cubic polynomial.

3. Explain what it means to factor a polynomial.

SECTION 10.1

Tell whether each expression is a polynomial. Write *Yes* or *No*. If not, explain why not.

4. $2m + 3n$

5. $5a^3 + 6$

6. 15

7. $7x^2 - 7x - 7$

8. $6x^{-2}$

9. $3x^2 - \dfrac{2}{x^2}$

Add. Give the degree of each sum.

10. $\begin{aligned}-3x^2 + 15x + 6 \\ \underline{3x^2 - 7x + 4}\end{aligned}$

11. $\begin{aligned}6a^2 - 6a - 7 \\ \underline{a^2 + 5a + 9}\end{aligned}$

12. $\begin{aligned}n^3 + 3n^2 - 2n + 2 \\ \underline{3n^2 - 2n + 2}\end{aligned}$

Add or subtract as indicated.

13. $(9x^2 - 2x + 5) + (-x^2 + 4x + 6)$

14. $(r^2 - 7r + 8) - (-3r^2 + 7r + 6)$

15. $(6x^3 - 9x^2 + 14) - (5x^2 + 10x)$

16. $(b + 2c - 6) - (-2b + 4c - 1)$

SECTION 10.2

Multiply.

17. $(x - 3)(x - 4)$

18. $(b - 9)(b + 5)$

19. $(r - 3)(5r - 6)$

20. $(3c + 8)(2c + 2)$

21. $3a(2a^2 + 9)$

22. $-d(d^2 - 3d - 6)$

23. $(2h + 7)(h^2 - 4h + 6)$

24. $(4x^2 + 3x + 2)(x - 8)$

25. Write a polynomial in standard form that represents the area of the blue region in the figure at the right.

SECTION 10.3

Factor each polynomial.

26. $13y^2 - 4y$ **27.** $6a^2 - 10a$ **28.** $6x^3 + 4x^2 - 8x$

Solve each equation.

29. $7y(y - 7) = 0$ **30.** $0 = 2x(4x + 1)$ **31.** $(3a - 3)(5a + 8) = 0$

32. $x^2 - 5x = 0$ **33.** $6m^2 + 6m = 0$ **34.** $8c = 4c^2$

35. Suppose the area of the shaded region in Question 25 is 12 in.^{2}. Find the value of x.

SECTIONS 10.4 *and* 10.5

Factor, if possible. If not, write *not factorable*.

36. $x^2 + 11x + 10$ **37.** $y^2 - 11y + 24$ **38.** $a^2 - 6a + 6$

Factor each polynomial.

39. $q^2 + 6q - 7$ **40.** $y^2 - 10y + 16$ **41.** $a^2 + a - 12$

42. $m^2 - 9$ **43.** $2x^2 - 3x + 1$ **44.** $4n^2 + 12n + 5$

Solve each equation by factoring.

45. $3x^2 - 8x - 11 = 0$ **46.** $7n^2 + n - 6 = 2$ **47.** $4a^2 - 3a = 10$

48. Writing Explain how to use the graph of the equation $y = 6x^2 + 30x + 36$ at the right to solve the equation $6x^2 + 30x = -36$. Describe two other methods you could use to solve the equation $6x^2 + 30x = -36$.

PERFORMANCE TASK

49. Suppose the equation $h = -16t^2 + 40t$ models the height h in feet of a ball t seconds after the ball has been hit. Will the ball reach a height of 24 ft? If so, when? Use methods you have learned in this chapter to support your answer.

11 | Rational Functions

SIZING things up in the air

Alex MacLean

Sometimes you have to look at things from a different perspective to understand them better. Alex MacLean sees things from a different perspective every time he goes to work. As an aerial photographer, he gets a bird's-eye view of his subjects as he flies over them. "From the air, it's possible to observe large-scale patterns and the connectedness of things," he explains. "That's a perspective you just can't get from the ground."

"It takes some luck to capture the perfect shot, but I also think I'm getting a little better at being lucky."

" From the air, it's possible to observe large-scale patterns and the connectedness of things. **"**

MacLean's photos show interesting patterns of human development. Some show the massive scarring of the land caused by the "clear-cutting" of trees in the Pacific Northwest. Others show vast landfills overflowing with paper trash. His photos do not just illustrate damage to the environment, however. "A lot of the things you see are really gorgeous—natural treasures that ought to be preserved," he says. "Still, by highlighting the negative things, the pictures can help you figure out what the positive things really are."

Patterns of Human Development

MacLean learned to fly while studying architecture and city planning. Instead of becoming an architect, he started an aerial photography company. At first, he provided photos to architects and city planners. Soon he ventured into other areas, taking pictures of such things as an archaeological site in New Mexico, wheat fields in Montana, and New York City's Central Park.

Getting the Right Angle

MacLean almost always flies alone. When he's ready to shoot, he banks his plane on its side and points the camera. For some shots, he points straight down. "If the camera is tilted at even a slight angle, the shot will come out distorted," he notes. He has to be at just the right height so that the subject completely fills the picture. "If you fly too low, you won't get it all in. If you fly too high, you'll have a lot of empty space on the edge of the picture."

Altitude and Image Size

MacLean can use a formula to determine how large the image of an object will be when he flies at a certain altitude. The formula says that flying height and image size show *inverse variation*. In other words, the higher MacLean flies, the smaller

the image size. In the photos of the river front in Manchester, New Hampshire, MacLean made the photographic image of the group of trees along the river appear to be large by flying the airplane at a low height. He made the image of the trees smaller and was able to include more of the city by flying higher. The graph shows the relationship between flying height and image size.

You can use the formula

$$\text{Image size} = \frac{k}{\text{Flying height}}$$

to help determine image size.

Image Size of the Group of Trees on a 35 mm Slide

Image Size (mm) — vertical axis: 0, 2, 4, 6, 8, 10, 12

Flying Height (feet) — horizontal axis: 1000, 2000, 3000, 4000, 5000, 6000

Alex MacLean takes his photographs from a four-seater Cessna 182.

1. Writing Use the graph to describe what it means for flying height and image size to show inverse variation. When flying height doubles, what happens to image size? How is this different from direct variation?

2. Research MacLean controls the size of the photographic image by changing his altitude. Some photographers use a telephoto lens to give the impression that the subject was close to them. Find out how a telephoto lens is different from a regular camera lens. Prepare a report of your findings.

3. Project Sketch a picture of your school as you think it would look from a plane. Compare your sketch with others made by students in your class. Which sketches look like they were made at the lowest heights? the greatest heights? Work together to put the sketches in order from lowest to highest.

Mathematics & Alex MacLean

In this chapter, you will learn more about how mathematics is related to photography.

Related Exercises

Section 11.1
• **Exercises 10 and 11**

Section 11.4
• **Exercises 20 and 21**

11.1

Learn how to...

- **recognize and describe data that show inverse variation**

So you can...

- **solve problems involving inverse variation, such as calculating driving times or planning aerial photographs**

Exploring Inverse Variation

These two rectangles have the same area, but different lengths and widths. What are some examples of other rectangles with the same area? If area is constant, how does the length of a rectangle affect its width?

3 ft

3 ft

4.5 ft

2 ft

EXPLORATION
COOPERATIVE LEARNING

Exploring Rectangles with Equal Areas

Work with a partner.
You will need:

- 36 algebra unit tiles
- graph paper

2 Use the algebra tiles to form as many rectangles as you can with area 36. Copy and extend the table to record the length *l* and the width *w* of each rectangle.

ℓ	w
12	3
3	12
2	?
?	?

1 Arrange the 36 tiles to form a rectangle. What are the length and the width of the rectangle?

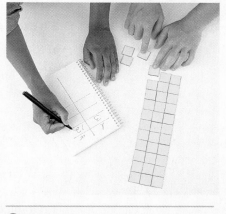

3 Graph the data pairs (*l*, *w*) from your table in a coordinate plane.

4 **a.** Could a rectangle with area 36 be 4.5 units wide? If so, what would the length be?

b. Draw a smooth curve passing through the points you graphed. Explain why it makes sense to connect the points.

5 Can a rectangle be 0 units long? What happens as the length of the rectangle gets very small? Adjust your graph if necessary.

6 Write an equation relating *l* and *w*. Describe how the length of a rectangle affects its width if the area is constant.

Inverse Variation

In the Exploration, the product of the rectangle's length and width was always constant. Whenever two variables have a constant, nonzero product, they are said to show **inverse variation**.

The constant product is called the **constant of variation**.

Inverse Variation

Inverse variation can be described by an equation in the form

$$xy = k \text{ or } y = \frac{k}{x}, \text{ where } k \neq 0 \longleftarrow$$ The letter k represents the constant of variation.

BY THE WAY

Between 1974 and 1987, states were required to enforce a 55 mi/h speed limit on all highways in order to receive federal aid for highways.

EXAMPLE 1   Application: Transportation

The time t in hours that it takes Derrick Le Marr to drive home from college varies inversely with his average speed s in miles per hour. If Derrick drives at an average speed of **50** mi/h, it takes him **6** h to get home.

a. Write and graph an equation expressing t in terms of s.

b. If Derrick wants to get home in **5** h, what must his average speed be?

SOLUTION

a. Since t varies inversely with s, find an equation in the form $t = \frac{k}{s}$.

You know that $t = 6$ when $s = 50$. Use this information to find k.

$$6 = \frac{k}{50}$$

$$k = 300$$

An equation expressing t in terms of s is $t = \frac{300}{s}$.

Graph the equation as $y = \frac{300}{x}$ on a graphing calculator.

Inverse variation graphs have two parts. To see both parts, view all four quadrants. One solution is (50, 6), so try the viewing window $-75 \leq x \leq 75$, $-75 \leq y \leq 75$.

Time and speed are always positive. Only the first-quadrant portion of the graph makes sense in this problem.

b. **Method 1** Substitute 5 for *t* in the equation from part (a). Solve for *s*.

$$5 = \frac{300}{s}$$

$$5s = 300$$

$$s = 60$$

Method 2 Use the graph.

X=60 Y=5

Derrick must drive at an average speed of 60 mi/h to get home in 5 h.

THINK AND COMMUNICATE

For Questions 1 and 2, use the information in Example 1.

1. The graph of $t = \dfrac{300}{s}$ never crosses either axis. Why?

2. What happens to *t* as *s* increases? as *s* decreases? How does the relationship between *t* and *s* compare with direct variation?

EXAMPLE 2 Application: Physics

A scuba diver 99 ft below the surface of the ocean exhales a small air bubble. As the bubble rises to the surface, the pressure on it decreases. The bubble grows in volume. The table describes the changes in pressure and volume. Does **volume** vary inversely with **pressure**?

SOLUTION

Find the product of pressure and volume at each depth. Pressure and volume vary inversely if the product is always the same.

Size of Rising Air Bubble

Water depth (ft)	Pressure (lb/in.²)	Volume of bubble (in.³)
99	59	0.22
66	44	0.30
33	29	0.45
0	15	0.87

$59 \cdot 0.22 = 12.98$ $44 \cdot 0.30 = 13.2$

$29 \cdot 0.45 = 13.05$ $15 \cdot 0.87 = 13.05$

The products are all very close to 13.

Volume varies inversely with pressure.

BY THE WAY

Scuba divers must exhale as they rise to the surface. Otherwise, expanding air in the lungs could cause the lungs to tear.

THINK AND COMMUNICATE

3. In Example 2, does bubble volume vary inversely with water depth?

Do the quantities described show inverse variation? Explain your thinking.

4. the length of a side of a square and the area of the square

5. the number of people sharing a pizza equally and the number of slices per person

☑ CHECKING KEY CONCEPTS

Tell whether the data show *inverse variation*, *direct variation*, or *neither*. If the data show direct variation or inverse variation, write an equation for *y* in terms of *x*.

1.

x	y
15	4
12	5
10	6

2.

x	y
6	18
4	12
3	9

3.

x	y
7	10
9	12
12	15

4. The time it takes each editor to proofread a chapter varies inversely with the editor's average rate of proofreading. Julia Poriet usually proofreads at a rate of about six pages per hour. She finished Chapter 11 in about 4.5 h. If Joan Kaplan proofreads at a rate of about five pages per hour, how long will it take her to finish Chapter 11?

5. Write and graph an equation showing that the product of two variables is always -150.

11.1 **Exercises and Applications**

Extra Practice exercises on page 579

Decide whether the data show inverse variation. If they do, find the missing value.

1.

x	y
12	4
15	5
18	6
?	7

2.

x	y
−15	−8
12	10
$\frac{2}{3}$	?
−8	−15

3.

x	y
12	?
−9	−16
8	18
−1	−144

4. TRANSPORTATION It takes Laurie Grandmaison $3\frac{1}{2}$ h driving at an average speed of 50 mi/h to go from home to her friend's house. How long will it take her to return if she drives at an average speed of 45 mi/h?

 Technology Graph each equation. Use a graphing calculator if you have one. Do *x* and *y* show inverse variation? Explain your answer.

5. $xy = 24$ **6.** $y = \dfrac{1}{x}$ **7.** $y = \dfrac{-24}{x}$ **8.** $\dfrac{y}{x} = 1$

9. SAT/ACT Preview If you want to earn a fixed amount of simple interest *I* in one year, the yearly interest rate *r* and the principal *P* you invest vary inversely. Which equation does *not* show this relationship?

A. $P = \dfrac{I}{r}$ **B.** $r = \dfrac{P}{I}$ **C.** $r = \dfrac{I}{P}$ **D.** $Pr = I$ **E.** none of these

Look back at the article on pages 460–462.

A

When Alex MacLean takes aerial photographs, he uses a formula to determine how high he must fly to achieve a desired image size.

$$\text{Altitude} = \frac{k}{\text{image size}}$$ ← The letter k represents a constant.

10. Investigation You will need a ruler. Alex MacLean took picture A from 1500 ft above the ground. At what heights were pictures B and C taken? How do you know?

11. Alex McLean was flying 1500 ft above the water when he took the picture at the right. At what altitude should he fly so the images of the boats are twice as long? half as long?

B

C

12. Writing You can use inverse variation to estimate the speed of the waves in the ocean. First estimate the horizontal distance *l* between the crests of the waves. Then count the time *t* between wave crests at a fixed point such as a buoy or pier pile. To find the wave speed *s*, use the formula $s = \frac{l}{t}$.

crest

a. Suppose you estimate the distance between wave crests to be about 7 m and the time between crests to be about 2 s. What is the speed of the wave?

b. Which two variables in the formula are inversely related?

c. If the distance between wave crests is constant, how does the speed of the waves affect the time between crests?

13. Open-ended Problem For each function graphed below, write an equation that could represent the function. Explain your thinking.

a.

b.

c.

d.

14. To loosen a bolt with a wrench with a 6 in. long handle, David Ruiz must exert a force of 250 lb. (That is like lifting a 250 lb person!) The force needed and the length of the handle vary inversely. If David slips a pipe over the wrench handle to make it 24 in. long, about how much force must he exert?

Bicycles in the late 1800s often had very large front wheels and no chains. The larger the wheel, the farther it traveled in one rotation and the faster a bicyclist could go.

Modern bicycles use a gear system to achieve similar speeds with much smaller wheels. The 24-speed bicycle below has gear numbers as large as 99. This means it can go as far with one turn of the pedals as an old-fashioned bicycle with a front wheel with diameter 99 in.

15. The bike at the right has a gear number of 40 when the chain is on the front chainwheel with 32 teeth and the rear sprocket with 21 teeth. If the chain remains on the same front chainwheel, then the gear number g is inversely related to the number of teeth n on the rear sprocket. Write an equation relating g and n.

16. What is the gear number when the chain is on the rear sprocket with 12 teeth?

17. What is the gear number when the chain is on the rear sprocket with 28 teeth?

18. Challenge If you always pedal at a constant rate, how does the number of teeth in the rear sprocket affect the speed of the bicycle? Explain.

Rear sprocket

Front chainwheel

19. Investigation How is the graph of an equation in the form $xy = k$ different when k is negative than when k is positive? Explain how you found your answer.

20. Writing List at least three things that all graphs of inverse variation functions have in common.

Find the mean, the median, and the mode(s) of each set of data. *(Section 1.3)*

21. 83, 88, 81, 87, 81, 85, 87, 87, 77

22. 101.25, 103.75, 102.25, 102.25

23. 11, 11, 11, 12, 12, 13, 13, 13, 13, 14, 14, 14, 15, 23

Solve each system of equations by adding or subtracting. *(Section 7.3)*

24. $3a - 2b = 2$
$4a + 2b = -16$

25. $x + y = 4$
$\frac{2}{3}x - y = 6$

11.2

Using Weighted Averages

Sometimes when you find an average, you may want to give more weight to one number than to the others. For example, your teacher may use a *weighted average* to calculate your semester grade average, so that your final exam grade counts more than each quiz grade. In the Exploration, you will see another use for weighted averages.

Learn how to...
- find weighted averages

So you can...
- summarize information such as grades or game scores

EXPLORATION
COOPERATIVE LEARNING

Comparing Game Scores

Work in a group of four.
You will need:
- a bowl
- a narrow cup or glass
- about two cups of dried beans

1 Place the cup upright inside the bowl on the floor.

2 Take turns trying to drop beans into the cup. Stand up straight and hold your arm straight out. You each get one minute. Count the number of beans that land:

- inside the cup (10 points)
- inside the bowl but outside the cup (5 points)
- outside the bowl (0 points)

Record your results in a table.

Name	Number in cup	×	10 points	Number in bowl	×	5 points	Number outside bowl	×	0 points
Lisa	18		180	6		30	3		0
?	?		?	?		?	?		?

3 Compute the total number of points for each person in your group. Can you tell who had the best aim by comparing total numbers of points? Explain.

4 Compute the average number of points per drop for each person in your group. Can you use this information to decide who had the best aim? Explain.

Average Points Per Bean

LISA $\dfrac{18(10)+6(5)+3(0)}{18+6+3} \approx 7.78$

Luke's Test Grades	
Test 1	75
Test 2	80
Midterm (worth 2 tests)	89
Test 3	62
Test 4	85
Final Exam (worth 3 tests)	?

To determine your final score for the semester, your teacher may look at many things, such as test grades, homework grades, and midterm grades. Some of these grades may carry more weight than others.

For example, the table at the left shows Luke's test grades so far in his algebra class. Luke found his average grade so far.

The red numbers are called weights.

The midterm counts as two tests.

$$\frac{\text{point total}}{\text{number of tests}} = \frac{1(75) + 1(80) + 2(89) + 1(62) + 1(85)}{1 + 1 + 2 + 1 + 1} = 80$$

The average is a **weighted average**.

EXAMPLE 1 **Application: Test Scores**

To earn a B for the semester, Luke's average grade must be at least 81. What must Luke earn on the final exam to get a B for the semester? The final exam counts as **three** tests.

SOLUTION

Write and solve an inequality.
Let x = Luke's grade on the final.

The final counts as 3 tests.

$$\frac{75 + 80 + 2(89) + 62 + 85 + 3x}{1 + 1 + 2 + 1 + 1 + 3} \geq 81$$

$$\frac{480 + 3x}{9} \geq 81$$

Luke's average grade must be at least 81.

$$480 + 3x \geq 729$$

$$3x \geq 249$$

$$x \geq 83$$

Luke needs a grade of 83 or better on the final to earn a B for the semester.

THINK AND COMMUNICATE

1. Find the mean of Luke's scores on the first five tests, including the midterm. How is a weighted average different from a mean?

2. You found a weighted average in Step 4 of the Exploration. What did your average describe? What were the weights?

EXAMPLE 2   **Application: Business**

Karen Arndt works for a development company that is designing a new shopping mall. For the project to be approved, Karen needs to make sure that the mall will bring in an average monthly rent of **$8/ft²**.

Two large department stores have each agreed to rent **100,000 ft²** at **$1.25/ft²**. Karen predicts that she can rent space to smaller stores for **$18.00/ft²**. How much space for small stores should be included in the mall's design to bring in an average monthly rent of $8/ft²?

SOLUTION

Step 1 Model the situation with an equation.

Let x = the amount of space (in square feet) designed for small stores.

$$\frac{\text{Average}}{\text{monthly rent}} = \frac{\text{Total monthly rent}}{\text{Total area}}$$

2 large stores are each renting **100,000 ft²**.

$$\$8/\text{ft}^2 = \frac{2 \cdot 100{,}000 \text{ ft}^2 \cdot \$1.25/\text{ft}^2 + x \cdot \$18.00}{2 \cdot 100{,}000 \text{ ft}^2 + x}$$

She needs to include x ft² for small stores.

$$8 = \frac{250{,}000 + 18x}{200{,}000 + x}$$

Step 2 Use a graph to solve the equation.

Graph the equations $y = 8$ and $y = \dfrac{250{,}000 + 18x}{200{,}000 + x}$ in the same coordinate plane.

```
Intersection
X=135000 ──── Y=8 ─────┘
```

Find the x-coordinate of the point where the graphs intersect.

The mall's design should include 135,000 ft² of space for small stores.

THINK AND COMMUNICATE

3. In Example 2, explain why the x-coordinate of the intersection of the two graphs is the solution of the equation $8 = \dfrac{250{,}000 + 18x}{200{,}000 + x}$.

4. If Karen provides more than 135,000 ft² of space for small stores, will the average rent go up or down? Explain. What problems might Karen run into if she provides more space for small stores than is necessary?

5. In Examples 1 and 2, what is being averaged? What are the weights?

1. The term paper in an English class counts as two tests. The table at the right shows three students' grades. Find each student's weighted average.

	Test 1	Test 2	Term paper
Daniel	85	85	75
Peter	75	85	75
Sarah	75	75	85

2. At the yearly Weld College clam bake, you can order either the lobster platter for $10 or the hamburger platter for $3. This year 124 people ordered the lobster platter and 95 ordered the hamburger platter. What was the average cost of a meal per person?

3. In Questions 1 and 2, what is being averaged? What are the weights?

4. **Technology** Look back at the Exploration on page 469. Instead of taking careful aim with each bean, Saul decided to pour all of the beans at once. He counted 66 beans in the cup and 83 in the bowl. Instead of counting the beans that spilled onto the floor, Saul estimated that his average score was about 4. About how many beans are on the floor if his estimate is correct?

11.2 Exercises and Applications

Extra Practice exercises on page 579

1. **BUSINESS** On a business trip, Michael Simms recorded information about his gasoline purchases.

   EXPENSE REPORT
   Name: *Michael Simms*

Date	Price/gallon	Amount (gal)
3/12	$1.20	18
3/14	$1.14	15
3/15	$1.16	10
3/18	$1.22	19

   a. What is the total cost of the gasoline?

   b. How many gallons of gasoline did Michael buy in all?

   c. Find the average cost per gallon.

2. **Test Scores** Look back at Example 1 on page 470. The minimum grade for an A average is 91. Is there any way Luke's final exam grade can bring his average up to an A? Explain.

3. **Test Scores** To earn an A for the semester in her Spanish class, Josie must have an average grade of at least 91. She earned grades of 87, 95, 92, and 88 on the tests during the semester. The final counts as two tests. What score does she need on the final to earn an A?

4. **Writing** A project in a math class involved comparison shopping. Students checked the prices of a gallon of low-fat milk at 12 stores. The results are given in the table at the left.

   a. Without calculating, estimate whether the average price will be closer to $1.80 or $2.65. Explain your reasoning.

   b. Calculate the average price per gallon.

Price of milk (dollars)	Number of stores
1.80	|| ②
1.95	|||| ④
2.35	ⅢⅢ ⑤
2.65		①

Connection SPORTS

Olympic figure skating judges use weighted scoring to rank skaters. Each skater is judged twice, in the technical program and in the free skating program. To find the final ranking, each skater's rank in the technical program is multiplied by 0.5 and added to the rank in the free skate. The skater with the lowest total wins the gold medal.

The table shows the rankings in both events of the 1994 Olympic Ladies' Figure Skating Championship in Hamar, Norway.

Skater COUNTRY	Ranking	
	Technical Program	Free Skate
Oksana Baiul UKR	2	1
Surya Bonaly FRA	3	4
Lu Chen CHN	4	3
Tonya Harding USA	10	7
Nancy Kerrigan USA	1	2
Yuka Sato JPN	7	5
Tanja Szewczenko GER	5	6
Katarina Witt GER	6	8

5. Find the final ranking.

6. Explain how weights are involved in the scoring.

7. How would the final ranking be different if the scores were not weighted?

---

8. **Technology** Look back at Example 2 on page 471. Suppose one of the larger stores decides not to rent space in the mall. The other agrees to a larger area of 150,000 ft^2. How much space should Karen Arndt provide for small stores as she revises her plans?

Technology For Exercises 9–11, use a graphing calculator to solve each equation.

9. $2 = \dfrac{x}{x - 3}$

10. $2 = \dfrac{x + 4}{x + 3}$

11. $3 = \dfrac{4x - 10}{x - 3}$

12. **BUSINESS** Martin Dionne sells peanuts for $1.70 per pound and cashews for $6 per pound. He decides to make a mixture of 70 lb of peanuts and 35 lb of cashews.

a. How much should he charge per pound for the mixture if he wants to earn the same amount as if he sold the nuts separately?

b. Write a variable expression to show how much Martin Dionne should charge if he adds x pounds of almonds to the mixture. He usually charges $4 per pound for almonds.

13. 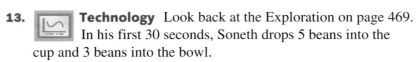 **Technology** Look back at the Exploration on page 469. In his first 30 seconds, Soneth drops 5 beans into the cup and 3 beans into the bowl.

a. What is Soneth's average score after the first 30 seconds?

b. Soneth suspects that if he drops the next few beans into the cup and then stops, he will get a higher average than his friend Tim. Tim's average score was 8.4. How many beans must Soneth drop into the cup to beat Tim's score?

Visual Learning For Exercises 14 and 15, graph each equation and compare the graphs. How can you predict what the graph of a rational equation will look like?

14. a. $y = \dfrac{1}{x}$ b. $y = \dfrac{1}{x + 1}$ c. $y = \dfrac{1}{x + 2}$ d. $y = \dfrac{1}{x - 2}$

15. a. $y = \dfrac{x}{x + 1}$ b. $y = \dfrac{2x}{x + 1}$ c. $y = \dfrac{3x}{x + 1}$ d. $y = \dfrac{-3x}{x + 1}$

16. **Challenge** Eiji Hirai and his friend Josh Smith share a two-bedroom apartment in a suburb of Boston. The rent is $950 per month. Both friends use common areas such as the kitchen and living room, so they each pay $225 toward those areas. They use weights to split the rest of the rent based on the sizes of their rooms. Eiji's room is 160 ft^2 and Josh's room is only 90 ft^2. How much rent should each pay? Explain how you found your answer.

17. a. **Open-ended Problem** Write a problem that can be solved using the weighted average $\dfrac{200(12) + x(20)}{200 + x} = 15$.

b. Solve the equation and tell what the answer means in terms of the situation you described in part (a).

18. **Cooperative Learning** Work with a partner to design and build a mobile. You need a lightweight dowel, string, and a collection of identical objects that are heavier than the dowel.

Step 1 Make a scale on the dowel as shown.

Step 2 Tie strings to several numbers. If you hang a single object from each string, the mobile will balance when it is hung from the average of the numbers. If you hang multiple objects from each string, the mobile will balance when it is hung from the weighted average of the numbers.

$$6\frac{7}{8} = \frac{1(1) + 2(4) + 3(8) + 2(11)}{1 + 2 + 3 + 2}$$

Step 3 Design your mobile first, using multiple objects on at least one string. Then test your design by making the mobile.

19. Writing Explain how weighted averages can be used in determining grades. When do you think it is to your advantage to have weighted averages used in calculating a grade? When is it to your disadvantage?

Solve each equation. *(Section 4.4)*

20. $\dfrac{x}{5} + \dfrac{x}{6} = 7$ **21.** $5 - \dfrac{2}{3}y = -\dfrac{7}{15}y$ **22.** $\dfrac{b}{3} - \dfrac{b}{9} = -\dfrac{1}{4}$

Tell whether each expression is a polynomial. Write *Yes* or *No*. If not, explain why not. *(Section 10.1)*

23. $2 + 3x^2 - 5x$ **24.** -5 **25.** $\dfrac{2}{x^3} - \dfrac{1}{x^2} + \dfrac{2}{5}$

ASSESS YOUR PROGRESS

VOCABULARY

inverse variation (p. 464) **weighted average** (p. 470)
constant of variation (p. 464)

Do the data show inverse variation? If they do, write an equation for *y* in terms of *x*. *(Section 11.1)*

1.

x	y
12	12
16	8
18	6

2.

x	y
5	24
6	20
8	15

3. GEOMETRY For a triangle with fixed area, the height and the length of the base vary inversely. One triangle has base length 32 cm and height 12 cm. A triangle with the same area has base length 24 cm. Find its height. *(Section 11.1)*

4. In Mark's statistics class the final is worth twice as much as each of the other tests. Mark earned the following grades on his tests: 80, 78, 96. Mark got a 90 on the final. What is his average grade? *(Section 11.2)*

5. Technology Jay Graydon is designing a hotel. He will include 50 economy rooms and at least 20 executive suites. If he plans to charge $80 per night for the economy rooms and $110 per night for the executive suites, how many executive suites must he include to bring in an average income per room of $90 per night? *(Section 11.2)*

6. Journal How are inverse variation and direct variation alike? How are they different? Include both equations and graphs in your answer.

11.3 Solving Rational Equations

Learn how to...
- solve rational equations

So you can...
- solve problems involving rational equations such as designing a mall, finding hiking speeds, and exploring weight at different altitudes

Look back at Example 2 on page 471 about designing a shopping mall. It showed how to solve the equation $8 = \dfrac{250,000 + 18x}{200,000 + x}$ by graphing. You can also solve this equation algebraically.

EXAMPLE 1

Solve the equation $8 = \dfrac{250,000 + 18x}{200,000 + x}$.

SOLUTION

> Eliminate the fraction by multiplying both sides of the equation by $(200,000 + x)$.

$$8 = \frac{250,000 + 18x}{200,000 + x}$$

$$(200,000 + x)8 = (200,000 + x)\frac{250,000 + 18x}{200,000 + x}$$

$$1,600,000 + 8x = 250,000 + 18x$$

$$1,350,000 = 10x$$

$$135,000 = x$$

The solution is 135,000.

Check

$$8 \overset{?}{=} \frac{250,000 + 18(135,000)}{200,000 + 135,000}$$

$$8 \overset{?}{=} \frac{2,680,000}{335,000}$$

$$8 = 8 ✔$$

Examples of Rational Expressions

$$\frac{300}{s}$$

$$\frac{1(75) + 2(89) + 1(x)}{1 + 2 + 1}$$

$$\frac{x - y}{x^2 - 2xy + y^2}$$

$$\frac{b}{b - \sqrt{2}}$$

The variable expression in Example 1 is an example of a *rational expression*. A **rational expression** is an expression that can be written as a fraction. The numerator and denominator must be polynomials.

A **rational equation** is an equation that contains only rational expressions.

Numerator is a polynomial. ⟶ $\dfrac{250,000 + 18x}{200,000 + x} = 8$ ⟵ Can be written as $\dfrac{8}{1}$.

Denominator ⟶ is a polynomial.

THINK AND COMMUNICATE

1. Decide whether each expression is a rational expression. Explain.

    a. $\dfrac{a + 2}{b - 1}$ **b.** $\dfrac{1}{\sqrt{y}}$ **c.** $x - 1$

EXAMPLE 2

Solve the rational equation $\dfrac{3}{2(t+1)} + \dfrac{5}{4t} = \dfrac{2}{t}$.

SOLUTION

Step 1 Find the least common denominator (LCD) of the fractions.

Look at the factors of the denominators.

$$\frac{3}{2\cdot(t+1)} + \frac{5}{2\cdot2\cdot t} = \frac{2}{t}$$

> Remember that the LCD is the least common multiple of the denominators.

The LCD is $2\cdot2\cdot t\cdot(t+1) = 4t(t+1)$.

Step 2 Multiply both sides of the equation by the LCD and solve.

$$\overset{2}{\cancel{4t}}(\cancel{t+1})\left(\frac{3}{\cancel{2(t+1)}}\right) + \cancel{4}t(t+1)\left(\frac{5}{\cancel{4t}}\right) = 4\cancel{t}(t+1)\left(\frac{2}{\cancel{t}}\right)$$

> Each denominator is a **factor** of the **LCD**.

$$6t + (5t+5) = 8t + 8$$

$$11t + 5 = 8t + 8$$

$$3t = 3$$

$$t = 1$$

The solution is 1.

Check

$$\frac{3}{2(1+1)} + \frac{5}{4} \overset{?}{=} \frac{2}{1}$$

$$\frac{8}{4} = 2 \checkmark$$

EXAMPLE 3

Solve the equation $\dfrac{5x-4}{2(x-4)} + \dfrac{9}{2} = \dfrac{8}{x-4}$.

SOLUTION

Multiply both sides of the equation by the LCD $2(x-4)$.

$$2(x-4)\frac{5x-4}{2(x-4)} + 2(x-4)\left(\frac{9}{2}\right) = 2(x-4)\left(\frac{8}{x-4}\right)$$

$$(5x-4) + (9x-36) = 16$$

$$14x = 56$$

$$x = 4$$

The equation has no solution.

Check

$$\frac{5\cdot2-4}{2(4-4)} + \frac{9}{2} \overset{?}{=} \frac{8}{4-4}$$

$$\frac{6}{0} + \frac{9}{2} \neq \frac{8}{0}$$

You can't have zero in the denominator, so 4 is not a solution.

◄ **WATCH OUT!**

Always check your answer! Sometimes you will get answers that are not solutions of the original equation.

EXAMPLE 4 Application: Recreation

Dan Patterson hikes downhill **twice as fast** as he hikes uphill. It took Dan **5.5 h** to hike from Lake Louise to the top of Saddleback and back. The trail up Saddleback is about **3.7 mi** long. What is Dan's average uphill hiking speed? his average downhill speed?

SOLUTION

Step 1 Organize the information in a table.

Let s = Dan's **uphill hiking speed** in miles per hour.

Then $2s$ = Dan's **downhill hiking speed** in miles per hour.

Use the formula **Time** $= \dfrac{\text{Distance}}{\text{Rate}}$ to find how long each part of the trip takes.

	Distance	*Rate*	*Time*
Uphill hike	3.7	s	$\dfrac{3.7}{s}$
Downhill hike	3.7	$2s$	$\dfrac{3.7}{2s}$

Distance traveled one way.

Step 2 Model the situation with an equation. The entire trip takes **5.5 h**.

$$\frac{\text{Time}}{\text{uphill}} + \frac{\text{Time}}{\text{downhill}} = 5.5$$

$$\frac{3.7}{s} + \frac{3.7}{2s} = 5.5$$

Step 3 Use a graph to solve the equation.

Graph the equations $y = \dfrac{3.7}{x} + \dfrac{3.7}{2x}$ and $y = 5.5$ on a graphing calculator.

Find the *x*-coordinate of the point where the graphs intersect.

Dan's average uphill hiking speed is about 1 mi/h. His average downhill speed is about $2(1) = 2$ mi/h.

Intersection
X=1.0090909 Y=5.5

THINK AND COMMUNICATE

2. Suppose Beth hikes uphill for 2 h and downhill twice as fast for 1 h, covering a total of 5 mi. Organize this information in a table like the one in Example 4 and write an equation. Let s = uphill speed.

3. Show how you can solve the equation in Example 4 algebraically.

Tell whether each equation is a rational equation. Explain your answer.

1. $y = \dfrac{36}{x}$

2. $xy = 36$

3. $\dfrac{16(0) + 13(5) + x(10)}{16 + 13 + x} = 7.5$

4. $x = \dfrac{-b \pm \sqrt{b^2 - 4ac}}{2a}$

Find the least common denominator of the fractions in each equation.

5. $\dfrac{x-2}{3} + \dfrac{3x}{8} = \dfrac{7}{6}$

6. $\dfrac{x}{4} + \dfrac{3}{x} = \dfrac{10x + 3}{4x}$

7. $\dfrac{6}{5x} = \dfrac{1}{10(x+1)}$

Solve each equation. If the equation has no solution, write *no solution*.

8. $\dfrac{1}{3} + \dfrac{2}{x} = \dfrac{3}{x}$

9. $\dfrac{9}{n} = 3 - \dfrac{4n-9}{n}$

10. $\dfrac{2}{6x} - \dfrac{1}{x-1} = \dfrac{3}{2x}$

11.3 Exercises and Applications

Tell whether each equation is a rational equation. If it is not, explain why not.

Extra Practice exercises on page 579

1. $\dfrac{3}{x+1} = \dfrac{2}{x}$

2. $P = 0.2\sqrt{L}$

3. $\dfrac{26.4 + 10x}{4.7 + x} = 7.3$

4. **Technology** Graph the equations $y = \dfrac{3}{x+1}$ and $y = \dfrac{2}{x}$ in the same coordinate plane. Use the graphs to solve the equation $\dfrac{3}{x+1} = \dfrac{2}{x}$. Check your answer by solving the equation algebraically.

Solve each equation. If the equation has no solution, write *no solution*.

5. $\dfrac{5}{x-1} = \dfrac{2}{3}$

6. $\dfrac{4}{x} + 2 = \dfrac{12}{x}$

7. $\dfrac{x}{4} + \dfrac{x}{6} = 2$

8. $\dfrac{6}{x+2} + 10 = \dfrac{1}{x+2}$

9. $\dfrac{6}{y} + \dfrac{3}{4} = 2$

10. $\dfrac{a+5}{a-6} = \dfrac{11}{a-6} + 4$

BY THE WAY

The Condor, a robotic spy plane, set an altitude record of 12.7 mi for piston-powered aircraft. It could fly for more than two and a half days without refueling.

11. PHYSICS On or near Earth, the weight w of an object depends on its altitude.

Let h = height in miles above sea level.
Let s = weight in pounds at sea level.

$$w = \dfrac{(4000)^2 s}{(4000 + h)^2}$$

   a. On long flights, passenger airplanes often fly at an altitude of about 6 mi. How much does a 150 lb passenger weigh 6 mi above sea level?

   b. How high above sea level must a plane fly for a 120 lb person to weigh 119 lb?

Cooperative Learning Work with a partner. One of you should do Exercise 12. The other should do Exercise 13. Work together on Exercise 14.

12. a. What is the LCD of the fractions in the equation $\dfrac{5}{x+3} = \dfrac{4}{x-1}$?

b. Solve the equation in part (a) by multiplying both sides by the LCD.

13. a. Is the equation $\dfrac{5}{x+3} = \dfrac{4}{x-1}$ a proportion? Explain.

b. Solve the equation in part (a) by using the means-extremes property.

14. Writing Compare the methods you used in Exercises 12 and 13. How are they alike? How are they different? Which do you prefer? Why?

Solve each equation. If the equation has no solution, write *no solution*.

15. $\dfrac{c-2}{5} = \dfrac{c-1}{4}$

16. $\dfrac{2}{x} + \dfrac{3}{5} = \dfrac{1}{4x}$

17. $\dfrac{t-1}{t} + \dfrac{3}{2t} = \dfrac{1}{3}$

18. $\dfrac{2(t-2)}{t} - 1 = \dfrac{-4}{t}$

19. SPORTS In his last NBA season, Larry Bird of the Boston Celtics made 52 of the 128 three-point shots he attempted, for an average of 40.6%. The Celtics lost in the semifinal round of the Eastern Conference Championship. If they had continued playing, how many consecutive three-point shots would Larry Bird have had to make to raise his average to 50%?

20. A game manufacturer is designing a dart board, as shown at the left.

a. Write an expression for the area of a bull's-eye with radius r. Write an expression for the area of the entire dart board.

b. The manufacturer wants the probability of a randomly thrown dart hitting the bull's-eye to be $\dfrac{1}{9}$. (Assume the dart hits the board.) What should the radius of the bull's-eye be?

Solve each equation. If the equation has no solution, write *no solution*.

21. $\dfrac{5}{z+2} = \dfrac{5}{z-4}$ **22.** $\dfrac{11}{5-b} = 4 + \dfrac{b+6}{5-b}$ **23.** $\dfrac{3}{4} + \dfrac{1}{6(t-5)} = \dfrac{5}{2(t-5)}$

24. a. **Technology** Solve the equation $\dfrac{x+10}{x-1} = \dfrac{12-x}{x-1} + 5$ algebraically.

b. Use a graphing calculator to solve the equation in part (a). Explain how the graph supports the answer you found in part (a).

25. Challenge The loudness of a sound decreases as the listener moves away from the sound. Suppose you are playing a radio at the beach with the volume very high. The intensity of the sound in Watts/m² can be modeled by the formula $I = \dfrac{0.01}{d^2}$, where d is the distance from the radio in meters. If your parents want to hear a sound only as intense as normal conversation (10^{-6} Watts/m²), how far from the radio should they be?

26. Writing How are rational expressions like rational numbers? How are they different?

In New Hampshire's White Mountains, there is a range of mountains named after presidents of the United States. Occasionally experienced hikers hike the 23 mi from one end of the range to the other.

27. Tanja and Adam Cook are hiking from Route 302 to Dolly Copp Campground along the path shown below. It took them 9 h to hike from Route 302 to the top of Mt. Washington. They usually hike downhill twice as fast as they hike uphill. What was their average uphill rate? their average downhill rate?

28. Predict how long it will take Tanja and Adam to hike from the top of Mt. Washington to Dolly Copp Campground. Explain your answer.

29. RESEARCH Which of the peaks do not have names of former presidents?

30. Open-ended Problem Write a rational equation that has no solution. Explain how you came up with your equation.

BY THE WAY

Normal summer temperatures on Mt. Washington are only about 47°F. Winds can reach up to 231 mi/h.

ONGOING ASSESSMENT

31. Writing How is finding the least common denominator of the fractions $\dfrac{3}{2(x-3)}$, $\dfrac{5}{4x}$, and $\dfrac{2}{2x}$ like finding the least common denominator of the fractions $\dfrac{3}{8}$, $\dfrac{5}{12}$, and $\dfrac{2}{3}$? How is it different?

SPIRAL REVIEW

Graph each equation. (*Section 3.4*)

32. $y = 2x - 5$

33. $y = -4x$

34. $y = \dfrac{2}{3}x + 3$

Solve each equation by factoring. (*Section 10.5*)

35. $2x^2 + 5x - 3 = 0$

36. $3y^2 - 15y + 12 = 0$

37. $3a^2 + 11a - 4 = 0$

11.4 Applying Formulas

You may think you never use formulas outside math class. But most people use formulas all the time. For example, you use a formula if you leave a 15% tip in a restaurant or estimate how long it will take to travel a given distance at a given speed.

Learn how to...

- solve a formula for one of its variables

So you can...

- perform multiple calculations with the same formula
- graph equations on graphing calculators
- make calculations with spreadsheets

EXAMPLE 1   **Application: Recreation**

Some hiking guidebooks estimate the time a hiker needs to hike a trail by using the formula $t = \frac{1}{2}m + v$, where t is the **time** in hours, m is the **length** of the trail in miles, and v is the **vertical rise** in thousands of feet. Jonathan wants to choose a trail which is not very strenuous. His guidebook says that the Tumbledown Mountain loop trail is **2 mi long** and takes about $2\frac{3}{4}$ **h** to hike. Use the formula to estimate the vertical rise of the trail.

SOLUTION

Method 1   Substitute **2.75** for t and **2** for m. Then solve for v.

$$t = \frac{1}{2}m + v$$

$$2.75 = \frac{1}{2} \cdot 2 + v$$

$$2.75 = 1 + v$$

$$1.75 = v$$

The vertical rise of the trail is about 1750 ft.

Method 2   Solve the formula for v. Then substitute **2.75** for t and **2** for m and simplify.

$$t = \frac{1}{2}m + v \qquad \text{Subtract } \tfrac{1}{2}m \text{ from both sides.}$$

$$t - \frac{1}{2}m = v$$

$$2.75 - \frac{1}{2} \cdot 2 = v$$

$$2.75 - 1 = v$$

$$1.75 = v$$

The vertical rise of the trail is about 1750 ft.

THINK AND COMMUNICATE

1. If you want to find the vertical rises of several trails, would you rather use the formula $t = \frac{1}{2}m + v$ or $t - \frac{1}{2}m = v$? Explain.

2. Do you think the Tumbledown Mountain loop trail is likely to be a strenuous hike? Explain.

If you need to use a formula many times, you may want to solve the formula for one of its variables first. You also solve an equation for a variable to graph it on a graphing calculator.

EXAMPLE 2

Graph the equation $x = \dfrac{y}{y + 2}$ on a graphing calculator.

SOLUTION

Solve the equation for y. Get y alone on one side of the equation.

$$x = \frac{y}{y + 2}$$

$$(y + 2)x = \frac{y}{y + 2}(y + 2) \qquad \begin{array}{l}\text{Multiply both} \\ \text{sides by } (y + 2) \\ \text{and simplify.}\end{array}$$

$$(y + 2)x = y \qquad \text{Note that } y \neq -2.$$

Move all terms containing **y** to the same side and factor out **y**.

$$yx + 2x = y$$

$$yx - y = -2x$$

$$y(x - 1) = -2x \qquad \begin{array}{l}\text{Divide both sides} \\ \text{by } x - 1 \text{ so } y \text{ is} \\ \text{alone. Note that} \\ x \neq 1.\end{array}$$

$$y = \frac{-2x}{x - 1}$$

Now you can graph the equation.

WATCH OUT!

The graph of your final equation may include a point that gives the original equation a zero denominator. If this happens, use an open dot to indicate that the point is not a part of the graph.

THINK AND COMMUNICATE

3. In the solution of Example 2, why must you factor out y?

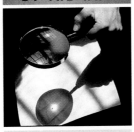
EXAMPLE 3 Application: Optics

If you look at a leaf with a magnifying glass, the distances between the leaf, the glass, and your eye are related by the formula below.

Let a = the distance from the magnifying glass to the leaf.

Let b = the distance from your eye to the magnifying glass.

Let f = the focal length of the magnifying glass.

$$\frac{1}{a} + \frac{1}{b} \approx \frac{1}{f}$$

$\longleftarrow b \longrightarrow$ $\leftarrow a \rightarrow$

Not drawn to scale

a. Solve the formula for a.

b. How far should you hold the magnifying glass from the leaf if your eye is **18 in.** from the glass and the focal length of the glass is **5 in**?

SOLUTION

a.
$$\frac{1}{a} + \frac{1}{b} \approx \frac{1}{f}$$
Multiply both sides by the least common denominator abf.

$$abf\left(\frac{1}{a}\right) + abf\left(\frac{1}{b}\right) \approx abf\left(\frac{1}{f}\right)$$

Move all terms containing a to the same side.
$$bf + af \approx ab$$

$$bf \approx ab - af$$
Factor out a.

$$bf \approx a(b - f)$$
Divide both sides by $(b - f)$.

$$\frac{bf}{b - f} \approx a$$

b. $a \approx \dfrac{bf}{b - f}$

$$\approx \frac{(18)(5)}{18 - 5}$$

$$\approx 6.9$$

To see the leaf clearly, you should hold the magnifying glass about 7 in. from the leaf.

CHECKING KEY CONCEPTS

Solve each equation for the variable in red.

1. $5z = 2y$

2. $x + y = 4$

3. $2m + d = 6$

4. $d = \frac{1}{2}at^2$

5. $\frac{1 - bc}{b} = c$

6. $\frac{f}{2} = \frac{1 - n}{n}$

Technology Graph each equation. Use a graphing calculator if you have one.

7. $3x + y = 7$

8. $y - 2x^2 - 1 = x$

9. Business Barbara Andrews travels frequently and needs to determine how long each of her trips will take. Express the formula $d = rt$ in the form she would find most useful.

11.4 **Exercises and Applications**

1. GEOMETRY The formula for the volume of a pyramid is $V = \frac{1}{3}Bh$, where B is the area of the base and h is the height.

Extra Practice exercises on page 580

  a. Solve the formula for h.

  b. Find h when $V = 140 \text{ cm}^3$ and $B = 42 \text{ cm}^2$.

2. RECREATION It takes 3 h to hike a trail that has a vertical rise of 750 ft. Use the formula from Example 1 to estimate the length of the trail.

Solve each equation for the variable in red.

3. $P = a + b + c$

4. $A = lw$

5. $P = C - S$

6. $y = a + bx$

7. $2x + y = 24$

8. $3x + 2y = 45$

9. $ax + by = c$

10. $A = 2x^2 + 4xy$

11. $y = \frac{1}{x}$

 Technology Graph each equation. Use a graphing calculator if you have one.

12. $x^2 + y - 9 = 0$

13. $x + 2y = 3$

14. $x^2 - \frac{1}{2}y = x$

15. $\frac{x - 1}{y} = 3$

16. $2(y - 3) = x^2$

17. $x(y + 1) = y - 1$

18. Writing Jen solved the formula $ry = a + y$ for y, as shown at the right. Describe what Jen did. What is wrong with her solution?

19. a. OPTICS Solve the equation in Example 3 on page 484 for f.

  b. Find the focal length of the magnifying glass if you can see the leaf clearly when $a = 40$ in. and $b = 10$ in.

$$ry = a + y$$
$$y = \frac{a + y}{r}$$

Look back at the article on pages 460–462.

Focal Lengths of Lenses
24 mm
28 mm
35 mm
50 mm
80–200 mm zoom
300 mm

Alex MacLean can adjust image sizes in aerial photographs by varying his flying height from about 0.2 mi to 2.5 mi or by choosing different lenses. The table at the left shows the focal lengths of the lenses Alex MacLean uses. Focal length f, flying height h, image size i, and actual size a are related by the formula $\frac{i}{a} = \frac{f}{h}$.

20. **a.** Solve the formula above for f.

b. Alex MacLean took a photograph of a $\frac{1}{2}$ mi long irrigator from a height of 1 mi. The image of the irrigator on 35 mm film is 14 mm long. Which lens did Alex use for this photograph?

A giant pivot irrigator stopping short of erosion gullies along the Teton River.

21. **Open-ended Problem** Look back at the map of New York's Central Park on page 87. Suppose you ask Alex MacLean to photograph Central Park. The park is about $\frac{1}{2}$ mi wide and $2\frac{1}{2}$ mi long.

a. Alex MacLean uses 35 mm film, which has frames that are 24 mm by 36 mm. Sketch how you would like Central Park to appear within the frame. How long will the image of the park be?

b. Choose a lens and flying height to produce the image you sketched in part (a).

22. **CARS** Pushing a bicycle pedal causes the wheels of the bicycle to turn. In a car, exploding gas pushes a piston which causes the wheels of the car to turn. The power of the car engine depends on many things.

Let H = the power of a car engine in *horsepower*.
Let p = the mean pressure on the piston in lb/in.^{2}.
Let l = the distance in feet the piston moves.
Let a = the area in square inches of the piston.
Let n = the number of explosions per minute.
Let k = the number of pistons in the engine.

exploding gas

p

a

l

$$H = \frac{plank}{33,000}$$

BY THE WAY

One horsepower is 33,000 foot-pounds per minute, which is about $1\frac{1}{3}$ times the actual power of a horse.

a. Solve the equation for p.

b. Find p for a 300 horsepower engine for which $l = 0.29$ ft, $a = 12.56$ in.^{2}, $n = 2500$/min, and $k = 8$.

Not drawn to scale

23. **Challenge** Solve the equation $\sqrt{y^2 - 1} = x - y$ for y.

24. **Spreadsheets** Suppose you are designing a spinning circular room for an amusement park. The faster the room spins, the more the riders are pressed against the wall. If the ride spins quickly enough, the riders are pressed into the wall more than they are pulled to the ground by gravity.

Let g = how much you feel "pulled" to the ground by gravity. (32 ft/s^2)
Let n = the number of g's with which you are pressed against the wall.
Let r = the radius of the circular room in feet.
Let v = the speed of the riders in feet per second.

Choose a reasonable value for r. Then use the formula $ng = \dfrac{v^2}{r}$ and a spreadsheet to find the speed v that will result in each value of n: 0.4, 0.8, 1.2, 1.6, 2.0, 2.4, 2.8, 3.2.

ONGOING ASSESSMENT

25. Writing When is it helpful to solve a formula for one of its variables? Give some examples to support your answer.

SPIRAL REVIEW

Find the reciprocal of each number. *(Section 4.2)*

26. 4

27. $\dfrac{1}{3}$

28. $-\dfrac{2}{3}$

Simplify each expression. *(Section 9.7)*

29. $\left(4x^3\right)^3$

30. $\left(\dfrac{a}{b^2}\right)^5$

31. $\left(\dfrac{m^5}{n^2 p^3}\right)^2$

ASSESS YOUR PROGRESS

VOCABULARY

rational expression (p. 476)   **rational equation** (p. 476)

Solve. *(Section 11.3)*

1. $\dfrac{x-1}{3} = \dfrac{x+1}{4}$

2. $\dfrac{5}{4x} + \dfrac{7}{8} = \dfrac{3}{x}$

3. $\dfrac{2}{x} + \dfrac{1}{3} = \dfrac{11}{3x}$

Solve each equation for the variable in red**.** *(Section 11.4)*

4. $x - 2y = 8$

5. $2(x - 3) = y^2 + 1$

6. $\dfrac{a+b}{2c} = b + d$

7. The formula $C = \dfrac{5}{9}(F - 32)$ gives the temperature C in degrees Celsius for a given temperature F in degrees Fahrenheit. Solve for F. Find F when $C = 20°$. *(Section 11.4)*

8. Journal What do you find most difficult about solving rational equations? How do you think you could make this less difficult?

11.5

Working with Rational Expressions

Joline Jacques wants to make 500 ice cream cones out of cookie dough for her ice cream shop. How can she tell how much dough to make?

Imagine cutting along the side of a cone and flattening it. The resulting shape is a sector of a circle. The area of the sector is the area of the dough needed to make the cone.

EXAMPLE 1   Application: Cooking

What is the area of the piece of dough Joline cuts out for each cone?

Let A = the area of the sector of dough as described above.
Let C = the circumference of the top of the cone.
Let s = the slant height of the cone.

SOLUTION

$$\frac{\text{area of sector of circle}}{\text{area of entire circle}} = \frac{\text{circumference of sector of circle}}{\text{circumference of entire circle}}$$

$$\frac{A}{\pi s^2} = \frac{C}{2\pi s} \qquad \text{To solve for } A,$$
$$\text{multiply both sides of the equation by } \pi s^2.$$

$$\pi s^2 \cdot \frac{A}{\pi s^2} = \pi s^2 \cdot \frac{C}{2\pi s}$$

$$\pi s^2 \cdot \frac{A}{\pi s^2} = \pi \overset{s}{s^2} \cdot \frac{C}{2\pi s} \qquad \text{Divide out common factors.}$$

$$A = \frac{sC}{2}$$

You need to cut out a piece of dough with area $A = \dfrac{sC}{2}$.

THINK AND COMMUNICATE

1. What is the area of the piece of dough Joline cuts out to make a cone with a circumference of 24 cm and a slant height of 10 cm?

2. About how thick should the sheet of dough described in Question 1 be? To make 500 cones, how much batter must Joline make?

You can multiply rational expressions just as you multiply fractions.

Multiplying fractions

$$\frac{4}{3} \cdot \frac{9}{2} = \frac{4 \cdot 9}{3 \cdot 2} = \frac{\overset{6}{\cancel{36}}}{\cancel{6}} = 6$$

$$\frac{\overset{2}{\cancel{4}}}{\cancel{3}} \cdot \frac{\overset{3}{\cancel{9}}}{\cancel{2}} = 2 \cdot 3 = 6$$

Multiplying rational expressions

$$\frac{x^2}{6y} \cdot \frac{10y}{x} = \frac{x^2 \cdot 10y}{6y \cdot x} = \frac{\overset{5}{\cancel{10}}\overset{x}{\cancel{x^2}}\cancel{y}}{\underset{3}{\cancel{6}}\cancel{x}\cancel{y}} = \frac{5x}{3}$$

$$\frac{\overset{x}{\cancel{x^2}}}{\underset{3}{\cancel{6y}}} \cdot \frac{\overset{5}{\cancel{10y}}}{\cancel{x}} = \frac{x}{3} \cdot 5 = \frac{5x}{3}$$

Multiply the numerators and **multiply the denominators**.

You can also divide out common factors before you multiply.

A rational expression is in **simplest form** when the greatest common factor of the numerator and denominator is 1.

EXAMPLE 2

Simplify $\dfrac{4b}{a^2} \cdot \dfrac{a^3}{14b^2}$.

SOLUTION

$$\frac{4b}{a^2} \cdot \frac{a^3}{14b^2} = \frac{4b \cdot a^3}{a^2 \cdot 14b^2}$$

Multiply numerators and denominators.

$$= \frac{\overset{2}{\cancel{4b}} \cdot \overset{a}{\cancel{a^3}}}{\underset{7b}{\cancel{a^2} \cdot \cancel{14b^2}}}$$

Divide out common factors.

$$= \frac{2a}{7b}$$

EXAMPLE 3

Simplify $\dfrac{2x + 8}{x + 1} \cdot \dfrac{x^2}{3x + 12}$.

SOLUTION

$$\frac{2x + 8}{x + 1} \cdot \frac{x^2}{3x + 12} = \frac{2(x + 4)}{x + 1} \cdot \frac{x^2}{3(x + 4)}$$

Factor the numerator and the denominator.

$$= \frac{2\cancel{(x + 4)}}{x + 1} \cdot \frac{x^2}{3\cancel{(x + 4)}}$$

$$= \frac{2x^2}{3(x + 1)}$$

You can divide rational expressions just as you divide fractions. To divide, multiply the first expression by the reciprocal of the second expression.

Dividing fractions

$$\frac{2}{3} \div \frac{7}{3} = \frac{2}{3} \cdot \frac{3}{7}$$

Dividing rational expressions

$$\frac{x^2}{2y} \div \frac{2x}{3} = \frac{x^2}{2y} \cdot \frac{3}{2x}$$

EXAMPLE 4

Simplify $\frac{m^2 - 1}{4m} \div (3m - 3)$.

SOLUTION

$$\frac{m^2 - 1}{4m} \div (3m - 3) = \frac{m^2 - 1}{4m} \div \frac{3m - 3}{1}$$

$$= \frac{m^2 - 1}{4m} \cdot \frac{1}{3m - 3}$$

$$= \frac{(m + 1)(m - 1)}{4m} \cdot \frac{1}{3(m - 1)}$$

$$= \frac{m + 1}{12m}$$

THINK AND COMMUNICATE

3. In Example 4, why was $3m - 3$ written as $\frac{3m - 3}{1}$?

4. Explain how to simplify the expression $\frac{b^3}{b^2}$.

☑ CHECKING KEY CONCEPTS

Simplify each expression. If it is already in simplest form, write *simplest form*.

1. $\frac{4}{8x + 16}$

2. $\frac{3z^2}{4z}$

3. $\frac{12x^3}{8x}$

4. $\frac{2y - 8}{3y - 12}$

5. $\frac{x^2 + 1}{x + 1}$

6. $\frac{n^2 + n}{2n + 2}$

7. $\frac{6x^2}{5y} \cdot \frac{10y}{3x}$

8. $\frac{x + 1}{12} \cdot \frac{8}{2x + 2}$

9. $\frac{6x}{y + 1} \cdot \frac{2y + 2}{3}$

10. $\frac{2xy}{3} \div \frac{4x^2}{3}$

11. $\frac{n + 1}{4} \div \frac{3n + 3}{16}$

12. $\frac{x + 2}{5} \div (x^2 + 2x)$

11.5 Exercises and Applications

Simplify each expression. If it is already in simplest form, write *simplest form*.

Extra Practice exercises on page 579

1. $\dfrac{b^2}{2b}$

2. $\dfrac{5x^3}{10x^2}$

3. $\dfrac{y^2}{12x}$

4. $\dfrac{4}{4a + 8}$

5. $\dfrac{c + 1}{c + 2}$

6. $\dfrac{c^2 + c}{c^2}$

7. $\dfrac{4x + 12}{x + 3}$

8. $\dfrac{n^2 + 2n}{2n + 2}$

9. Writing Jim and Ted disagree about whether $\dfrac{3x + 5}{2x + 5}$ can be simplified. Jim says no. Ted says yes. Ted's work is shown at the right. Who is correct? Give an argument to support your answer.

$$\frac{3\cancel{x}+5}{2\cancel{x}+5} = \frac{3\cancel{x}}{2\cancel{x}} = \frac{3}{2}$$

Connection ▸ PROBABILITY

The probability of spinning both spinners and having the first pointer land on the red region and the second pointer land on "4" is the product of the individual probabilities: $\dfrac{1}{3} \cdot \dfrac{1}{4} = \dfrac{1}{12}$.

You can extend this principle to other situations. In general:

Let a = the probability that event A will happen.
Let b = the probability that event B will happen.

If events A and B do not depend on each other, then the probability that both events will happen is $a \cdot b$.

10. Suppose that one of two darts thrown randomly hits the green target and the other hits the red target. Write a variable expression for the probability that both darts land in the bonus region of the target. Simplify your expression.

11. When you fly into Mexico City from the United States, you must go through customs. At customs you press a button which randomly activates a light. If the light is green you may pass immediately, but if the light is red you must wait as your bags are searched.

  a. If the bags of about one in every n people are searched, what is the probability that your bags will be searched?

  b. If you are traveling with a friend, what is the probability that you will both get a green light?

Simplify each expression.

12. $\dfrac{6x}{4y^2} \cdot \dfrac{y^3}{x^2}$

13. $\dfrac{4m - 8}{10y} \cdot \dfrac{15y^2}{5m - 10}$

14. $(2a + 14)\dfrac{1}{3a + 21}$

15. $\dfrac{9y^2}{x + 1} \cdot \dfrac{2x + 2}{6y}$

16. $\dfrac{n^2 - 4}{n + 2} \cdot \dfrac{n}{3n - 6}$

17. $\dfrac{a + 1}{a + 2} \cdot \dfrac{2}{a}$

18. a. Writing Evaluate $\dfrac{3x + 6}{4x + 8}$ when $x = 5$ and when $x = -3$.

   b. Predict the value of $\dfrac{3x + 6}{4x + 8}$ when $x = 9678$. Explain how you know.

Simplify each expression.

19. $\dfrac{8z}{9k} \div \dfrac{4z}{15k^2}$

20. $\dfrac{5t^2(r - 1)}{4r} \div \dfrac{r - 1}{tr^2}$

21. $\dfrac{n}{n + 3} \div \dfrac{n - 2}{n + 3}$

22. $\dfrac{ax + bx}{x^2} \div \dfrac{a + b}{2x}$

23. $\dfrac{n^2 - 1}{2n} \div \dfrac{n - 1}{4n^3}$

24. $\dfrac{6x^2 - 6x}{10} \div (6x - 6)$

25. Open-ended Problem Write three different rational expressions that all equal $\dfrac{2x}{5}$ when simplified.

26. SAT/ACT Preview Which variable expression is equivalent to $\dfrac{x^2}{8} \div \dfrac{2}{x}$?

   A. $\dfrac{2x^2}{8x}$ **B.** $\dfrac{x}{4}$ **C.** $\dfrac{x^3}{8}$ **D.** $\dfrac{x}{2}$ **E.** none of these

GEOMETRY Write a ratio in simplest form comparing the area of the shaded region to the total area of the figure.

27.

28.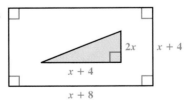

29. Writing Kris and Sandra disagree about whether the two expressions $\dfrac{x^2 - 2x - 35}{x - 7}$ and $(x + 5)$ are equivalent for all values of x.

Kris says they are. Sandra says they are not. Who is correct? Explain.

30. Challenge Complete each simplification.

   a. $\dfrac{15a^2}{7b} \div \underline{\ ?\ } = \dfrac{3b}{14}$

   b. $\underline{\ ?\ } \div \dfrac{w^2}{a^2 - 4} = \dfrac{aw}{4(a + 2)}$

If a train goes around a curve like the one shown, each wheel on the inside of the curve travels a distance of πT, and each outer wheel travels a distance of $\pi(T + t)$. The outer wheels must travel farther, but they cannot turn any faster than the inside wheels because the wheels are connected by solid axles.

Train wheels are designed as shown at the left. The radius of each wheel varies from R on one side to $R + r$ on the other.

As the train goes around a curve, the train shifts slightly to the outside of the turn so the wheels are off center on the rail as shown. Where the outer wheel contacts the rail, the radius is large. Where the inside wheel contacts the rail, the radius is small. This allows the outer wheels to travel farther with each rotation.

In Exercises 31 and 32, you will find the radius T of the sharpest turn a train can make without its wheels slipping.

31. a. Let $n =$ the number of turns the inner wheel makes on the curve shown. With each turn, the wheel travels a distance of $2\pi R$. So $n(2\pi R) = \pi T$. Explain this equation and solve it for n.

b. Write and solve an equation similar to the one in part (a) for the outer wheel.

c. Write an equation relating T, $T + t$, R, and $R + r$. Use the fact that n must be the same for both wheels.

d. Solve the equation from part (c) for T.

32. Suppose a train's wheels vary from a radius of 14 in. to a radius of 14.2 in. On turns, the distance between the rails is 57 in. Find T. Do you think trains ever turn sharply enough for the wheels to slip? Explain.

ONGOING ASSESSMENT

33. Open-ended Problem Find a product that equals $\dfrac{x + 3}{x}$ when simplified. Find a quotient that equals $\dfrac{x + 3}{x}$ when simplified.

SPIRAL REVIEW

Use the distributive property to find each product. (*Section 6.4*)

34. $2x(3 - 5x)$ **35.** $(4x + 2)(5x - 4)$ **36.** $\left(4\sqrt{3} + 5\right)\left(3\sqrt{3} + 1\right)$

Simplify each expression. (*Section 9.6*)

37. $a^2 \cdot a^3 \cdot a^4$ **38.** $\dfrac{a^{12}}{a^3}$ **39.** $\dfrac{c^{-2}}{c^2}$

11.6 Exploring Rational Expressions

Learn how to...
- add and subtract rational expressions

So you can...
- explore patterns with fractions
- solve problems such as investigating car accidents and reducing fuel costs

Mathematics involves the study of patterns. The Exploration will help you discover a pattern for subtracting rational expressions.

EXPLORATION
COOPERATIVE LEARNING

Subtracting Rational Expressions

Work with a partner.
You will need:
- a ruler
- graph paper

1 Look at the diagram. Explain how it shows that $\frac{1}{2} - \frac{1}{3} = \frac{1}{6}$.

2 Draw a diagram to show that $\frac{1}{3} - \frac{1}{4} = \frac{1}{12}$.

Questions

1. Look for a pattern in the equations from Steps 1 and 2. Use this pattern to complete the following equations.

   a. $\frac{1}{4} - \frac{1}{5} = \underline{?}$ b. $\frac{1}{5} - \frac{1}{6} = \underline{?}$ c. $\frac{1}{6} - \frac{1}{7} = \underline{?}$

2. Now use the pattern you found to complete this equation.

   $\frac{1}{x} - \frac{1}{x+1} = \underline{?}$

You can add and subtract rational expressions the same way you add and subtract fractions.

Fractions

$$\frac{5}{8} + \frac{2}{8} = \frac{5+2}{8} = \frac{7}{8}$$

Rational expressions

$$\frac{7}{a+b} - \frac{1}{a+b} = \frac{7-1}{a+b} = \frac{6}{a+b}$$

When the denominators of the expressions are different, use the **LCD** to rewrite each expression.

Fractions

$$\frac{7}{8} - \frac{3}{10} = \frac{35}{40} - \frac{12}{40}$$

$$= \frac{35 - 12}{40}$$

$$= \frac{23}{40}$$

Rational expressions

$$\frac{1}{x} - \frac{1}{x+1} = \frac{x+1}{x(x+1)} - \frac{x}{x(x+1)}$$

$$= \frac{x+1-x}{x(x+1)}$$

$$= \frac{1}{x(x+1)}$$

THINK AND COMMUNICATE

1. Terence simplified the expression $\frac{7}{8} - \frac{3}{10}$ as shown at the right.

Compare Terence's method to the method above. Which do you prefer?

$$\frac{7}{8} - \frac{3}{10} = \frac{7 \cdot 10}{8 \cdot 10} - \frac{3 \cdot 8}{10 \cdot 8}$$

$$= \frac{70 - 24}{80}$$

$$= \frac{46}{80}$$

$$= \frac{23}{40}$$

EXAMPLE 1

Simplify each expression.

a. $\dfrac{7n}{n+2} + \dfrac{14}{n+2}$

b. $\dfrac{2x+5}{2x} - \dfrac{x-1}{2x}$

SOLUTION

a. $\dfrac{7n}{n+2} + \dfrac{14}{n+2} = \dfrac{7n+14}{n+2}$

$$= \frac{7(n+2)}{n+2}$$

$$= 7$$

Always write your answer in simplest form.

b. $\dfrac{2x+5}{2x} - \dfrac{x-1}{2x} = \dfrac{(2x+5) - (x-1)}{2x}$

$$= \frac{2x + 5 - x + 1}{2x}$$

$$= \frac{x + 6}{2x}$$

◀ **WATCH OUT!**

When you subtract expressions, use parentheses to avoid confusion.

EXAMPLE 2

Simplify the expression $\dfrac{5}{6x} - \dfrac{x-3}{4x^2}$.

SOLUTION

First find the LCD. The LCD of $\dfrac{5}{6x}$ and $\dfrac{x-3}{4x^2}$ is $12x^2$.

$$\frac{5}{6x} - \frac{x-3}{4x^2} = \frac{2x \cdot 5}{2x \cdot 6x} - \frac{3(x-3)}{3 \cdot 4x^2}$$

$$= \frac{10x}{12x^2} - \frac{3x-9}{12x^2}$$

$$= \frac{10x - 3x + 9}{12x^2}$$

$$= \frac{7x+9}{12x^2}$$

Notice that $\dfrac{2x}{2x}$ and $\dfrac{3}{3}$ are both equal to 1. Multiplying by 1 does not affect the value of the fractions.

Types of Functions

The table below summarizes the different kinds of functions that have been described in this book.

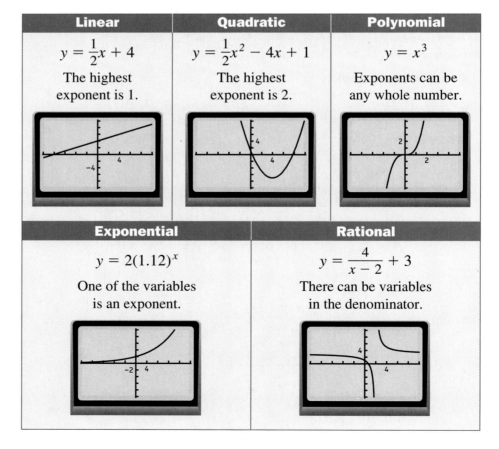

Linear	Quadratic	Polynomial
$y = \dfrac{1}{2}x + 4$	$y = \dfrac{1}{2}x^2 - 4x + 1$	$y = x^3$
The highest exponent is 1.	The highest exponent is 2.	Exponents can be any whole number.

Exponential	Rational
$y = 2(1.12)^x$	$y = \dfrac{4}{x-2} + 3$
One of the variables is an exponent.	There can be variables in the denominator.

Simplify each expression.

1. $\dfrac{1}{y} + \dfrac{3}{y}$

2. $\dfrac{3}{x-2} - \dfrac{4x}{x-2}$

3. $\dfrac{a}{a+1} - \dfrac{3a+2}{a+1}$

4. $\dfrac{1}{x} - \dfrac{1}{x^2}$

5. $\dfrac{4}{5m} + \dfrac{1}{3m^2}$

6. $\dfrac{p}{10p^2} - \dfrac{1}{15p}$

State whether each equation can be best described as *linear*, *quadratic*, *polynomial*, *rational*, or *exponential*.

7. $p = \dfrac{1}{q} - \dfrac{1}{q^2}$

8. $d = 2 + 3t - 16t^2$

9. $\dfrac{1}{2}a - \dfrac{1}{3}b = 7$

11.6 **Exercises and Applications**

Simplify each expression.

Extra Practice
exercises on
page 581

1. $\dfrac{2t}{t-1} - \dfrac{t+2}{t-1}$

2. $\dfrac{b+a}{ab} + \dfrac{b-a}{ab}$

3. $\dfrac{4f+3}{f-1} - \dfrac{2f+5}{f-1}$

4. $\dfrac{3}{m} + \dfrac{2}{m^2}$

5. $\dfrac{5}{4xy^2} - \dfrac{3}{10x^2y}$

6. $\dfrac{a}{a+5} + \dfrac{2}{a}$

State whether each equation can be best described as *linear*, *quadratic*, *polynomial*, *rational*, or *exponential*.

7. $s = 15t^2 - 4$

8. $m = 4r + 2$

9. $y = 2.9(0.94)^x$

10. $3x + 2y = -7$

11. $b = \dfrac{1}{a} + \dfrac{1}{a-5}$

12. $V = d^4 - \dfrac{1}{2}d^2 + \dfrac{1}{4}$

Investigation In the Exploration on page 494, you used a diagram to discover a pattern for subtracting fractions. In Exercises 13–16, you will use the diagram to find a pattern for adding fractions.

13. Place a blank sheet of paper over the diagram at the right. Trace bars of length $\dfrac{1}{2}$ and $\dfrac{1}{3}$ end to end. Find a bar as long as the sum of the two bars you traced. Complete the equation $\dfrac{1}{2} + \dfrac{1}{3} = \underline{\ ?\ }$.

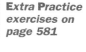

14. Complete the equation $\dfrac{1}{3} + \dfrac{1}{4} = \underline{\ ?\ }$.

15. **Writing** Compare the equations in Exercises 13 and 14. What pattern do you notice? Use the pattern to simplify each expression below.

a. $\dfrac{1}{4} + \dfrac{1}{5}$

b. $\dfrac{1}{5} + \dfrac{1}{6}$

c. $\dfrac{1}{42} + \dfrac{1}{43}$

d. $\dfrac{1}{x} + \dfrac{1}{x+1}$

16. Verify the pattern you discovered in Exercise 15 by simplifying the expression $\dfrac{1}{x} + \dfrac{1}{x+1}$.

17. a. CARS A car's *gas mileage* tells how far the car goes on one gallon of gasoline. Let x = your car's gasoline mileage in miles per gallon. Write a variable expression for the amount you spend on gasoline each year if gasoline averages $1.20 per gallon and you drive about 15,000 mi each year.

    b. Write and simplify an expression for the amount of money you will save each year if you can increase your car's gas mileage by 10 mi/gal. (*Hint:* First write an expression for the cost of gasoline if the mileage increases. Then find the difference.)

    c. Maeghan Maloney estimates that she can save $200 each year on gasoline costs. Use your variable expression from part (b) to estimate her car's present gas mileage to the nearest mile per gallon.

Simplify each expression.

18. $\dfrac{2}{5cd} - \dfrac{3}{10cd}$ **19.** $\dfrac{2}{x+8} + \dfrac{6}{x-3}$ **20.** $\dfrac{y+6}{9y^2} - \dfrac{7}{6y}$

21. $\dfrac{2}{x+1} + 3$ **22.** $5 - \dfrac{1}{x}$ **23.** $\dfrac{b^2}{4a} - \dfrac{c}{a}$

24. a. Simplify each expression. Describe any patterns you see.

$$\frac{1}{3} - \frac{1}{6} \qquad \frac{1}{4} - \frac{1}{8} \qquad \frac{1}{5} - \frac{1}{10} \qquad \frac{1}{6} - \frac{1}{12}$$

    b. Write and simplify a variable expression to verify your pattern.

25. TRANSPORTATION Michael and Danielle Evans are driving 250 mi to the Super Bowl. Michael wants Danielle to drive faster to save time.

    a. Write an expression that gives the time saved if the car's average speed s is increased by 5 mi/h.

    b. How much time will they save if Danielle averages 55 mi/h rather than 50 mi/h?

26. CARS If a car goes around a turn too quickly, it will create *yaw marks* on the pavement. A yaw mark looks like an arc of a circle. Accident investigators can use the radius r of the circle and the drag factor f of the road surface to tell how fast the car was going.

To find the radius r of a yaw mark, investigators pick two points A and B on the arc and measure the distance d between them. Then they measure the perpendicular distance m from the midpoint of segment AB to the arc.

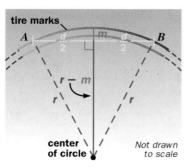

    a. Use the Pythagorean Theorem to write an equation relating d, m, and r. Solve for r.

    b. Find the radius r if $d = 35$ ft and $m = 3.5$ ft.

    c. For a level road, the equation $S \approx 3.86\sqrt{rf}$ gives the speed of the car in miles per hour. Find S if $f = 0.76$. Use your value for r from part (b).

27. a. Open-ended Problem Simplify the expression $\dfrac{1}{x} + \dfrac{1}{4x}$.

b. Find two more rational expressions whose sum is the same as your answer to part (a).

28. SAT/ACT Preview If $A = \dfrac{1}{x} + \dfrac{1}{y} + 1$ and $B = \dfrac{x + y}{xy}$ for positive integers x and y, then:

   A. $A > B$ **B.** $B > A$ **C.** $A = B$ **D.** relationship cannot be determined

29. Challenge Shannon added two rational expressions with different denominators. After simplifying, her answer was $\dfrac{2x + 5}{x^2 + 5x}$. What might the original expressions have been? How did you get your answer?

ONGOING ASSESSMENT

30. a. Open-ended Problem Use a pattern to create several numerical expressions as in Exercises 15 and 24. Simplify each expression and describe any patterns you see.

b. Write and simplify a variable expression to verify your pattern.

SPIRAL REVIEW

Solve each equation. If the equation has no solution, write *no solution*.
(Section 11.3)

31. $\dfrac{3}{x} - 8 = \dfrac{5}{x}$ **32.** $\dfrac{5}{y} + 1 = \dfrac{1}{2}$ **33.** $\dfrac{5r - 4}{2(r - 4)} + \dfrac{9}{2} = \dfrac{8}{r - 4}$

Solve each equation for the variable in red. *(Section 11.4)*

34. $p = 2l + 2w$ **35.** $x = 2xy + y$ **36.** $\dfrac{1}{a} + \dfrac{1}{b} = c$

ASSESS YOUR PROGRESS

VOCABULARY

simplest form of a rational expression (p. 489)

Simplify each expression. *(Sections 11.5 and 11.6)*

1. $\dfrac{2t}{5t + 15} \cdot \dfrac{t + 3}{t}$ **2.** $\dfrac{x + 1}{3x^2} \cdot \dfrac{6x}{5x + 5}$ **3.** $\dfrac{3y - 3}{y + 1} \div \dfrac{y - 1}{y}$

4. $\dfrac{y}{y + 1} + \dfrac{5}{y + 1}$ **5.** $\dfrac{3z}{z + 1} - \dfrac{5}{2(z + 1)}$ **6.** $\dfrac{3}{m + 2} + \dfrac{6}{m(m + 2)}$

7. Journal How is subtracting rational expressions like subtracting fractions? How is dividing rational expressions like dividing fractions? How are they different?

Balancing Weights

Have you ever ridden a seesaw or teeter-totter? Often you must adjust your position on the board to balance with the person on the other end. In the picture, which way should the person on the left move to make the seesaw balance?

PROJECT GOAL Explore the relationship between weights and their distances from the center of a balance.

Making a balance

Work with a partner to make a balance, as shown.

1. USE loops of string to hang plastic bags from a yardstick.

2. HANG the balance using string and a binder clip attached to the middle of the yardstick.

3. PUT different numbers of marbles in the bags.

4. SLIDE the bags until they balance.

Exploring Inverse Variation

1. Let b = the distance from bag B to the center of the balance. Record the distance b to the nearest $\frac{1}{4}$ in., and record the number of marbles in bag B.

2. Use tape to hold bag A in place. Put a different number of marbles in bag B and move it until both bags are balanced again. Record the distance b and the number of marbles in bag B. Repeat this step at least six more times.

3. Make a scatter plot of your data. Write an equation for the distance b as a function of the number of marbles in bag B.

Exploring Rational Expressions

4. Add bag C to the balance as shown at the right.

5. Adjust the three bags until they balance. Record the number of marbles in each bag and measure the distances *a*, *b*, and *c*. Repeat Step 4 at least 7 times, putting different numbers of marbles in the bags for each trial.

6. For each trial, make the calculations shown at the right. What do you notice?

7. Suppose the bags are balanced and contain different numbers of marbles. If you add *x* marbles to each bag, how far must you move bag A so they balance again? Write and simplify a variable expression.

Bag A

Trial	Number of marbles × Distance *a* (in.) = Product		
1	12	16	192
2			
3			
4			
5			
6			
7			
8			

Bag C

Trial	Number of marbles × Distance *c* (in.) = Product		
1	11	12	132
2			
3			
4			

Bag B

Trial	Number of marbles × Distance *b* (in.) = Product		
1	20	2.75	55
2			
3			
4			
5			

Trial	Product for A	Product for B	+	Product for C	=	?
1	192	55	+	132	=	187
?	?	?		?		?

Summarizing Your Results

Write a report about your experiments. Include your data, calculations, and results.

Remember to:

- Explain how the number of marbles in bag B affects the distance at which it balances.

- Explain how you constructed the variable expression in Step 7, and show how you simplified it.

You may want to extend your report to include some of the ideas below:

- Find the weight of an object by balancing it with available objects of known weight.

- Ask your science teacher to show you a beam scale. Explain how it works.

Self-Assessment

In your report, describe what you learned while completing this project. You may want to mention things you learned about balances as well as things you learned about how you approach problems and how you work with others.

11 | Review

STUDY TECHNIQUE

Look over the sections in the chapter. List the key skills presented in each section. For each skill, solve a related exercise from the text. If possible, choose an exercise that you have not already worked on.

VOCABULARY

inverse variation (p. 464)
constant of variation (p. 464)
weighted average (p. 470)

rational expression (p. 476)
rational equation (p. 476)
simplest form of a rational expression (p. 489)

SECTION 11.1

If two variables have a constant, nonzero product, they show **inverse variation**.

x	y
12	16
32	6
4	48
−24	−8

$12 \cdot 16 = 192$

$32 \cdot 6 = 192$

$4 \cdot 48 = 192$

$-24(-8) = 192$

$xy = 192, \quad y = \dfrac{192}{x}$

$$y = \frac{192}{x}$$

The graph does not cross the axes.

SECTIONS 11.2 and 11.3

In a **weighted average**, the data values have different weights.

Quiz	82
Midterm	90
Quiz	70
Final	80

The midterm and final count **twice** as much as each quiz. Weighted average:

$$\frac{1(82) + 2(90) + 1(70) + 2(80)}{1 + 2 + 1 + 2} = 82$$

You can solve a **rational equation** by multiplying both sides of the equation by the **least common denominator (LCD)**.

$$\frac{5}{x} = \frac{1}{2-x} + \frac{3}{2(2-x)}$$

$$2x(2-x)\frac{5}{x} = 2x(2-x)\frac{1}{2-x} + 2x(2-x)\frac{3}{2(2-x)}$$

$$20 - 10x = 2x + 3x$$

$$x = \frac{4}{3}$$

You can also use a graph to solve the rational equation.

$$y = \frac{1}{2-x} + \frac{3}{2(2-x)} \qquad y = \frac{5}{x}$$

Intersection
X=1.3333333 Y=3.75

Remember to check your answer.

The solution is about 1.33.

SECTION 11.4

To solve a formula for a variable, you need to get the variable alone on one side of the equation. Perform the same operations on both sides.

To solve for y, first multiply both sides by the LCD, **2xy**, and simplify.

$$\frac{1}{x} + \frac{1}{y} = \frac{1}{2}$$

$$2y + 2x = xy$$

$$2y - xy = -2x$$

Move all of the y-terms to one side.

$$y(2 - x) = -2x$$

Factor out y.

$$y = \frac{-2x}{2-x}$$

SECTIONS 11.5 *and* 11.6

1. To divide by a rational expression, multiply by the reciprocal.

2. Multiply the numerators and multiply the denominators.

3. Simplify if possible.

$$\frac{2(n+1)}{3} \div \frac{n+1}{2} = \frac{2(n+1)}{3} \cdot \frac{2}{n+1}$$

$$= \frac{2 \cdot 2(n+1)}{3(n+1)}$$

$$= \frac{4}{3}$$

1. To add or subtract rational expressions, rewrite each using the **LCD**.

2. Then add or subtract the numerators. Simplify if possible.

$$\frac{1}{3r^2} + \frac{1}{2r} = \frac{2}{6r^2} + \frac{3r}{6r^2}$$

$$= \frac{2 + 3r}{6r^2}$$

11

Assessment

Vocabulary Review

1. Explain what is meant by a weighted average. Give an example of a situation in which you might use a weighted average.

2. Write an equation that is an example of inverse variation.

3. Give an example of a rational expression that is not in simplest form.

Sections 11.1 *and* 11.2

CONSUMER ECONOMICS The math department at Stanley High School received a $6000 grant to buy calculators. The number of calculators the department can buy varies inversely with the cost per calculator, as shown in the table. Use the table for Questions 4–8.

4. Complete the table.

5. **Writing** Explain why this situation is an example of inverse variation.

6. Write an equation showing how the number of calculators the math department can buy varies inversely with the cost per calculator.

Cost per calculator	Number of calculators
$100	?
$60	?
$50	?
$30	?

7. Graph the equation from Question 6. Be sure to show all quadrants. Which parts of the graph make sense in the situation described above?

8. If the math department wants to buy 250 graphing calculators, how much can the department afford to pay for each one?

9. Jeanne's grades in French class are shown at the left. Each test is worth two quizzes, and the final is worth three quizzes. What must Jeanne earn on the final to get at least a 92 average for the semester?

Jeanne's French Grades	
Quiz 1	92
Test 1	96
Quiz 2	89
Test 2	87
Quiz 3	91
Final	?

10. **Technology** Peder Anderson is designing an apartment complex. He plans to devote 11,500 ft^2 to two-bedroom apartments that bring in $.93/ft^2 in rent each month. Peder will also include one-bedroom apartments that have a monthly rent of $1.03/ft^2. How many square feet of space should Peder allow for one-bedroom apartments if he wants to receive an average rent of $1.00/ft^2?

Solve each equation. If the equation has no solution, write *no solution*.

11. $7 = \dfrac{4 - x}{2 + 3x}$

12. $\dfrac{3}{2} - \dfrac{x + 4}{x + 2} = \dfrac{5}{2(x + 2)}$

13. $\dfrac{4}{x - 5} - \dfrac{3}{x} = \dfrac{20}{x(x - 5)}$

14. $\dfrac{2}{r - 1} - \dfrac{3}{2r} = \dfrac{5}{4(r - 1)}$

Solve each equation for the variable in red.

15. $3x + 4y = 12$

16. $\dfrac{1}{m} + \dfrac{1}{n} = 2$

17. **Technology** Rewrite the equation $2x = \dfrac{4y}{y + 2}$ so it can be graphed on a graphing calculator. Then graph it.

18. GEOMETRY Triangles and squares are examples of *polygons*. The figure at the right is a 5-sided polygon. The formula $S = (n - 2)180$ gives the sum, S, of the measures of the angles inside any polygon with n sides. Solve for n and find the value of n for which $S = 1260$.

150°
60°
80°
140°
110°
n = 5

Simplify each expression.

19. $\dfrac{15ab}{4} \cdot \dfrac{2b^2}{3a^2}$

20. $\dfrac{8c}{6c + 2} \cdot \dfrac{3c + 1}{5}$

21. $\dfrac{2t + 7}{9t} \div (6t + 21)$

22. $\dfrac{p^2 - 4}{8} \div \dfrac{p + 2}{p - 2}$

23. $\dfrac{5t - 1}{3t} + \dfrac{t + 1}{3t}$

24. $\dfrac{3k + 4}{k + 2} - \dfrac{2k - 1}{k + 2}$

25. $\dfrac{y + 1}{3y^2} + \dfrac{1}{6y}$

26. $\dfrac{3}{z + 1} - \dfrac{5}{z}$

27. $\dfrac{n}{n - 7} - \dfrac{n + 7}{n}$

28. Open-ended Problem Give an example of each type of equation: linear, quadratic, polynomial, rational, and exponential.

PERFORMANCE TASK

29. Ask your teacher how he or she assigns final grades. Determine what your final grade will be as a function of your grade on the final exam. You may need to make some assumptions about your grades for class participation, homework, etc.

12 Discrete Mathematics

Animating the Imagination

INTERVIEW **Phil Roman**

Growing up in Fresno, California, during the Great Depression of the 1930s, Phil Roman picked grapes with his family to earn money for food. His life changed when he turned eleven and saw the Disney movie *Bambi* for the first time. "I was captured by those images on the screen," Roman recalls. "Here were drawn animals that you could relate to. It really struck a chord in me." After he left the movie theater, he began to draw figures of Bambi everywhere—on school notebooks and any scrap of paper he could lay his hands on. "That was the start of it all," he says. "I read books on animation and gradually pointed myself in that direction."

> "I love to draw, to bring characters to life, to animate."

A Dream Come True

Roman arrived in Hollywood at the age of 18 with $60 in his pocket and an urge to draw. He took a job as an usher at a movie theater, using half his life savings as a down payment for art school. His dream came true a year later when he landed a job at Disney. In 1984, after almost three decades working at several different studios, he started his own company. Initially just "three people in a small room," the company is now the largest independent animation studio in Hollywood. Its productions include popular television shows, such as *Garfield and Friends* and *The Simpsons.*

> " **Animation takes ... an understanding of math to make it look natural.** "

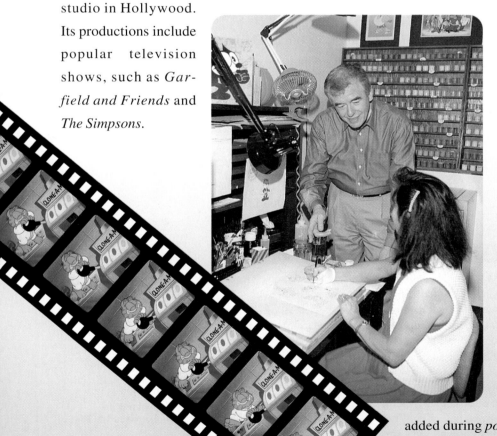

From Script to Film

As studio head, Roman oversees every aspect of producing a film. The starting point is a script that describes the action and characters. Next to be created is a *storyboard*— an elaborate comic strip that tells the story in a visual way. Animators then create thousands of drawings that are photographed by a movie camera, frame by frame. Music and sound effects are added during *post-production*, with some last-minute cutting and trimming done in the editing stage. Finally, the lights go dim, the projector rolls, and the characters come to life.

Creating an Illusion

When you watch an animated film, the action on the screen looks smooth and seamless. But if you examine the film itself, you will see a string of pictures that are each slightly different. The frames are *discrete*. When the film is projected at a speed of 24 frames per second, your eye is tricked into seeing the images as continuous movement. The challenge for an animator is to choose the series of pictures that will model continuous motion. In *discrete mathematics*, you will use such techniques as choosing the correct series of steps for solving a problem and counting the possibilities at each stage in a process. "When I got into animation, I had no idea that the stuff I learned in high school would be

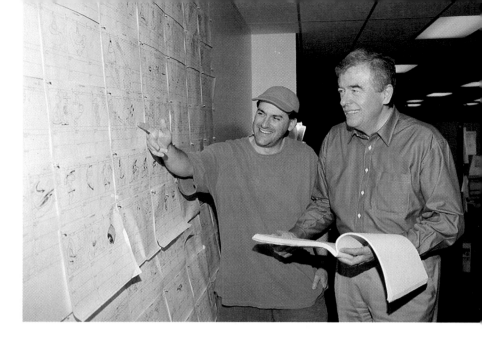

so important," Roman says. "But the math comes in very handy. I use it every day." Although the movement in animated films seems natural, that doesn't just happen, he explains. "It takes hard work, a lot of planning, and an understanding of math to make it look natural."

EXPLORE AND CONNECT

Phil Roman with two animated characters.

1. Research There are many different techniques used to create animated films. Some animators create thousands of drawings. Some use puppets, clay figures, and other three-dimensional objects. Some use computer animation. Find out more about one of these or another animation technique. Report your findings.

2. Project Create an animated flip book. Each page of a flip book shows a picture that is slightly different from the one before it. When you flip through the pages quickly, you see a short piece of animation.

3. Writing Have you noticed any differences between animation in movies and animation in television shows? Describe the differences.

Mathematics & Phil Roman

In this chapter, you will learn more about how mathematics is related to animation.

Related Explorations and Exercises

Section 12.1
• **Exploration**
• **Exercises 14–16**

12.1

Exploring Algorithms

Learn how to...

- **write and use algorithms**

So you can...

- **understand real-world problems such as how to animate an action; or**

- **assign shelf keeping units**

Animators like Phil Roman break simple motions into discrete steps creating sequences of pictures that give the appearance of natural motion on film. You can write an *algorithm* to break a motion into steps. An **algorithm** is a step-by-step method for accomplishing a goal.

Writing an Algorithm

Work in a group of 3–5 students.

1 Work together to write an algorithm that explains how to sit down in a chair. Make each step as clear and precise as possible.

2 Have one of the members of your group follow your algorithm exactly. Did each step work the way you planned?

3 If necessary, revise your algorithm. Have someone in your group test your new algorithm.

4 Compare your algorithm with those of other groups. Are there different steps you could use, or are all of the algorithms basically the same?

Algorithms break a complicated task into discrete steps. If you have ever followed a recipe, you have used an algorithm. If you have ever programmed a calculator or computer, you have written an algorithm.

EXAMPLE   Application: Fundraising

The drama club took orders for 204 plants to raise money. Danielle needed to determine how many plants each of the 17 club members should deliver. She tried to use her calculator, but found that the division key was broken. Write an algorithm Danielle could use to find a solution without using the division key. Use your algorithm to find a solution.

SOLUTION

Method 1 Write an algorithm to use the subtraction key.

Step 1 Subtract 17 from 204.

Step 2 Subtract 17 from the remainder.

Step 3 Repeat Step 2 until the number left is less than 17.

Step 4 Count how many times 17 was subtracted.

You need to subtract 12 times. There is no remainder. Each student should deliver 12 plants.

Method 2 Write an algorithm using multiplication and guess and check.

In Steps 3–5, if your answer is *no*, skip to the next step.

	Algorithm	First try	Second try	Third try
Step 1	Make a guess.	Guess 15.	Guess 10.	Guess 12.
Step 2	Multiply your guess by 17.	$15 \cdot 17 = 255$	$10 \cdot 17 = 170$	$12 \cdot 17 = 204$
Step 3	Is the result correct? You are done.	No	No	12 is correct. Done.
Step 4	Is the result too big? Try again with a smaller number.	15 is too big. Try again.	No	
Step 5	Is the result too small? Try again with a larger number.		10 is too small. Try again.	

BY THE WAY

The word *algorithm* comes from Persia. It is derived from the name of the mathematician Muhammad ibn Mūsā al-Khwārizmī, who lived around 780–850 A.D.

Each student should deliver 12 plants.

THINK AND COMMUNICATE

For Questions 1–3, look at the Example on pages 510 and 511.

1. Which of the two algorithms do you prefer? Why?

2. Suppose there were 1230 plants to distribute. Which algorithm would you prefer? Why?

3. How could you change the algorithm in Method 2 to make it work for a problem with a remainder?

☑ CHECKING KEY CONCEPTS

For Questions 1–3, refer to the algorithm for making scrambled eggs.

Step 1 Take the eggs out of the refrigerator and put them in a bowl.

Step 2 Mix them all up together and pour them in a pan.

Step 3 Stir them around as they cook.

1. What could go wrong if you follow this algorithm exactly?

2. Which steps in this algorithm need to be clarified? What steps would you add to this algorithm?

3. **Cooperative Learning** Work with a partner to write a clear, concise, and complete algorithm for making scrambled eggs.

4. Look back at the Example on page 510. Write an algorithm that uses the calculator's addition key to solve Danielle's problem.

12.1 Exercises and Applications

Extra Practice exercises on page 581

For Exercises 1–3, write an algorithm that could be used to program a robot or computer to complete each task.

1. Make your favorite sandwich.

2. Use two given points to write an equation of a line.

3. Graph the equation $y = \dfrac{1}{x + 2}$ on a graphing calculator.

4. Look back at Exercises 20–23 on page 163. Write an algorithm for the method described by the Ahmes papyrus.

5. **Writing** How would you change the two algorithms in the Example on pages 510 and 511 if 19 students were going to help deliver plants?

6. **FINE ARTS** Santana Martinez makes coil pots in the traditional Pueblo style. The photographs show some of the key steps she follows. Put the steps in the correct order. Write an algorithm for the process.

A.   B.   C.

D.   E.   F.

BY THE WAY

Santana Martinez's mother-in-law, Maria Martinez, rediscovered techniques used by the Pueblos to make black-on-black pots eight centuries earlier. Maria Martinez was the first Pueblo potter to sign her work.

For Exercises 7–9, write an algorithm that explains how to use a calculator to solve each problem.

7. Divide 326 by 52 without using the division key.

8. Multiply 13 by 34 without using the multiplication key.

9. Compute 4^8 without using the exponent key.

10. **Open-ended Problem** Write an algorithm for a common activity, such as tying your shoe or climbing a flight of stairs.

11. **Challenge** Write an algorithm for playing tic-tac-toe.

12. **CARS** The algorithm below describes a method for computing fuel mileage.

Step 1   Drive the car until it runs out of fuel.

Step 2   Put exactly one gallon of fuel in the tank.

Step 3   Record the mileage on the odometer.

Step 4   Drive until you run out of fuel again.

Step 5   Record the mileage on the odometer. Subtract the old figure from the new figure. This is the rate of fuel consumption in miles per gallon.

a. Comment on this algorithm. Will it work? Is it practical?

b. Write a better algorithm to compute a car's fuel mileage.

13. **Open-ended Problem** Suppose you have a computer program that can factor numbers into primes. Write an algorithm that uses this computer program to find the least common multiple of two numbers.

INTERVIEW **Phil Roman**

Look back at
the interview
on pages 506–508.

Cooperative Learning *Suppose an animation studio wishes to include a sequence on making a cake in an upcoming film. You are part of the team of animators working on this segment. Work in a group of three students.*

14. Write an algorithm for making a cake. Think about how the process appears to someone watching the action.

15. Illustrate each step in your algorithm to make a storyboard.

16. Estimate the time each step will take in the film. At a rate of 24 frames per second, how many frames will be required for each step? for the entire sequence?

Phil Roman reviews the storyboards for an animated film.

Your storyboard can be made by drawing pictures, or you might want to use photos from magazines.

Bob Brill is an algorithmic artist. Instead of drawing, he writes computer programs to create art. Brill repeatedly applied the P2 algorithm to transform the original image (shown at the right) into the images below. Use this algorithm for Exercises 17–19.

Bob Brill's original image

P2 Algorithm

Step 1 Divide the *starting* rectangle into rows and columns of squares.

Step 2 Use the rules below to move the squares from the *starting* rectangle to a *new* rectangle with the same dimensions as the starting rectangle.

Rule 1 Start at the top left of the *starting* rectangle. Remove the pieces of the top row from left to right. When a row is empty, start at the left end of the next row.

Rule 2 Start at the top right of the *new* rectangle. Place the squares by column from top to bottom. When a column is full, start filling the column to its left.

Algorithm applied 254 times.

Algorithm applied 341 times.

17. The images below show the result of applying the P2 algorithm to a rectangle several times. Sketch the results of applying the P2 algorithm to the numbered rectangle three, four, five, and six times.

18. Use the patterned rectangle. Sketch the results of applying the P2 algorithm three, four, five, and six times.

19. Create your own pattern in a rectangle. Cut it into equally sized squares. Apply the P2 algorithm to your pattern until it repeats. Sketch the original image and the result of several applications of the algorithm.

20. **Open-ended Problem** Create your own algorithm for changing a pattern. Explain your algorithm. Does the pattern return to its original form if you repeat the algorithm enough times?

By the Way shows the algorithm applied 512 times.

BUSINESS A mail order company is organized to fill customers' orders quickly. Each item is assigned a shelf keeping unit (SKU). The letters and digits of the SKU tell the packer where to find the item.

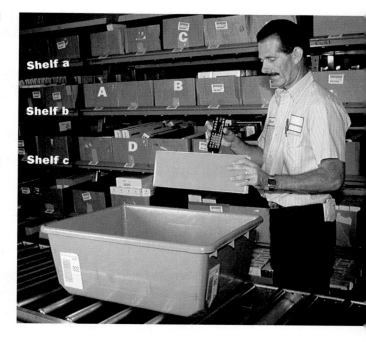

Shelf a

C

A   B

Shelf b

D

Shelf c

21. In one company, the first two digits of the SKU indicate a shelving unit. The lower case letter indicates a shelf. The last two digits indicate the position on the shelf. Match each SKU with a lettered package on shelving unit 03 in the photo at the right.

03-b-03   03-c-02

03-a-03   03-b-01

22. **Writing** When assigning SKUs, the company considers how well the item is expected to sell, as well as the weight and dimensions of the package. Explain how each of these would affect where you would shelve an item. Write an algorithm for assigning SKUs.

O N G O I N G   A S S E S S M E N T

23. **Open-ended Problem** The results of using an algorithm to solve the equation $12x + 9 = 45$ are given below:

Try $x = 10$.	$12(10) + 9 = 129$	Too big!
Try $x = 1$.	$12(1) + 9 = 21$	Too small!
Try $x = 5$.	$12(5) + 9 = 69$	Too big!
Try $x = 3$.	$12(3) + 9 = 45$	That's it!

a. Write an algorithm that could have given these results. Do you think this is a good algorithm to use?

b. Write two different algorithms that you could use to solve the same equation.

S P I R A L   R E V I E W

Graph each inequality. *(Section 7.5)*

24. $y > 5$

25. $y \leq 2x - 4$

26. $4x - 5y < -10$

Simplify each expression. *(Section 11.5)*

27. $-\dfrac{3x}{2y} \cdot \dfrac{2}{5x}$

28. $\dfrac{x}{y} \div 2x$

29. $\dfrac{21r}{20s^2} \cdot \dfrac{15s^3}{7r}$

30. $\dfrac{x^2 + 2}{x} \div \dfrac{x + 2}{3x}$

31. $\dfrac{xy}{4x - 4y} \cdot \dfrac{8x + 2y}{x^2}$

32. $\dfrac{3x + 1}{x - 2} \div \dfrac{x^2}{x^2 - 4}$

12.2 Finding Paths and Trees

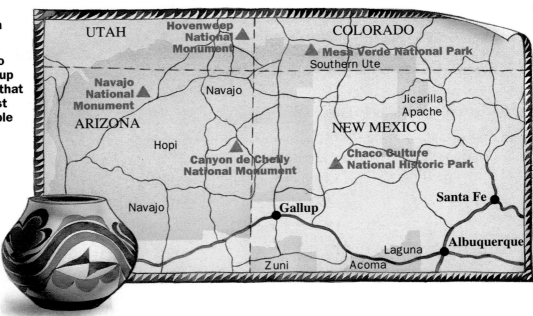

Learn how to...

- **find the best path or tree for a graph**

So you can...

- **minimize the distance you must travel on a trip**
- **find the way to connect a group of computers that takes the least amount of cable**

How do you decide where to go first when you have several places to go? You might want to arrange your trip so that you travel the shortest distance. You can use discrete math to find a good *path*.

A group of students is planning a trip to visit five Anasazi sites. The students found information about the sites and located them on a map. They want to find a good route to use to visit all of the sites.

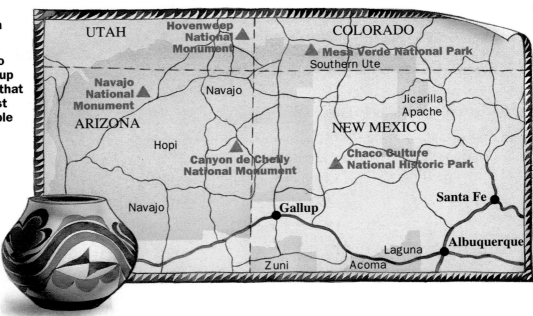

The students decided to simplify the map by modeling it with a **graph** consisting of points, called **vertices**, and lines, called **edges**. The graph below models the Anasazi sites and the distances between them.

Each site is modeled by a **vertex** of the graph.

The route between any two sites is modeled by an **edge** of the graph. Edges are not drawn to scale.

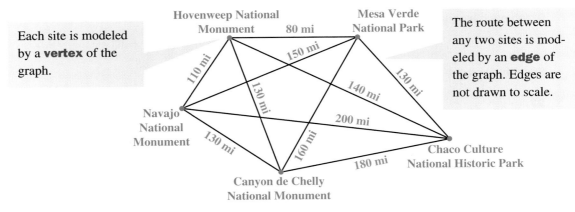

516 Chapter 12 *Discrete Mathematics*

A **path** among the vertices on a graph is a collection of edges that connect one vertex to another without repeating any vertex or edge. Example 1 uses a *greedy algorithm* to find a short path. A **greedy algorithm** always chooses the best edge from the ones currently available.

EXAMPLE 1

Use a greedy algorithm to find a short path for the students to use to visit each of the sites in the graph on page 516. How far will the students travel if they use this path?

SOLUTION

Step 1 Choose a site from which to start.

Step 2 Go to the closest site that you have not visited.

Step 3 Continue choosing the next closest site until all of the sites have been visited.

The algorithm suggests visiting the sites in this order: Hovenweep, Mesa Verde, Chaco Culture, Canyon de Chelly, Navajo.

Find the sum of the lengths of the edges used.

$$80 + 130 + 180 + 130 = 520$$

The students will travel 520 miles if they use this path.

The only known way to find the *best* path is to try all the possible paths. A greedy algorithm will usually find a good path.

THINK AND COMMUNICATE

1. Use the graph in Example 1. If you start at Canyon de Chelly, there are two sites that are the same distance from your starting point. Use a greedy algorithm to determine which of the two choices is better.

2. Why is *greedy algorithm* a good name for the steps used to find a good path through a graph?

3. Why is the algorithm used in Example 1 a greedy algorithm?

Trees

A **tree** connects all the vertices in a graph with the smallest possible number of edges. You can use a greedy algorithm to find the best tree for a graph.

This is not a tree. One vertex is not connected to an edge.

This is a tree. All of the vertices are connected. There are no extra edges.

This is not a tree. The extra edge between the vertices creates a loop.

EXAMPLE 2

In a small office, five employees need to connect their computers to each other and to a shared printer. The map shows the lengths of all of the possible connections between the computers and the printer. Find the shortest tree to connect the computers to the printer.

SOLUTION

Model the map with a graph. Use a greedy algorithm.

Step 1 Choose the shortest edge. If there is a tie, choose one.

Step 2 Choose the next shortest edge. Do not make any loops.

Step 3 Repeat Step 2 until all the vertices have been connected.

The length of the shortest tree is
$6 + 7 + 8 + 10 + 12 = 43$ ft.

4. Compare Example 1 to Example 2. How is a path different from a tree? Explain why the graph at the right is both a path *and* a tree.

5. How is the algorithm in Example 1 different from the one given in Example 2? How are they alike?

☑ CHECKING KEY CONCEPTS

Tell whether each graph is a *path*, a *tree*, *both*, or *neither*.

1.

2.

For Questions 3–5, copy the graph at the right.

3. Use a greedy algorithm to find a short path from vertex *A* to vertex *B*. Your path does not need to visit all the vertices. How long is the path?

4. Start at vertex *A*. Find a path through all of the vertices. How long is it? Try starting at other vertices. Can you always find a path?

5. Find a tree for this graph. How long is it? Is it the shortest tree? How do you know?

12.2 | **Exercises and Applications**

Tell whether each graph is a *path*, a *tree*, *both*, or *neither*. Explain why.

Extra Practice exercises on page 581

1.

2.

3.

4.

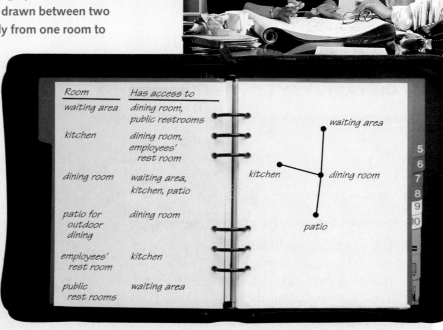

Architects may use an *access graph* to model the floor plan of a building. In an access graph, each vertex represents a room. An edge is drawn between two vertices if you can walk directly from one room to the other.

5. A client gives an architect a list of preferences for the design of a restaurant.

   a. Copy and complete the access graph for the restaurant.

   b. Is there more than one possible graph? Explain your reasoning.

Room	Has access to
waiting area	dining room, public restrooms
kitchen	dining room, employees' rest room
dining room	waiting area, kitchen, patio
patio for outdoor dining	dining room
employees' rest room	kitchen
public rest rooms	waiting area

For Exercises 6 and 7, draw an access graph to model each plan of a traditional dwelling unit in Burkina Faso in west Africa.

6.

7.

a traditional home in Segenega, Burkina Faso

For Exercises 8 and 9, draw an access graph to model each house plan.

8.

9.

10. **Writing** How are the access graphs for Exercises 8 and 9 alike? Do you see any patterns? How are the house plans different?

For Exercises 11–14, copy the graph below.

11. Find a short path from vertex *A* to vertex *B*. How long is this path?

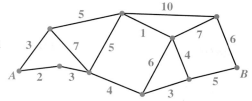

12. Start at vertex *A*. Find a path through all of the points. How long is this path?

13. Find the shortest tree for the graph.

14. Write a greedy algorithm to find the *longest* tree for a graph. Use your algorithm to find the longest tree for the graph.

15. **SAT/ACT Preview** Evaluate the expression $7^2 + 2(7)(3) + 3^2$.

   A. 42   **B.** 100   **C.** 1764   **D.** 1000   **E.** 84

BUSINESS **For Exercises 16–18, use the map of the camping area.**

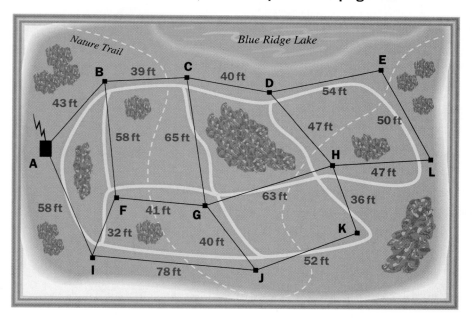

16. Dave Stapp owns a camping area. He is installing power lines to provide electricity to each site. He plans to locate the power source at *A*. The power lines will run along existing roads.

   a. Does Dave need to find a *path* or a *tree* to connect the sites?

   b. Use a greedy algorithm to decide where Dave should put his power lines. How many feet of power line does Dave need?

17. **Writing** The power source could also be located at site *I*. Would changing the location of the power source affect the amount of power line Dave needs? Explain your answer.

18. **Challenge** Dave checks each camp site every day. Find a short path through all of the sites. Does it matter where the path begins and ends?

Mary Longacre is installing a new computer system at six regional offices of her company. Her travel agent gave her prices for routes between each of the cities she will visit.

Airfare Between Cities (dollars)						
	Bos	Dal	Phnx	Atl	Chi	S F
Boston	0	654	671	461	304	734
Dallas		0	456	361	461	576
Phoenix			0	556	565	644
Atlanta				0	356	669
Chicago					0	712
San Francisco						0

19. Draw a graph that models the cost of traveling between the cities.

20. a. Does Mary need to find a *path* or a *tree* for her trip?

  b. Use a greedy algorithm to find a low-cost route through all six of the cities.

  c. What is the total cost for the trip?

21. Open-ended Problem Do you think that the route with the lowest cost will always be the same as the route that takes the shortest time? Explain your reasoning.

BY THE WAY

People in sales and industry are often concerned with finding the best route through a city or group of cities. Mathematicians often call this type of problem a *Traveling Salesman Problem.*

ONGOING ASSESSMENT

22. Open-ended Problem Think of a situation where you have a choice of paths between several places. Model the situation with a graph. How do you decide which is the shortest route? Does a greedy algorithm give you the same route?

SPIRAL REVIEW

Solve each equation. (*Section 4.4*)

23. $11.5x - 3.8 = 12.3$ **24.** $\dfrac{3}{0.7}x + \dfrac{4}{1.4}x = 20$ **25.** $\dfrac{8}{11} - \dfrac{5}{3}y = y$

Solve each equation by factoring. (*Section 10.5*)

26. $x^2 + 2x - 35 = 0$ **27.** $2x^2 - 5x - 3 = 0$ **28.** $3a^2 + 5a = 2$

29. $4y^2 - 25 = 0$ **30.** $6a^2 - 11a + 3 = 0$ **31.** $4y^2 - 4y = 3$

12.3

Voting and Fair Division

Learn how to...

- recognize fairness in elections and divisions

So you can...

- choose the best voting method for school elections

You are constantly faced with deciding how to be fair. You may need to decide how to divide something up among several people. You may be part of a group that has to make a choice.

Rocky Mount School is having a student council election. The dress code is the biggest topic in the campaign. Two of the candidates, Jack and Denzel, conducted a poll using the questionnaire shown below.

What kind of dress code do you think the school should have?
(Check one.)

1. No dress code ❑
2. Minimal dress code ❑
3. Moderate dress code ❑
4. Detailed dress code ❑
5. Strict dress code ❑
6. School uniform ❑

Jack and Denzel made a bar graph of the poll results. They asked the other candidates which type of school dress code they would support and added this information to their bar graph. Then they predicted the election results.

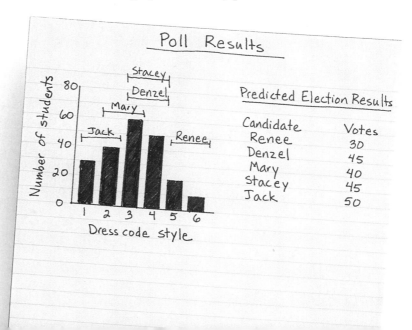

Poll Results

Predicted Election Results	
Candidate	Votes
Renee	30
Denzel	45
Mary	40
Stacey	45
Jack	50

THINK AND COMMUNICATE

For Questions 1–3, use the results of Jack and Denzel's poll on page 523.

1. Explain why Jack and Denzel predicted that Renee will receive 30 votes. Why do they predict that Mary will receive 40 votes?

2. Based on the bar graph, which candidates represent the views of the most students? Explain your reasoning.

3. Jack and Denzel predict that Jack will win the election. Why might he get more votes than candidates who represent the views of more students?

In a **one person-one vote** election, each person can vote for one candidate. In an **approval voting** election, each person can vote for any number of candidates. In the event of a tie, there is a tiebreaking election.

EXAMPLE 1

Writing Use the results from Jack and Denzel's poll to predict the results of an approval voting election. Do you think the results of an approval voting election would better reflect the opinion of the voters?

SOLUTION

To predict the number of votes for each candidate, I found the number of students that support each candidate's position. For example, Denzel would receive 60+50= 110 votes. Renee would get 30 votes. I predict the election results shown below.

Candidate	No. of votes
Denzel	110
Stacey	110
Mary	100
Jack	70
Renee	30

I think this election will show a tie, since Denzel and Stacey each appeal to the same voters.

This type of voting shows which candidates appeal to the largest number of students. More students will be satisfied with the outcome. I think it is better than one person - one vote in this situation.

THINK AND COMMUNICATE

4. Explain why Renee is predicted to receive 30 votes in both an approval voting election and a one person-one vote election.

5. The student who wrote the solution in Example 1 suggests that an approval voting election would result in a tie. What led the student to make this conclusion? In this situation, do you think that a tie is likely to occur in an approval voting election? Explain why or why not.

6. Discuss which candidate you think *should* win the election.

Voting is a way for a large group of people to decide among a few choices. In a **fair division** problem a group of objects needs to be divided among a group of people. Example 2 shows one type of fair division strategy.

EXAMPLE 2

Amy decided not to take all of her CDs with her to college. She gave the rest of her collection to two of her friends to split. Pam and Maria like all of the CDs. They want to be friendly about the division. How can they split the collection fairly?

SOLUTION

First, Pam should divide the CDs into two groups so that she will be equally happy with either group.

Second, Maria picks the group of CDs that she likes best.

Third, Pam gets the remaining group.

Pam and Maria used the **cut-and-choose** algorithm.

Step 1 The first person decides on a fair way to divide the item(s) into two parts.

Step 2 The second person chooses one of the two parts.

Step 3 The first person gets the remaining part.

BY THE WAY

Dividing things fairly in a large group is much harder than dividing into two pieces. In 1994, mathematicians Steven J. Brams and Alan D. Taylor solved the problem of *envy-free division* for any number of people.

THINK AND COMMUNICATE

7. Why is this way of dividing something fair for the first person? Why is it fair for the second person?

8. In Example 2, Pam might choose to split the CDs so that one group has fewer CDs than the other. Explain why she might call this a fair division of the CDs.

9. Will the cut-and-choose algorithm always work for two people? Explain why or why not.

☑ CHECKING KEY CONCEPTS

Jack and Denzel discovered that they missed a stack of questionnaires when they tallied the results shown on page 523. For Questions 1–3, combine the results from these questionnaires with those used on page 523.

Dress code	1	2	3	4	5	6
Number of votes	0	2	9	8	3	0

1. Predict the results of a one person-one vote election.

2. Predict the results of an approval voting election.

3. Suppose a new candidate who supports the same type of dress code as Jack enters the campaign. Predict the results of both a one person-one vote election and an approval voting election with this new candidate.

12.3 Exercises and Applications

Extra Practice exercises on page 582

For Exercises 1 and 2, tell how to apply the cut-and-choose algorithm in each situation.

1. There is one piece of pie left. You and your sister both want some.

2. You and a friend are being paid to weed a neighbor's flower garden.

3. Writing Explain the difference between a one person-one vote election and an approval voting election.

4. In major league baseball, the Cy Young Award is given to the best pitcher in each league. The winner is chosen by 28 members of the Baseball Writers Association of America. The electors rank their top three choices.

BY THE WAY

The Cy Young Award winner is chosen using a Borda Count. In a Borda Count, voters rank their choices. The Borda Count is named after Jean-Charles de Borda (1733–1799), the French naval captain who invented it.

a. Give each pitcher 5 points for each first-place vote, 3 for each second-place vote, and 1 for each third-place vote. Rank the pitchers based on total points.

b. Suppose the number of points awarded is changed to 3 points for first-place, 2 for second-place, and 1 for third-place. Explain how this affects the order of the ranking of the pitchers.

Cy Young Award			
Pitcher, Team	1	2	3
John Burkett, San Francisco	—	3	—
Tom Glavine, Atlanta	4	7	8
Tommy Greene, Philadelphia	—	—	2
Bryan Harvey, Florida	—	—	1
Greg Maddux, Atlanta	22	2	3
Randy Myers, Chicago	—	—	1
Mark Portugal, Houston	—	—	2
Jose Rijo, Cincinnati	—	1	5
Bill Swift, San Francisco	2	15	6

BUSINESS **For Exercises 5–7, the diagram shows the layout of a small restaurant. The number of seats at each table is indicated.**

5. Open-ended Problem How do you think the manager should decide on a fair distribution of tables among the people waiting on the tables?

6. a. How would you divide up the tables fairly between two people?

   b. What is a fair division of the tables among three people?

7. Challenge Suppose twelve people want to sit together. How can the restaurant seat them? Does this change the division of the tables among three people? Justify the fairness of the division.

Connection   VOLUNTEERING

8. Cooperative Learning Consider each volunteer opportunity.

   a. Rank each type of volunteer work in order of your preference. Give your first choice 4 points, your second choice 3 points, your third choice 2 points, and your fourth choice 1 point.

   b. Predict the ranking of each type of volunteer work in order of your class's preference. Explain your reasoning.

   c. Collect the rankings of other students in your class. Add up the scores. Compare the class ranking to your prediction.

Hospital

Food Bank

9. Open-ended Problem Three students decide to work as a group to do volunteer work. They each rank their interest in the four volunteer opportunities shown in the table. They rank their first choice with a 1. Create a schedule for them to divide their time fairly. Explain your reasoning.

Animal Shelter

Each student can volunteer four hours per week.

Volunteer Opportunity	Tomoko	Samuel	Rita
Hospital	2	2	4
Food Bank	3	3	1
Animal Shelter	1	4	3
Local Cleanup	4	1	2

Local Cleanup

Each volunteer opportunity needs students to help three hours per week.

The President of the United States is not elected directly by the votes of citizens. The final vote is made by the Electoral College. In most states, the candidate who gets the most votes in the state receives all of the Electoral College votes for that state. For Exercises 10 and 11, use the table of election results.

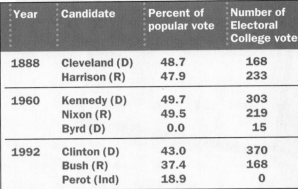

Year	Candidate	Percent of popular vote	Number of Electoral College votes
1888	Cleveland (D)	48.7	168
	Harrison (R)	47.9	233
1960	Kennedy (D)	49.7	303
	Nixon (R)	49.5	219
	Byrd (D)	0.0	15
1992	Clinton (D)	43.0	370
	Bush (R)	37.4	168
	Perot (Ind)	18.9	0

1960 Electoral College Votes

- Kennedy (Democrat)
- Nixon (Republican)
- Byrd (Democrat)

10. For each of the elections in the table, calculate the percent of the Electoral College votes each candidate received. In each election, how does the percent of the Electoral College vote compare to the percent of the popular vote?

11. **Open-ended Problem** Do you think the popular vote or the Electoral College vote better indicates the opinion of the voters? Why?

12. **Research** Find out more about one of the topics below.

Topic A
How is the number of Electoral College votes decided for each state?

Topic B
Dividing areas into voting districts is a complicated fair division problem. What factors are considered in making this division?

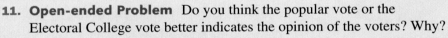

ONGOING ASSESSMENT

13. **Writing** Write a letter to your student council explaining the different voting methods you have learned about in this section. Which method do you think your school should use in student council elections? Explain why the method you suggest is best for your student council elections.

Find the theoretical probability that a randomly thrown dart that hits each target hits the shaded area. *(Section 5.6)*

14.

15.

Write each number in scientific notation. *(Section 9.5)*

16. 10,025

17. 0.0034

18. 54,000,000

19. 0.0010078

20. 9.6 million

21. 3 ten-thousandths

ASSESS YOUR PROGRESS

VOCABULARY

algorithm (p. 509)

graph (p. 516)

vertex (p. 516)

edge (p. 516)

path (p. 517)

greedy algorithm (p. 517)

tree (p. 518)

one person-one vote (p. 524)

approval voting (p. 524)

fair division (p. 525)

cut-and-choose (p. 525)

1. Write an algorithm for adding two trinomials. *(Section 12.1)*

VOLUNTEERING For Exercises 2 and 3, use the graph below. The graph models the locations where donations for a food bank need to be picked up today.

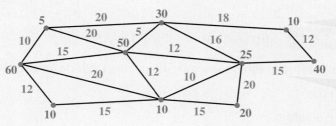

weight in pounds of the donation at this location

time in minutes to drive between two locations

2. Open-ended Problem Find a short path for one person to use to visit each of the locations. *(Section 12.2)*

3. Suppose two people will each pick up some of the donations. Copy the graph and suggest a fair division of the locations. Explain why your division is fair. *(Section 12.3)*

4. Journal Explain what is *discrete* about each of the topics covered in Sections 12.1–12.3.

12.4 Permutations

In Section 12.2, you learned that the only known way to be sure that you have found the best path between two vertices is to try all of the possibilities. How do you know that you have found them all? In this section, you will learn a way of counting all the possibilities.

Learn how to...

- **count all the ways groups of objects can be arranged**

So you can...

- **find all the possible outfits of clothing; or**
- **all possible computer passwords; or**
- **all possible Braille characters**

A fashion designer is planning a new collection of clothing that will look good together in any combination. The collection includes two pairs of pants, four shirts, and three sweaters.

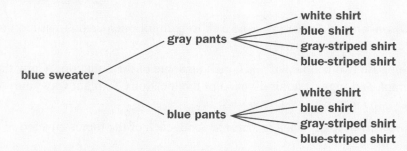

EXAMPLE 1   Application: Fashion Design

How many different three-piece outfits include the blue sweater?

SOLUTION

Use a tree diagram.

```
blue sweater ─┬─ gray pants ─┬─ white shirt
              │               ├─ blue shirt
              │               ├─ gray-striped shirt
              │               └─ blue-striped shirt
              │
              └─ blue pants ─┬─ white shirt
                              ├─ blue shirt
                              ├─ gray-striped shirt
                              └─ blue-striped shirt
```

There are eight different outfits that include the blue sweater.

For Questions 1–3, use the information about the clothing collection on page 530.

1. How would the tree diagram in Example 1 look if you chose a shirt first and then a pair of pants to go with the blue sweater?

2. How many different three-piece outfits include the gray sweater?

3. How many different three-piece outfits are possible if you use all the items planned for the collection?

A tree diagram models all the possible ways to choose items so that you have one item from each of several categories. You can use a tree diagram to count the total number of choices, as you did in Example 1. Another way to find the number of possible choices is to use the *multiplication counting principle.*

Multiplication Counting Principle

When you have *m* choices in one category and *n* choices in a second category, you have *m · n* possible choices.

When you have *r* choices in a third category, you have a total of *m · n · r* possible choices.

EXAMPLE 2 **Application: Fashion Design**

The fashion designer in Example 1 decides to add one pair of pants and two shirts to the collection in colors that coordinate well with those already included. There are now 3 sweaters, 3 pairs of pants, and 6 shirts. What is the total number of different three-piece outfits in the collection?

SOLUTION

Use the multiplication principle.

Sweaters × Pants × Shirts = Number of outfits

3 · 3 · 6 = 54

There are 54 different three-piece outfits.

EXAMPLE 3

Kevin decides to rearrange the letters of his name to use as a computer password. How many different arrangements can he use?

SOLUTION

1st letter	2nd letter	3rd letter	4th letter	5th letter
5 choices	4 choices	3 choices	2 choices	1 choice

$$5 \cdot 4 \cdot 3 \cdot 2 \cdot 1 = 120$$

There are 120 different arrangements of the letters in *Kevin*. One of them is *Kevin*, so he can use 119 of them as passwords.

In Example 3, the product $5 \cdot 4 \cdot 3 \cdot 2 \cdot 1$ can be written $5!$, which is read as *five factorial*. The symbol $!$ indicates a **factorial**. The number $n!$ is the product of all the integers from 1 to n. By definition, $0! = 1$.

In Example 3, each password is a *permutation* of the letters in Kevin's name. A **permutation** is an arrangement of a group of items in a definite order. The order of the items is important. In Example 3, *ekniv* is different from *vnkei*.

For more information on using a calculator to find factorials, see the *Technology Handbook*, p. 602.

Permutations

A group of n objects has $n!$ permutations.	A group of 5 objects has $5!$ permutations.
$n! = n(n-1)(n-2) \ldots (3)(2)(1)$	$5! = 5 \cdot 4 \cdot 3 \cdot 2 \cdot 1$

☑ CHECKING KEY CONCEPTS

1. Use the multiplication counting principle to solve Example 1. How does the multiplication counting principle relate to the tree diagram?

2. Suppose the designer in Example 2 decides to add two more pairs of pants to her collection for a total of five pairs. How many three-piece outfits are now possible?

3. How many different paths are possible through all five Anasazi sites in Example 1 on page 517?

4. **a.** Evaluate $4!$, $7!$, and $10!$. How can you use $7!$ to help find $10!$?

   b. Try finding other factorials with your calculator. How large a factorial can your calculator evaluate?

5. How many different passwords can be made by rearranging the letters in *Claire*?

WATCH OUT! ▶

Your calculator cannot evaluate large factorials. If you get an error message when you try to evaluate a large factorial, write your answer as a factorial.

12.4 Exercises and Applications

The ninth grade at Central High School is about to elect class officers for next year. For Exercises 1 and 2, use the list of candidates.

Extra Practice exercises on page 582

President	Vice President	Secretary	Treasurer
Susan	Leith	Trevor	Bryan
Naomi	Denise	Jeremy	Zara
Anna		Chelsea	
Miguel			

1. Make a tree diagram to show all the possible groups of officers if Susan is elected president. How many different groups are possible?

2. How many possible groups of officers are there in all, including all the presidential candidates?

3. **a. Writing** Make a table giving the value of $n!$ for $n = 1$ to $n = 12$.

   b. Do you think the last digit of every factorial larger than 4! is 0? Explain your reasoning.

4. **Challenge** Show why $6! \times 7! = 10!$.

Connection   BRAILLE

Each character in Braille is a grouping of raised dots. The dots are arranged in a 3-by-2 rectangle. In the photo, the large black circles represent raised dots. The small black circles represent positions in the rectangle that do not contain raised dots. Characters are used to represent letters, words, and punctuation marks.

5. **a.** How many different ways can the raised and unraised dots be grouped to make a Braille character? Explain how you know.

   b. One of these groupings is not used as a Braille character. Why?

6. Braille uses some pairs of characters for numbers and commonly used words. How many different pairs of Braille characters are possible? Explain how you know.

BY THE WAY

Louis Braille was blinded in an accident when he was 3. He invented a method of reading with raised dots in 1824, when he was 15 years old.

The picture at the left shows Helen Keller during a presentation she gave in Paris, France.

King High School is hosting a Model United Nations (MUN). Each participating school will represent one of the countries listed below.

North and Central America	South America	Africa	Europe	Asia
Canada	Brazil	Algeria	Estonia	China
Costa Rica	Chile	Botswana	France	Indonesia
Cuba	Colombia	Chad	Germany	Japan
El Salvador	Peru	Egypt	Spain	Pakistan
Guatemala	Uruguay	Somalia		Thailand
Mexico		Togo		Vietnam
United States		Zaire		
		Zimbabwe		

BY THE WAY

The United Nations was created after World War II. Countries that had cooperated during the war hoped that they could maintain international peace by continuing to work together.

7. The MUN will have an economic council consisting of one representative from each continent. The representative can come from any country. How many different councils are possible?

8. **Writing** After the economic council is chosen, the security and development councils will be picked. Each council will consist of one representative from each continent. No country can be on more than one council. How many different ways can the security council be chosen? the development council? Explain your reasoning.

9. a. **Open-ended Problem** The MUN will have an assembly of 100 members. At least one representative from each country will be in the assembly. The remaining seats must be distributed among the countries. Write an algorithm for fairly distributing assembly seats in any MUN.

   b. Discuss your algorithm with other students. What do the algorithms have in common? How are they different?

10. **Research** Find out about one of the programs run by the United Nations.

11. SPORTS There are nine players on an intramural softball team. In how many different orders can the players go to bat? What factors might limit the number of orders the team actually uses?

12. SAT/ACT Preview Pauline read 24 pages of an assigned book in 20 min before lunch. How long will it take her to read the remaining 60 pages of her assignment if she continues reading at the same rate?

   A. 30 min   **B.** 40 min   **C.** 50 min   **D.** 60 min   **E.** 70 min

 Technology Your calculator has a function $_nP_r$ that will find the number of permutations of *r* items chosen from a group of *n* items. **For Exercises 13–15, use the function $_nP_r$.**

13. The *Mind Rockers* have 15 songs that they perform in concert. At the upcoming Battle of the Bands, they will play three songs. In how many different orders could the band perform three of their songs?

14. Challenge On a sales trip, Jane Gupta needs to visit 10 locations in 2 days. She decides to visit 5 locations each day. How many possible paths can she take the first day? How many possible paths remain on the second day? Explain your reasoning.

15. Open-ended Problem How many seven-digit numbers are possible if no digit is repeated? if digits are repeated? How many seven-digit numbers could be telephone numbers in the United States? Explain how you decided which numbers could *not* be telephone numbers.

ONGOING ASSESSMENT

16. Open-ended Problem Automatic teller machines (ATMs) usually require customers to use a four-digit password. Many computer systems require a password that is at least six characters long and can include letters and numbers.

   a. How many passwords are possible in each system?

   b. What are the advantages of requiring a longer password? What are the disadvantages?

   c. Why do you think banks choose to require a password with only four digits for ATM use?

SPIRAL REVIEW

Solve each system of equations by adding or subtracting. *(Section 7.3)*

17. $3x - y = -3$
$4x + y = -11$

18. $4p - 2q = -8$
$4p - 5q = -17$

19. $5m - n = 17$
$m + n = 1$

Solve each equation using the quadratic formula. *(Section 8.5)*

20. $x^2 + 2x - 9 = 0$ **21.** $y^2 - 5y - 1 = 0$ **22.** $3z^2 + 5z + 1 = 0$

23. $5x^2 + 3x = 5$ **24.** $11n^2 - 3n + 4 = 0$ **25.** $2a^2 - 7a = 8$

12.5

Combinations

How many different ways can you choose two people from a group? How many distances do you need to know to find the best path through a group of cities? In the Exploration, you will investigate a related problem.

Learn how to...

- **count the number of possible pairs of objects in a group**

So you can...

- **find the number of edges in a complete graph**
- **find the number of games in a tournament**

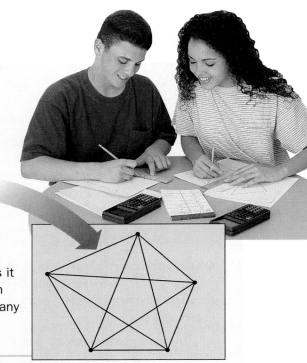

EXPLORATION
COOPERATIVE LEARNING

Making Connections

Work in a group of 2–4 students.
You will need:

● a graphing calculator

SET UP A **complete graph** has a single edge between every pair of vertices.

1 Draw a complete graph with 2 vertices. How many edges does it have? Draw complete graphs with 3, 4, 5, 6, and 7 vertices. How many edges does each graph have? Organize your results in a table.

Put the number of vertices on the horizontal axis.

◀2 Use a graphing calculator to make a scatter plot showing how the number of edges is related to the number of vertices. Choose a good viewing window.

3 Predict the number of edges in a complete graph with 8 vertices. Explain your reasoning. Check your prediction.

4 Write an equation for the number of edges in a complete graph as a function of the number of vertices. Explain your reasoning.

5 Graph the function from Step 4 on your scatter plot from Step 2. Does it seem to fit the data?

If you choose three students from your class in a particular order, you are creating a permutation. You can also choose the students without paying any attention to order. A selection of items chosen from a group in which order is not important is called a **combination**.

In the Exploration, you wrote an equation that allows you to count combinations of two items chosen from a group.

Combinations

The number of **combinations** of 2 items chosen from a group of n items is given by the formula:

$$_nC_2 = \frac{n(n-1)}{2!}$$

For example:

$$_5C_2 = \frac{5 \cdot 4}{2 \cdot 1}$$

The number of **combinations** of 3 items chosen from a group of n items is given by the formula:

$$_nC_3 = \frac{n(n-1)(n-2)}{3!}$$

For example:

$$_5C_3 = \frac{5 \cdot 4 \cdot 3}{3 \cdot 2 \cdot 1}$$

EXAMPLE 1

Paul Green is leading a workshop for 20 people. He wants each person to meet every other person. If each pair of people shake hands when they meet, how many handshakes will occur?

SOLUTION

There is one handshake for each pair in a group of 20 people. Count the number of combinations of 2. Use the formula for $_nC_2$ and let $n = 20$.

$$_{20}C_2 = \frac{n(n-1)}{2}$$

$$= \frac{20(20-1)}{2}$$

$$= 190$$

There will be 190 handshakes.

THINK AND COMMUNICATE

1. Explain the difference between a *combination* and a *permutation*. Give an example of each.

2. Explain how you can use a complete graph to model the number of handshakes in a group of people.

EXAMPLE 2

Cecilia Jordan plans to choose a different group of three students each day to lead the discussion of the assigned reading from the previous day. Can she choose a different combination of three students from her class of 25 students each day of the 180-day school year?

SOLUTION

Find the number of ways to choose 3 students from a group of 25.

$$_{25}C_3 = \frac{n(n-1)(n-2)}{3!}$$

$$= \frac{25(25-1)(25-2)}{3 \cdot 2 \cdot 1}$$

$$= 2300$$

There are 2300 possible groups of three. Not only could Cecilia Jordan choose a different group each day of the school year, she could choose a different group each school day for more than twelve years.

THINK AND COMMUNICATE

3. How many diagonals are there in a polygon with 5 sides? How is this related to the number of edges in a complete graph with 5 vertices?

4. Explain how to change the formula for the number of edges in a complete graph to get a formula for the number of diagonals of a polygon as a function of the number of sides.

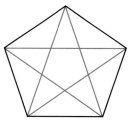

☑ CHECKING KEY CONCEPTS

Tell how many handshakes there would be in a group of each size if each pair of people shook hands.

1. 8   **2.** 15   **3.** 100   **4.** 1000

Evaluate.

5. $_7C_2$ **6.** $_6C_3$ **7.** $_{11}C_2$ **8.** $_{15}C_3$

9. Maya must read two of the six books on the summer reading list at her school. She can pick any two of the books. How many different combinations of two books can she choose?

10. How many edges does a complete graph with 18 vertices have?

11. How many ways can you choose a combination of 3 students from a class of 28 students?

12.5 Exercises and Applications

GOVERNMENT **For Exercises 1–4, use the 12 members of the student council listed below.**

Extra Practice *exercises on* *page 583*

Julie	Karen	Nathan	Susan
Booker	Eric	Tamai	Ivan
Mariah	Paula	Isabel	Greg

1. The council is sending a two-person delegation to the state model legislature. How many two-person delegations are possible?

2. How many of the possible delegations have either Nathan or Tamai on them? (*Hint:* Count the Nathan-Tamai delegation only once.)

3. How many possible delegations will not have Isabel, Karen, or Booker? Explain your reasoning.

4. How many three-person delegations are possible?

Connection ▶ SOCIAL STUDIES

Many countries have flags with three horizontal stripes. The colors of the stripes have meanings for each country. For Exercises 5 and 6, use the colors red, blue, orange, white, green, black, and yellow.

5. Both Austria and Belarus have flags with red and white horizontal stripes. How many different pairs of colors could you choose for a two-color flag? How many different two-color flags with three horizontal stripes of alternating colors could be made?

BY THE WAY

When the Olympic flag was first flown in 1920, every nation's flag contained at least one of the five colors on the Olympic flag.

Ethiopia   Gabon   Bolivia   Netherlands

Germany   Austria   Yemen   Belarus

6. a. How many different combinations of three colors could you choose for a three-color flag?

   b. How many different flags with three horizontal stripes can be made with green, red, and blue? How many different flags with three horizontal stripes can be made with three colors?

Evaluate.

7. $_{13}C_2$ **8.** $_9C_3$ **9.** $_{17}C_3$ **10.** $_{21}C_2$

11. a. Jill Keyes asked her students to answer two of the five essay questions on an English test. How many different choices do her students have?

 b. For extra credit the students can answer a third essay question. How many combinations of three questions can the students answer?

12. **MANUFACTURING** A book publisher selects two books at random from each lot for a check of production quality. How many different combinations of two books can be selected from a lot of 25 books?

 Spreadsheet Virginia set up the spreadsheet below to calculate the number of different handshakes possible for a group of people. She assumed that each pair of people shakes hands once.

13. Cooperative Learning Work in a group of 2 or 3 students. Discuss how the formula for Column B works. Explain in words what the formula A7 + B7 in cell B8 means. Why does it give the correct number of handshakes?

B8		=A7+B7
	A	**B**
1	People	Handshakes
2	1	
3	2	1
4	3	3
5	4	6
6	5	10
7	6	15
8	7	21

14. Make a spreadsheet of your own. Use it to find the number of handshakes in a group of 35 people.

15. **GEOMETRY** Draw three points on a circle. How many triangles can you draw connecting the points? How many triangles can you draw connecting four points on a circle? Make a table for the number of triangles connecting 3, 4, 5, and 6 points on a circle. How is this related to the formula for $_nC_3$?

16. At a local restaurant, you can choose from nine pizza toppings.

 a. How many different two topping pizzas can you choose?

 b. How many different three topping pizzas can you choose?

17. **Technology** The $_nC_r$ function on your calculator will calculate the number of ways you can choose r items from a list of n possibilities. Let $n = 9$ pizza toppings. Make a table of the number of pizzas with r toppings for values of r from 1 to 9. Explain any patterns you see.

18. a. Challenge Write a formula for $_nC_4$. You may want to think about how many ways there are to choose 4 of the pizza toppings in Exercise 17. Explain your reasoning.

 b. Write a general formula for $_nC_r$. Test your formula by using the $_nC_r$ function on your calculator and choosing values for n and r.

19. **SPORTS** North Central High is planning a tournament for the city's intramural volleyball league. Each team will play a three-game match against each of the other teams. There are enough volleyball courts to play about ten matches each day of the tournament. If nine teams participate, how many days will the tournament take to complete? Explain your reasoning.

20. **Open-ended Problem** Write a counting problem that could be solved by the equation $\dfrac{15 \cdot 14}{2} = 105$.

21. **Writing** Melissa tried to find how many ways she could choose two cassette tapes to listen to in the car. What do the numbers 1190 and 595 represent?

I have 35 tapes, so there are 35 choices for the first tape. Then there will be 34 choices for the second tape. Because 35 · 34 = 1190, I can choose two tapes 1190 ways. But if I use the formula $\dfrac{n(n-1)}{2}$,

I get $\dfrac{35 \cdot 34}{2} = 595$ choices for my pair of tapes.

The equations in Exercises 22–27 describe quantities that grow or decay exponentially. For each equation, tell whether the quantity described *grows* or *decays*. Then give the growth or decay factor for the quantity.
(Sections 9.2 and 9.3)

22. $y = 2.1(1.32)^x$

23. $y = 0.15(0.52)^x$

24. $y = 3.2(0.91)^x$

25. $y = 0.02(2)^x$

26. $y = A(1 + r)^x$

27. $y = ab^x$

Decide whether the data show inverse variation. If they do, write an equation for *y* in terms of *x*. *(Section 11.1)*

28.

x	y
−2.5	6.00
2.5	−6.00
6.5	−2.31
10.5	−1.43

29.

x	y
−1.5	−2.79
−0.5	−0.93
1.0	1.86
3.5	6.51

30.

x	y
−14	−7.2
6	16.7
8	12.5
12	8.4

12.6 Connecting Probability and Counting

Learn how to...
- **use counting strategies to find probabilities**

So you can...
- **find the probability of winning a random drawing**

Nancy Greenleaf has a deal for her algebra class. She agrees to cancel all of the tests for the rest of the year and give everyone an A for the class. But she will do this only if she rolls five 1's when she rolls five dice with different numbers of sides. How likely is that to happen?

You can use the counting strategies you have learned in this chapter to solve many probability problems. Recall the meaning of *theoretical probability* from Chapter 5.

$$\text{Theoretical probability} = \frac{\text{Number of favorable outcomes}}{\text{Number of possible outcomes}}$$

EXAMPLE 1

Find the probability that Nancy Greenleaf will roll five 1's when she rolls a 6-sided die, an 8-sided die, a 10-sided die, a 12-sided die, and a 20-sided die.

SOLUTION

The number of favorable outcomes is the number of ways to roll five 1's. There is one favorable outcome.

Use the multiplication counting principle to find the number of possible outcomes.

six sides	eight sides	ten sides	twelve sides	twenty sides

$$6 \;\cdot\; 8 \;\cdot\; 10 \;\cdot\; 12 \;\cdot\; 20 \;=\; 115{,}200$$

$$\frac{\text{Favorable outcomes}}{\text{Possible outcomes}} = \frac{1}{115{,}200}$$

The probability of rolling five 1's is $\frac{1}{115{,}200}$.

EXAMPLE 2

Deb and Kyle are two of the 15 people who submitted correct solutions to a puzzle contest. Two prize winners will be chosen in a random drawing.

a. What is the probability that both Deb and Kyle will win a prize?

b. What is the probability that Deb will be one of the prize winners?

SOLUTION

a. The number of favorable outcomes is the number of ways to choose both Deb and Kyle. There is one favorable outcome.

The number of possible outcomes is the number of combinations of two winners. Use the formula for $_nC_2$ and let $n = 15$.

$$_{15}C_2 = \frac{n(n-1)}{2}$$

$$= \frac{15 \cdot 14}{2}$$

$$= 105$$

There are 105 ways to choose 2 winners out of 15 people.

$$\frac{\text{Favorable outcomes}}{\text{Possible outcomes}} = \frac{1}{105}$$

The probability that both Deb and Kyle will be chosen is $\frac{1}{105}$.

b. Use the multiplication counting principle. If Deb is one winner, there are $15 - 1 = 14$ choices for the second winner. There are $1 \cdot 14 = 14$ favorable outcomes.

$$\frac{\text{Favorable outcomes}}{\text{Possible outcomes}} = \frac{14}{105} = \frac{2}{15}$$

The probability that Deb will be a winner is $\frac{2}{15}$.

THINK AND COMMUNICATE

1. How do you decide when to use the multiplication counting principle and when to use the number of combinations to determine the number of possible results?

2. What is the probability of Kyle being selected as one of the two prize winners in Example 2?

3. What is the probability of getting four heads if you flip one coin four times? if you flip four coins one time?

4. When Robyn put her new license plate on her car, she was surprised to notice that her license plate had the same four digits in a different order as her sister's license plate. Explain how you would find the probability of this happening.

For Questions 1–4, suppose that you toss a penny four times.

1. List all of the possible results.

2. What is the probability that you will get four heads?

3. What is the probability that you will get four tails?

4. What is the probability that you will get two heads and two tails?

5. POLITICS Ben and Lara want to be the delegates for a local political convention. Two people will be chosen at random from a group of 11 students. What is the probability of Ben and Lara being chosen?

12.6 | Exercises and Applications

Extra Practice
exercises on
page 583

For Exercises 1 and 2, find the probability of getting the result shown when each group of coins is tossed.

1.

2.

For Exercises 3–5, suppose you toss six pennies.

3. What is the total number of possible results?

4. What is the probability that the pennies will all come up heads?

5. Writing Explain why the probability of getting one tail and five heads is the same as the probability of getting five tails and one head.

6. a. SPORTS There are 12 players on Elisa's basketball team. Each team member plays a one-on-one game with each of the other members of the team. How many games are played?

b. The school's video club is going to pick one game at random to film for the video yearbook. What is the probability that Elisa and Bridget's game will be filmed? What is the probability that one of Elisa's games will be filmed?

7. Suppose you throw a 4-sided die, a 20-sided die, and a 100-sided die at the same time. What's the probability of rolling three 4's?

8. Suppose you roll three 4-sided dice. What is the probability that you will roll three of the same number? Explain your reasoning.

Cooperative Learning For Exercises 9–12, work with a partner. Use the game explained below. You will need a calculator with a random number generator.

Take turns until one player wins. On each turn:

Step 1 Start with a score of zero.

Step 2 Generate a random number, _n_.
If $n \leq 0.5$, subtract 1 from your score.

If $n > 0.5$, add 1 to your score.

Step 3 Repeat Step 2 five times.

a. If your score = 3 or −3, you win the game.

b. If your score is anything else, it is your partner's turn.

9. Play at least six games. Record how many turns it takes to win the game. Record the ending score after each turn during the games. You may want to use bar graphs to organize this information.

10. What is the probability of winning the game on any given turn? Explain your reasoning.

11. **Open-ended Problem** Choose a way to change the rules in the game described above. Explain how to play and win your new game.

12. **Challenge** Play the game you described in Exercise 11 several times. Can you determine the probability of winning the game on any given turn? Explain why or why not.

13. **Technology** You can program your calculator to model the roll of a 6-sided die. See your calculator's manual for information on how to program your calculator.

iPart tells the calculator to ignore everything to the right of the decimal point.

```
PROGRAM:SIXSIDED
:iPart (rand*6+1
)→A
:Disp A
:Stop
:
```

rand generates a random number between 0 and 1.

a. **Writing** Explain how this model works. How can you use it without the iPart function? Why do you need to add 1 to the random number?

b. How can you model the roll of a 20-sided die?

c. Program your calculator to model the roll of a 6-sided die. Run your program 36 times. Do you get the results you would expect from a fair die?

14. **PHOTOGRAPHY** Luisa removed the used batteries from her camera. She accidently dropped the two used batteries into her pile of eight new ones. The batteries all look the same. What is the probability that she will pick up at least one used battery when she chooses two to put into her camera? What is the probability that she will pick up two new batteries?

Connection ▶ LITERATURE

Rosencrantz and Guildenstern are Dead by Tom Stoppard is a twentieth century play about two characters from Shakespeare's *Hamlet*. When the play begins, Rosencrantz and Guildenstern are carrying large leather money bags. Guildenstern's (Guil) is almost empty. Rosencrantz's (Ros) is almost full.

15. Guildenstern tosses 92 coins, all of which come up heads. What is the number of possible outcomes of tossing 92 coins? Explain how you know.

16. What is the probability of getting 92 heads in a row?

17. Guildenstern makes another wager later in the play. "*Bet* me then. . . . Year of your birth. Double it. Even numbers I win, odd numbers I lose." What is the probability that Guildenstern will win this wager?

ROSENCRANTZ AND GUILDENSTERN ARE DEAD

Guil: *takes a coin out of his bag, spins it, letting it fall.*
Ros: Heads. *He picks it up and puts it in his bag. The process is repeated.*

    Heads. *Again.*

    Heads. *Again.*

    Heads. *Again.*

    Heads.

Guil: . . . *flips over two more coins.* . . . *Ros announces each of them as heads."*

Ros: . . . Eighty-five in a row—beaten the record!

Guil: . . . And if you'd lost? If they'd come down against you, eighty-five times, one after another, just like that?

Ros: . . . Eighty-five in a row? *Tails?*

Guil: Yes! What would you think?

Ros: (*doubtfully*): Well . . . . Well, I'd have a good look at your coins for a start! . . .

Guil: . . . It must be indicative of something, besides the redistribution of wealth. . . . each individual coin spun individually (*he spins one*) is as likely to come down heads as tails and therefore should cause no surprise each individual time it does. . . . The equanimity of your average tosser of coins depends upon a law, or rather a tendency, or let us say a probability, or at any rate a mathematically calculable chance, which ensures that he will not upset himself by losing too much or upset his opponent by winning too often.

18. a. **Writing** George, Helen, and Inez agree to toss a coin twice to decide who will go first at batting practice. George goes first if the result is two heads. Helen does if the result is two tails. Inez does if the result is one head and one tail. Explain why Inez is most likely to go first.

    b. Suggest a fair way to decide who goes first using coins or dice. Why is your suggestion fair to all three people?

SPIRAL REVIEW

Add. *(Section 10.1)*

19. $(x + 2) + (4x - 12)$

20. $(4a^2 - 3a + 1) - (3a^2 + 7a - 9)$

21. $(4 - 3n - 5n^2) + (n - 3n^2)$

22. $(3x - 7y - 5z) - (x - 4y - 5)$

Simplify each equation. *(Section 11.6)*

23. $\dfrac{3}{2ab} - \dfrac{5}{3ab}$

24. $\dfrac{6}{m + 1} + 7$

25. $\dfrac{2t + 3}{5} + \dfrac{5 - t}{4}$

26. $\dfrac{m^2}{7n} - \dfrac{5}{n}$

27. $\dfrac{3}{2a} - \dfrac{1 + a}{b}$

28. $\dfrac{a + b}{3} + \dfrac{a + b}{5} + \dfrac{a + b}{10}$

ASSESS YOUR PROGRESS

VOCABULARY

multiplication counting principle (p. 531)
factorial (p. 532)
permutation (p. 532)
complete graph (p. 536)
combination (p. 537)

1. How many ways can you select 1 shirt, 1 sweater, and 1 hat from 3 shirts, 5 sweaters, and 4 hats? *(Section 12.4)*

2. A company has 713 employees. Explain why at least two of them must have the same pair of first and last initials. For example, Ann Martin has the same initials as Alan Monchick. *(Section 12.4)*

3. a. **ENTERTAINMENT** Martha and Gwen have 150 movies to choose from at the video store. They plan to rent 2 movies. How many different pairs of movies could they choose? *(Section 12.5)*

   b. They decide to rent 1 of 36 comedies and 1 of 23 new releases. How many different pairs could they choose? *(Section 12.4)*

4. If you throw two six-sided dice, what is the probability that you will roll at least one 6? *(Section 12.6)*

5. **Journal** Give examples of probability problems that help you to remember the difference between permutations and combinations. Explain how permutations or combinations are used in each problem.

Playing the Game

For centuries, people in Africa and Asia have been developing strategies for winning games that involve moving counters around a board. Some versions of this ancient game are *mancala*, *adi*, *kala*, and *oware*. Finding strategies for winning many different kinds of games is an important part of a field of mathematics known as *game theory*.

This mancala board is from the Neolithic Age (7000–5000 B.C.).

PROJECT GOAL **Develop strategies for winning a game of wari from Ghana, a country in Africa.**

How to Play Wari

MATERIALS The Wari board has 12 playing cups, in two rows of six, and two scoring cups. The game is played with 48 counters—stones or seeds are often used.

1. SET UP Work with a partner. Each of you should choose one side of the board. Your scoring cup is on your right. Place 4 stones in every cup but the scoring cups.

2. PLAY On your turn, choose any cup on your side of the board. Take all the stones out of this cup. Moving counter-clockwise drop one stone in each cup (not including the scoring cups) until you run out of stones.

3. CAPTURE If the last stone is placed in one of your partner's cups and makes the total number of stones in that cup 2 or 3, capture those stones and put them in your scoring cup.

If your partner's side of the board is empty, you must play so that you drop *at least one stone* into your partner's cups. If you cannot, you capture all remaining stones.

WINNING The game ends when no captures are possible. The player who has captured the most stones is the winner.

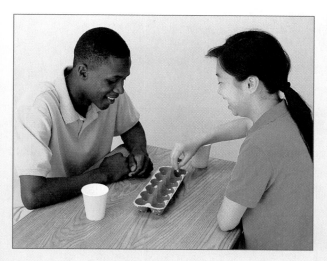

Developing Strategies

Play several games of *wari*. Take turns going first. While you are playing try different strategies for collecting as many stones as possible. Discuss your strategies. Which seem to work best? Do any of your strategies depend on whether you play first?

You may want to use an egg carton for your board. You may want to use dried beans or pumpkin seeds instead of stones.

Presenting Your Results

Give an oral report on this project to your class. Demonstrate the strategies you developed. Explain how they work and how you developed them.

You may want to extend your report to explore other ideas:

Senet

• Research and explore some of the variations of this game. Develop a strategy for winning a different version of the game. How does this new strategy compare with the strategy that you developed for *wari?*

• Research the history and directions for another game. Where and when was the game developed? Explore the game and develop a strategy for winning that game. Present your findings to your classmates. Some games you might research are *Go*, *Nim*, *Senet*, and *Nyout*.

Self-Assessment

What general advice would you give someone who is trying to develop strategies for winning a new game? How did you and your partner work together to develop strategies? Based on your experience playing wari, does the same strategy for winning always work? How do your partner's moves affect your strategy?

Go

12 Review

STUDY TECHNIQUE

Look back at the study techniques you used in other Review and Assessment sections. Decide which technique(s) helped you most. Why was it helpful? Choose a technique you think would be most helpful in studying Chapter 12.

VOCABULARY

algorithm (p. 509)
graph (p. 516)
vertices (p. 516)
edges (p. 516)
path (p. 517)
greedy algorithm (p. 517)
tree (p. 518)
one person-one vote (p. 524)
approval voting (p. 524)

fair division (p. 525)
cut-and-choose (p. 525)
multiplication counting principle (p. 531)
factorial (p. 532)
permutation (p. 532)
complete graph (p. 536)
combination (p. 537)

SECTIONS 12.1 *and* 12.2

A **graph** consists of points called **vertices** and lines called **edges**. A **path** is a group of edges that connect one vertex to another without repeating any vertices or edges. A **tree** uses as few edges as possible to connect all the vertices in a graph.

This is a graph.

This is a path from *A* to *E*.

This is a tree.

An **algorithm** is a step-by-step method for accomplishing a goal.

You can use a **greedy algorithm** to find a short path.

   Step 1 Choose a vertex.

   Step 2 Choose the nearest vertex you have not already chosen.

   Step 3 Repeat Step 2 until you've connected all the vertices.

SECTION 12.3

In a **one person-one vote** election, each person can vote for one candidate. In an **approval voting** election, each person can vote for any number of candidates. The table shows poll results based on both types of voting.

•••• Type of Election ••••		
Candidate	One person-one vote	Approval voting
Phuong	48	73
Gioberti	62	68
Wallach	41	82

The poll results show that Gioberti would probably win a one person-one vote election. Wallach would probably win an approval voting election.

You can use a **cut-and-choose** algorithm to create a **fair division** of a collection into two groups.

SECTIONS 12.4, 12.5, *and* 12.6

A **permutation** is an arrangement of a group of items in a definite order. A **combination** is a selection of items from a group in which the order of selection does not matter.

Problem	Type	Number
Hang 4 flags in a row.	permutation	$4! = 4 \cdot 3 \cdot 2 \cdot 1 = 24$
Hang 2 of 4 flags in a row.	permutation	$4 \cdot 3 = 12$
Choose 2 of 4 flags.	combination	$_4C_2 = \dfrac{4 \cdot 3}{2!} = 6$

To find the probability of getting exactly 2 heads when 5 coins are tossed, you can use the formula for theoretical probability.

The number of favorable outcomes is the number of combinations of 2 heads.

$$_5C_2 = \frac{5 \cdot 4}{2!} = 10$$

There are two possible outcomes for each coin toss. By the multiplication counting principle, the number of possible outcomes when 5 coins are tossed is $2 \cdot 2 \cdot 2 \cdot 2 \cdot 2 = 32$.

$$\text{Theoretical probability} = \frac{\text{Number of favorable outcomes}}{\text{Number of possible outcomes}} = \frac{10}{32}.$$

The probability of getting exactly 2 heads when 5 coins are tossed is $\frac{5}{16}$.

12 Assessment

VOCABULARY REVIEW

For Questions 1–3, complete each paragraph.

1. Describe the difference between a permutation and a combination.

2. Give an example of a greedy algorithm.

3. Give an example of a situation in which you would find a *tree* to solve a problem. Give an example of a situation in which you would find a *path*.

SECTION 12.1

For Questions 4 and 5, refer to the algorithm for doing laundry.

Step 1 Divide the laundry evenly into loads.

Step 2 Put one load in the washing machine.

Step 3 Turn the machine on.

Step 4 When the cycle is done, remove the laundry from the machine.

Step 5 Repeat Steps 2 through 4 until the laundry is finished.

4. What could go wrong if you follow this algorithm exactly?

5. How could you clarify this algorithm?

6. Write an algorithm to add two fractions with different denominators.

SECTION 12.2

For Questions 7–9, use the graph at the right.

7. Find the shortest path from vertex *A* to vertex *D*.

8. Start at vertex *D*. Find a short path through all the points.

9. Find the shortest tree for the graph.

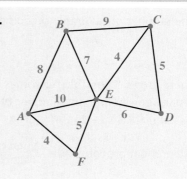

10. **Writing** Ballots are being printed for an election. Explain how the instructions for a one person-one vote election would be different from those for an approval voting election.

11. **Open-ended Problem** Describe a situation in which the cut-and-choose algorithm could be used to create a fair division. Explain how to apply the algorithm in this situation.

SECTIONS 12.4 *and* 12.5

12. In a class election, there are four candidates for president, five candidates for vice-president, and six candidates for secretary. How many possible groups of officers are there?

13. In how many ways can a three-member finance committee be chosen from among ten students?

14. There are 14 baseball teams in a community league. Suppose the league champion is determined by having each team in the league play every other team. The team with the best record wins. How many games are necessary to determine the league champion?

SECTION 12.6

15. Three coins are tossed. Find the probability of getting exactly two heads.

16. Luisa and Keith each prepared an individual project for their history class. Each day, 2 of the 17 students in the class will present their projects. Find the probability that both Luisa and Keith will be chosen to present their projects on the first day.

PERFORMANCE TASK

17. Choose one story of a building to model with a graph. Let each vertex represent a room. Connect two vertices with an edge if you can walk from one room to the other without going through another room. Give an example of a path through all the rooms that does not visit any room twice, or explain why such a path is not possible. In how many different orders can you visit all of the rooms in the building? (You may pass through the same room more than once. Only count the first time you visit each room.)

Cumulative Assessment
CHAPTERS $10-12$

CHAPTER 10

Add, subtract, or multiply each pair of polynomials as indicated. Write each answer in standard form and give the degree of the polynomial.

1. $(4x - 3) + (-2x^2 - x + 1)$ **2.** $(z^2 - 6z + 5) - (3z^2 + 10z + 7)$

3. $(4y^3 - 7y^2 + 3) + (5y^2 - 2y + 9)$ **4.** $(7s - 2)(6s - 4)$

5. $8x(10x^2 - 4x + 7)$ **6.** $(2y - 9)(4y^2 - 7y + 5)$

Factor, if possible. If not, write *not factorable*.

7. $x^2 + 4x - 21$ **8.** $x^2 - 49$ **9.** $a^2 - 9a + 18$

10. $y^2 - 10y - 56$ **11.** $3z^2 + 25z - 18$ **12.** $4x^2 - 16x + 15$

Solve each equation.

13. $x^2 + x - 12 = 0$ **14.** $r^2 + 16r + 64 = 0$ **15.** $81y^2 - 25 = 0$

16. $9z^2 + 9z + 5 = 3$ **17.** $3s^2 - 8s - 1 = 2$ **18.** $6x^2 = 11x - 5$

19. If $4x^2 + bx + 9$ is a perfect square trinomial, what is the value of b?

20. Writing When can you combine terms in a polynomial? Write a few sentences explaining your answer.

21. Open-ended Problem Write a quadratic equation that cannot be solved by factoring. Show how to solve it using another method. Explain why you chose that method.

CHAPTER 11

Tell whether the data show *direct variation*, *inverse variation*, or *neither*. If the data show direct or inverse variation, write an equation describing the situation.

22.

x	y
1	3
3	1
6	$\frac{1}{2}$

23.

x	y
6.2	1.55
7.0	1.75
8.4	2.10

24.

x	y
1	2
3	6
6	12

Solve each equation.

25. $\dfrac{3}{a + 3} + 2 = \dfrac{7}{a + 3}$ **26.** $\dfrac{5}{6y} + \dfrac{2}{3} = \dfrac{1}{y}$

27. Here are scores for nine students in a math contest. Find the mean score.

Score	75	83	87	88
Number of students	1	2	4	2

28. a. In a trapezoid with height h and bases b_1 and b_2, the area A can be calculated using the formula $A = \frac{1}{2}h(b_1 + b_2)$. Solve the formula for b_1.

b. Find b_1 for a trapezoid with an area of 60 cm^2, a height of 4 cm, and $b_2 = 17$ cm.

Simplify each expression.

29. $\dfrac{7}{3ab} - \dfrac{5}{9a}$

30. $\dfrac{5y}{3y^2} \div \dfrac{15}{9y^4}$

31. $\dfrac{z-1}{2(z+1)} + \dfrac{3z+5}{2(z+1)}$

32. Writing Describe how to simplify a rational expression.

33. Open-ended Problem Write two rational expressions. Use the same variable(s) in each expression. Multiply your expressions and simplify the resulting expression.

CHAPTER 12

34. Write an algorithm that could be used to program a computer to follow the order of operations. Be sure to include all of the operations you have learned this year.

35. Use a greedy algorithm to find the shortest tree for the graph at the right.

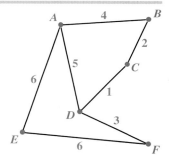

Calculate.

36. $7!$

37. $9!$

38. $_8C_2$

39. $_9C_3$

40. $_{13}C_2$

41. Each day, the cafeteria offers a lunch special that includes a sandwich, a salad, and a dessert. The menu includes 10 sandwiches, 4 salads, and 6 desserts. How many different lunch specials can the cafeteria offer?

42. a. Suppose you toss a penny and a nickel. What is the probability that the penny shows heads and the nickel shows tails?

b. When you toss a penny and a nickel, what is the probability of getting exactly one head and one tail?

43. Writing Suppose every pair of people in a group shakes hands once. Explain why the number of possible handshakes in a group of people is related to the number of combinations of two items chosen from a group.

44. Open-ended Problem Give an example of a situation where you might want to find the number of combinations. Give an example of a situation where you might want to find the number of permutations.

Contents of Student Resources

Extra Practice

CHAPTER 1

For Exercises 1–6, use the histogram at the right. *Section 1.1*

1. How many days does the histogram represent?

2. For how many days was the high temperature 90°F to 94°F?

3. a. If a high temperature of 80°F to 89°F is considered normal, how many days had normal temperatures?

   b. How many days had above-normal temperatures?

4. For how many days was the temperature $t \le 89$?

5. Write two inequalities that describe a daily high temperature of at least 90°F.

6. Write two inequalities that describe the range of daily high temperatures for the month of August.

Daily High Temperatures for August

Simplify each expression. *Section 1.2*

7. $4(10 - 8) + 3$

8. $15 - 3(2 - 1.5)$

9. $28 - 15 \div 3 \cdot 5$

10. $\dfrac{16 + 8}{2 \cdot 6}$

11. $\dfrac{4(8)}{2(6)}$

12. $2 + \dfrac{7}{20} \cdot 4$

13. $0.8(25 - 3.7)$

14. $3 + (82 - 8) \div 7$

15. $\dfrac{8.9 - 5}{6 + 7}$

16. $\dfrac{3}{2} + 4(10 - 2)$

17. $\dfrac{5 + 7 \cdot 32}{17 \cdot 4}$

18. $\dfrac{(19 + 18 \div 2) \div 2}{5 + 2}$

Evaluate each variable expression when $r = 2$, $s = 5$, and $t = 6$. *Section 1.2*

19. $s^2 - 4 + t$

20. $r^2 + t^2$

21. $6t - r$

22. $3s + 2r$

23. $\dfrac{7t}{r + s}$

24. $7(r + s) + t$

25. $\dfrac{t - s}{5r}$

26. $3(t - r) + 7$

27. $\dfrac{t + s - r}{(s - r)^2}$

28. $s^2 - rt + r^2 - s$

29. $t^2 + 7t - st$

30. $\dfrac{3r^2 - t}{2s}$

Find the mean, the median, and the mode(s) of each set of data. *Section 1.3*

31. Kilometers hiked each day
    10, 18, 15, 13, 19, 6, 13

32. Number of people in computer classes
    22, 18, 25, 21, 17, 17, 18, 20

33.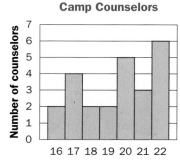

Camp Counselors

34. Terri's quiz scores
99, 85, 94, 100, 82, 76, 94, 88, 95, 81

35. Number of people who swim laps each day
12, 8, 5, 9, 8, 15, 8, 8, 8, 9, 9, 10, 14, 10, 11, 12, 1

Simplify each expression. *Section 1.4*

36. $-3 + 10$

37. $9 + (-14)$

38. $5 - 17$

39. $-5 + (-16)$

40. $4 - (-2)$

41. $-8 - (-3)$

42. $15 - 19$

43. $-2.1 + 5.3$

44. $|-2| + 5$

45. $|-6| - |-5|$

46. $3|-7|$

47. $|-3 - 2|$

48. $4 - |10| + 2$

49. $-8 + |5 - 9|$

50. $|4 - 1| + |3 - 10|$

51. $\dfrac{|-5 - 2|}{14}$

52. $\dfrac{4|2 - 5|}{3}$

53. $|7.4 - 6| - (-7)$

54. $\dfrac{|-25|}{|-15|}$

55. $\dfrac{2(-1 + 5)}{12|-2|}$

Evaluate each variable expression when $a = 3$, $b = -2$, and $c = -4$. *Section 1.4*

56. $b + a$

57. $b + (-5)$

58. $a - c$

59. $4 - c + 1$

60. $a^2 + b + c$

61. $\dfrac{2a^2}{b + 8}$

62. $|c - b|$

63. $\dfrac{|b + c|}{2}$

64. $|b| - |a + c|$

65. $a|b| - (-10)$

66. $2|a + b|$

67. $7|3 - a + c|$

Simplify each expression. *Section 1.5*

68. $-7(8)$

69. $\dfrac{-18}{2}$

70. $(-9)(-8)$

71. $\dfrac{-42}{-7}$

72. $(-1)(3 - 7)$

73. $-8(5) + (-8)$

74. $|6 - 4 - (-4)|$

75. $\left(\dfrac{1}{2} \cdot 5\right) \cdot (-2)$

76. $\dfrac{|(-4)(6)|}{-3}$

77. $-6 + (-3) \cdot 7 + 4$

78. $0.5 \cdot (-15) \cdot 10$

79. $\dfrac{-1 + (-3)(-9)}{-7}$

80. $\dfrac{7}{8}(-5)(16)$

81. $4 - (5 - 8) \cdot (-3)$

82. $(-72 \div 6) - 25(-2)$

Evaluate each variable expression for the given values. *Section 1.5*

83. $4x$, when $x = -3$

84. $-5a + 3$, when $a = 4$

85. $-3p - 7$, when $p = -5$

86. $\dfrac{|x|}{2}$, when $x = -6$

87. $4r + s$, when $r = -5$ and $s = -4$

88. $5xyz$, when $x = -4$, $y = 2$, and $z = -1$

89. $-2|d - e| + d$, when $d = -3$ and $e = 7$

90. $\dfrac{-16 - 9}{2n - p}$, when $n = -2$ and $p = 1$

Write a variable expression for each situation. *Section 1.6*

91. The rate for a long-distance call is $.35 for the first minute and $.10 for each additional minute. Write a variable expression that represents the total cost of a call.

92. The cost of a basic platter at a fast-food restaurant in the mall is $3.75. Each additional item costs $.75. Write a variable expression that represents the cost of any platter that is ordered.

93. The estimated deer population in a certain area is 2200. It is estimated that the population increases by 130 deer each year. Write a variable expression that represents the predicted population for each future year.

Simplify each variable expression. *Section 1.7*

94. $5(2a + 3b)$

95. $-8(-4 + 2x)$

96. $-(7y - z)$

97. $7a + 3b - 3a$

98. $3m^2 - 4 + 9m^2 + 6$

99. $18n - 3n^2 - 5n^2 + n$

100. $5c + 2cd + d$

101. $9 - (6f + 8) + 1$

102. $-4(g + 8) - (5g + 3)$

103. $3(y - 4y) - 4(2 + y)$

104. $4t + 9t^2 - (2t^2 - t)$

105. $2(a^2 - 5a + 2a) + 9a^2 + 6a$

106. $-9x - 7y + 3(-x + 5y)$

For Exercises 107–109, use matrices *E*, *F*, *G*, and *H*. *Section 1.8*

$$E = \begin{bmatrix} -4 & 2 & 1 & 9 \\ 3 & -8 & 7 & -9 \\ 0 & 1 & -2 & 6 \end{bmatrix}$$

$$F = \begin{bmatrix} -2 & 5 & -15 & 3 \\ 10 & -4 & 13 & -9 \\ 3 & 7 & -1 & -1 \end{bmatrix}$$

$$G = \begin{bmatrix} -1 & 9 & 3 \\ 0 & -11 & 1 \\ -6 & 14 & -4 \\ 12 & 30 & -6 \end{bmatrix}$$

$$H = \begin{bmatrix} 16 & -12 & 8 \\ -2 & 8 & -2 \\ -3 & 4 & 0 \\ -13 & -20 & -1 \end{bmatrix}$$

107. Give the dimensions of each matrix.

108. Add each pair of matrices.

    a. *E* and *F*
        b. *G* and *H*

109. Which pairs of matrices cannot be added? Explain.

CHAPTER 2

Solve each equation. *Section 2.1*

1. $3.1 + a = 5$

2. $n - 15 = 40$

3. $0.9 - b = 2.2$

4. $-4 = x + (-9)$

5. $4s = -64$

6. $22 = \dfrac{v}{4}$

7. $15 = 5 - q$

8. $\dfrac{r}{-2} = 9$

9. $-4 = -b$

10. $-7w = -3.5$

11. $\dfrac{m}{3} = 0.12$

12. $s + 9 = -45$

13. $-8a = 0.56$

14. $k + \dfrac{2}{3} = \dfrac{5}{6}$

15. $5 - x = \dfrac{1}{2}$

16. $\dfrac{n}{-9} = -7$

Solve each equation. *Section 2.2*

17. $3x + 5 = 23$

18. $-22 = 4n + 2$

19. $12 - \dfrac{x}{2} = -3$

20. $9 + 4b = -7$

21. $13 - 6c = -17$

22. $-17 = -13 - 6c$

23. $-1 = 9 + \dfrac{c}{6}$

24. $\dfrac{s}{3} - 8 = 7$

25. $2m + 0.5 = 6.5$

26. $-42 = 3v - 6$

27. $27 = -9 - \dfrac{n}{3}$

28. $-3 = 42 - a$

29. $-10 + \dfrac{a}{6} = -2.7$

30. $-82 + 9b = -19$

31. $32 = -5f - 6$

32. $\dfrac{c}{21} + 2 = 0$

Write and solve an equation for each problem. *Section 2.2*

33. Chad bought a bottle of fruit juice that cost $2.29 and 9 juice boxes. He paid a total of $5.17. How much did each juice box cost?

34. The perimeter of a rectangle is 22 in. If the length of the rectangle is 7 in., what is the width of the rectangle?

For Exercises 35–37:

a. **Write an equation for the function.**

b. **Describe the domain and the range of the function.**

c. **Solve the problem.** *Section 2.3*

35. Connie Nardelli rents a car for $27 per day. The total cost is a function of the number of days she rents the car. How much will she pay for an 8-day rental?

36. A city parking lot charges $4.50 for the first hour and $1.50 for each additional hour. The total cost is a function of the amount of time spent in the lot. How much does it cost to park for 12 hours?

37. Calligraphy pens cost $3 each. A package of special paper costs $5.75. The total cost of a package of paper and some pens is a function of the number of pens purchased. How much will the paper and 6 pens cost?

For Exercises 38 and 39, use the coordinate graph. *Section 2.4*

38. Write an ordered pair for each labeled point.

39. Name the quadrant (if any) in which each point lies.

Graph each point in a coordinate plane. Label each point with its letter.
Section 2.4

40. $A(2, 0)$

41. $B(1, -2)$

42. $C(3, 3)$

43. $D(-3, 2)$

44. $E(4, 2)$

45. $F(-1, 4)$

Graph each equation. Name an ordered pair that is a solution of the equation.
Section 2.5

46. $y = x + 5$

47. $y = -2x + 2$

48. $y = 4 - 3x$

49. $y = 3x$

50. $y = -x + 4$

51. $y = x + 2$

52. $y = 3 + 3x$

53. $y = -3x - 1$

For Exercises 54 and 55, show how to use a graph to solve the problem. Then show how to use an equation. *Section 2.6*

54. Alison is trying to raise money for her soccer team by selling 100 raffle tickets. She has already sold 12 tickets. She estimates that she can sell an average of 5 tickets per day. About how many days will it take her to sell all 100 tickets?

55. Mike Ellis is paying back a loan from his parents. The original amount of the loan was $7500. He pays $125 per month, and has already paid $1625. How many more payments must he make?

CHAPTER 3

Copy and complete each equation. *Section 3.1*

1. $\dfrac{50 \text{ cycles}}{1 \text{ s}} \cdot \dfrac{60 \text{ s}}{1 \text{ min}} = \underline{\ ?\ }$

2. $\dfrac{\$450}{1 \text{ week}} \cdot \underline{\ ?\ } = \dfrac{\$23{,}400}{1 \text{ year}}$

3. $\dfrac{\$1.53}{1 \text{ gal}} \cdot \dfrac{18 \text{ gal}}{1 \text{ tank}} = \underline{\ ?\ }$

4. $\dfrac{\$200}{1 \text{ h}} \cdot \dfrac{1 \text{ h}}{60 \text{ min}} \approx \underline{\ ?\ }$

5. $\dfrac{8 \text{ pages}}{15 \text{ min}} \cdot \underline{\ ?\ } = \dfrac{32 \text{ pages}}{1 \text{ h}}$

6. $\dfrac{22{,}000 \text{ rotations}}{1 \text{ h}} \cdot \dfrac{1 \text{ h}}{60 \text{ min}} \approx \underline{\ ?\ }$

7. $\dfrac{1170 \text{ mi}}{3 \text{ h}} \cdot \dfrac{1 \text{ h}}{60 \text{ min}} = \underline{\ ?\ }$

8. $\dfrac{2 \text{ ft}}{1 \text{ min}} \cdot \underline{\ ?\ } \cdot \dfrac{1 \text{ mi}}{5280 \text{ ft}} = \dfrac{1 \text{ mi}}{44 \text{ h}}$

Express each rate in the given units. For conversion rates, see the Table of Measures on page 611. *Section 3.1*

9. $88 \text{ km/h} \approx \underline{\ ?\ } \text{ m/s}$

10. $1270 \text{ ft/h} \approx \underline{\ ?\ } \text{ mi/day}$

11. $2 \text{ km/h} = \underline{\ ?\ } \text{ cm/h}$

12. $\$30{,}000/\text{min} = \underline{\ ?\ } \text{ per second}$

13. $\$290/\text{month} = \underline{\ ?\ } /\text{year}$

14. $14{,}700 \text{ papers/week} = \underline{\ ?\ } \text{ papers/day}$

Tell whether the data show direct variation. If they do, give the constant of variation and write an equation that describes each situation. *Section 3.2*

15.

Tickets sold	Money received (dollars)
20	150.00
50	375.00
75	562.50
100	750.00
120	900.00

16.

Old height	New height
15 cm	31.5 cm
24 cm	50.4 cm
30 cm	63.0 cm
45 cm	94.5 cm
50 cm	105.0 cm

17.

Hours bicycling	Distance traveled
3	45 km
4	64 km
5	70 km
6	95 km
7	105 km

18.

Number of bagels	Cost (dollars)
1	0.45
2	0.90
3	1.35
6	2.70
8	3.60

19.

Number of counties	State population (millions)
4	1.1
8	3.3
16	1.2
58	29.8

20.

Building height (ft)	Number of stories
400	33
550	40
600	44
900	64
1050	77

Graph each equation. Find the slope of the line. *Section 3.3*

21. $y = -2x + 3$

22. $y = \frac{1}{3}x - 1$

23. $y = 3x$

24. $y = \frac{1}{4}x + 5$

25. $y = 2$

26. $y = x - 6$

27. $y = 4 - x$

28. $y = -4x - 3$

29. $y = 9 + 2x$

30. $y = 2 + x$

31. $y = 5x - 2$

32. $y = -3 - x$

Find the slope of the line through each pair of points. *Section 3.3*

33. $(2, 3)$ and $(5, 6)$

34. $(-1, 5)$ and $(2, -1)$

35. $(3, -3)$ and $(5, 7)$

36. $(5, -2)$ and $(6, -4)$

37. $(8, 2)$ and $(3, -4)$

38. $(-4, -4)$ and $(3, 3)$

39. $(-6, 2)$ and $(-2, 3)$

40. $(10, -7)$ and $(15, 8)$

41. $(-7, -9)$ and $(1, 1)$

42. $(14, 6)$ and $(-8, 9)$

43. $(8, -12)$ and $(-11, 3)$

44. $(20, 2)$ and $(4, -14)$

45. $(4, -13)$ and $(-3, -3)$

46. $(25, 35)$ and $(20, 50)$

47. $(-21, -7)$ and $(-2, 11)$

Give the slope and *y*-intercept of each graph. Then write an equation in slope-intercept form. *Section 3.4*

48.

49.

50.

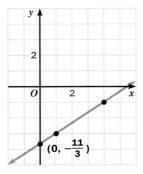

Graph each equation. Give the slope and *y*-intercept of the graph if they exist.

Section 3.4

51. $y = x - 2$

52. $y = 4x - 2$

53. $y = -3x + 1$

54. $y = -x + 5$

55. $y = 8x$

56. $x = 7$

57. $y = \frac{1}{2}x + 6$

58. $y = x + 1$

59. $y = \frac{1}{3}x - 2$ **60.** $y = \frac{2}{3}x + 4$ **61.** $y = \frac{3}{2}x$ **62.** $y = 5x - 10$

63. $y = -5x - (-1)$ **64.** $y = -1$ **65.** $y = -7x - 8$ **66.** $y = \frac{5}{2} - \frac{3}{2}x$

Find an equation of the line with the given slope and through the given point.
Section 3.5

67. slope $= 2$; $(1, 3)$ **68.** slope $= 5$; $(-2, 1)$ **69.** slope $= -3$; $(4, -5)$

70. slope $= \frac{1}{3}$; $(-8, 2)$ **71.** slope $= 0$; $(-4, 2)$ **72.** slope $= 6$; $(8, -12)$

73. slope is undefined; $(5, -14)$ **74.** slope $= -\frac{2}{3}$; $(9, 4)$ **75.** slope $= \frac{1}{4}$; $(3, -2)$

76. slope $= \frac{7}{2}$; $(6, 1)$ **77.** slope $= -11$; $\left(\frac{1}{2}, 0\right)$ **78.** slope $= -4$; $\left(-\frac{1}{5}, \frac{3}{10}\right)$

Find an equation of the line through the given points. *Section 3.5*

79. $(1, 5), (2, 4)$ **80.** $(-2, 3), (4, 9)$ **81.** $(4, -3), (-5, -5)$ **82.** $(-1, -7), (2, 2)$

83. $(2, -1), (-1, -9)$ **84.** $(0, 0), (3, -6)$ **85.** $(3, 4), (6, 0)$ **86.** $(-4, 10), (-5, 7)$

Find an equation that models the data in each scatter plot. *Section 3.6*

87.   **88.**   **89.**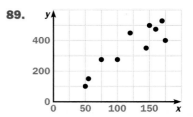

CHAPTER 4

For Exercises 1–4, use the graph at the right. *Section 4.1*

1. a. What is the cost of one round-trip ride?

   b. Write an expression for the cost of x round-trip rides.

2. Suppose you want to spend as little as possible on bus fares. If you plan to make 15 round-trip rides this month, should you buy a monthly pass or pay each day? Why?

3. Which costs more, 23 daily rides or a monthly pass? How much more?

4. When does it become more economical to buy a monthly bus pass? Explain.

Bus Fares

Rewrite each expression with a fraction as the coefficient. Then give the reciprocal of the coefficient. *Section 4.2*

5. $\dfrac{3a}{-2}$

6. $\dfrac{h}{12}$

7. $\dfrac{-5b}{8}$

8. $\dfrac{-r}{9}$

Solve each equation. *Section 4.2*

9. $\dfrac{2}{3}n = 14$

10. $\dfrac{-1}{8}a = -3$

11. $\dfrac{2x}{5} = 6$

12. $15 = -\dfrac{1}{4}h$

13. $\dfrac{-7c}{3} = -14$

14. $-9 = \dfrac{2}{7}y$

15. $\dfrac{-z}{6} = 5$

16. $-\dfrac{3t}{8} = -24$

17. $\dfrac{-4x}{3} = -8$

18. $\dfrac{3}{4}x + 1 = 16$

19. $8 = \dfrac{2r}{3} - 4$

20. $\dfrac{3}{5}n + 4 = -5$

21. $7 = \dfrac{1}{2}x - 5$

22. $-\dfrac{3}{2}y + 2 = -13$

23. $-3 + \dfrac{1}{4}s = 19$

24. $-12 = -\dfrac{5n}{2} + 3$

25. $\dfrac{-a}{7} - 6 = 2$

26. $-\dfrac{5}{6}n + 1 = 0$

27. $\dfrac{-b}{3} - 1 = -5$

28. $3 = \dfrac{2x}{5} + 6$

Solve each equation. *Section 4.3*

29. $2x + 7 = 3x$

30. $9a = 6a - 6$

31. $5b - 2 = 8b - 14$

32. $-8n + 3 = -11 - n$

33. $3r - 10 = -5r - 14$

34. $2n + 6 = 5n$

35. $8y = 3y - 10$

36. $5(a + 3) - 6 = 2a - 6$

37. $\dfrac{1}{2}(4n + 6) = 3n + 2(9 - 2n)$

Solve each equation. If an equation is an *identity* or there is *no solution*, say so.
Section 4.3

38. $5n - 4 = 10n + 6$

39. $3(x - 7) = 15 + 3x$

40. $7a + 3(a - 9) = 3$

41. $2(n - 3) = 2n - 6$

42. $4(x + 4) = 3(x - 1)$

43. $-7(m + 2) = 5(m + 3)$

44. $3(2x - 1) = 5x + 18$

45. $7(x - 2) = 3 + 7x$

46. $3d + 7 = 6 - d$

47. $\dfrac{4}{5}x - 1 = \dfrac{4}{5}x + 1$

48. $-5 - \dfrac{1}{2}c = 6 + \dfrac{1}{2}c$

49. $2n + 6 - 6n = 2(3 - 2n)$

50. $2(-5 - 3a) = 7a + 3$

51. $4(x - 5) = 3(6 - 2x)$

52. $8g - 12 = 3(4g - 4)$

53. $-1 - t = -t + 1$

54. $6n + 3 = 2n + 1$

55. $4(h - 3) = h - 11 + 3h$

56. $44 - 5x = 4(-3x - 3)$

57. $\dfrac{3}{2}(6b - 10) = 4b$

58. $9c - 15 = -3(5 - 3c)$

Find the least common denominator of the fractions in each equation. Then solve each equation. *Section 4.4*

59. $\dfrac{r}{4} - \dfrac{r}{5} = 6$

60. $\dfrac{1}{2}a + 6 = \dfrac{1}{3}a$

61. $\dfrac{1}{5}n - 2n = 1$

62. $2 - \dfrac{2}{3}s = \dfrac{5}{6}s$

63. $\dfrac{x}{3} - \dfrac{x}{5} = \dfrac{4}{3}$

64. $\dfrac{3}{4}b - \dfrac{2}{3}b = \dfrac{1}{6}$

65. $\dfrac{1}{8}n - \dfrac{1}{2} = \dfrac{1}{4}n$

66. $\dfrac{3c}{5} - \dfrac{c}{3} = -24$

Tell what power of ten you would multiply by to make each equation easier to solve. Then solve each equation. *Section 4.4*

67. $3 = 0.1n + 1.2$

68. $5 - 1.5m = 0.5$

69. $0.5x - 0.02 = 1.48$

70. $0.15 + 1.8c = 1.95$

71. $2.1 = 0.5x + 0.2x$

72. $1.5 = 6n - 1.5n$

73. $0.002y - 0.018 = 0.7$

74. $0.02t = 4.83 - 0.01t$

75. $1.8 - 0.3b = 0.3$

76. $0.09d + 0.11 = 0.2d$

77. $3x - 0.02 = 1.8x$

78. $0.2 - 0.3a = 1.2a - 1$

For Exercises 79–81, graph each equation. Then write an inequality for each graph that represents the *y*-values when *x* < 1. *Section 4.5*

79. $y = x - 2$

80. $y = -3x$

81. $y = -x + 3$

For Exercises 82–84, use the graph. *Section 4.5*

82. The graph shows the costs of buying videos through two video clubs. Which club charges a membership fee? What is the fee?

83. Write an inequality that shows when it is more economical to join Club B. Let v = the number of videos purchased.

84. Write an inequality that shows when it is more economical to join Club A. Let v = the number of videos purchased.

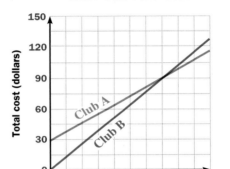

Video Club Purchases

Solve each inequality. Graph the solution on a number line. If there is no solution, write *no solution*. *Section 4.6*

85. $x - 3 > 1$

86. $-4a > -20$

87. $\frac{1}{2}x < 5$

88. $6 \le -3d$

89. $-3t + 4 < -8$

90. $3x - 1 \ge 2$

91. $7n + 6 > 6n$

92. $2t + 4 \le -7 + 2t$

93. $32 - 5a > -22$

94. $1 - 3s < -8 - s$

95. $4r + 3 \le 5r - (-3)$

96. $5c \ge 6(c - 1)$

97. $\frac{2n}{3} > \frac{n}{4} - 5$

98. $-4k + 3.5 \le 2.3 - 4k$

99. $4(3x + 4) < -2(-2 + x)$

Write and solve an inequality for each problem. *Section 4.6*

100. Dawn's neighbor will pay her $13 to mow the front lawn and $4 per hour to take care of the garden. If she mows the front lawn, what is the least number of hours she has to work on the garden in order to earn at least $25?

101. The weight of a carton of books to be shipped must be under 44 lb. If each book weighs 0.6 lb and the carton weighs 2 lb, what is the greatest number of books that can go in the carton?

CHAPTER 5

Find the ratio of the first quantity to the second quantity. Write your answer as a fraction in lowest terms. *Section 5.1*

1. 150 m; 300 m
2. 6 in.; 2 ft
3. 1 m; 1 km
4. 200 mL; 2 kL
5. 1 mi; 2000 ft
6. 1 mm; 2 m
7. 192 oz; 3 lb
8. 20 g; 4 kg
9. 0.72 m; 40 cm

Solve each proportion. *Section 5.1*

10. $\dfrac{5}{12} = \dfrac{10}{n}$

11. $\dfrac{1}{4} = \dfrac{b}{10}$

12. $\dfrac{v}{8} = \dfrac{63}{18}$

13. $\dfrac{12}{6} = \dfrac{a}{18}$

14. $\dfrac{1.2}{4.8} = \dfrac{0.7}{k}$

15. $\dfrac{2}{5.6} = \dfrac{3}{n}$

16. $\dfrac{2}{3} = \dfrac{12}{2a}$

17. $\dfrac{x}{3.5} = \dfrac{2}{7}$

18. $\dfrac{21}{1.5} = \dfrac{1.4}{v}$

19. $\dfrac{x+2}{3} = \dfrac{9}{2}$

20. $\dfrac{9}{4a} = \dfrac{1}{6}$

21. $\dfrac{1}{x+3} = \dfrac{6}{7}$

22. $\dfrac{c}{1+c} = \dfrac{5}{9}$

23. $\dfrac{3p-2}{4p} = \dfrac{3}{8}$

24. $\dfrac{d+3}{6} = \dfrac{d-1}{2}$

25. A floor plan of a house is drawn to a scale of 1 in. = $2\frac{1}{2}$ ft. Find the dimensions of these rooms in the floor plan. *Section 5.2*

    a. living room: 15 ft by 20 ft

    b. bathroom: 8 ft by 10 ft

    c. kitchen: 12 ft by 13 ft

    d. bedroom: 14 ft by 16 ft

26. Give the scale factor used in each situation.

    a. enlarge a 24 cm long picture to 64 cm long

    b. reduce a 3 ft high poster to 15 in. high

    c. reduce a 9 ft long banner to 9 in. long

    d. enlarge a 7.5 ft wide billboard photo to 4 yd wide

27. Fred is using a copy machine. Give the scale factor he should use to make each change.

    a. enlarge a 3 in. wide photo to $4\frac{1}{2}$ in. wide

    b. enlarge a 9 cm wide chart to 21 cm wide

    c. reduce an 11 in. by 14 in. paper to approximately $8\frac{1}{2}$ in. by 11 in.

    d. reduce a 7 in. wide diagram to $3\frac{1}{2}$ in. wide

For Exercises 28–33, use figures *ABCD* and *PQRS*.
Figure *ABCD* ~ figure *PQRS*. *Section 5.3*

28. Find the measure of ∠ *B*.

29. Find the measure of ∠ *P*.

30. $\overline{DC}$ and __?__ are corresponding sides.

31. What is the ratio of *BC* to *QR*?

32. Find *RS*.

33. Find *PS*.

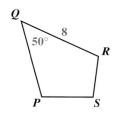

For Exercises 34–36, tell whether the triangles are similar. If they are, list each pair of corresponding angles and sides. If they are not similar, explain why. *Section 5.3*

34.

35.

36.

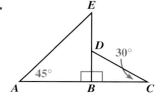

37. Suppose quadrilateral *DEFG* is similar to quadrilateral *HIJK*, and ∠ *F* = 120°. What is the measure of ∠ *J*?

38. **a.** Suppose △*ABC* ~ △*DEF* and *AB* = 20, *DE* = 5, and *AC* = 16. How long is $\overline{DF}$?

   b. Suppose *EF* = 3. How long is $\overline{BC}$?

Find the missing values for each pair of similar figures. *Section 5.4*

39.

$P = \underline{\ ?\ }$
$A = 42$

$P = 48$
$A = \underline{\ ?\ }$

40.

$P = 40$
$AB = \underline{\ ?\ }$
$A = \underline{\ ?\ }$

$P = \underline{\ ?\ }$
$EF = \underline{\ ?\ }$

41.

$P = \underline{\ ?\ }$
$A = 200$

$P = 140$
$A = \underline{\ ?\ }$

For Exercises 42–46, use the spinner shown. Write each probability as a fraction and as a percent. *Section 5.5*

42. a. What is the probability of spinning a 6?

   b. What is the probability of spinning a number other than 6?

43. What is the probability of spinning an even number?

44. What is the probability of spinning a number greater than 8?

45. What is the probability of spinning a number less than 5?

46. a. Suppose that for 120 spins, an odd number came up 40 times. What is the experimental probability of spinning an odd number?

   b. What is the theoretical probability of spinning an odd number?

For Exercises 47–52, the targets are a regular hexagon, squares, an equilateral triangle, and a circle. Find the probability that a randomly thrown dart that hits each target hits the shaded area. *Section 5.6*

47.

48.

49.

50.

51.

52.

CHAPTER 6

Find the unknown length for each right triangle. *Section 6.1*

1.

2.

3.

Tell whether the given lengths can be the sides of a right triangle. *Section 6.1*

4. 8 cm, 15 cm, 17 cm

5. 5 in., 8 in., 9 in.

6. 20 ft, 48 ft, 52 ft

7. 1 m, 2.4 m, 2.6 m

8. 7 yd, 7 yd, 13 yd

9. 0.3 cm, 0.4 cm, 0.5 cm

10. 33 cm, 544 cm, 545 cm

11. 3.5 m, 8.4 m, 9.1 m

12. 56 ft, 105 ft, 119 ft

The legs of each right triangle have lengths *a* and *b*. The hypotenuse has length *c*. Find the unknown length for each right triangle. *Section 6.1*

13. $a = 9, b = 12$

14. $a = 15, c = 25$

15. $b = 36, c = 39$

16. $a = 7, c = 25$

17. $a = 16, b = 30$

18. $b = 60, c = 61$

Write each decimal as a fraction in lowest terms. *Section 6.2*

19. $0.111\ldots$

20. 0.02

21. $0.7272\ldots$

22. $0.012012\ldots$

23. $0.\overline{6}$

24. $0.\overline{396}$

25. $1.\overline{7}$

26. $0.011011\ldots$

27. $0.\overline{006}$

Tell whether each number is *rational* or *irrational*. *Section 6.2*

28. $1.666\ldots$

29. $-\sqrt{25}$

30. $\sqrt{10}$

31. $\dfrac{22}{7}$

32. $0.0202202220\ldots$

33. $\sqrt{121}$

Estimate each square root within a range of two integers. *Section 6.3*

34. $\sqrt{30}$

35. $\sqrt{95}$

36. $\sqrt{41}$

37. $\sqrt{135}$

Simplify each expression. *Section 6.3*

38. $\sqrt{32}$

39. $\sqrt{45}$

40. $\sqrt{63}$

41. $\sqrt{80}$

42. $\sqrt{28}$

43. $\sqrt{90}$

44. $\sqrt{144}$

45. $\sqrt{126}$

46. $\sqrt{68}$

47. $3\sqrt{75}$

48. $3\sqrt{500}$

49. $\sqrt{\dfrac{3}{4}}$

50. $\sqrt{\dfrac{9}{49}}$

51. $\sqrt{\dfrac{1}{36}}$

52. $9\sqrt{3} + 3\sqrt{3}$

53. $2\sqrt{27} + \sqrt{12}$

54. $\sqrt{50} - 2\sqrt{2}$

55. $3\sqrt{20} + 3\sqrt{45}$

56. $\sqrt{48} - \sqrt{75}$

57. $\sqrt{5} \cdot \sqrt{5}$

58. $2\sqrt{2} \cdot \sqrt{8}$

59. $\sqrt{12} \cdot 2\sqrt{6}$

60. $2\sqrt{6^2}$

61. $3\sqrt{10} \cdot 2\sqrt{9}$

For Exercises 62–64, write an equation for each tile model. *Section 6.4*

62.

63.

64.

Find each product. *Section 6.4*

65. $2(3x - 6)$

66. $6(x + 1)$

67. $5x(x)$

68. $x(9x)$

69. $4x(2x + 7)$

70. $(4 - 3x)7$

71. $5x(2 + 4x)$

72. $(x - 2)(x + 3)$

73. $(x + 8)(5x - 6)$

74. $(12x - 15)(-3x)$

75. $(5x - 3)(2x + 8)$

76. $(-4)(5 + \sqrt{5})$

77. $4(6 + \sqrt{5})$

78. $\sqrt{2}(1 + \sqrt{2})$

79. $(5 - \sqrt{14})(3 - \sqrt{2})$

80. $(4\sqrt{2} + 2)(3\sqrt{3} - 1)$

Find each product. *Section 6.5*

81. $(x + 4)^2$

82. $(y + 6)(y - 6)$

83. $(a - 8)^2$

84. $(g - 4)(g + 4)$

85. $(2w + 1)(2w - 1)$

86. $(4x - 2)^2$

87. $(7k + 0.4)^2$

88. $(r - s)^2$

89. $(6b - 2c)^2$

90. $(x + 4y)(x - 4y)$

91. $(2 - \sqrt{7})^2$

92. $(7 + \sqrt{5})(7 - \sqrt{5})$

Find each product using mental math. *Section 6.5*

93. 31^2

94. $27 \cdot 33$

95. $(5.2)^2$

96. $\left(4\frac{1}{2}\right)^2$

Solve. *Section 6.5*

97. The side of a square is $3 + \sqrt{5}$ units long. What is the area of the square? What is the perimeter?

98. The length of a rectangle is $n + 7$ units and the width is $n - 7$ units. What expression represents the area of the rectangle?

CHAPTER 7

Graph each equation by making a table and graphing each point. *Section 7.1*

1. $x + 3y = 3$

2. $4x - 2y = 12$

3. $-3x + y = -6$

4. $2y = -3x + 7$

5. $8x + y = -8$

6. $-6 + 3x = 2y$

7. $-1 - 2x = 5y$

8. $-x + 7y = 14$

9. $0.6x + 0.1y = 2$

10. $\frac{1}{2}x + \frac{2}{3}y = 2$

11. $\frac{2}{5}x = \frac{1}{3}y - 3$

12. $0.6x - 1.5y = 3$

For Exercises 13–15, match each equation with its *x*- and *y*-intercepts. *Section 7.1*

13. $y = \frac{1}{2}x - 3$

A. The *x*-intercept is 2. The *y*-intercept is 2.

14. $x + y = 2$

B. The *x*-intercept is 6. The *y*-intercept is 4.

15. $y = -\frac{2}{3}x + 4$

C. The *x*-intercept is 6. The *y*-intercept is -3.

Use substitution to solve each system of equations. *Section 7.2*

16. $x + y = 3$
$x - y = 1$

17. $2x - y = 1$
$x - 2y = 1$

18. $x - y = 2$
$3x + y = 6$

19. $y = -2x + 3$
$y = 3x - 1$

20. $x = y$
$x = 3$

21. $3x + y = 2$
$x + 2y = -1$

22. $x + y = 8$
$x - y = -5$

23. $y = -x + 3$
$y = 2x - 3$

 Technology Solve each system of equations by graphing. Use a graphing calculator if you have one.

24. $2x + y = -2$
$-5x + 2y = 3$

25. $4x = y$
$\frac{1}{2}x = y$

26. $2x - 2y = 1$
$x + 3y = 6$

27. $5 - 2y = x$
$6 + y = 2x$

Solve each system of equations by adding or subtracting. *Section 7.3*

28. $2x + y = 8$
$x - y = 1$

29. $4a - 6b = 2$
$-4a + b = 5$

30. $3n - 2p = 6$
$3n - 7p = -4$

31. $3 = -2x + 2y$
$-7 = 3x - 2y$

32. $6x - 2 = -4y$
$-2x + 10 = 4y$

33. $a + 2b = -6$
$7a + 2b = 6$

34. $a + b = 6$
$a - b = 4$

35. $x - y = -4$
$x = 6 - y$

36. $2n + p = 8$
$-2n + 3p = 0$

37. $3x + y = 13$
$2x = y + 2$

38. $5y = 3x + 12$
$7x - 5y = 8$

39. $-8 = 3a - b$
$-6 = -3a - b$

For each system of equations, give the solution or state whether there are
no solutions or many solutions. *Section 7.4*

40. $x + y = 3$
$2x + 2y = 2$

41. $x + 2y = 5$
$2x + y = 7$

42. $3a + 2b = 8$
$6a + 4b = 12$

43. $-3 = r - 2s$
$9 = 3r - 3s$

44. $2x + y = 2$
$2y = -4x + 4$

45. $2a + b = 7$
$3a - 2b = 7$

46. $4 = 2x + 3y$
$2 = 3x + 4y$

47. $b = -3a - 6$
$2b = a + 2$

48. $y = -3x + 4$
$2y = -x + 3$

49. $y = 4x + 2$
$8x + 2y = 4$

50. $3r + 4s = -7$
$2r + s = -3$

51. $2y = 8x + 6$
$y = 2x + 9$

52. $x + y = -5$
$2x + 2y = -10$

53. $3x - 2y = 0$
$3x - 2y = 5$

54. $2a + 3b = -7$
$3a - 2b = -4$

55. $2x + 5y = 10$
$15y = -6x + 30$

Match each inequality with its graph. *Section 7.5*

56. $y \le 2$

57. $x \ge 2$

58. $y \ge -2$

59. $x \le 2$

A.

B.

C.

D.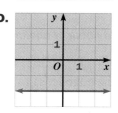

Graph each inequality. *Section 7.5*

60. $y \le x - 1$

61. $y > 2x + 6$

62. $x + y \ge -2$

63. $y \le 4x$

64. $x - y < -1$

65. $2x - 3y < 9$

66. $y \ge -x + 3$

67. $y > 3x$

68. $x \ge 5$

69. $x < y$

70. $\frac{1}{2}x + 2y \ge 3$

71. $3x - 4y > 12$

72. $6 < y - 2x$

73. $0.4x - y < 3$

74. $-\frac{x}{3} + \frac{y}{5} \le \frac{2}{5}$

75. $y \le 1.6x + 2$

For Exercises 76–79, tell which region, *A, B, C,* or *D*, is the solution of each system of inequalities. *Section 7.6*

76. $y < -2x + 2$
    $y > x - 1$

77. $y < x - 1$
    $y < -2x + 2$

78. $y < x - 1$
    $y > -2x + 2$

79. $y > -2x + 2$
    $y > x - 1$

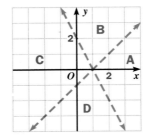

Graph each system of inequalities. *Section 7.6*

80. $y < 3$
    $x > -2$

81. $y \geq x - 2$
    $y \geq -x + 1$

82. $x + y > 5$
    $y < x - 1$

83. $2x + y \geq 2$
    $x - 2y \geq 3$

84. $y \geq \frac{1}{2}x - 1$
    $y \leq 2x$

85. $3x - y < 1$
    $x + 2y \leq 2$

86. $4 < 2x - y$
    $-2y > x$

87. $2y < -x - 5$
    $3x + y \geq -3$

CHAPTER 8

Graph each equation. Tell whether each equation is *linear* or *nonlinear*. *Section 8.1*

1. $y = 4x - 2$

2. $y = |x| - 2$

3. $y = 2x^2$

4. $y = \frac{1}{2}x$

5. $y = -2x^2 + 1$

6. $y = -2|x|$

7. $y = 2x + 1$

8. $x^2 - 2 = y$

9. $y = \frac{1}{3}x^3$

10. $y = 6x$

11. $y = -x + 3$

12. $y = -4x$

Tell whether each relationship is *linear* or *nonlinear*. *Section 8.1*

13. the relationship between any positive number x and its reciprocal, $\frac{1}{x}$

14. the relationship between the edge length s of a cube and its surface area, $6s^2$

15. the relationship between any number x and its opposite, $-x$

 Technology For Exercises 16–23, use a graphing calculator to check your predictions. *Section 8.2*

a. **Predict how the graph of the equation will compare with the graph of $y = x^2$.**

b. **Sketch the graph of each equation.**

16. $y = -2x^2$

17. $y = \frac{1}{3}x^2$

18. $y = -x^2$

19. $y = -\frac{1}{4}x^2$

20. $y = 7x^2$

21. $y = 0.1x^2$

22. $y = \frac{5x^2}{2}$

23. $y = -1.75x^2$

For Exercises 24 and 25, match each graph with its equation. *Section 8.2*

24.

25.
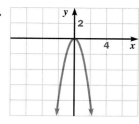

A. $y = -\dfrac{1}{4}x^2$

B. $y = -2x^2$

C. $y = \dfrac{1}{2}x^2$

D. $y = 4x^2$

Solve each equation algebraically or by using a graph. *Section 8.3*

26. $x^2 - 25 = 0$

27. $144 = n^2$

28. $4a^2 = 64$

29. $-3x^2 = -27$

30. $\dfrac{1}{3}p^2 = 12$

31. $42 = n^2 + 6$

32. $b^2 - 15 = 1$

33. $4x^2 = 196$

34. $196 = x^2$

35. $3x^2 + 2 = 245$

36. $11 = 2n^2 + 7$

37. $g^2 = 0.01$

38. $\dfrac{2}{5}x^2 - 3 = 1$

39. $\dfrac{1}{2}c^2 - 2 = -1$

40. $12a^2 - 5 = 55$

41. $\dfrac{3}{4}x^2 - 4 = 2$

Solve each equation using the given graph. *Section 8.4*

42. $x^2 + x - 6 = 0$

43. $x^2 + 6x + 9 = 0$

44. $-x^2 + 5x - 4 = 0$

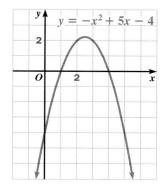

Technology Use a graphing calculator to solve each equation. If an equation has no solution, write *no solution*. *Section 8.4*

45. $x^2 + 3x - 8 = 0$

46. $x^2 - 8x = -15$

47. $2x^2 - 8x + 6 = 0$

48. $-x^2 + x - 3 = 0$

49. $2x^2 - 3x = 5$

50. $-x^2 + x = 2$

51. $42x^2 = 10$

52. $-5x^2 - 2x + 18 = 0$

53. $3x - 5x^2 = 6$

54. $3x^2 + 6x + 4 = 0$

55. $-x^2 + x = 5$

56. $2x^2 - 2x + 6 = 0$

Solve each equation using the quadratic formula. *Section 8.5*

57. $x^2 - 9x + 20 = 0$

58. $3x^2 - 3x - 6 = 0$

59. $x^2 - 2x - 15 = 0$

60. $-2x^2 + 4x = 0$

61. $-n^2 + 5n - 6 = 0$

62. $a^2 - 6a + 9 = 0$

63. $2k^2 + 9k - 5 = 0$

64. $p^2 + 2p + 1 = 0$

65. $-x^2 + 6x = -16$

66. $x^2 + 2x + 1 = 0$

67. $9x^2 - 18x + 9 = 0$

68. $5x^2 + 10x - 7 = 0$

69. $-2m^2 + 4m = -6$

70. $16g^2 + 16g = 32$

71. $x^2 - x = 42$

72. $3x^2 - 23x = -30$

For each equation, find the number of solutions. *Section 8.6*

73. $x^2 + 7x + 12 = 0$ **74.** $y^2 + y + 1 = 0$ **75.** $2c^2 + 5 = 0$ **76.** $6x^2 - x - 2 = 0$

77. $-2x^2 = 2x + 12$ **78.** $3x^2 + 6x + 3 = 0$ **79.** $x^2 + 8 = 7x$ **80.** $3x^2 + x + 2 = 0$

For each equation, find the number of solutions. Then solve using the quadratic formula. *Section 8.6*

81. $x^2 + 4x + 4 = 0$ **82.** $x^2 - 3x = 18$ **83.** $x^2 - 5x - 24 = 0$ **84.** $3x^2 + 13x - 10 = 0$

85. $-x^2 + x + 20 = 0$ **86.** $x^2 - 4x = 0$ **87.** $x^2 + 10x = -25$ **88.** $-x^2 + 14x - 48 = 0$

89. $5x^2 + 3x + 2 = 0$ **90.** $4x^2 - 17x = -4$ **91.** $x^2 + 16x + 64 = 0$ **92.** $x^2 + 4x = 1$

CHAPTER 9

Rewrite each product using exponents. *Section 9.1*

1. $ab \cdot ab \cdot ab \cdot ab \cdot ab \cdot ab$

2. $-4(6 \cdot 6 \cdot 6 \cdot 6 \cdot 6 \cdot 6 \cdot 6)$

3. $\left(-\frac{1}{2}\right)\left(-\frac{1}{2}\right)\left(-\frac{1}{2}\right)\left(-\frac{1}{2}\right)\left(-\frac{1}{2}\right)$

4. $-b \cdot b \cdot b$

5. $3\left(\frac{2}{5}\right)\left(\frac{2}{5}\right)$

6. $cde \cdot cde \cdot cde \cdot cde$

Evaluate each power. *Section 9.1*

7. 7^3 **8.** 3^5 **9.** $(-5)^2$ **10.** $(1.1)^4$

11. $\left(\frac{1}{3}\right)^3$ **12.** $\left(\frac{1}{2}\right)^6$ **13.** $(-1.5)^3$ **14.** $-\left(\frac{1}{6}\right)^2$

Evaluate each expression when $x = 2$ and $y = 4$. *Section 9.1*

15. x^2y **16.** $2y^2$ **17.** $-5x^2y$ **18.** $x^2 - y^3$

19. $(y - x)^3$ **20.** y^x **21.** $4(x + y)^3$ **22.** y^y

The equations in Exercises 23–28 model the growth of a quantity y over a certain number of intervals. For each equation:

a. Give the initial amount of the quantity.

b. Give the growth factor and the growth rate.

c. Give the number of intervals. *Section 9.2*

23. $y = 26(1.25)^3$ **24.** $y = 2(1.5)^7$ **25.** $y = 300(1.35)^{10}$

26. $y = 750(1.01)^5$ **27.** $y = 1500(1.8)^4$ **28.** $y = 50(2.1)^9$

For Exercises 29–34, find the value of y when $x = 6$. Round decimal answers to the nearest hundredth. *Section 9.2*

29. $y = 30(1.05)^x$ **30.** $y = 33(2.01)^x$ **31.** $y = 52(12)^x$

32. $y = 2.5(1.04)^x$ **33.** $y = 600(1.15)^x$ **34.** $y = 14(2.5)^x$

35. A population of 250 fish grows at the rate of 20% per year. What is the expected population in five years?

Find the value of y when $x = 4$. Round decimal answers to the nearest hundredth. *Section 9.3*

36. $y = 25(0.5)^x$ **37.** $y = 200(0.1)^x$ **38.** $y = 50\left(\dfrac{1}{4}\right)^x$

Find the value of y when $x = 3$. Round decimal answers to the nearest hundredth. *Section 9.3*

39. $y = 4(0.75)^x$ **40.** $y = 100(0.9)^x$ **41.** $y = 25(0.8)^x$ **42.** $y = 450(0.3)^x$

43. $y = 325\left(\dfrac{1}{2}\right)^x$ **44.** $y = 900\left(\dfrac{2}{3}\right)^x$ **45.** $y = 0.7(0.95)^x$ **46.** $y = 1250(0.65)^x$

The equations in Exercises 47–52 describe quantities that grow or decay exponentially. For each equation:

a. Tell whether the quantity described *grows* or *decays*.

b. Give the initial amount of the quantity.

c. Give the growth or decay factor for the quantity. *Section 9.3*

47. $y = 455(1.15)^x$ **48.** $y = 20(0.6)^x$ **49.** $y = 335(1.01)^x$

50. $y = 75(0.87)^x$ **51.** $y = 100(2.3)^x$ **52.** $y = 575(0.1)^x$

Rewrite each expression using only positive exponents. *Section 9.4*

53. x^{-3} **54.** a^{-10} **55.** $b^{-2}c^{-5}$ **56.** $7m^{-4}$

57. $y^{-4}z^7$ **58.** $6p^4q^{-2}$ **59.** $w^2x^{-1}z^3$ **60.** $\left(\dfrac{x}{y}\right)^{-6}$

61. $\dfrac{b^{-5}}{4}$ **62.** $10a^{-2}b^{-6}$ **63.** $\dfrac{m^{-5}}{n^6}$ **64.** $\dfrac{a^7}{b^{-3}}$

65. $(3c)^{-2}$ **66.** $(ab)^{-4}$ **67.** $2x^{-5}y^{-4}z^8$ **68.** $(pqr)^{-3}$

Evaluate each expression when $a = 2$, $b = 3$, and $c = 4$. *Section 9.4*

69. a^{-5} **70.** bc^{-1} **71.** $a^{-2}c^0$ **72.** $\left(\dfrac{1}{c}\right)^{-3}$

73. a^4b^{-2} **74.** $(a + c)^{-4}$ **75.** $b^{-1} + c^{-2}$ **76.** $a^{-1}b^3c^{-3}$

77. $\left(\dfrac{a}{c}\right)^{-3}$ **78.** $4^{c - b}$ **79.** $a^{-5} - b^{-1}$ **80.** $b^{-a}c^a$

Write each number in decimal notation. *Section 9.5*

81. 3×10^5 **82.** 1.4×10^{-4} **83.** 5.341×10^0 **84.** 3.97×10^6

85. 4.503×10^{-5} **86.** 5.7×10^{-8} **87.** 3.001×10^{10} **88.** 6.517×10^{-7}

89. 9.9×10^{-6} **90.** 6.5×10^{13} **91.** 6.0032×10^{11} **92.** 7.03×10^{-9}

Write each number in scientific notation. *Section 9.5*

93. 9675   **94.** 3,000,000   **95.** 0.0058   **96.** 0.000045

97. 2.7 million   **98.** 0.000000862   **99.** 87,492,000,000   **100.** 0.0004

101. 9 millionths   **102.** 4.92 hundredths   **103.** 0.0004000562   **104.** 455 hundredths

105. 4506 millionths   **106.** 475 million   **107.** 6.8 million   **108.** 543,570,000,000,000

Write each expression as a single power. *Section 9.6*

109. $10^6 \cdot 10^3$ **110.** $2^8 \cdot 2^3$ **111.** $4^4 \cdot 4^{-1}$ **112.** $\dfrac{5^8}{5^4}$

113. $\dfrac{9^{-4}}{9^3}$ **114.** $7^{-5} \cdot 7^{-7}$ **115.** $\dfrac{3^5}{3^{-6}}$ **116.** $\dfrac{6^3 \cdot 6^{-2}}{6^4}$

117. $\dfrac{2^{-2}}{2^{-6}}$ **118.** $8^{-3} \cdot 8^6 \cdot 8^{-5}$ **119.** $\dfrac{2^9}{2^5 \cdot 2^{-2}}$ **120.** $\dfrac{3^{-4}}{3^6 \cdot 3^{-2}}$

Simplify each expression. Write your answers using positive exponents.

Section 9.6

121. $a^3 \cdot a^6$ **122.** $x^{-4} \cdot x^{-3}$ **123.** $a^2 \cdot b^{-3}$ **124.** $\dfrac{x^8}{x^6}$

125. $\dfrac{c^{-2}}{c^5}$ **126.** $\dfrac{y^5}{3y^3}$ **127.** $3b^5 \cdot 4b^{-2}$ **128.** $(5a^{-7}b)(3a^{-2}b^{-4})$

129. $(7c^8)(2c^{-3})$ **130.** $(2x^3y)(4x^{-6}y^4)$ **131.** $\dfrac{5bc^5}{b^2c^{-3}}$ **132.** $(5 \times 10^2)(1 \times 10^8)$

133. $\dfrac{6 \times 10^9}{2 \times 10^6}$ **134.** $\dfrac{4 \times 10^6}{8 \times 10^3}$ **135.** $\dfrac{4x^{-1}y^7}{2x^5y^{-3}}$ **136.** $(3 \times 10^6)(4 \times 10^5)$

Evaluate each power. Write your answers in scientific notation. *Section 9.7*

137. $(4 \times 10^6)^2$ **138.** $(2 \times 10^7)^3$ **139.** $(1.5 \times 10^{-4})^2$

140. $(1.0 \times 10^4)^{-2}$ **141.** $(3 \times 10^{-3})^4$ **142.** $(2.21 \times 10^{-8})^2$

Simplify each expression. *Section 9.7*

143. $(b^5)^3$ **144.** $(a^2c^3)^4$ **145.** $(3x)^3$ **146.** $\left(\dfrac{x}{y}\right)^8$

147. $\left(\dfrac{a^2}{b^4}\right)^3$ **148.** $(2c^4)^3$ **149.** $4(y^3)^4$ **150.** $(3a^3)^2$

151. $\left(\dfrac{b^3}{c^4}\right)^4$ **152.** $(x^3y^6)^5$ **153.** $(7a^2b^3c^5)^2$ **154.** $(2xy^4)^4$

CHAPTER 10

Write each polynomial in standard form. Then classify the polynomial as *linear*, *quadratic*, *cubic*, or *none of these*. *Section 10.1*

1. $2x - 3x^2 + 7$ **2.** $4a^2 + 2a - 3 - 3a^3$ **3.** $5c - 4c^2$

4. $2 - 3y^4 + 6y$ **5.** $\dfrac{1}{2}b^3 - 4 + 2b$ **6.** $1.6x + 4.5x^2 - 7.2$

7. $22 + \dfrac{1}{2}y$ **8.** $r^2 - 3r^4 + 7r + 1 - r^3$ **9.** $5t^2 - 8 + 9t$

Add or subtract. *Section 10.1*

10. $(5x^2 - 4x + 6) + (3x^2 + 6x - 2)$

11. $(4a^2 - 3a + 5) - (2a^2 + 5a + 7)$

12. $(6x^2 - 7xy + 5y^2) - (x^2 + 2xy + 6y^2)$

13. $(3b^2 - 4b + 5) - (4b^2 + 15)$

14. $(a^2 - 2b^2) + (8a^2 - 3ab - 2b^2)$

15. $(7x + y - 4) - (2x + 7 - 3y)$

16. $(3st^2 - 8st + 12) + (9st - 14)$

17. $(-a - 4b + 2) + (3a^2 + 7b - 8)$

18. $(8s^3 + 3s^2 - 5) + (s^3 - 3s^2 - 4)$

19. $(-r^3 - 3r - 7) - (-7r^3 + 2r^2 - 8r - 9)$

Multiply. *Section 10.2*

20. $2a(2a^2 - 3a)$

21. $6n(3n + 8)$

22. $-s(2s^2 + 5s + 4)$

23. $(x + 3)(x + 5)$

24. $(t + 9)(t - 4)$

25. $(4b - 3)(b - 7)$

26. $(3c - 8)(4c + 6)$

27. $(-4x + 5)(-4x - 5)$

28. $5n(2n^2 + 3n - 9)$

29. $0.2a(4a^2 - 3a - 5)$

30. $(n - 1)(n^2 + 3n)$

31. $(3d + 10)(-2d + 7)$

32. $(c - 3)(c^2 + 5c + 2)$

33. $(s + 4)(3s^2 - 5s - 12)$

34. $(7r^2 + 2r - 4)(3r - 5)$

35. $(-3y - 3)(6y^2 + y - 4)$

36. $(3a + 2b)(4a - 7b + 5)$

37. $(x + 5y)(x + 3y - 1)$

38. $(2x^2 - 4x + 5)(x - 5)$

39. $(2t - 3)(t^2 + 4t + 2)$

40. $(7x - 4)(x^2 - x - 8)$

Write a polynomial in standard form to represent the volume of each figure.
Section 10.2

41.

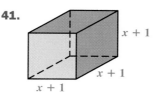

$x + 1$
$x + 1$
$x + 1$

42.

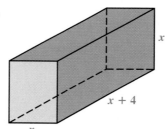

$x + 1$
$x + 4$
x

43.

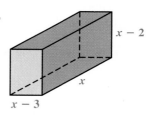

$x - 2$
x
$x - 3$

Factor each polynomial. *Section 10.3*

44. $a^2 + 8a$

45. $7n + 14$

46. $16 + 4b^2$

47. $x^3 - 5x$

48. $4c^2 + 8c$

49. $24t^2 - 6t$

50. $12y^2 + 8y$

51. $3v^2 - 6v$

52. $30b^2 - 25b$

53. $15a^2 + 9a - 9$

54. $16b^2 - 2b + 4$

55. $8z^3 - 4z$

56. $y^3 - 3y^2 - 3y$

57. $18x^3 + 6x^2 + 12x$

58. $2p^2 - 4p + 4$

59. $18x^3 - 14x^2 + 9x$

Solve each equation. *Section 10.3*

60. $b(b - 5) = 0$

61. $0 = 3x(x + 9)$

62. $8a(3a - 12) = 0$

63. $0 = 4x(6x + 3)$

64. $7y(5y - 8) = 0$

65. $0 = 0.9c(4c - 24)$

66. $(x - 4)(x - 9) = 0$

67. $(s - 8)(s + 3) = 0$

68. $(2t + 8)(t - 6) = 0$

69. $0 = (6x + 6)(3x + 4)$

70. $(-r + 5)(2r + 7) = 0$

71. $0 = (4k + 3)(7k - 5)$

72. $6t^2 - 2t = 0$

73. $0 = 4x^2 - 12x$

74. $15y^2 = -y$

75. $9a = 18a^2$

76. $20d^2 + 15d = 0$

77. $x^2 - 16x = 0$

For each set of algebra tiles, write a polynomial that represents the area. Then express the area as the product of two binomials. *Section 10.4*

78.

79.

80.

Factor, if possible. If not, write *not factorable*. *Section 10.4*

81. $x^2 + 4x + 4$

82. $a^2 + 7a + 12$

83. $b^2 - 8b + 7$

84. $y^2 + 13y + 13$

85. $c^2 - 7c + 10$

86. $n^2 + 11n + 28$

87. $b^2 - 14b + 49$

88. $t^2 + 13t + 30$

89. $g^2 - 12g + 24$

90. $3n^2 + 7n + 3$

91. $5n^2 - 7n + 2$

92. $7n^2 + 8n + 2$

93. $4c^2 + 8c + 3$

94. $4y^2 + 13y + 3$

95. $6n^2 - 5n + 1$

96. $9x^2 + 9x + 2$

97. $9t^2 - 19t + 2$

98. $10n^2 - 19n + 6$

Factor each polynomial. *Section 10.5*

99. $x^2 + 3x - 4$

100. $n^2 - n - 12$

101. $y^2 + y - 6$

102. $a^2 - 64$

103. $t^2 + 3t - 28$

104. $b^2 - 15b + 36$

105. $3x^2 + x - 2$

106. $4x^2 - 5x - 6$

107. $7d^2 - 25d + 12$

108. $8b^2 - 2b - 15$

109. $3x^2 - 12x - 15$

110. $6x^2 - 19x + 3$

111. $15c^2 + c - 2$

112. $10n^2 - 3n - 4$

113. $3r^2 - 8r - 3$

Solve each equation by factoring. *Section 10.5*

114. $a^2 - a - 12 = 0$

115. $x^2 - 11x + 30 = 0$

116. $0 = y^2 - 7y - 8$

117. $n^2 + 10n + 21 = 0$

118. $c^2 - 3c - 10 = 0$

119. $b^2 + 13b = -42$

120. $x^2 + 5x - 30 = 6$

121. $3n^2 - n - 4 = 0$

122. $4z^2 - 9 = 0$

123. $2t^2 - 5t = 3$

124. $8y^2 = 32$

125. $12x^2 - 4x - 21 = 0$

Solve each equation using whichever method you prefer. *Section 10.5*

126. $y^2 - 81 = 0$

127. $(3a - 2)(a + 3) = 0$

128. $0 = 9b^2 - 6b$

129. $12x^2 - x - 6 = 0$

130. $0 = x^2 + 16x + 64$

131. $(5x - 7)(2x + 9) = 0$

132. $8 = 3b^2 - 2b$

133. $9x^2 + 7x = 14 - 8x$

134. $10n^2 - 20n = 14n + 24$

CHAPTER 11

Decide whether the data show inverse variation. If they do, find the missing value. *Section 11.1*

1.

x	y
6	18
12	9
?	72
36	3

2.

x	y
14	12
4	42
−6	?
−7	−24

3.

x	y
$\frac{2}{3}$	?
3	112
4	84
8	42

4. It takes Roy four hours to travel to North Lake when he drives at an average speed of 45 mi/h. The time it takes him to drive to the lake varies inversely with his average speed.

   a. How fast must he drive in order to return home in $3\frac{1}{2}$ hours?

   b. If he averages 42 mi/h on the return trip, how long will it take him to get home?

Solve. *Section 11.2*

5. The average scores for the five teams in the Eastern Division of a mathematics competition are shown at the right. What is the average score for the students in the Eastern Division?

Team	Average score	Number of students
A	46	5
B	47	4
C	42	6
D	49	4
E	44	6

6. Gina's test grades in history are 92, 89, 89, 85, and 90. Her final exam for the marking period counts as two tests. What score does she need on the exam in order to get an average of at least 90 for the marking period?

7. The drama club has 350 adult tickets and 150 children's tickets for a performance. The club plans to sell the children's tickets for $3 and wants to earn an average price of $6 per ticket.

   a. What is the minimum amount the club needs to charge for the adult tickets?

   b. Suppose the club decides to charge $7 for an adult. What must it charge for a child's ticket in order to keep the average price at $6 per ticket?

Find the least common denominator of the fractions in each equation. *Section 11.3*

8. $\dfrac{1}{4x} + \dfrac{1}{8} = \dfrac{1}{2x}$

9. $\dfrac{1}{3x^2} + \dfrac{1}{4x^2} = \dfrac{1}{x+1}$

10. $\dfrac{2}{3(a+1)} = \dfrac{a}{3(a-1)}$

11. $\dfrac{4t}{2(t+1)} - \dfrac{1}{3t} = \dfrac{5}{5(t+1)}$

12. $\dfrac{8}{b^2} = \dfrac{3}{b(b+1)} + \dfrac{b}{b+1}$

13. $\dfrac{1}{4y} - \dfrac{3y}{2(y-2)} = \dfrac{7}{6(y+2)}$

Solve each equation. If the equation has no solution, write *no solution*.

Section 11.3

14. $\dfrac{5}{x} + 3 = \dfrac{2}{x}$ **15.** $\dfrac{2}{x} + \dfrac{2}{3} = \dfrac{2}{3}$ **16.** $\dfrac{a}{5} + \dfrac{a}{2} = 3$

17. $\dfrac{1}{x-2} + \dfrac{1}{2} = \dfrac{2}{x-2}$ **18.** $\dfrac{b-2}{b+3} = 2 + \dfrac{5}{b+3}$ **19.** $\dfrac{x+1}{6} = \dfrac{x+1}{4}$

20. $\dfrac{a-2}{a} + \dfrac{4}{3a} = \dfrac{1}{3}$ **21.** $\dfrac{3}{y} + \dfrac{2}{3} = \dfrac{1}{3y}$ **22.** $\dfrac{3}{x+2} = \dfrac{2}{x-2}$

23. $\dfrac{4}{x+3} = \dfrac{4}{x-6}$ **24.** $\dfrac{4}{a^2} + \dfrac{3}{a} = \dfrac{5}{a^2}$ **25.** $\dfrac{2}{x-2} + \dfrac{3}{(x+2)(x-2)} = \dfrac{1}{x+2}$

26. $\dfrac{3}{4(c-2)} + \dfrac{1}{3(c-2)} = \dfrac{2}{c+3}$ **27.** $\dfrac{1}{x(x-1)} - \dfrac{2}{x^2} = \dfrac{1}{x-1}$ **28.** $\dfrac{3}{b^3} + \dfrac{3}{b^2} = \dfrac{6}{b}$

Solve each equation for the variable in red. *Section 11.4*

29. $t = v + w + x$ **30.** $2a + b = c$ **31.** $cd = fg$

32. $4m - 5n = 25$ **33.** $y = cx + b$ **34.** $2m + 3n = 4p$

35. $t + st = 2$ **36.** $3 - 5m = 4n$ **37.** $a = \dfrac{b+2}{a}$

38. $\dfrac{2}{x} - \dfrac{1}{y} = z$ **39.** $4 - 6a = 15$ **40.** $\dfrac{2 - ab}{2} = c$

41. $\dfrac{y}{3} = \dfrac{2+x}{7}$ **42.** $a(b + 2) = b - 2 + c$ **43.** $2ax + 4ay = 5$

44. $4(a - 5) = b^2 - 2$ **45.** $\dfrac{a}{b} + \dfrac{c}{d} = a$ **46.** $a + 3b = 4a - 8$

Simplify each expression. If it is already in simplest form, write *simplest form*.

Section 11.5

47. $\dfrac{x^3}{5x}$ **48.** $\dfrac{7a^2}{3a}$ **49.** $\dfrac{a^2b}{a+b}$

50. $\dfrac{5y + 10}{5}$ **51.** $\dfrac{9t^2}{12y}$ **52.** $\dfrac{c^2 - 4}{c + 2}$

53. $\dfrac{n-6}{4n^2 - 24n}$ **54.** $\dfrac{(x+1)^2}{x-1}$ **55.** $\dfrac{4b^2 + 2b}{2b}$

Simplify each expression. *Section 11.5*

56. $\dfrac{4a^3}{3b} \cdot \dfrac{9b^2}{2a}$ **57.** $\dfrac{6x}{5y} \div \dfrac{2x^2}{15y^2}$ **58.** $(3n + 12) \cdot \dfrac{2}{2n + 8}$

59. $\dfrac{4}{n+1} \div \dfrac{2}{3n+3}$ **60.** $\dfrac{3x+6}{x-3} \cdot \dfrac{x^2}{x+2}$ **61.** $\dfrac{6a+9}{9b^2} \cdot \dfrac{3b}{4a+6}$

62. $\dfrac{1}{3x^2} \div \dfrac{x-2}{12x}$ **63.** $\dfrac{y}{y-4} \div \dfrac{2y^2 + y}{y-4}$ **64.** $\dfrac{x^2 - 9}{2(x+1)} \cdot \dfrac{6x+1}{x+3}$

Simplify each expression. *Section 11.6*

65. $\dfrac{1+x}{xy} + \dfrac{x-1}{xy}$

66. $\dfrac{2a}{a-1} + \dfrac{a-4}{a-1}$

67. $\dfrac{3x}{3x^2y} - \dfrac{3xy}{3x^2y}$

68. $\dfrac{4a^2}{a^2} + \dfrac{1}{a^2}$

69. $\dfrac{3}{y} - \dfrac{2}{y+2}$

70. $\dfrac{4}{2a^2b} + \dfrac{5}{3ab^2}$

71. $\dfrac{x}{x-2} + \dfrac{1}{x}$

72. $\dfrac{5}{x+3} + \dfrac{3}{x-4}$

73. $2 - \dfrac{1}{x+1}$

74. $\dfrac{y}{3x} - \dfrac{2}{x}$

75. $\dfrac{3}{x-5} + \dfrac{2}{x}$

76. $\dfrac{1}{2a} + \dfrac{3}{10a}$

77. $\dfrac{x}{x+1} - \dfrac{2}{x+1}$

78. $\dfrac{6}{8y} - \dfrac{2}{y^2}$

79. $\dfrac{1}{xy^2} - \dfrac{1}{x^2y}$

80. $\dfrac{3}{4n} - \dfrac{5}{4n+3}$

CHAPTER 12

For Exercises 1–3, write an algorithm that explains how to use a calculator to solve each problem. *Section 12.1*

1. Multiply 15 by 52 without using the multiplication key.

2. Divide 451 by 79 without using the division key.

3. Estimate the value of $\sqrt{30}$ without using the square root key.

For Exercises 4–7, write an algorithm for each task. *Section 12.1*

4. Make change from a $5 bill for a $3.52 purchase.

5. Find 30% of 132 without using a calculator.

6. Make soup from a can of condensed soup.

7. Graph the equation $y = 3x + 2$ in a coordinate plane.

For Exercises 8–12, copy the graph at the right. *Section 12.2*

8. Use a greedy algorithm to find a short path from vertex A to vertex B.

9. Find another path from vertex A to vertex B through all the vertices. How long is it? How does it compare to the path you found in Exercise 8?

10. Find a path from vertex C to vertex B through all the vertices. How long is the path?

11. Find the shortest tree for this graph. How long is it?

12. Suppose each vertex represents a home in a new neighborhood. Vertex C represents the main entry for the telephone lines for the neighborhood. Would you use a path or a tree to bring phone lines to all the homes in the neighborhood? Explain.

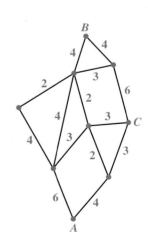

Four students voted for their favorite after-school activities, with 1 being their first choice. The results are shown in the table below. *Section 12.3*

13. Rank the activities in order of popularity. Explain how you decided on your ranking.

14. If two additional students both chose newspaper, community service, yearbook, and computer club, in that order, how would the group's preferences change?

● ● ● ● After-school Activities ● ● ● ●				
	Anna	**Glenn**	**Justin**	**Jenna**
Newspaper	2	4	3	2
Yearbook	1	3	1	4
Community Service	4	1	2	3
Computer Club	3	2	4	1

For Exercises 15–17, tell how to apply the cut-and-choose algorithm in each situation. *Section 12.3*

15. You and your sister both want the remaining ear of corn-on-the-cob.

16. You and a friend share a newspaper delivery route.

17. You and a friend volunteer to pick up litter in front of your school.

Find the number of possibilities for each situation. *Section 12.4*

18. ordering a three-course meal from a menu with 6 appetizers, 23 main dishes, and 8 desserts

19. sandwiches made with a choice of 8 different fillings on a choice of 4 types of bread

20. fabric for a chair in 7 designs, 2 fabric weights, and 6 colors

21. phones with or without built-in answering machines, with or without speaker-phone capability, and available in 7 different colors

For Exercises 22–26, find the number of permutations. *Section 12.4*

22. the number of ways four students can arrange themselves in a line

23. the number of ways you can arrange the letters in the word *pencil*

24. the number of different codes that can be programmed into a lock that has 4 digits, ranging from 0 through 9

25. the number of orders in which 8 different award winners can be announced

26. the number of different license plates that can be made using 3 different letters followed by 3 different digits

For Exercises 27 and 28, use the cards at the right. *Section 12.5*

27. **a.** How many pairs of cards can be made from the pictured cards?

    b. List the pairs that can be made.

    c. How many of the possible pairs have either a 2 or a 3 in them?

28. **a.** How many sets of three cards can be made from the pictured cards?

    b. List the possible sets.

    c. How many sets will not have a 4, a 5, or a 6 in them?

29. How many different ways can a committee of 3 students be selected from a class of 22 students?

30. For a history test, students must select 3 out of 8 questions to answer. How many different sets of questions can be chosen?

31. **a.** A jukebox at a diner has 50 songs. If you pay $.50, you can choose any 3 different songs. How many different combinations of 3 songs are available?

    b. In how many different orders could you choose 3 songs?

For Exercises 32 and 33, suppose three spinners are spun at the same time.
Section 12.6

32. **a.** If the spinners have equally sized sections numbered 1–8, 1–5, and 1–6, respectively, what is the probability of spinning three 4's?

    b. What is the probability of spinning three 6's?

33. If the spinners each have 10 equally sized sections numbered 1–10, what is the probability of spinning three 7's?

34. **a.** Suppose you are among 200 people who have entered a contest given by the local supermarket. If one name is drawn at random to be the winner, what is the probability that you will win?

    b. Suppose two names will be drawn at random as winners, and your friend also has entered the contest. What is the probability that you both will win?

Toolbox

FACTORS AND MULTIPLES

A common factor of two numbers is a number that is a factor of both numbers. The *greatest common factor* (GCF) of two numbers is the *largest* number that is a common factor of both numbers.

EXAMPLE 1

Find the GCF of 24 and 180.

SOLUTION

Find the prime factorization of each number.

$$24 = 2 \cdot 2 \cdot 2 \cdot 3$$
$$180 = 2 \cdot 2 \cdot 3 \cdot 3 \cdot 5$$

> Find the factors that appear in both lists. These are the **common factors**.

The GCF is the product of the common factors.

$$GCF = 2 \cdot 2 \cdot 3 = 12$$

A common multiple of two numbers is a number that is a multiple of both numbers. For example: $8 \cdot 10 = 80$ and $20 \cdot 4 = 80$, so 80 is a common multiple of 8 and 20. The *least common multiple* (LCM) of two numbers is the *smallest* number that is a common multiple of both numbers.

EXAMPLE 2

Find the LCM of 8 and 20.

SOLUTION

First find the GCF of the numbers.

$$8 = 2 \cdot 2 \cdot 2$$
$$20 = 2 \cdot 2 \cdot 5$$

The LCM is the product of the GCF and the other factors.

$$LCM = 4 \cdot 2 \cdot 5 = 40$$

> The GCF is 4.

PRACTICE

Find the GCF and LCM of each set of numbers.

1. 10, 45   **2.** 28, 49   **3.** 15, 70   **4.** 20, 24

5. 54, 72   **6.** 18, 25   **7.** 16, 64   **8.** 14, 42, 70

9. Pencils are sold in packages of 10, and pencil-top erasers are sold in packages of 6. Find the smallest number of each type of package you can buy so that you have the same number of pencils as erasers.

POWERS OF TEN

Numbers like the ones in the table below are **powers of ten**.

$10 = 10$	$\frac{1}{10} = 0.1$
$10 \cdot 10 = 100$	$\frac{1}{10} \cdot \frac{1}{10} = \frac{1}{100} = 0.01$
$10 \cdot 10 \cdot 10 = 1000$	$\frac{1}{10} \cdot \frac{1}{10} \cdot \frac{1}{10} = \frac{1}{1000} = 0.001$

The examples below show you some shortcuts to use when multiplying or dividing with powers of ten.

EXAMPLE 1

Multiply.

a. 1.2×100

b. 37×0.001

SOLUTION

a. $1.2 \times 100 = 1.20 = 120$

b. $37 \times 0.001 = .037 = 0.037$

There are **2** zeros in 100. Move the decimal point **2** places to the right.

There are **3** decimal places in 0.001. Move the decimal point **3** places to the left.

Insert a zero as a place holder.

Divide.

a. $23 \div 10,000$

b. $9.45 \div 10$

SOLUTION

a. $23 \div 10,000 = .0023 = 0.0023$

b. $9.45 \div 10 = 9.45 = 0.945$

There are **4** zeros in 10,000. Move the decimal point **4** places to the left.

Insert zeros as place holders.

There is **1** zero in 10. Move the decimal point **1** place to the left.

PRACTICE

Multiply.

1. 25×1000

2. 18.4×0.01

3. $1.94 \times 100,000$

4. 0.003×100

5. 3.68×0.001

6. 0.05×10

7. 0.781×0.0001

8. 9346×0.00001

Divide.

9. $29.3 \div 100$

10. $108.3 \div 1000$

11. $468 \div 10$

12. $413 \div 1000$

13. $0.06 \div 100$

14. $0.09 \div 10$

15. $1995 \div 10,000$

16. $21.973 \div 100,000$

RATIOS AND RATES

You can use *ratios* to compare two numbers using division. A ratio can be written in three ways:

$$a \text{ to } b \qquad a : b \qquad \frac{a}{b}$$

EXAMPLE 1

Write the ratio 6 to 8 in lowest terms.

SOLUTION

To write a ratio in lowest terms, write it as a fraction in lowest terms.

$$\frac{6}{8} = \frac{3}{4} \quad \text{or} \quad 3 \text{ to } 4 \quad \text{or} \quad 3 : 4$$

EXAMPLE 2

Write the ratio $\dfrac{6 \text{ h}}{3 \text{ days}}$ in lowest terms.

SOLUTION

When the quantities being compared are in different units, first write the quantities in the same unit and then write the ratio.

Convert days to hours.

$$\frac{6 \text{ h}}{3 \text{ days}} = \frac{6}{72} = \frac{1}{12} \quad \text{or} \quad 1 \text{ to } 12 \quad \text{or} \quad 1 : 12$$

A *rate* is a ratio that compares two different quantities, such as dollars and hours. A *unit rate* is a rate per one unit of a given quantity.

EXAMPLE 3

What is the unit rate if you earn $420 in 21 hours?

SOLUTION

Write the comparison as a fraction: $\dfrac{\text{dollars}}{\text{hours}} \longrightarrow \dfrac{420}{21} = \dfrac{20}{1}$

The rate is $20 per hour.

PRACTICE

Write each ratio in lowest terms.

1. 12 to 4

2. 62 : 279

3. $\dfrac{42}{64}$

4. 85 to 15

5. 20 min to 3 h

6. 75 mg : 65 g

7. 9 in. : 4 ft

8. 12 oz to 1 lb

Write each unit rate.

9. 750 mi in 15 h

10. $12.50 for 5 hats

11. 840 vibrations in 10 s

12. 63 hours in 7 days

13. $522 in 12 months

14. $3.60 for 8 pens

FRACTIONS, DECIMALS, AND PERCENTS

A percent is a ratio that compares a number to 100. Remembering that *percent* means *per 100* can help you convert fractions, decimals, and percents.

EXAMPLE 1

a. Write 28% as a decimal and as a fraction in lowest terms.

b. Write 0.375 as a percent and as a fraction in lowest terms.

SOLUTION

a. $28\% = \dfrac{28}{100} = 0.28$ and $28\% = \dfrac{28}{100} = \dfrac{7}{25}$

b. $0.375 = 37.5\%$ and $0.375 = \dfrac{375}{1000} = \dfrac{375 \div 125}{1000 \div 125} = \dfrac{3}{8}$

Move the decimal point two places to the right. Include the % symbol.

EXAMPLE 2

Write each fraction as a decimal and as a percent.

a. $\dfrac{5}{4}$

b. $\dfrac{1}{12}$

SOLUTION

a. $\dfrac{5}{4} = \dfrac{5 \times 25}{4 \times 25} = \dfrac{125}{100} = 1.25 = 125\%$

b. $\dfrac{1}{12} = 1 \div 12 = 0.08333\ldots = 8.\overline{3}\% \approx 8.3\%$

Equivalent Percents, Decimals, and Fractions		
$1\% = 0.01 = \dfrac{1}{100}$	$33\tfrac{1}{3}\% = 0.\overline{3} = \dfrac{1}{3}$	$60\% = 0.6 = \dfrac{3}{5}$
$10\% = 0.1 = \dfrac{1}{10}$	$40\% = 0.4 = \dfrac{2}{5}$	$66\tfrac{2}{3}\% = 0.\overline{6} = \dfrac{2}{3}$
$25\% = 0.25 = \dfrac{1}{4}$	$50\% = 0.5 = \dfrac{1}{2}$	$100\% = 1$

You may want to memorize the information in the chart.

PRACTICE

Write each percent as a decimal and as a fraction in lowest terms.

1. 4% **2.** 87% **3.** 175% **4.** 82.4% **5.** $\dfrac{1}{2}\%$

Write each decimal as a percent and as a fraction in lowest terms.

6. 0.75 **7.** $0.\overline{6}$ **8.** 0.025 **9.** 1.6 **10.** 0.513

Write each fraction as a decimal and as a percent.

11. $\dfrac{7}{20}$ **12.** $\dfrac{4}{5}$ **13.** $\dfrac{27}{80}$ **14.** $\dfrac{14}{25}$ **15.** $\dfrac{8}{15}$

INTEGERS

The *integers* are the numbers . . . , $-3, -2, -1, 0, 1, 2, 3, ...$ You can use integer chips to model integers.

| +1, or 1 | −1 | 0 | −2 | 0 + 3, or 3 |

Since $1 + (-1) = 0$, a positive chip and a negative chip form a *zero pair*. The value of a zero pair is 0.

EXAMPLE 1

Add.

a. −2 + (−5)

b. 4 + (−3)

SOLUTION

Model the first number with chips and add chips for the second number. Count the chips.

a.

$$-2 + (-5) = -7$$

b.

Add 3 negative chips to 3 positive chips to form zero pairs. 1 positive chip remains.

$$4 + (-3) = 1$$

EXAMPLE 2

Subtract.

a. −3 − (−1)

b. −3 − 1

c. −2 − (−4)

SOLUTION

Model the first number with chips. You may need to add zero pairs so you are able to subtract.

a.

Subtract 1 negative chip. 2 negative chips remain.

b.

Add a zero pair so you have a positive chip to subtract. Subtract 1 positive chip. 4 negative chips remain.

c.

Add zero pairs until you have 4 negative chips to subtract. Subtract 4 negative chips. 2 positive chips remain.

$$-3 - (-1) = -2$$

$$-3 - 1 = -4$$

$$-2 - (-4) = 2$$

Add or subtract.

1. $7 + (-2)$
2. $-7 + (-2)$
3. $-7 - (-2)$
4. $-2 - (-7)$
5. $0 - 8$
6. $-3 + (-3)$
7. $-10 + (-3)$
8. $-7 - 4$
9. $2 - (-9)$
10. $-6 - (-1)$
11. $5 - (-5)$
12. $-6 - (-15)$
13. $-6 - (-8)$
14. $2 - (-4)$
15. $-10 - 2$

EXAMPLE 3

Multiply -4×2.

SOLUTION

Two groups of -4 chips are the same as one group of -8 chips.

$$-4 \times 2 = -8$$

EXAMPLE 4

Divide $-9 \div 3$.

SOLUTION

One group of -9 chips is the same as three groups of -3 chips.

$$-9 \div 3 = -3$$

PRACTICE

Multiply or divide.

16. $6 \times (-3)$
17. $-6 \div 3$
18. -6×3
19. -6×0
20. 4×5
21. $1 \times (-1)$
22. $-15 \div 5$
23. -1×3
24. $4 \times (-4)$
25. $4 \div 4$
26. $0 \div 2$
27. $-8 \div 8$
28. $2 \times (-5)$
29. $7 \times (-3)$
30. $-18 \div 9$

INEQUALITIES

A mathematical statement that has an *inequality symbol* between two numbers or quantities is called an *inequality*.

You can use a *variable* such as n to represent numbers or quantities that vary. The inequality $n < 7$ means "all numbers less than 7."

Inequality symbol	Meaning
$<$	less than
$\leq$	less than or equal to
$>$	greater than
$\geq$	greater than or equal to

EXAMPLE 1

Write an inequality to describe the numbers less than or equal to 50.

SOLUTION

Choose a variable and tell what it represents. Let n = any number.

$n \leq 50$ Use the $\leq$ symbol.

EXAMPLE 2

Write an inequality to describe the phrase "at least $10."

SOLUTION

Choose a variable and tell what it represents. Let d = number of dollars.

$d \geq 10$ Use the $\geq$ symbol.

EXAMPLE 3

What integers make the inequality $x < 2$ a true statement?

SOLUTION

The inequality $x < 2$ describes all the integers that are less than 2.

The integers $1, 0, -1, -2, -3, \ldots$ make $x < 2$ a true statement.

PRACTICE

Write a phrase to describe each inequality.

1. $a > 6$ **2.** $x \leq 9$ **3.** $h < -2$ **4.** $p \geq 0$

Write an inequality for each statement. Tell what the variable represents.

5. numbers greater than or equal to 7   **6.** numbers less than 20

7. numbers greater than 3   **8.** numbers greater than or equal to 24

9. a price less than $30   **10.** a hike less than or equal to 12 km

Describe the integers that make each inequality true.

11. $n > 7$ **12.** $r \geq -1$ **13.** $v < 5$ **14.** $t \leq 3$

GRAPHING INEQUALITIES

You can use a number line to *graph* an inequality.

EXAMPLE 1

Graph each inequality on a number line.

a. $x > -2$ **b.** $x \leq 5$

SOLUTION

Step 1 Draw a number line.

Step 2 Draw an open or closed circle at the number. Use an open circle for < and >, and a closed circle for ≤ and ≥.

Step 3 Decide which direction the arrow should point, to the left for < and ≤, or to the right for > and ≥. Then draw the line and the arrow.

The open circle means −2 is not included in the graph.   The closed circle means that 5 is included in the graph.

a. $x > -2$

b. $x \leq 5$

EXAMPLE 2

Write an inequality for each graph.

a.

b.

SOLUTION

The arrow is pointing to the left. The inequality is either < or ≤.   The closed circle includes the value 1. Choose ≤.

The open circle does not include the value −3. The inequality is either > or <.   The arrow is pointing to the right. Choose >.

a.

$x \leq 1$

b.

$x > -3$

PRACTICE

Graph each inequality on a number line.

1. $n \leq 6$ **2.** $r > -1$ **3.** $a < 0$ **4.** $t \geq 4$

5. $x \geq -5$ **6.** $k < 7$ **7.** $h > 3$ **8.** $b \leq -5$

Write an inequality for each graph.

9.

10.

SYSTEMS OF MEASUREMENT

The **United States customary system** of measurement uses these units for length, capacity, and weight.

Length	Capacity	Weight
12 in. (inches) = 1 ft (foot)	8 fl oz (fluid ounces) = 1 c (cup)	16 oz = 1 lb (pound)
3 ft = 1 yd (yard)	2 c = 1 pt (pint)	2000 lb = 1 T (ton)
5280 ft = 1 mi (mile)	2 pt = 1 qt (quart)	
	4 qt = 1 gal (gallon)	

The **metric system** of measurement uses these units for length, capacity, and weight. The prefixes *milli-* (0.001), *centi-* (0.01), and *kilo-* (1000) are added to indicate different amounts.

Length (meter)	Capacity (liter)	Mass (gram)
1 mm = 0.001 m	1 mL = 0.001 L	1 mg = 0.001 g
1 cm = 0.01 m	1 cL = 0.01 L	1 cg = 0.01 g
1 m = 0.001 km	1 L = 0.001 kL	1 g = 0.001 kg
1 km = 1000 m	1 kL = 1000 L	1 kg = 1000 g

EXAMPLE

Convert each measurement.

a. 10 quarts to gallons

b. 0.7 kg to grams

SOLUTION

a. $10 \text{ qt} = \underline{\ ?\ } \text{ gal}$

$\dfrac{10}{4} = 2\dfrac{2}{4}$

$10 \text{ qt} = 2\dfrac{1}{2} \text{ gal}$

> Since 4 qt = 1 gal, divide by 4.

> When you convert to a larger unit (gal), you will have fewer gallons than quarts.

b. $0.7 \text{ kg} = \underline{\ ?\ } \text{ g}$

$(0.7)(1000) = 700$

$0.7 \text{ kg} = 700 \text{ g}$

> Since 1 kg = 1000 g, multiply by 1000.

> When you convert to a smaller unit (g), you will have more grams than kilograms.

PRACTICE

Convert each measurement.

1. $87 \text{ in.} = \underline{\ ?\ } \text{ ft } \underline{\ ?\ } \text{ in.}$

2. $2.5 \text{ T} = \underline{\ ?\ } \text{ lb}$

3. $1 \text{ qt} = \underline{\ ?\ } \text{ fl oz}$

4. $550 \text{ m} = \underline{\ ?\ } \text{ cm}$

5. $3400 \text{ g} = \underline{\ ?\ } \text{ kg}$

6. $7\dfrac{1}{4} \text{ yd} = \underline{\ ?\ } \text{ ft}$

7. $5\dfrac{1}{2} \text{ qt} = \underline{\ ?\ } \text{ pt}$

8. $250 \text{ L} = \underline{\ ?\ } \text{ mL}$

PERIMETER AND CIRCUMFERENCE

The *perimeter* of a figure is the distance around it.

The *circumference* of a circle is the distance around it.

Formulas for finding perimeter and circumference are listed on page 615.

EXAMPLE

Find the perimeter or circumference of each figure.

a.

b.

c.

SOLUTION

a. Add the measures of each side.

$6 + 6 + 5 + 3 = 20$

The perimeter is 20 cm.

b. Use a formula.

$P = 2l + 2w$

$P = 2(9) + 2(4)$

$ = 18 + 8$

$ = 26$

The perimeter is 26 in.

c. Use a formula.

$C = 2\pi r$ Use 3.14 for π.

$ \approx 2(3.14)(4)$

$ \approx 25.12$

The circumference is about 25.12 ft.

PRACTICE

Find the perimeter or circumference of each figure.

1.

10.2 m

15.3 m

2.

2.1 ft

3.

2 cm

2 cm

3 cm

4 cm

2 cm

5 cm

4.

8 in.

5. a square with a side of 4.2 cm

6. a circle with radius of 3.3 mm

7. an equilateral triangle with sides of $5\frac{1}{2}$ in.

8. a circle with diameter of 14 yd

9. The diameter of Josh's bicycle wheel is 22 inches. About how far will the wheel travel in one turn?

AREA AND VOLUME

The *area* of a figure is the number of square units it encloses.

The *volume* of a space figure is the amount of space it encloses. Volume is measured in cubic units.

Formulas for finding area and volume are listed on page 615.

Formulas for finding area and volume are listed on page 615.

EXAMPLE

a. Find the area.

b. Find the volume.

SOLUTION

Use 3.14 for π.

a. $A = \pi r^2$

$\approx (3.14)(5)^2$

$\approx (3.14)(5)(5)$

≈ 78.5

The area is about 78.5 m².

b. $V = lwh$

$= (3)(2)(2)$

$= 12$

The volume is 12 in.³.

PRACTICE

Find the area of each figure.

1. circle

2. rectangle

3. parallelogram

4. triangle

Find the volume of each rectangular prism.

5.

6.

7. A cube with sides of length 25 in.

8. A box with length 7.5 cm, width 9.5 cm, and height 3.5 cm

9. A rectangular hole 7 yd long, 4 yd wide, and 2 yd deep was dug for a swimming pool. How many cubic yards of dirt were removed from the hole?

10. A pizza restaurant charges $9 for a large round pizza that has a diameter of 14 in. This week the rectangular pizza, which is 15 in. by 12 in., also costs $9. Which pizza has the greater area?

ANGLES

An angle is formed by two rays (called *sides*) with a common endpoint (called a *vertex*).

An angle is measured in degrees.

∠ is the symbol for angle.

This angle is named ∠*ABC*, ∠*CBA*, or ∠*B*.

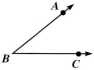

Some angles have special names based on their measures.

right angle: 90°

acute angle: less than 90°

obtuse angle: greater than 90° and less than 180°

straight angle: 180°

EXAMPLE 1

Measure the angle and tell whether it is *right*, *acute*, *obtuse*, or *straight*.

SOLUTION

Read the number of degrees from zero that the other side passes through.

Line up one side of the angle with 0°.

The measure of the angle is 120°. It is obtuse.

PRACTICE

Find the measure of each angle. Use a protractor if necessary. Then tell whether the angle is *right*, *acute*, *obtuse*, or *straight*.

1.

2.

3.

4.

5. Draw an angle that has a measure of 85°.

6. Draw an angle that has a measure of 160°.

7. Draw a figure showing two intersecting lines forming angles of 75° and 105°.

STATISTICAL GRAPHS

A *bar graph* is a visual display of data that fall into distinct categories.

EXAMPLE 1

Use the bar graph at the right.

a. What is the most popular car color among the people surveyed?

b. How many people in the survey plan to buy a red car?

SOLUTION

a. The teal bar is the tallest. The most popular color is teal.

b. The bar for red is 5 units high. Five people plan to buy a red car.

Choice of Color for Next Car

PRACTICE

1. How many people are represented by the bar graph?

2. What is the least popular color?

3. How many more people choose blue than choose white?

A *line graph* is a graph that shows the amount and the direction of change in data over a period of time.

EXAMPLE 2

The line graph shows the average precipitation in Portland, Oregon, and Burlington, Vermont.

a. Estimate the average precipitation during May and June in Burlington.

b. During January and February, about how much more precipitation is there in Portland than in Burlington?

Average Precipitation

SOLUTION

a. Locate May/June on the horizontal axis. Move vertically up the grid line to the bullet and then move horizontally to the vertical axis. Estimate the number.

The average precipitation during May and June is about 7 inches.

b. Estimate the average precipitation for **Portland** and **Burlington** during January and February. Then subtract.

$$10 - 4 = 6$$

During January and February, Portland gets about 6 more inches of precipitation than Burlington.

Use the line graph in Example 2.

4. Estimate the average precipitation for Burlington and the average precipitation for Portland during November and December.

5. When does Burlington get the most precipitation?

6. During July and August about how much more precipitation is there in Burlington than in Portland?

A *pictograph* can be used to display data that fall into distinct categories. A symbol is used to represent a given number of items.

EXAMPLE 3

Use the pictograph below.

a. About how many Super Trekker bicycles does Super Cycles manufacture?

b. About how many more Mountain Blazers are manufactured than Super Trekkers?

Type of Bicycle	Number of Bicycles Manufactured by Super Cycles
Super Trekker	🚲 🚲 🚲 🚲 🚲 🚲
Path Master	🚲 🚲 🚲 🚲 🚲
Tandem Tiger	🚲 🚲 🚲
Mountain Blazer	🚲 🚲 🚲 🚲 🚲 🚲 🚲 🚲 🚲

🚲 = 1000 bicycles

SOLUTION

a. Each symbol represents 1000 bicycles.

There are $5\frac{1}{2}$ symbols for Super Trekker.

Multiply.
$(5.5)(1000) = 5500$

Super Cycle manufactures about 5500 Super Trekkers.

b. Compare the number of symbols. The Mountain Blazer has about 3 more symbols than the Super Trekker.

Multiply.
$3(1000) = 3000$

There are about 3000 more Mountain Blazers manufactured than Super Trekkers.

PRACTICE

Use the pictograph in Example 3.

7. About how many bicycles are manufactured altogether?

8. For which bicycle is the amount manufactured the smallest?

9. How many symbols would you use to represent 9500 bicycles?

A *circle graph* shows how parts relate to a whole and to each other.

The conductor of a community orchestra analyzed the ideal makeup of the musicians in the orchestra. The results are shown in the *circle graph* at the right.

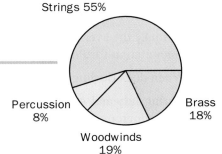

Strings 55%

Percussion 8%

Brass 18%

Woodwinds 19%

EXAMPLE 4

a. Suppose there are 65 musicians in the orchestra. About how many should be string players?

b. Suppose the conductor would like to have 22 woodwind players. About how big should the orchestra be to maintain an ideal balance?

SOLUTION

a. 55% of the orchestra's musicians are string players. Let s = the number of string players.

$$s = 0.55(65)$$

$$= 35.75$$

There should be about 36 string players.

b. 19% of the orchestra are woodwind players. Let N = the number of orchestra members.

$$0.19N = 22$$

$$\frac{0.19N}{0.19} = \frac{22}{0.19} \approx 115.8$$

There should be about 116 members in the orchestra.

PRACTICE

Use the circle graph in Example 4.

10. The conductor decides to expand the orchestra for a performance. To maintain balance, about how many players should be in each section in a 140 piece orchestra?

11. The conductor decides to have 65 string players. About how big should the orchestra be to maintain an ideal balance?

A *stem-and-leaf plot* displays data grouped into intervals. For example, the stem-and-leaf plot at the right groups daily protein intake of health club members for one day into intervals of 10. Each data item consists of two parts, a *stem* and a *leaf*. For example the number 57 is written

$$5 \mid 7$$

5 is the stem.   7 is the leaf.

Protein Consumed (grams)

4	3 4
5	2 6 7
6	3 3 4 8 8 9
7	0 1 2 5 6
8	2 4 4 7

EXAMPLE 5

Use the stem-and-leaf plot showing protein consumed.

a. How many different amounts of protein are represented?

b. What is the highest amount of protein? the lowest?

SOLUTION

a. Count each leaf. Do not count duplicates. There are 17 different amounts represented.

b. Find the highest stem, then find the highest leaf in that row. The highest amount of protein consumed is 87 grams.

Find the smallest stem, then find the smallest leaf in that row. The lowest amount of protein consumed is 43 grams.

PRACTICE

Use the stem-and-leaf plot showing finishing times of racers.

12. How many racers finished in under 50 minutes?

13. What was the fastest time? the slowest?

Finishing Times (minutes)

```
3 | 2  4
4 | 1  3  4  5  5  6  7
5 | 0  1  1  2  2  8
6 | 2  3  5  6
7 | 1
```

A *histogram* is a bar graph that shows how many data items occur in each of one or more intervals. The height of each bar shows the frequency, or number of items in each interval. Note that there are no spaces between the bars of a histogram.

EXAMPLE 6

Use the histogram at the right.

a. How many students walked from 80 to 89 laps?

b. How many students in all are represented by the histogram?

SOLUTION

a. Find the height of the bar for the interval 80–89. About 5 students walked from 80 to 89 laps.

b. Find the height of each bar. The sum of these heights is the total number of students.

$$3 + 5 + 4 + 7 + 6 + 10 + 5 + 2 = 42$$

The histogram represents 42 students.

Walkathon Results

PRACTICE

Use the histogram in Example 6.

14. How many students walked 40 or more laps?

15. How many students walked fewer than 40 laps?

16. Which interval has the smallest frequency? What is the frequency in this interval?

17. Change the intervals to 20–39, 40–59, 60–79 and 80–99. Redraw the histogram using these intervals.

PROBLEM SOLVING

Often there is more than one way to solve a problem. If you are not sure how to solve a problem, consider using one or more of the following strategies.

Strategy	Application
Break the problem into parts	When intermediate steps are needed
Guess, check, and revise	When you do not seem to have all the needed information
Identify a pattern	When you can identify several cases
Make an organized list	When organizing information helps you solve the problem
Make a diagram	When words describe a picture
Solve a simpler problem	When simpler numbers help you solve the problem
Use a formula	When you recognize a known relationship
Use logical reasoning	When you use the facts to reach a conclusion
Work backward	When you are looking for a fact leading to a known result

---

EXAMPLE 1

How many segments can be drawn to connect ten points equally spaced around a circle?

SOLUTION

Problem Solving Strategy: Find a pattern.

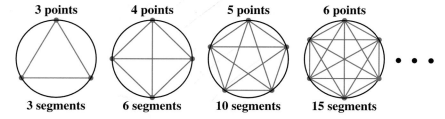

Compare the numbers of points and segments in one circle to the numbers in the previous circle to find a pattern. Then continue the pattern until the number of points equals 10.

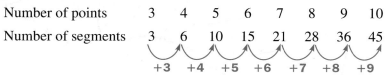

A circle with 10 evenly spaced points can have 45 drawn segments.

EXAMPLE 2

Marva's savings account statement for May showed a deposit of $250, two withdrawals of $50 and $80, an interest payment of $1.29, and an ending balance of $637.52. Find the starting balance.

SOLUTION .

Problem Solving Strategy: Work backward.

$637.52	Ending balance
− 1.29	Subtract the interest paid.
$636.23	
+ 80.00	Add back the withdrawals.
+ 50.00	
$766.23	
− 250.00	Subtract the deposit.
$516.23	

Marva's starting balance was $516.23.

PRACTICE

Solve each problem.

1. Each digit from 1 to 9 is used only once in the problem at the right. Find the missing digits.

```
    1   2   ?
+   ?   ?   7
─────────────
    4   ?   ?
```

2. List all the ways you can pay a bridge toll of $1.50 if you have only quarters and dimes.

3. Complete the pattern.

Days	1	2	3	4	5	6	7
Length of vine (inches)	12	24	36	?	?	?	?

4. The perimeter of a rectangular garden is 20 ft and the width is 4 ft. Find the area of the garden.

5. The Changs drove for 2 h at a speed of 50 mi/h, and then drove for 45 min at 40 mi/h. If their car gets about 25 miles per gallon of gasoline, about how many gallons of gasoline did the trip require?

6. A stack of cans in a grocery store has 1 can on top, 3 cans in the second row, 5 in the third, and so on. If there are 20 rows of cans altogether, how many cans are there in the stack?

7. Lois wants to buy a compact disk player that costs $145 plus 5% sales tax. She has already saved $35. Lois earns $4.75 an hour working at a department store. How many more hours does Lois need to work to be able to buy the compact disk player?

8. The sum of the measures of the angles of a triangle is 180°. Suppose the three angles of a triangle are equal in measure. Find the measure of each angle.

Technology Handbook

This handbook introduces you to the basic features of most graphing calculators. Check your calculator's instruction manual for specific keystrokes and any details not provided here.

PERFORMING CALCULATIONS

The Keyboard

Look closely at your calculator's keyboard. Notice that most keys serve more than one purpose. Each key is labeled with its primary purpose, and labels for any secondary purposes appear somewhere near the key. You may need to press 2nd , SHIFT , or ALPHA to use a key for a secondary purpose.

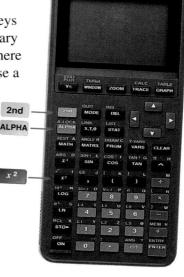

Examples of using the x^2 key:

Press x^2 to square a number.

Press 2nd and then x^2 to take a square root.

Press ALPHA and then x^2 to get the letter I.

The Home Screen

Your calculator has a *home screen* where you can do calculations. You can usually enter a calculation on a graphing calculator just as you would write it on a piece of paper.

Shown below are some things to remember as you enter calculations on your graphing calculator.

The calculator has a subtraction key, − , and a negation key, (−) . If you use these incorrectly, you will get an error message.

You may need to use parentheses to avoid confusion. For example, enter $|5 - 9|$ as **ABS (5 − 9)** so the calculator does not interpret the expression as $|5| - 9$.

To perform an operation that is not often used, such as to calculate 8! or $_{20}C_2$, you may need to press MATH to display a menu of operations.

```
3--2
                    5
abs (5-9)
                    4
20 nCr 2
                  190
```

Use your calculator to find the value of each expression.

1. $-5 - 3$ **2.** $|-30|$ **3.** 2π **4.** $\sqrt{12.25}$

5. 2^5 **6.** 2.34×5^{-3} **7.** $8!$ **8.** $_7C_2$

DISPLAYING GRAPHS

Entering and Graphing a Function

To graph a function, enter its equation in the form $y = \ldots$. Rewrite the equation with x and y if necessary. For example, rewrite $t = 3 - s$ as

$y = 3 - x$ and rewrite $y - 3 = \frac{1}{2}x$ as $y = \frac{1}{2}x + 3$.

The graph of $y = \frac{1}{2}x + 3$ is shown.

Use parentheses. If you enter $y = 1/2x + 3$, the calculator may interpret the equation as $y = \frac{1}{2x} + 3$.

The Viewing Window

Think of the calculator screen as a *viewing window* that lets you look at a portion of the coordinate plane.

On many calculators, the *standard window* shows values from -10 to 10 on both the x- and y-axes. You can adjust the viewing window by pressing WINDOW or RANGE and entering new values for the window variables.

The x-axis will be shown for $-10 \le x \le 10$.

The y-axis will be shown for $-10 \le y \le 10$.

With scale variables set to equal 1, tick marks will be 1 unit apart on both axes.

Squaring the Screen

A *square screen* is a viewing window with equal unit spacing on the two axes. For example, the graph of $y = x$ is shown for two different windows.

Standard Viewing Window

Square Screen Window

On a square screen, the line $y = x$ makes a 45° angle with the x-axis.

On many graphing calculators, the ratio of the screen's height to its width is about 2 to 3. Your calculator may have a feature that gives you a square screen. If not, choose values for the window variables that make the "length" of the y-axis about two-thirds the "length" of the x-axis:

$$(\text{Ymax} - \text{Ymin}) \approx \frac{2}{3}(\text{Xmax} - \text{Xmin})$$

The *Zoom* Feature

To get a closer look at a point of interest on a graph, you can use the *zoom* feature. Your calculator may have more than one way to zoom. A common way is to put a *zoom box* around the point.

Place one corner of the zoom box. Then place the opposite corner.

The calculator draws what is inside the box at full-screen size.

CALCULATOR PRACTICE

9. Enter and graph each equation separately. Use the standard viewing window.

a. $y = 3x + 1$ **b.** $y = x^2 - 4x + 2$ **c.** $y = \dfrac{5}{x}$ **d.** $y = 2^x$

e. Zoom in on part of your graph from part (d) and sketch it.

10. Find a good viewing window for the graph of $y = 65 - 3x$. Be sure your window shows where the graph crosses both axes. Be sure the window is square.

READING A GRAPH

The *Trace* Feature

After a graph is displayed, you can use the calculator's *trace* feature. When you press TRACE, a flashing cursor appears on the graph. The *x*- and *y*-coordinates of the cursor's location are shown at the bottom of the screen. Press ◀ and ▶ to move the *trace* cursor along the graph.

The *trace* cursor is at the point (2.2340426, 2.2340426) on the graph of $y = x$.

Suppose the equation $y = 7x - 70$ gives your profit *y* from selling *x* T-shirts. You can use the *trace* feature to find how many T-shirts you must sell to make a profit of $800.

Graph the equation $y = 7x - 70$. Using TRACE, move the cursor along the graph until the *y*-value is approximately equal to 800. This gives you the value of *x*, the number of T-shirts you must sell.

Another way to solve the problem is to graph two equations on the same screen: $y = 7x - 70$ and $y = 800$. Use *trace* to move the cursor to the intersection of the two lines to find your answer. On many graphing calculators you can use ▲ and ▼ to move between graphs. This helps you find the intersection.

Trace along one line until you are near the intersection.

Press ▲. Now you are on the other line.

The *x*-coordinates are the same. The *y*-coordinates are close.

You can *zoom* in and repeat the process described above until the *y*-coordinates are the same to the nearest tenth, hundredth, or any other decimal place.

Friendly Windows

As you press while tracing a graph, the *x*-coordinate may increase by "unfriendly" increments, giving "unfriendly" *x*-values.

Your calculator may allow you to control the *x*-increment, ΔX, directly. If not, you can control it indirectly by choosing an appropriate Xmax for a given Xmin.

For example, on a TI-82 graphing calculator, choose Xmin and Xmax so that

$$\text{Xmax} - \text{Xmin} = 94\Delta X \qquad \text{or} \qquad \text{Xmax} = \text{Xmin} + 94\Delta X.$$

This number depends upon the calculator you are using.

Suppose you want ΔX to equal 0.1 and Xmin to equal -5. Then:

$$\text{Xmax} = -5 + 94(0.1)$$
$$= 4.4$$

Set Xmax to 4.4 for a "friendly window" in which the *trace* cursor's *x*-coordinate will increase by 0.1 each time you press ▶.

Calculating Coordinates

Some calculators have a *calculate* feature. You can use this feature to find the coordinates of a point of interest automatically.

For example, on a TI-82   Enter the *x*-value.   Press ENTER.
use *value* to find the value
of *y* for a given value of *x*.

You can use the *intersection* feature to find the intersection of two graphs.

Use ▲ and ▼ to choose the two graphs.

If there is more than one point of intersection, use ▶ and ◀ to choose one of them.

First curve?
X=5   Y=-10

Guess?
X=3.0851064 Y=-9.822091

Intersection
X=3.1043561 Y=-10

You can use the *root* feature to find the point where a graph crosses the *x*-axis.

Choose a root by marking a point to the left of it and a point to the right of it.

Then guess the location of the root to help the calculator find it more quickly.

The calculator approximates the root. The number 1E−13 means 10^{-13} or about zero.

Upper Bound?
X=-1.276596 Y=-.8171118

Guess?
X=-1.489362 Y=.19692168

Root
X=-1.44949   Y=1E-13

CALCULATOR PRACTICE

11. Graph the equation $y = x + \dfrac{1}{7}$.

    a. Choose a friendly window where $\Delta X = 0.1$. Use the *trace* feature to determine, to the nearest tenth, the value of *x* for which $y = 5$.

    b. Try zooming in on the point where $y = 5$. Between what two values, to the nearest hundredth, does the *x*-coordinate of the point lie?

12. Graph $y = 2\pi x$. Use the *calculate* feature to find an approximate value of *y* when $x = 9$.

13. Graph the equations below. Use the *calculate* feature to find the coordinates of the point of intersection.

$$9x + 3y = 14$$
$$-3x + 2y = 8$$

14. Use the *calculate* feature to find the roots of the equation $y = -4x^2 + 2x + 15$.

WORKING WITH STATISTICS

Adding Matrices

Suppose you want to use matrices to find each baseball team's total wins and losses for the 1992 season. Use your calculator's *matrix* feature to enter (or *edit*) the data into matrices.

AMERICAN LEAGUE

Games Before the All-Star Game

EAST	W	L
Toronto	53	34
Baltimore	49	38
Milwaukee	45	41
Boston	42	43
New York	42	45
Detroit	41	48
Cleveland	36	52
WEST	**W**	

AMERICAN LEAGUE

Games After the All-Star Game

EAST	W	L
Toronto	43	32
Baltimore	40	35
Milwaukee	47	29
Boston	31	46
New York	34	41
Detroit	34	39
Cleveland	40	34
WEST	**W**	

Give the dimensions of matrix *A* and enter the data for games before the All-Star Game.

Enter the data for games after the All-Star Game in matrix *B*.

From the home screen, add the two matrices. To type [A] you may need to press MATRX 1 or 2nd 1.

The answer is a matrix with the same dimensions as the original matrices. Its elements are the sums of elements in the same positions in [A] and in [B].

Scatter Plots and Finding a Line of Fit

Some calculators use linear regression to find a line of fit for a set of ordered pairs. First enter the two lists of data and then graph the paired data in a scatter plot.

On a TI-82 graphing calculator, press STAT to edit, or enter, the data.

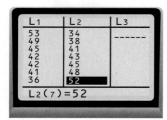

Turn STAT PLOT on and GRAPH the data. See your manual or your teacher if you need help.

Use linear regression in the statistics calculations menu to fit a line to your data.

The equation $y = -x + 90$ models the data.

The *Table* Feature

Some calculators have a *table* feature. You can use this feature to examine the values of a function. The screen shows a table of values for $y = x^2 - 6x + 7$.

Some calculators have a table set-up feature that allows you to change the *x*-values.

Notice that the *y*-values change sign between $x = 1.5$ and $x = 1.6$.

You can also use the table feature like a spreadsheet. Suppose you want to change the units of the data at the right from centimeters to inches.

Heights of Children (cm)				
102	106	96	106	102
101	100	106	112	102
105	106	100	100	100
100	94	94	106	97

Use **TblSet** to instruct the calculator to ask for *x*-values.

Enter the *x*-values in the **TABLE**.

Using **Y= .** , enter a formula to change the units of the data to inches.

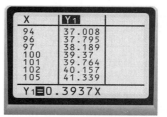

Now you can use **Y-VARS** to perform calculations on Y_1.

CALCULATOR PRACTICE

15. Add matrices *A* and *B*.

$$A = \begin{bmatrix} 1 & -2 & 7 \\ -1 & 5 & 0 \end{bmatrix} \quad B = \begin{bmatrix} -3 & 1 & -2 \\ 0 & -1 & 5 \end{bmatrix}$$

16. Graph the data about can sizes. Find a line of fit. If a can has a diameter of 3.1 in., what will the circumference be?

17. Create a table of values for the equation $y = x^2 - 6x + 7$. As the *x*-values increase, at approximately what *x*-value do the *y*-values stop decreasing and begin increasing?

18. Use the *table* feature to transform the data at the right from hours to minutes.

Sizes of Cans	
Diameter (inches)	Circumference (inches)
2.1	6.7
2.7	8.4
3.0	9.4
3.2	10.0
3.4	10.8
4.1	12.8
4.3	13.4
6.2	19.4

Time Spent on Homework	
Day	Time (hours)
Monday	1
Tuesday	1.5
Wednesday	2
Thursday	1
Friday	0
Saturday	4.75
Sunday	0

USING A SPREADSHEET

Suppose you want to buy a CD player that costs $195, including tax. You already have $88 and can save $9 per week. After how many weeks can you buy the player? You can use a computer spreadsheet to solve this problem.

row numbers column letters

CD Savings

	A	B
1	Week number	Total saved
2	0	88
3	= A2 + 1	= B2 + 9
4	= A3 + 1	= B3 + 9
5	= A4 + 1	= B4 + 9
6	= A5 + 1	= B5 + 9
7	= A6 + 1	= B6 + 9
8	= A7 + 1	= B7 + 9
9	= A8 + 1	= B8 + 9
10	= A9 + 1	= B9 + 9
11	= A10 + 1	= B10 + 9
12	= A11 + 1	= B11 + 9
13	= A12 + 1	= B12 + 9
14	= A13 + 1	= B13 + 9
15	= A14 + 1	= B14 + 9

A spreadsheet is made up of cells named by a column letter and a row number, like A3 or B4. Cells can contain labels, numbers, or formulas.

Cell B1 contains the label "Total saved."

Cell B2 contains the number 88.

Cell B3 contains the formula " = B2 + 9." This formula tells the computer to take the number in cell B2, add 9 to it, and put the result in cell B3.

Instead of typing a formula into each cell individually, you can use the spreadsheet's *series*, *fill down*, and *copy* commands.

The computer replaces all the formulas with calculated values. You can have the computer draw a scatter plot with a line connecting the plotted points.

CD Savings

	A	B
1	Week number	Total saved
2	0	88
3	1	97
4	2	106
5	3	115
6	4	124
7	5	133
8	6	142
9	7	151
10	8	160
11	9	169
12	10	178
13	11	187
14	12	196
15	13	205

You will have enough money to buy the CD player after 12 weeks.

See pages 41–42 of Section 1.8 for more information about spreadsheets.

SPREADSHEET PRACTICE

19. Suppose you borrowed $150 from your parents and are paying back $7 each week. Use a spreadsheet to find out when you will owe only $45. How long will it take to pay back the loan completely?

Table of Measures

Time

60 seconds (s)	= 1 minute (min)
60 minutes	= 1 hour (h)
24 hours	= 1 day
7 days	= 1 week
4 weeks (approx.)	= 1 month

$\left.\begin{array}{l} 365 \text{ days} \\ 52 \text{ weeks (approx.)} \\ 12 \text{ months} \end{array}\right\} = 1 \text{ year}$

$\left.\begin{array}{l} 10 \text{ years} \\ 100 \text{ years} \end{array}\right.$ 10 years = 1 decade
100 years = 1 century

Metric

Length

10 millimeters (mm) = 1 centimeter (cm)

$\left.\begin{array}{l} 100 \text{ cm} \\ 1000 \text{ mm} \end{array}\right\} = 1 \text{ meter (m)}$

1000 m = 1 kilometer (km)

Area

100 square millimeters = 1 square centimeter
(mm^2)   (cm^2)

10,000 cm^2 = 1 square meter (m^2)

10,000 m^2 = 1 hectare (ha)

Volume

1000 cubic millimeters = 1 cubic centimeter
(mm^3)   (cm^3)

1,000,000 cm^3 = 1 cubic meter (m^3)

Liquid Capacity

1000 milliliters (mL) = 1 liter (L)

1000 L = 1 kiloliter (kL)

Mass

1000 milligrams (mg) = 1 gram (g)

1000 g = 1 kilogram (kg)

1000 kg = 1 metric ton (t)

Temperature — Degrees Celsius (°C)

0°C = freezing point
of water

37°C = normal body
temperature

100°C = boiling point
of water

United States Customary

Length

12 inches (in.) = 1 foot (ft)

$\left.\begin{array}{l} 36 \text{ in.} \\ 3 \text{ ft} \end{array}\right\} = 1 \text{ yard (yd)}$

$\left.\begin{array}{l} 5280 \text{ ft} \\ 1760 \text{ yd} \end{array}\right\} = 1 \text{ mile (mi)}$

Area

144 square inches (in.^{2}) = 1 square foot (ft^2)

9 ft^2 = 1 square yard (yd^2)

$\left.\begin{array}{l} 43,560 \text{ ft}^2 \\ 4840 \text{ yd}^2 \end{array}\right\} = 1 \text{ acre (A)}$

Volume

1728 cubic inches (in.^{3}) = 1 cubic foot (ft^3)

27 ft^3 = 1 cubic yard (yd^3)

Liquid Capacity

8 fluid ounces (fl oz) = 1 cup (c)

2 c = 1 pint (pt)

2 pt = 1 quart (qt)

4 qt = 1 gallon (gal)

Weight

16 ounces (oz) = 1 pound (lb)

2000 lb = 1 ton (t)

Temperature — Degrees Fahrenheit (°F)

32°F = freezing point
of water

98.6°F = normal body
temperature

212°F = boiling point
of water

Table of Symbols

Symbol		Page	Symbol		Page
$\geq$	is greater than or equal to	3	$^\circ$	degree(s)	172
$\leq$	is less than or equal to	3	$a : b$	the ratio of a to b	192
$>$	is greater than	3	$\triangle ABC$	triangle ABC	205
$<$	is less than	3	$\overline{AB}$	segment AB	205
$=$	equals, is equal to	3	AB	the length of $\overline{AB}$	205
$(\ )$	parentheses—a grouping symbol	8	$\angle A$	angle A	205
$\cdot$	multiplication, times ($\times$)	9	$\sim$	is similar to	206
$\approx$	is approximately equal to	14	π	pi, an irrational number approximately equal to 3.14	215
$\ldots$	and so on	19	$\sqrt{a}$	the nonnegative square root of a	240
$\lvert x \rvert$	absolute value of x	19	$\pm$	plus or minus	240
$-a$	opposite of a	20	$0.\overline{27}$	the repeating decimal $0.272727\ldots$	246
$[\ ]$	brackets—a grouping symbol	26	a^n	nth power of a	370
$\stackrel{?}{=}$	is this statement true?	28	a^{-n}	$\frac{1}{a^n}, a \neq 0$	391
$\begin{bmatrix} 1 & 0 \\ 0 & 1 \end{bmatrix}$	matrix	40	$!$	factorial	532
(x, y)	ordered pair	73	${}_nP_r$	permutation	535
m	slope	120	${}_nC_r$	combination	537
$\frac{1}{a}$	reciprocal of a, $a \neq 0$	153			
$\neq$	is not equal to	154			

Table of Squares and Square Roots

No.	Square	Sq. Root	No.	Square	Sq. Root	No.	Square	Sq. Root
1	1	1.000	51	2,601	7.141	101	10,201	10.050
2	4	1.414	52	2,704	7.211	102	10,404	10.100
3	9	1.732	53	2,809	7.280	103	10,609	10.149
4	16	2.000	54	2,916	7.348	104	10,816	10.198
5	25	2.236	55	3,025	7.416	105	11,025	10.247
6	36	2.449	56	3,136	7.483	106	11,236	10.296
7	49	2.646	57	3,249	7.550	107	11,449	10.344
8	64	2.828	58	3,364	7.616	108	11,664	10.392
9	81	3.000	59	3,481	7.681	109	11,881	10.440
10	100	3.162	60	3,600	7.746	110	12,100	10.488
11	121	3.317	61	3,721	7.810	111	12,321	10.536
12	144	3.464	62	3,844	7.874	112	12,544	10.583
13	169	3.606	63	3,969	7.937	113	12,769	10.630
14	196	3.742	64	4,096	8.000	114	12,996	10.677
15	225	3.873	65	4,225	8.062	115	13,225	10.724
16	256	4.000	66	4,356	8.124	116	13,456	10.770
17	289	4.123	67	4,489	8.185	117	13,689	10.817
18	324	4.243	68	4,624	8.246	118	13,924	10.863
19	361	4.359	69	4,761	8.307	119	14,161	10.909
20	400	4.472	70	4,900	8.367	120	14,400	10.954
21	441	4.583	71	5,041	8.426	121	14,641	11.000
22	484	4.690	72	5,184	8.485	122	14,884	11.045
23	529	4.796	73	5,329	8.544	123	15,129	11.091
24	576	4.899	74	5,476	8.602	124	15,376	11.136
25	625	5.000	75	5,625	8.660	125	15,625	11.180
26	676	5.099	76	5,776	8.718	126	15,876	11.225
27	729	5.196	77	5,929	8.775	127	16,129	11.269
28	784	5.292	78	6,084	8.832	128	16,384	11.314
29	841	5.385	79	6,241	8.888	129	16,641	11.358
30	900	5.477	80	6,400	8.944	130	16,900	11.402
31	961	5.568	81	6,561	9.000	131	17,161	11.446
32	1,024	5.657	82	6,724	9.055	132	17,424	11.489
33	1,089	5.745	83	6,889	9.110	133	17,689	11.533
34	1,156	5.831	84	7,056	9.165	134	17,956	11.576
35	1,225	5.916	85	7,225	9.220	135	18,225	11.619
36	1,296	6.000	86	7,396	9.274	136	18,496	11.662
37	1,369	6.083	87	7,569	9.327	137	18,769	11.705
38	1,444	6.164	88	7,744	9.381	138	19,044	11.747
39	1,521	6.245	89	7,921	9.434	139	19,321	11.790
40	1,600	6.325	90	8,100	9.487	140	19,600	11.832
41	1,681	6.403	91	8,281	9.539	141	19,881	11.874
42	1,764	6.481	92	8,464	9.592	142	20,164	11.916
43	1,849	6.557	93	8,649	9.644	143	20,449	11.958
44	1,936	6.633	94	8,836	9.695	144	20,736	12.000
45	2,025	6.708	95	9,025	9.747	145	21,025	12.042
46	2,116	6.782	96	9,216	9.798	146	21,316	12.083
47	2,209	6.856	97	9,409	9.849	147	21,609	12.124
48	2,304	6.928	98	9,604	9.899	148	21,904	12.166
49	2,401	7.000	99	9,801	9.950	149	22,201	12.207
50	2,500	7.071	100	10,000	10.000	150	22,500	12.247

Properties of Algebra

Identity Property of Multiplication, p. 24

For every number a:

$$1 \cdot a = a \qquad\qquad 1 \cdot 3 = 3$$

Multiplicative Property of −1, p. 24

For every number a:

$$-1 \cdot a = -a \qquad\qquad -1 \cdot 3 = -3$$

Commutative Property, p. 26

You can add or multiply numbers in any order.

$$a + b = b + a \qquad -3 + 7 = 7 + (-3)$$
$$ab = ba \qquad\qquad (-3)(7) = (7)(-3)$$

Associative Property, p. 26

When you add or multiply three or more numbers, you can group the numbers without affecting the result.

For example: $(7 + 8) + 2 = 7 + (8 + 2)$

$$(7 \cdot 8) \cdot 2 = 7 \cdot (8 \cdot 2)$$

Distributive Property, p. 35

For all numbers a, b, and c:

$$a(b + c) = ab + ac$$

For example: $3(10 + 2) = 3(10) + 3(2)$

Property of Reciprocals, p. 153

For every nonzero number a, there is exactly one number $\frac{1}{a}$ such that

$$a \cdot \frac{1}{a} = 1 \quad \text{and} \quad \frac{1}{a} \cdot a = 1$$

For example: $2 \cdot \frac{1}{2} = 1 \quad \text{and} \quad \frac{1}{2} \cdot 2 = 1$

Means-Extremes Property, p. 193

For any proportion, the product of the extremes is equal to the product of the means.

$$\text{If } \frac{a}{b} = \frac{c}{d}, \text{ then } ad = bc.$$

Properties of Square Roots, p. 252

For all nonnegative numbers a and b:

$$\sqrt{a^2} = a$$
$$\sqrt{ab} = \sqrt{a} \cdot \sqrt{b}$$
$$\sqrt{\frac{a}{b}} = \frac{\sqrt{a}}{\sqrt{b}}, b \neq 0$$

Zero and Negative Exponents, p. 391

For any nonzero number a and any integer n:

$$a^0 = 1 \quad \text{and} \quad a^{-n} = \frac{1}{a^n}$$

Rules for Products and Quotients of Powers, p. 402

For any nonzero number a and any integers m and n:

$$a^m \cdot a^n = a^{m+n} \quad \text{and} \quad \frac{a^m}{a^n} = a^{m-n}$$

Rules for Powers of Products and Quotients, p. 408

For any nonzero numbers a and b and any integer m:

$$(ab)^m = a^m b^m \quad \text{and} \quad \left(\frac{a}{b}\right)^m = \frac{a^m}{b^m}$$

Power of a Power Rule, p. 408

For any nonzero number a and any integers m and n:

$$(a^m)^n = a^{m \cdot n}$$

Zero-Product Property, p. 437

If the product of factors is equal to zero, then one or more of the factors is equal to zero.

For example: If $(x - 3)(x - 4) = 0$,

$$\text{then } (x - 3) = 0 \quad \text{or} \quad (x - 4) = 0$$
$$x = 3 \quad \text{or} \qquad x = 4$$

Formulas from Geometry

To find the perimeter P of any plane figure, find the sum of the lengths of each side of the figure.

For the formulas below, let A = area, C = circumference, V = volume, and B = the area of the base of a space figure. Let $\pi \approx 3.14$.

Parallelogram

$A = $ base $\times$ height

$A = bh$

Triangle

$A = \dfrac{1}{2} \times$ base $\times$ height

$A = \dfrac{1}{2}bh$

Trapezoid

$A = \dfrac{1}{2} \times$ sum of bases $\times$ height

$A = \dfrac{1}{2}(b_1 + b_2)h$

Circle

$C = 2\pi r$ or $C = \pi d$

$A = \pi r^2$

Rectangular Prism

$V = $ area of base $\times$ height

$V = Bh = lwh$

Cylinder

Area of base $B = \pi r^2$

$V = $ area of base $\times$ height

$V = Bh = \pi r^2 h$

Pyramid

$V = \dfrac{1}{3} \times$ area of base $\times$ height

$V = \dfrac{1}{3}Bh$

Cone

Area of base $B = \pi r^2$

$V = \dfrac{1}{3} \times$ area of base $\times$ height

$V = \dfrac{1}{3}Bh = \dfrac{1}{3}\pi r^2 h$

Sphere

$V = \dfrac{4}{3}\pi r^3$

1 Domains and Ranges of Functions

Learn how to...
- **find the domain and range of a function**

So you can...
- **determine reasonable values to use in real-world situations, such as the height of a ball after it is thrown**

Functions can be used to model many real-world situations. In such cases, the *domain* of a function consists of all input values that it makes sense to use. The *range* of the function consists of all resulting output values. The range depends on the domain of the function.

EXAMPLE 1

A function is defined by the equation $y = 5 - 2x$.

a. If the domain is the whole numbers $0, 1, 2, 3, \ldots$, what is the range?

b. If the domain is the integers $\ldots, -2, -1, 0, 1, \ldots$, what is the range?

SOLUTION

Problem Solving Strategy: Make a table.

a.

x	$5 - 2x = y$
0	$5 - 2 \cdot 0 = 5$
1	$5 - 2 \cdot 1 = 3$
2	$5 - 2 \cdot 2 = 1$
3	$5 - 2 \cdot 3 = -1$
...	...

The range consists of the numbers $5, 3, 1, -1, \ldots$.

b.

x	$5 - 2x = y$
...	...
-2	$5 - 2 \cdot -2 = 9$
-1	$5 - 2 \cdot -1 = 7$
0	$5 - 2 \cdot 0 = 5$
1	$5 - 2 \cdot 1 = 3$
...	...

The range consists of the numbers $\ldots, 9, 7, 5, 3, \ldots$.

THINK AND COMMUNICATE

1. Describe in words the range in both parts of Example 1.

2. A newspaper vending machine holds 30 papers when full. The papers cost $.50 each.

a. Give an equation for the amount of money collected in the machine as a function of papers sold.

b. Describe the domain and range of the function.

c. Would the graph of this function be a line? Explain why or why not.

EXAMPLE 2 **Application: Sports**

The equation $y = -16x^2 + 20x + 6$ gives the height y in feet of a ball as a function of x, the number of seconds after it is thrown. Graph the equation on a graphing calculator. Give the domain and range of the function.

SOLUTION

Trace the graph to find the value of x when $y = 0$.

X=1.5   Y=0

Trace the graph to find the maximum value of y.

X=.625   Y=12.25

The ball is in the air from time $x = 0$ to time $x = 1.5$. The domain of the function is $x \geq 0$ and $x \leq 1.5$.

The ball reaches a maximum height of $y = 12.25$ and then falls to the ground ($y = 0$). The range of the function is $y \geq 0$ and $y \leq 12.25$.

Exercises and Applications

1. A function is defined by the equation $y = 4x - 10$.
   a. If the domain is the numbers $-3, -1.5, 1.5$, and 3, what is the range?
   b. If the domain is the integers $\ldots, -2, -1, 0, 1, \ldots$, what is the range?

2. **BIOLOGY** A person's height in feet is a function of age in years.
   a. Describe the domain and range of the height function.
   b. Would the graph of this function be a line? Explain why or why not.

For Exercises 3–6, describe the domain and range of each function.

3. The cost C of an item including sales tax is related to the price P in dollars of the item by the equation $C = 1.06P$.

4. The distance d in miles remaining on a 200-mile trip is related to the time h in hours by the equation $d = 200 - 50h$.

5.

x	−2	−1	0	1	2	3	4
$(1-x)^2$	9	4	1	0	1	4	9

6. An object falls to the ground from a height of 30 m. The distance d in meters that the object falls is related to the time t in seconds since it began falling by the equation $d = 4.9t^2$.

7. **BIOLOGY** The formula $w = 446l^3$ gives the approximate weight w in grams of a hognose snake whose length is l meters. Give a reasonable domain and range for this function. Explain your reasoning.

2

Graphing Linear and Quadratic Equations

The slope-intercept form of a linear equation is $y = mx + b$. In this equation, m is the slope of the line and b is the y-intercept. The numbers m and b are called *parameters*. When $m = 1$ and $b = 0$, you get the *parent* function $y = x$, whose graph is shown in black.

THINK AND COMMUNICATE

1. Give the values of m and b for each equation graphed above.

2. How is the graph of $y = mx + b$ with $m > 0$ different from the graph of $y = mx + b$ with $m < 0$?

3. What happens to the graph of $y = mx + b$ as $|m|$ increases?

4. Describe the graphs of linear equations all having the same value of m.

The parameter m determines how steep the graph of $y = mx + b$ is and whether the line slants up or down as x increases. The parameter b determines where the line crosses the y-axis.

EXAMPLE 1

Describe how the graph of $y = 5x - 3$ compares with the graph of $y = x$. Then graph $y = x$ and $y = 5x - 3$.

SOLUTION

In $y = 5x - 3$, $m = 5$ and $b = -3$.
In $y = x$, $m = 1$ and $b = 0$. Because $-3 < 0$, the graph of $y = 5x - 3$ crosses the y-axis below the point where the graph of $y = x$ does. Because $|5| > |1|$, the graph of $y = 5x - 3$ is steeper than the graph of $y = x$.

Parameters in Quadratic Equations

The parameters in the quadratic equation $y = ax^2 + c$ are a and c. The parameter a determines whether the graph (a parabola) opens up or down, and $|a|$ determines whether the parabola is wider than or narrower than the graph of the parent function $y = x^2$. The parameter c is the y-intercept of the parabola. The graph of $y = ax^2 + c$ always has its vertex at $x = 0$, so c is also the y-coordinate of the vertex.

THINK AND COMMUNICATE

5. Identify the value of a for each equation graphed at the right.

6. How is the graph of $y = ax^2$ with $a > 0$ different from the graph of $y = ax^2$ with $a < 0$?

7. What happens to the graph of $y = ax^2$ as $|a|$ increases? as $|a|$ decreases?

8. Predict what the graph of the equation $y = -4x^2$ looks like. Graph the equation to check.

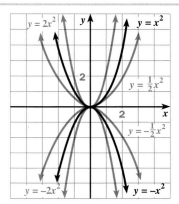

EXAMPLE 2

Compare the graph of $y = -0.4x^2 + 3$ with the graph of $y = x^2$. Then graph $y = x^2$ and $y = -0.4x^2 + 3$.

SOLUTION

In $y = -0.4x^2 + 3$, $a = -0.4$ and $c = 3$. In $y = x^2$, $a = 1$ and $c = 0$. Because $-0.4 < 0$, the graph of $y = -0.4x^2 + 3$ opens down. Because $1 > 0$, the graph of $y = x^2$ opens up. Because $|-0.4| < |1|$, the graph of $y = -0.4x^2 + 3$ is wider than the graph of $y = x^2$. Because $3 > 0$, the vertex of the graph of $y = -0.4x^2 + 3$ lies above the vertex of the graph of $y = x^2$.

Plot the vertex at $(0, 3)$.

If $x = -4$,
$y = -0.4(-4)^2 + 3 = -3.4$.
Plot the point $(-4, -3.4)$.

If $x = 4$,
$y = -0.4(4)^2 + 3 = -3.4$.
Plot the point $(4, -3.4)$.

Exercises and Applications

Describe how the graph of each equation compares with the graph of $y = x$. Then graph $y = x$ and the equation.

   1. $y = 4x + 3$ **2.** $y = 0.2x - 4$ **3.** $y = -2x + 1.5$

   4. Graph the equations in parts (a)–(c) in the same coordinate plane.

      a. $y = 3x - 2$ **b.** $y = -3x + 2$ **c.** $y = -3x - 2$

      d. How are the graphs in parts (a)−(c) alike? How are they different?

For each line:

a. Tell whether m is *greater than*, *equal to*, or *less than* zero.

b. Tell whether b is *greater than*, *equal to*, or *less than* zero.

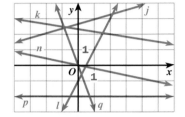

   5. j **6.** k

   7. l **8.** n

   9. p **10.** q

   11. CARS Use the information in the table. Assume that each vehicle starts with a full tank of gas. Let g = the fuel remaining in gallons and d = the miles driven.

Fuel Efficiency (mi/gal)		
Vehicle	City	Highway
Car with a 12.7 gal gas tank	28	37
Truck with a 20 gal gas tank	17	25

      a. Write an equation giving g as a function of d for driving the car in the city. Write an equation for driving the car on the highway.

      b. Write an equation giving g as a function of d for driving the truck in the city. Write an equation for driving the truck on the highway.

      c. Explain how gas tank capacity and fuel efficiency for each vehicle affect the graphs of the equations you wrote in parts (a) and (b).

Compare the graph of each equation with the graph of $y = x^2$. Then graph $y = x^2$ and the equation.

   12. $y = 2x^2 + 1$ **13.** $y = 0.2x^2 - 3$ **14.** $y = -4x^2 + 5$

   15. Graph the equations in parts (a)–(c) in the same coordinate plane.

      a. $y = 3x^2 - 2$ **b.** $y = -3x^2 + 2$ **c.** $y = -3x^2 - 2$

      d. How are the graphs in parts (a)–(c) alike? How are they different?

PHYSICS The equation $d = 16t^2$ gives the distance d in feet that an object falls in t seconds of free fall on Earth.

   16. The equation $d = 2.7t^2$ gives the distance d in feet that an object falls in t seconds of free fall on the moon. Compare the graph of this equation with the graph of $d = 16t^2$.

   17. If an object falls from a height of 10 ft on Earth, the equation $h = 10 - 16t^2$ gives the height h of the object t seconds after it begins to fall. Compare the graph of this equation with the graph of $d = 16t^2$.

3

Solving Systems Using Tables and Tiles

Learn how to...

- **solve systems of linear equations using tables and algebra tiles**

So you can...

- **have multiple methods for solving linear systems**

In many situations, you will need to find solutions that balance two different needs. You may be able to set up two different equations and find a solution that works for both equations. A pair of equations for which you try to find a single solution is called a *system of equations*.

A *solution* of a system of equations in two variables is a pair of numbers that satisfy both equations of the system. There are several methods that you can use to solve systems of equations. Example 1 shows one way you can set up a table to find a solution.

EXAMPLE 1

Use a table to solve the system of equations.

$$3x + y = 11$$
$$y - 2x = 1$$

SOLUTION

Problem Solving Strategy: Guess, check, and revise.

Solve both equations for y.

$$3x + y = 11 \qquad\qquad y - 2x = 1$$
$$y = 11 - 3x \qquad\qquad y = 1 + 2x$$

Choose a value of x, and use the equations to calculate values of y. Then check whether the values of y are the same.

x	$y = 11 - 3x$	$y = 1 + 2x$	Are the values of y the same?
0	$y = 11 - 3 \cdot 0$ $= 11$	$y = 1 + 2 \cdot 0$ $= 1$	No. Try another value of x.
1	$y = 11 - 3 \cdot 1$ $= 8$	$y = 1 + 2 \cdot 1$ $= 3$	No, but they are closer than before. Try a larger value of x.
3	$y = 11 - 3 \cdot 3$ $= 2$	$y = 1 + 2 \cdot 3$ $= 7$	No. This time the second value is larger. Try a smaller value of x.
2	$y = 11 - 3 \cdot 2$ $= 5$	$y = 1 + 2 \cdot 2$ $= 5$	Yes. The solution is $x = 2$ and $y = 5$.

The solution of the system of equations is (2, 5).

Example 2 shows how you can use algebra tiles to solve a system of equations.

EXAMPLE 2

Use algebra tiles to solve the system of equations.

$$y - 2x = -1$$
$$2y - x = 4$$

SOLUTION

Step 1 Model each equation using algebra tiles. Use red tiles for negative numbers and terms with negative coefficients.

$y - 2x = -1$

$2y - x = 4$

Step 2 Choose one of the equations, for example $y - 2x = -1$. Solve for y.

Add two x-tiles to each side of the equation. A positive tile and a negative tile of the same type have a sum of 0.

Remove pairs of tiles that have a sum of 0.

One y-tile is equal to two x-tiles and one negative 1-tile.

Step 3 Substitute the result from Step 2 for each y-tile in the other equation. Solve for x.

Remove pairs of tiles that have a sum of 0.

Add two 1-tiles to each side of the equation.

Only x-tiles remain on one side of the equation.

There are two 1-tiles for each x-tile, so $x = 2$.

Step 4 Find the value of y by using the result of Step 2. Substitute two 1-tiles for each x-tile.

There are three 1-tiles for each y-tile, so $y = 3$.

The solution of the system of equations is $(2, 3)$.

THINK AND COMMUNICATE

1. Look at the two equations in Example 2. Why might you choose the first equation instead of the second equation to solve for y in Step 2?

2. a. In Step 2 of Example 2, you could choose to solve an equation for x instead of y. Which equation would you choose? Why?

   b. If you solved the second equation for x in Step 2, what would you substitute for each negative x-tile in the first equation in Step 3?

Exercises and Applications

Solve each system of equations using a table.

1. $x - y = 0$
$2 - y = x$

2. $x + y = 5$
$2x - 3y = 5$

3. $2x - y = 8$
$3x + y = 17$

4. a. Use a table to solve the system of equations. Try 4 values for x.

$y = 2x - 1$

$-2x = 3 - y$

   b. Writing Does this system of equations have a solution? Explain your reasoning.

5. a. Use a table to solve the system of equations. Try 4 values for x.

$y = 2x - 4$

$6x = 3y + 12$

   b. Writing What do you notice about the last column of your table? What does this tell you about the system of equations?

Solve each system of equations using algebra tiles.

6. $2x + y = 7$
$x - y = 2$

7. $x - 2y = -2$
$3x - y = 9$

8. $x - 2y = 3$
$x + 2y = 3$

Glossary

absolute value (p. 19) The distance that a number x is from zero on a number line, written $|x|$.

algorithm (p. 509) A step-by-step method for accomplishing a goal.

approval voting (p. 524) A voting method in which each person can vote for any number of candidates.

base of a power (p. 370) *See* power.

binomial (p. 260) A variable expression with two unlike terms.

coefficient (p. 36) A number multiplied by a variable in a term of a variable expression. For example, the coefficient of $6x^2$ is 6.

combination (p. 537) A selection made from a group of items. The order of the selected items does not matter.

complete graph (p. 536) In discrete mathematics, a graph with a single edge between every pair of vertices. *See also* graph.

constant of variation (pp. 107, 464) *See* direct variation *and* inverse variation.

converse of the Pythagorean theorem (p. 241) If the lengths of a triangle's sides have the relationship $a^2 + b^2 = c^2$, then the triangle is a right triangle. *See also* Pythagorean theorem.

coordinate plane (p. 73) A grid formed by two *axes* that intersect at the *origin*. The axes divide the coordinate plane into four *quadrants*.

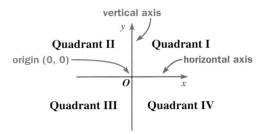

cubic polynomial (p. 423) A polynomial of degree three.

cut-and-choose (p. 525) An algorithm used for fair division in which one person makes a division, another person chooses the part he or she wants, and the first person gets the remaining part.

data (p. 3) A collection of numerical facts.

decay factor (p. 384) The expression $1 - r$ in an exponential decay function. *See also* exponential decay.

degree of a polynomial (p. 423) The largest degree of all the terms.

degree of a term (p. 423) The exponent of the variable part of a term. If there is more than one variable, the degree is the sum of the exponents of the variables.

dense (p. 246) A group of numbers is dense if between any two numbers there is always a third number. For example, the real numbers are dense.

difference of two squares (p. 266) The product of the sum and the difference of two terms:
$$(a + b)(a - b) = a^2 - b^2$$

dimensions of a matrix (p. 40) *See* matrix.

direct variation (p. 107) A relationship between two variables in which the variables have a constant ratio k, $k \neq 0$. If x and y are the variables, then $\frac{y}{x} = k$, or $y = kx$.

The number k is called the *constant of variation*.

discriminant (p. 355) The value of the expression $b^2 - 4ac$ in the quadratic formula.

domain of a function (p. 67) *See* function.

edge (p. 516) In discrete mathematics, a route connecting two vertices of a graph. *See also* graph.

element of a matrix (p. 40) *See* matrix.

equation (p. 55) A mathematical sentence that says that two numbers or expressions are equal.

equivalent expressions (p. 35) Two expressions that are equal.

evaluate a variable expression (p. 9) Finding the value of a variable expression for given values of the variable(s).

event (p. 218) *See* probability.

experimental probability (p. 219) The ratio of the number of times an event actually occurs to the number of times an experiment is performed.

exponent (pp. 10, 370) *See* power.

exponential decay (p. 384) A decreasing exponential function modeled by the equation $y = a(1 - r)^x$, where r is the fixed rate of decrease and a represents the initial amount before any decrease.

exponential growth (p. 377) An increasing exponential function modeled by the equation $y = a(1 + r)^x$, where r is the fixed rate of growth and a represents the initial amount before any growth.

extremes of a proportion (p. 193) In the proportion $\frac{a}{b} = \frac{c}{d}$, a and d are the extremes.

factorial (p. 532) The product of all integers from 1 to n, represented by the symbol $n!$.

factoring a polynomial (p. 437) Rewriting a polynomial as a product of polynomial factors.

fair division (p. 525) A way of dividing something among a group of people so that everyone feels he or she has received a fair portion.

favorable outcome (p. 220) *See* probability.

function (p. 66) A relationship between input and output. For each input, there is exactly one output. The *domain* of a function consists of all possible input values. The *range* consists of all possible output values. The range depends on the domain.

geometric probability (p. 224) A probability based on area.

graph (p. 516) In discrete mathematics, a collection of points, called *vertices*, and lines, called *edges*, that connect the vertices.

vertex ⟶ ⟵ edge

greedy algorithm (p. 517) A method for finding a good path or tree by choosing the best edge from the ones currently available on a graph. *See also* graph, path, and tree.

growth factor (p. 377) The expression $1 + r$ in an exponential growth function. *See also* exponential growth.

growth rate (p. 377) The constant rate at which an exponential function increases. *See also* exponential growth.

histogram (pp. 4, 599) A bar graph that shows how many data items occur in each of one or more intervals. The height of each bar shows the frequency, or number of items, in each interval.

horizontal axis (p. 72) *See* coordinate plane.

horizontal coordinate (p. 73) *See* ordered pair.

horizontal intercept (p. 282) *See* x-intercept.

hypotenuse (p. 239) In a right triangle, the side opposite the right angle. *See also* Pythagorean theorem.

identity (p. 161) An equation that is true for all numbers.

inequality (p. 171) A mathematical statement formed by placing an inequality symbol between numerical or variable expressions.

integers (p. 19) The numbers . . . , $-3, -2, -1, 0, 1, 2, 3, ...$

inverse operations (p. 56) Operations that undo each other. Addition and subtraction are inverse operations. Multiplication and division are inverse operations.

inverse variation (p. 464) A relationship between variables in which the variables have a constant product k, $k \neq 0$. If x and y are the variables, then $xy = k$, or $y = \frac{k}{x}$, $x \neq 0$. The number k is called the *constant of variation*.

irrational number (p. 246) A real number that cannot be written as a ratio of two integers; an irrational number written as a decimal never repeats or terminates.

leg (p. 239) In a right triangle, one of two sides that are shorter than the hypotenuse. *See also* Pythagorean theorem.

like terms (p. 36) In a variable expression, terms that have identical variable parts.

line of fit (p. 130) A line on a scatter plot that is as close as possible to all the data points on the scatter plot. A way to model linear data.

line of symmetry (p. 333) *See* symmetric.

linear data (p. 131) Paired data that come close to forming a line when displayed in a scatter plot.

linear equation (p. 119) An equation whose graph is a line.

linear inequality (p. 307) An inequality whose graph on a coordinate plane is a region bounded by a line.

linear polynomial (p. 423) A polynomial of degree one.

mathematical model (p. 130) A table, a graph, an equation, a function, or an inequality that describes a real-world situation.

matrix, matrices (p. 40) A group of numbers arranged in rows and columns. Each number is an *element* of the matrix. The number of rows and columns in a matrix are its *dimensions*. Matrix A has dimensions 2×3.

$$A = \begin{bmatrix} 3 & 11 & 0 \\ -1 & 4 & 7 \end{bmatrix}$$

mean (p. 14) In a data set, the sum of all the data divided by the number of data items.

means of a proportion (p. 193) In the proportion $\frac{a}{b} = \frac{c}{d}$, b and c are the means.

median (p. 14) In a data set, the middle number when you put the data in order from smallest to largest. When the number of data items is even, the median is the mean of the two middle numbers.

mode (p. 14) In a data set, the most frequently appearing data item, or items.

modeling (p. 130) Using a mathematical model.

monomial (p. 260) An expression with a single term.

multiplication counting principle (p. 531) When you have m choices in one category and n choices in a second category, you have $m \cdot n$ possible choices. When you have r choices in a third category, then you have a total of $m \cdot n \cdot r$ possible choices.

nonlinear equation (p. 326) An equation whose graph is not a line.

one person-one vote (p. 524) A voting method in which each person can vote for one candidate.

opposites (p. 20) Two numbers whose sum is zero. For example, the numbers 3 and -3 are opposites.

order of operations (pp. 8, 371) A set of rules people agree to use so that an expression has only one value.

ordered pair (p. 73) A pair of numbers, such as $(5, -3)$, that can be graphed on a coordinate plane. The first number is the *horizontal coordinate*, or *x-coordinate*, and the second number is the *vertical coordinate*, or *y-coordinate*.

origin (p. 73) On a coordinate plane, the point with coordinates $(0, 0)$. *See also* coordinate plane.

outcome (p. 218) One possible result of a probability experiment. *See also* probability.

parabola (p. 331) A nonlinear graph that is U-shaped. *See also* quadratic function.

path (p. 517) In discrete mathematics, a collection of edges on a graph that allow you to travel from one vertex to another without repeating any vertex or edge. *See also* graph.

perfect square (p. 240) A number that is the square of an integer.

perfect square trinomial (p. 265) A trinomial that is the product of squaring a binomial. *See also* trinomial.

permutation (p. 532) An arrangement of a group of items in a definite order.

polynomial (p. 423) An expression that can be written as a sum of monomials with exponents that are whole numbers.

polynomial equation (p. 436) An equation in which both sides are polynomials.

power (p. 370) The product when a number or expression is used as a factor a given number of times. In the power 3^5, 3 is the *base* and 5 is the *exponent*. $3^5 = 3 \cdot 3 \cdot 3 \cdot 3 \cdot 3$.

probability (p. 218) A ratio that measures how likely an *event* is to happen. For example, suppose you want to get an even number when you roll a six-sided die. There are six possible *outcomes* when you roll the die, only three of which are *favorable* to making this event happen. The probability of rolling an even number is $\frac{3}{6}$ or $\frac{1}{2}$.

proportion (p. 193) An equation that shows two ratios to be equal.

Pythagorean theorem (p. 239) If the length of the hypotenuse of a right triangle is c and the lengths of the legs are a and b, then $a^2 + b^2 = c^2$.

quadrant (p. 73) *See* coordinate plane.

quadratic equation (pp. 342, 345) An equation that can be written in the form $ax^2 + bx + c = 0$, when $a \neq 0$.

quadratic formula (p. 350) The formula $x = \dfrac{-b \pm \sqrt{b^2 - 4ac}}{2a}$, for the solutions of the equation $ax^2 + bx + c = 0$, when $a \neq 0$.

quadratic function (p. 332) A function or an equation whose graph is a parabola.

quadratic polynomial (p. 423) A polynomial of degree two.

radical form (p. 252) When an expression is written using the symbol $\sqrt{}$, it is in radical form.

range of a function (p. 67) *See* function.

rate (p. 99) A ratio that compares the amounts of two unlike quantities, for example, miles per hour.

rate of decrease (p. 384) The constant rate at which an exponential function decreases. *See also* exponential decay.

ratio (p. 192) A quotient that compares two quantities.

rational equation (p. 476) An equation that contains only rational expressions.

rational expression (p. 476) A variable expression that can be written as a fraction, the numerator and denominator of which are polynomials.

rational number (p. 246) A real number that can be written as a ratio of two integers; when written as a decimal, a rational number repeats or terminates.

real number (p. 246) Any number that is either rational or irrational.

reciprocals (p. 153) Two numbers whose product is 1.

right triangle (p. 239) A triangle that has a right, or 90°, angle.

scale (p. 198) The ratio of the size of a representation of an object to the actual size of the object.

scale factor (p. 199) The ratio of a length on an enlarged or reduced image to the corresponding length on the original figure.

scatter plot (p. 106) A graph that shows a relationship between two sets of data.

scientific notation (p. 396) A number expressed in the form $a \times 10^n$, where $1 \leq a < 10$ and n is an integer.

similar figures (p. 206) Two figures are similar if corresponding angles have equal measure and the ratios of the lengths of corresponding sides are equal.

simplest form of an expression (p. 36) A variable expression with no parentheses and with all like terms combined.

simplest form of a rational expression (p. 489) A rational expression that consists of a single fraction whose numerator and denominator have no common factors other than 1.

slope (pp. 113, 115) The measure of the steepness of a line.

slope-intercept form (p. 120) The equation of a line in the form $y = mx + b$, where m is the slope of the line and b is the vertical intercept.

solution of an equation (p. 56) Values of variables that make an equation true.

solution of an equation in two variables (p. 79)
An ordered pair that makes the equation true.

solution of an inequality (p. 177) The set of values of a variable that make an inequality true.

solution of a system of inequalities (p. 312) Any ordered pair of numbers that satisfy all the inequalities of the system.

solution of a system of linear equations (p. 288) Any ordered pair of numbers that satisfy all the equations of the system.

solving algebraically (p. 62) Using symbols to solve an equation.

solving an equation (p. 55) Finding all values of variables that make the equation true.

spreadsheet (p. 41) A computer-generated arrangement of data in rows and columns. A spreadsheet can be used to perform calculations on large groups of data.

square root (p. 240) One of two equal factors of a number.

standard form of a linear equation (p. 282) A linear equation written in the form $ax + by = c$, where a, b, and c are integers and x and y are variables.

standard form of a polynomial (p. 423) A polynomial that has been simplified and whose terms are written in order from greatest exponent to smallest exponent.

symmetric (p. 333) When a graph or shape can be folded so that one side fits exactly over the other side, the graph or shape is symmetric about the fold line, called the *line of symmetry*.

system of inequalities (p. 312) Two or more inequalities that state relationships between the same variables.

system of linear equations (p. 288) Two or more linear equations that state relationships between the same variables.

terms (p. 36) The parts of a variable expression that are added together.

theoretical probability (p. 220) When all possible outcomes of an experiment are equally likely, the ratio of the number of favorable outcomes to the total number of possible outcomes.

tree (p. 518) In discrete mathematics, a path that connects all the vertices in a graph with the smallest possible number of edges. *See also* graph.

trinomial (p. 260) A variable expression with three unlike terms.

unit rate (p. 99) A rate per one given unit.

value of a variable (p. 3) A number represented by a variable.

variable (p. 3) A letter used to stand for a quantity that varies.

variable expression (p. 9) An expression made up of variables, numbers, and operations.

vertex, vertices (p. 516) In discrete mathematics, a point used to model a location on a graph. *See also* graph.

vertex of a parabola (p. 332) The point where the function reaches a minimum or maximum value.

vertical axis (p. 72) *See* coordinate plane.

vertical coordinate (p. 73) *See* ordered pair.

vertical intercept (pp. 120, 282) *See* y-intercept.

weighted average (p. 470) An average of data values in which each value is multiplied by a weight, so that some values count more than others.

x-axis (p. 73) The horizontal axis of a coordinate plane. *See also* coordinate plane.

x-coordinate (p. 73) *See* ordered pair.

x-intercept (p. 282) The x-value of a point where a graph crosses the x-axis. *Also called* horizontal intercept.

y-axis (p. 73) The vertical axis of a coordinate plane. *See also* coordinate plane.

y-coordinate (p. 73) *See* ordered pair.

y-intercept (pp. 120, 282) The y-value of a point where a graph crosses the y-axis. *Also called* vertical intercept.

Index

210, 211, 215, 217, 221,
223, 226, 229, 234, 235,
243, 249, 250, 257, 263,
268, 269, 271, 275, 285,
293, 296, 298, 304, 305,
310, 314, 315, 317, 324,
328, 334, 336, 342, 347,
353, 358, 361, 365, 368,
375, 382, 387, 389, 393,
395, 399, 400, 413, 426,
427, 433, 434, 435, 439,
440, 441, 445, 451, 453,
455, 501, 512, 515, 520,
521, 526, 528, 533, 534,
541, 544, 545, 547

Commutative property
of addition, 26
of multiplication, 26

Comparing
quantities with ratios, 191
radicals, 251
riser/tread ratios, 316–317
tables and graphs, 147–148
theoretical values to experimental values, 271, 361, 413

Computer
spreadsheets, 41–42, 43, 44, 45,
83, 170, 249, 256, 381, 428,
452, 487, 540
See also Calculator, Graphing
calculator, *and* Technology
Handbook.

Connections
architecture, 117, 520
art, 111, 514
astronomy, 335
biology, 88
Braille, 533
business, 292, 440, 522
cars, 64
chemistry, 28
computers, 374
consumer economics, 394
cooking, 34, 297, 329, 388
crafts, 433
dance, 33
ecology, 17
economics, 70
energy use, 81, 151
engineering, 358, 493
genetics, 268
geography, 75, 128
geometry, 243

government, 196
highway safety, 348
history, 59, 76, 123, 163, 210,
262, 303, 410
literature, 304, 404, 546
meteorology, 174
music, 103, 156
physics, 12
politics, 528
probability, 491
recreation, 468, 481
social studies, 222, 534, 539
sports, 23, 227, 342, 354, 452,
473
statistics, 133, 216
transportation, 179
volunteering, 527
See also Multicultural connections.

Constant of variation, 107
Converse
of Pythagorean Theorem, 241
Cooperative learning
Exercises, 17, 34, 39, 60, 83,
111, 118, 123, 150, 170,
180, 203, 211, 223, 228,
229, 243, 255, 263, 285,
298, 309, 330, 336, 341,
359, 387, 428, 434, 441,
474, 480, 513, 527, 540,
545
Explorations, 8, 30, 61, 78, 106,
119, 132, 147–148, 158,
176, 191, 205, 218, 251,
259, 299, 306, 325, 331,
355, 369, 383, 401, 429,
436, 442, 463, 469, 494,
509, 536
Portfolio Projects, 46–47, 90–91,
136–137, 182–183,
230–231, 270–271,
316–317, 360–361,
412–413, 454–455,
500–501, 548–549
Coordinate axes, 72–73
Coordinate graphs, 72–73, 85,
90–91, 106, 113–115, 119,
121, 136–137, 147–148,
172, 183, 282, 301, 317,
331, 345, 355, 357, 361,
378, 383, 413, 434, 436,
462, 536
Coordinate plane, 73

Corresponding angles, 205–207
Corresponding sides, 205–207
Critical thinking, *See* Reasoning
and Visual thinking.
Cross-curriculum connections,
See Connections.
Cubic polynomial, 423
Cumulative Review, *See* Review.
Cut-and-choose algorithm, 525

Data analysis
collecting data, 46, 90, 106,
144–146, 182, 230, 270,
316, 360, 454
comparing tables and graphs,
147
finding possible sources of error,
455
making a graph, 91, 106
organizing data, 2, 30, 40, 46,
90, 106, 183, 230–231,
270–271, 317, 361, 412,
454–455
portfolio project using, 46–47,
90–91, 182–183, 230–231,
270–271, 316–317,
360–361, 412–413,
454–455, 500–501,
548–549
using spreadsheets, 41–42, 46,
90, 412
using a table to find a pattern, 30,
106
Decay factor, 384
Decimal
repeating, 246, 247
terminating, 246
See also Toolbox, 587.
Degree of a polynomial, 423
Dense, 246
Decision making, 287,
523–525
Deviation from the mean, 23
Difference of two squares, 266
Dimensional analysis, 99–100
Dimensions of a matrix, 40
Direct variation, 106–108
Discrete mathematics
access graph, 520
combinations, 536–537

Factoring a polynomial, 436–437, 442, 447

Fair division, 523, 525

FOIL method of multiplying binomials, 431

Function
  definition of, 66
  domain of, 67, 91, 616–617
  exponential, 366–368, 376, 383–385, 496
  linear, 54, 130–132, 281, 496, 616
  as mathematical model, 130–132, 183, 281, 337, 344, 361, 376–378, 383–385, 413, 423, 429, 500, 616–617
  polynomial, 423, 429, 496
  quadratic, 324, 331–332, 337–338, 344, 442–444, 447–450, 496, 617
  range of, 67, 91, 616–617
  rational, 496
  table, 66, 67, 68, 616–617

Game theory, 548–549

Geometric probability, 224–225

Geometry exercises and applications, 39, 65, 70, 112, 179, 243, 256, 293, 328, 357, 374, 427, 434, 451, 475, 485, 492, 540

Graphing calculator
  exercises and applications, 43, 77, 85, 88, 106, 119, 121, 134, 147–148, 168, 172, 175, 203, 228, 254, 271, 284, 286, 290, 291, 297, 301, 314, 329, 330, 331, 334, 341, 345, 347, 348, 353, 355, 357, 359, 381, 383, 387, 413, 434, 436, 441, 462, 466, 472, 473, 474, 478, 483, 510, 535, 536, 540, 545
  fitting a line, 134
  Trace feature, 85
  Viewing window, setting, 119
  Zoom feature, 85
  See also Technology Handbook, 602.

Graphs
  access graph, 520
  bar graphs, 6, 47
  with calculator, 85, 106, 119, 121, 147–148, 172, 271, 283, 301, 331, 345, 355, 357, 378, 383, 413, 434, 436, 462, 536, 617
  on coordinate plane, 72–73, 85, 90–91, 106, 113–114, 119, 121, 136–137, 147–148, 172, 182–183, 283, 301, 317, 331, 345, 355, 357, 361, 378, 383, 413, 434, 436, 462, 510, 536, 618–619
  data, 2, 46–47, 90–91, 182–183, 317, 361, 413
  of direct variation, 108
  in discrete mathematics, 516
  of exponential equations, 378
  finding intercepts, 120, 282
  finding rates from, 100–101
  histogram, 4, 14–16, 47
  of inequalities, 171, 177–178, 182–183, 280, 306–307, 317
  intersection of, 172, 288, 301
  linear, 113–114, 119–121, 136–137, 281–283, 618
  as mathematical models, 2, 47, 91, 130–131, 171, 182–183, 317, 361, 413
  nonlinear, 325–327
  of a parabola, 331–333, 619
  pictograph, 216
  scale of, 90–91, 200
  scatter plot, 108, 130, 270–271, 361, 500
  slope of a line, 113–115, 136–137
  solving quadratic equations with, 338
  solving systems of linear equations with, 288
  of system of inequalities, 280, 317
  See also Toolbox, 596–599.

Greatest common factor, 437
  See also Toolbox, 584.

Greedy algorithm, 517

Group activities
  Exercises, 17, 34, 39, 60, 83, 111, 118, 123, 150, 180, 203, 211, 223, 229, 243, 255, 263, 285, 298, 309, 330, 336, 341, 359, 387, 428, 434, 441, 513, 527, 540, 545
  Explorations, 8, 24, 30, 61, 78, 106, 119, 132, 147–148, 158, 176, 191, 205, 218, 251, 259, 299, 306, 325, 331, 355, 369, 383, 401, 429, 436, 442, 463, 469, 494, 509, 536
  Portfolio Projects, 46–47, 90–91, 136–137, 182–183, 230–231, 270–271, 316–317, 360–361, 412–413, 454–455, 500–501, 548–549

Growth factor, 376–377

Growth rate, 377

Histogram, 4, 14, 15, 16

Historical notes, 53, 59, 63, 76, 111, 123, 154, 163, 172, 210, 221, 262, 268, 293, 297, 303, 344, 353, 360–361, 370, 410, 440, 511, 517, 533, 539

Horizontal axis, 72

Hypotenuse, 239

Identity, 161

Identity property of multiplication, 24

Indirect measurement, 207

Inequality
  adding to, 176
  definition, 3, 171
  describing data, 3
  graphing linear, 280, 306, 317
  linear, 306–307
  as mathematical model, 130, 183
  as solution to problem, 172
  solving, 177
  systems of, 280, 311–312, 317
  writing from graphs, 171

Integers, 19

Similar triangles, 205–207
Similarity
  areas of similar figures, 212–214
  of figures, 205–207
  perimeters of similar figures,
    212–213
Simplifying
  exponents, 10
  expressions, 8
Slope of a line, 113, 618
  positive and negative, 114
  undefined, 115
**Slope-intercept form of a linear
    equation,** 120, 283, 618
Solutions, *See* Functions *and*
  Graphs.
Solving equations, *See*
  Equations.
Solving inequalities, 177
Solving problems, *See*
  Problem-solving strategies.
Spatial reasoning, *See*
  Manipulatives *and* Visual
  thinking.
Special products of binomials,
    264–266
Spiral review, *See* Review.
Spreadsheet
  Fill down command for spread-
    sheets, 42
  problem solving with, 43, 44, 45,
    83, 148–149, 170, 249, 256,
    381, 428, 452, 487, 540
  using, 41–42
  See also Technology Handbook,
    610.
Square, perfect, 240
Square roots
  defined, 240
  estimating, 253
  properties of, 252
Squaring a binomial, 265
Standard form
  of linear equation, 281–282
  of polynomial equation, 436
Statistics
  deviation from mean, 23
  mean, median, mode, 14
  See also Data analysis.
Study technique, 48, 92, 138, 184,
    232, 272, 318, 362, 414,
    456, 502, 550

Subtraction
  from inequalities, 177
  of linear equations to solve
    system, 294–295
  of negative integers, 20
  of rational expressions, 494
**Sum and difference of two
    terms,** 266
Symmetry, line of, 333
Systems of inequalities, 280,
    311–312, 317
Systems of linear equations
  solving by adding or subtracting,
    287–289, 294–295
  solving by algebra tiles, 622–623
  solving by graphing, 288
  solving by multiplication,
    299–300
  solving by substitution, 289
  solving by using a table, 621

Table
  defined, 30
  function, 66–68
  as mathematical model, 2, 46,
    90, 130, 147–149, 183, 231,
    317, 412
  writing as matrix, 40
Technology
  calculator, 8, 12, 13, 22, 39, 64,
    77, 88, 119, 121, 134, 147,
    148, 168, 172, 175, 203,
    228, 245, 254, 271, 284,
    286, 290–291, 297, 314,
    329, 330, 334, 341, 345,
    347, 348, 353, 357, 359,
    373, 381, 387, 393, 400,
    413, 434, 436, 441, 464,
    466, 472, 473, 474, 475,
    479, 480, 485, 510, 535,
    536, 540, 545
  computer, 41–42, 43, 44, 45, 83,
    148, 149, 170, 249, 256,
    381, 428, 452, 487, 540
Technology Handbook, 602–611
Temperature
  converting between Fahrenheit
    and Celsius, 28, 87
Test for similar triangles, 207

Testing, *See* Assessment.
Theoretical probability, 220
Time
  calculating cost of, 182
  measurement of, 360
Toolbox, examples and practice of
    previously learned algebra
    and arithmetic skills,
    584–601
Tree, 516, 518
Triangles
  right, 239
  similar, 205–207
Trinomial
  defined, 260
  factoring, 442
  perfect square, 265
Turning radius, 54

Undefined slope, 115
Unit rate, 99

Value of a variable, 3–4
Variable
  defined, 3, 9
  moving to one side of an
    equation, 158–161
Variable expression, 9, 30
Variation
  constant of, 107, 462
  direct, 106–108
  inverse, 462, 500–501
Vertex
  of a graph, 516
  of a parabola, 332
Vertical axis, 72
Vertical intercept, 120
Visual thinking, 2, 4, 6, 14–16,
    30, 32, 46–47, 72–73, 85,
    90–91, 100–101, 106, 108,
    113–115, 118–121, 130,
    136–137, 147–148, 162,
    171–172, 200, 209, 216,
    270–271, 281–283, 288,
    299, 301, 305, 306, 317,
    325, 331, 332, 339, 345,

Credits

Cover

Front cover   Earth Imaging/Tony Stone Images/Chicago Inc.(tl) James Andrew Bareham/Tony Stone Images/Chicago Inc.(r) **Back cover**   Magnus Rietz/The Image Bank

Chapter Opener Writers

Linda Borcover interviewed Carrie Teegardin (pp. xxx–2), Mae Jemison (pp. 96–98), Bill Pinkney (pp. 322–324), Linda Pei (pp. 420–422);   Laird Harrison interviewed Daniel Galvez (pp. 188–190);   Yleana Martinez interviewed Lori Arviso Alvord (pp. 366–368);   Steve Nadis interviewed Bob Hall (pp. 52–54), Barbara Romanowicz (pp. 144–146), Nathan Shigemura (pp. 236–238), Nancy Clark (pp. 278–280), Alex MacLean (pp. 460–462), Phil Roman (pp. 506–508).

Acknowledgements

xxx–2 From "Where Children Live" (map) and masthead logo from the June 6, 1993, issue of *The Atlanta Journal/The Atlanta Constitution*. Reprinted with permission from The Atlanta Journal and The Atlanta Constitution.   **96** From "Music of the Spheres" from the July 10, 1993, issue of *The Boston Globe*. Reprinted by permission of The Boston Globe.   **210** From *Was Pythagoras Chinese?: An Examination of Right Triangle Theory in Ancient China* by Frank J. Swetz and T.I. Kao. University Park: The Pennsylvania State University Press, 1977. **285** From "Counting Carbohydrates" by Nancy Clark from the January 1992 issue of *American Fitness*. Copyright ©1992 by Aerobics and Fitness Association of America. Reprinted by permission of the author.   **303** From *The Crest of the Peacock: Non-European Roots of Mathematics*, by George Gheverghese Joseph. London: I.B. Tauris & Co. Ltd., 1991.   **304** "The Two Parallels" from *Songs From the Gallows*, by Christian Morgenstern. Copyright ©1993 by Yale University. Reprinted by permission of Yale University Press.   **404** From *The Rolling Stones*, by Robert A. Heinlein. New York: Charles Scribner's Sons, 1952, 1980. **546** From *Rosencrantz and Guildenstern Are Dead*, by Tom Stoppard. Copyright ©1967 by Tom Stoppard. Used by permission of Grove/Atlantic, Inc.

Stock Photography

i James Andrew Bareham/Tony Stone Images/Chicago Inc.;   **iv** Courtesy of Arlene Blum/Expedition photographer unknown(t);   **iv** David Weintraub/Stock Boston(m);   **vi** Manoj Shah/Tony Stone Images/Chicago Inc.(t);   **vi** NASA(b);   **viii** Jeffrey Dunn(t);   **viii** Jeffrey Dunn(b);   **ix** Superstock(t);   **ix** Clifford R. Mueller/Illinois State Police. Courtesy of Nathan S. Shigemura(b);   **xi** M. & E. Bernheim/Woodfin Camp and Associates(t);   **xi** Paul Elledge (b);   **xii** Clive Brunskill/Allsport(t);   **xii** Courtesy of *The Chicago Tribune*(b);   **xiii** Courtesy of Linda Pei/Photo by Gus Bower;   **xiv** Alex S. MacLean/Landslides(b);   **xv** Bob Brill(t);   **xv** Bob Brill(m);   **xv** Courtesy of Film Roman(b);   **xvii** Craig T. Mathew/Courtesy of Film Roman(tl);   **xvii** Michael Heller(tr);   **xvii** Clifford R. Mueller/Illinois State Police. Courtesy of Nathan S. Shigemura(br);   **xxi** David Ulmer/Stock Boston(bl);   **xxi** Mark Segal/Tony Stone Images/Chicago Inc.(br);   **xxiii** Michel Hans/Allsport;   **xxiv** Lonny Kalfus/Tony Stone Images/Chicago Inc.(l);   **xxv** Tom McHugh/Photo Researchers, Inc.(t);   **xxv** Warren Morgan/Westlight(b);   **xxix** Jeffrey Dunn(bl);   **6** Superstock(br); **7** Negative # 314993, Courtesy of Department of Library Services, American Museum of Natural History;   **12** Norm Thomas/Photo Researchers, Inc.(tl);   **12** Terry Raines(m);   **12** Ian Murphy/Tony Stone Images/Chicago Inc.(bl);   **12** Terry Raines(br);   **15** Tony Duffy/Allsport;   **16** Paul Kenward/Tony Stone Images/Chicago Inc.; **17** ©Tom Van Sant/Geosphere Project, Santa Monica/Science Photo Library/Photo Researchers, Inc.(tr);   **17** Gregory Silber(ml);   **17** C. Faesi/Projecto Vaquita/Marine Mammal Images(bl);   **19** David Weintraub/Stock Boston(tr);   **19** Courtesy of Arlene Blum/Expedition photographer unknown(l);   **19** David Weintraub/Stock Boston(br);   **22** Mike Yamashita/Woodfin Camp and Associates(tl);   **22** Robert Frerck/Odyssey Productions/Chicago(lower ml);   **22** Loren McIntyre/Woodfin Camp and Associates(upper ml);   **22** Michael J. Howell/The Picture Cube(mr);   **22** Giraudon/Art Resource(bl);   **23** Paul J. Sutton/Duomo(t);   **23** Duomo(b);   **26** David Harding/Tony Stone Images/Chicago Inc.;   **28** Craig Newbauer/Peter Arnold, Inc.(tl); **28** Dennis MacDonald/PhotoEdit(ml);   **28** Stephen Frisch/Stock Boston(bl);   **32** Runk/Schoenberger/Grant Heilman Photography(bl);   **32** Runk/Schoenberger/Grant Heilman Photography(bm);   **32** Barry L. Runk/Grant Heilman Photography(br);   **34** Cuisine Studio, 400(t);   **34** Felicia Martinez/PhotoEdit(tm);   **34** Cuisine Studio, 400(b);   **34** Felicia Martinez/PhotoEdit(bm);   **38** Miro Vintoniv/Stock Boston(l);   **41** Peter Tarry/Action-Plus;   **44** Phil McCarten/PhotoEdit(t);   **44** Richard Hutchings/Photo Researchers, Inc.(b);   **46** Warren Morgan/Westlight(tl);   **53** Courtesy of Boston Athletic Association/Fay Foto(b);   **55** Neal Boenzi/New York Times Pictures **58** Focus on Sports, Inc.;   **59** UPI/Bettmann;   **60** Courtesy of Bob Hall(r);   **63** The Bettmann Archive;   **64** Ferguson & Katzman/Tony Stone Images/Chicago Inc.;   **70** Lawrence Migdale/Stock Boston;   **72** Kevin Schafer/Tony Stone Images/Chicago Inc.;   **75** Mark Segal/Stock Boston(l);   **75** David Burnett/Contact Press Images(r);   **76** Michael J. Howell/The Stock Shop(r);   **76** *"The Region of The Great Wall of China"*. #15261e2. By permission of The British Library(b);   **80** Letraset(bkgrnd); **81** Gabe Palmer/The Stock Market;   **87** Jack Gescheidt;   **88** Joe McDonald/Animals Animals(inset);   **96-98** NASA;   **100** Mark Wagner/Tony Stone Images/Chicago Inc.;   **102** David Nunuk/Science Photo Library/Photo Researchers, Inc.;   **104** NASA; **105** ©Tom Van Sant/The Geosphere Project/The Stock Market(tr); **105** Dorothy Littell/Stock Boston(br);   **110** Brooks Air Force Base(bl);   **110** NASA(br);   **111** Art Resource;   **113** Zefa Germany/The Stock Market;   **116** Mark Antman/Stock Boston; **117** Michael Newman/PhotoEdit(tl);   **117** Terry Qing/FPG International(br);   **123** The Bettmann Archive(tl);   **123** Photographs courtesy of The Time Museum, Rockford, Illinois(r, bl); **125-126** Ferne Saltzman/Albuquerque International Balloon Fiesta; **128** William Johnson/Stock Boston;   **130** David Ulmer/Stock Boston(t);   **145** U.S. Geological Survey/Albuquerque Seismological Laboratory(t);   **145** AP/Wide World Photos(b);   **146** AP/Wide World Photos(t);   **146** AP/Wide World Photos(l);   **146** Natural History Museum, London(r);   **156** Bob Daemmrich/Uniphoto; **162** Mark C. Burnett/Stock Boston;   **163** Michael Holford;   **165** AP/Wide World Photos;   **165** U.S. Geological Survey/Albuquerque Seismological Laboratory(t);   **171** George Hunter/Tony Stone Images/Chicago Inc.;   **172** Ulrike Welsch/Photo Researchers, Inc.;

Selected Answers

CHAPTER 1

Page 5 Checking Key Concepts

1. Answers may vary. **3.** Answers may vary. An example is given. I disagree. It is true that $h > 0$ and $h < 100$, but these inequalities apply to the heights of all human beings and give no useful description of this particular group. **5.** 17 students **7.** that you spend at most 6 h visiting

Pages 5–7 Exercises and Applications

1. 0 **3.** 13 states **5.** 20%; 40% **7.** Let e = weekly earnings; $e \le 600$. **9.** Let l = price of lunch; $l \ge 3.95$. Let d = price of dinner; $d \ge 4.95$. **13.** about 15% **19.** If w = the weight in grams, then $w \ge 2$ and $w \le 30,000,000$; if w = the weight in kilograms, then $w \ge 0.002$ and $w \le 30,000$. **21.** Answers may vary. An example is given. 29.95, 49.99, 119.99 **27.** United States **29.** 2.67 **31.** 5.56 **33.** 0.13

Page 11 Checking Key Concepts

1. 9 **3.** 5 **5.** 5.1 **7.** $\frac{1}{2}$ **9.** 0 **11.** 4 **13.** 10.7 **15.** $\frac{7}{17}$

Pages 11–13 Exercises and Applications

1. a. 0 **b.** 4 **3. a.** 42 **b.** 42 **5. a.** 8 **b.** 11.5 **9.** 3.5 **11.** 5.88 **13.** $6\frac{1}{5}$ **15.** $\frac{2}{11}$ **17.** $32\frac{1}{2}$ **19.** $46\frac{1}{8}$ **23.** $25g$ mi **25.** 36 in. **27.** 53 **29.** $4\frac{3}{8}$ **31.** $\frac{11}{6}$ **33.** 1 **35.** 0 **37.** about 256 ft **41.** Answers may vary. Example: $(\!(2 \times 5 - 7)\!) \div 15 =$ **43.** C **47.** Choices of variables may vary. An example is given. Let s = the number of species counted; $s \le 80$.

Page 16 Checking Key Concepts

1. about 3.2; 3; 2 **3.** about 21.9, 22, 22

Pages 16–18 Exercises and Applications

1. 2; 2; 0 **3.** about 234.9; 240; 244; Answers may vary. An example is given. The three values are fairly close, so I think no one average describes the data better than the others. **11.** $\frac{1}{20}$; 5% **13.** $\frac{1}{4}$; 25% **15.** $\frac{7}{4}$; 175%

Page 18 Assess Your Progress

1. 16 **2.** 2 **3.** $\frac{1}{5}$ **4.** $18\frac{2}{3}$ **5.** 10 **6.** $\frac{1}{5}$ **7.** 24 students **8.** 15 students **9.** For example, $n \ge 0$ and $n \le 6$; the variable n represents the number of movies seen by each student. **10.** 2.75; 3, 3 and 4

Page 21 Checking Key Concepts

1. –24 **3.** $-\frac{1}{4}$ **5.** $-n$ **7. a.** 6 **b.** 6 **9. a.** –6 **b.** –2 **11. a.** 2 **b.** 8

Pages 21–23 Exercises and Applications

1. –10 **3.** –13 **5.** –8 **7.** –4 **9.** 3 **13.** 1050 years **15.** 503 years **17.** 1100 years **19.** –12 **21.** 78 **23.** 4.3 **25.** –3 **27.** $\frac{9}{4}$ **29.** Answers may vary. Examples are given. **a.** $x = 5, y = 2$ **b.** $x = 2, y = 5$ **c.** $x = 5, y = 5$ **31.** –1 **33.** 8 **35.** 0 **37.** 12 **41.** D **47.** 8; 4 **49.** $4\pi \approx 12.56$; $4\pi \approx 12.56$ **51.** 7; 2.5

Page 27 Checking Key Concepts

1. –6 **3.** –6 **5.** 18 **7.** $\frac{13}{16}$; The quotient of two numbers is positive if both numbers are negative. **9.** $-\frac{12}{7}$; The quotient of two numbers is positive if both numbers are negative, so $\frac{-12}{-7} = \frac{12}{7}$. The opposite of a positive number is negative, so $-\left(\frac{12}{7}\right) = -\frac{12}{7}$. **11.** Answers may vary. An example is given. You can multiply 7 and 8 and then multiply the product by 0, or you can multiply 8 and 0 and then multiply the product by 7. I prefer the second way because the product of 8 and 0 is 0, and the product of 0 and 7 is 0; this is easier than remembering the product of 7 and 8.

Pages 27–29 Exercises and Applications

3. –4 **5.** 70 **7.** 7 **9.** 7 **15.** 5°F **17.** –12 **19.** –30 **21.** 6 **23.** 12 **27.** equal; associative property **29.** equal; associative property **31.** not equal **33.** equal; both properties **39.** 8; 120 **41.** 1.2; 120% **43.** $0.\overline{8}$; about 88.9%

Page 29 Assess Your Progress

1. 9 **2.** 0 **3.** –3 **4.** 5 **5.** 12 **6.** 8 **8.** 42 **9.** –8 **10.** 5 **11.** 7 **12.** $-\frac{8}{3}$ **13.** 12

Page 32 Checking Key Concepts

1. $0.75(g - 5)$ dollars **3.** Let f = the number of friends she invites to the party; $5 + 2f$ people. The expression represents the total number of people she invites to the party, which is 5 more than twice the number of friends she invites.

Pages 32–34 Exercises and Applications

1. $10 + 1.5t$ dollars; Answers may vary. An example is given. whole numbers from 0 to 4 **3.** Let n = the number of ballerinas; $200n$ pairs of toe shoes are ordered.

7. Let d = the number of days Chris is on vacation; $2d + 2$ books (if "a couple" of books means two books); the number of books Chris wants to take along is two more than twice the number of vacation days.   **15.** Let n = the number of people to be served; $1.5n$ pounds.   **19.** 1   **21.** 195   **23.** 48   **25.** 9

Page 37   **Checking Key Concepts**

1. $2(2x + 1) = 4x + 2$ **3.** $2x - 8$ **5.** $40x - 16$ **7.** $13 + 8x$
9. $2x - 7$ **11.** $-3x - 20$

Pages 37–39   **Exercises and Applications**

1. $2x + 2y$ **3.** $-10t - 14$ **5.** $-\frac{5}{3}a + 2b$ **7.** $24 - 12y$
9. $15z + 20$ **11.** $-1.75 + x$ **13.** Answers may vary. An example is given. The teacher can give each student a book and a calculator at the same time (which is like adding, then multiplying). The result will be the same if the teacher gives a book to each student and then a calculator to each student (which is like multiplying first). In symbols, $(1 + 1)n = n + n$, where n = number of students.   **17.** $3m + 4$
19. $3x + 4y^2$ **21.** 15   **23.** $-21u + 21uv - 2v$ **25.** If he walks m miles, then the amount of money he raises is $m + 1.5m + 2m + m + m$, or $6.5m$ dollars. I prefer to evaluate the expression $6.5m$ because it requires multiplying once. Evaluating the other expression requires multiplying twice and then adding five numbers.   **27.** $36 + 4x$ **29.** $13 - 3x$
31. $6x + 15y - 7$ **33.** 0   **35. a.** evaluating: $-5(6 + 12(5)) + 3(20(5) + 10) = -5(66) + 3(110) = -330 + 330 = 0$; simplifying first: $-30 - 60x + 60x + 30 = 0$; It is easier to simplify the expression first because you can see that the expression is equal to 0.   **b.** Answers may vary. An example is given. $4(3x - 6y) - 3(-8y + 4x) = 12x - 24y + 24y - 12x = 0$ **41.** 20
43. -2 **45.** 22

Page 42   **Checking Key Concepts**

1. A: 1×3; B: 3×1; C: 1×3; D: 3×1
3. a. 12 * B3 + C3; 88   **b.** 82.25; The number is the mean height of the four athletes in inches.

Pages 43–45   **Exercises and Applications**

1. A: 2×3; B: 2×3; C: 3×2; D: 3×2 **5. a.** The sales manager used the fill down command to divide each person's expenses for the month by that person's mileage for the month.   **b.** Milner, \$.04; Dubois, \$.05; Perez, \$.04; Chen, \$.04   **9.** The number of males in the work force and the number of females in the work force increased steadily from 1960 to 1980 and then decreased from 1980 to 1990.
11. Answers may vary. An example is given. The number and percent of 16- and 17-year-olds in the work force increased between 1960 and 1980 and then declined. I think the percents give a better sense of how many teenagers were in the work force. They tell how likely it is a teenager was in the work force in a given year.   **17.** 0   **19.** 7   **21.** $\frac{1}{2}$

Page 45   **Assess Your Progress**

1. $60h$ min   **2.** $m + 250$ mi   **3.** $3x - 6$ **4.** $5x^2 + 2y - x$
5. $x - 1$ **6.** X: 3×3; Y: 3×3; Z: 2×2;

$$X + Y = \begin{bmatrix} 6.6 & 2.25 & -2 \\ -2 & 18 & 3 \\ 5 & 3 & 11 \end{bmatrix}$$ **7.** B6 = B2 + B3 + B4 + B5

Pages 50–51   **Chapter 1 Assessment**

1. Answers may vary. An example is given. $3x + 2y - 5x + 6 = -2x + 2y + 6$ **2.** The absolute value of a number is its distance from zero on the number line. For example, $|-7| = 7$, $|3| = 3$, and $|0| = 0$.   **3.** The mean is the sum of the data divided by the number of data items. The median is the middle number when you put the data in order from smallest to largest. When the number of data items is even, the median is the mean of the two middle numbers.   **4.** 10   **5.** 25
6. $g \geq 76$ and $g \leq 100$ **7.** 24   **8.** 8   **9.** 1   **10. a.** 19.6; 19; 19   **11.** -3 **12.** 1   **13.** -7 **14.** 30   **15.** -5 **16.** 5
17. 1   **18.** $-11\frac{2}{3}$ **19.** -11 **20.** -20 **21.** -10 **22.** $24d$; 168 h   **23.** Let f = the number of friends; $2.90(f + 1)$ dollars; \$23.20.   **24.** $11x + 11$ **25.** $5 + 5x + 7y$ **26.** y
27. $-6x^2 + 6x$ **28.** amount; price per item; total cost; D2 = B2*C2   **29.** A: 3×2; B: 2×2; C: 3×1; D: 3×2

30. $A + D = \begin{bmatrix} 2 & 8 \\ 7 & 0 \\ 0 & 6 \end{bmatrix}$

CHAPTER 2

Page 58   **Checking Key Concepts**

1. -23 **3.** 0.75   **5.** -16 **7.** 140   **9.** -10 **11.** 4.1
13. about 10.31 m/s

Pages 58–60   **Exercises and Applications**

1. -415 **3.** 6.5   **5.** $-\frac{3}{8}$ **7.** -5 **9.** 0   **11.** 0   **13.** 6.25
15. $-\frac{1}{2}$ **17.** -4.5 **21.** A   **25.** about 135.97 mi/h
27. $29{,}887 = 100d$; about 299 days   **29.** A; The rate per day, 15 mi, times the number of days d equals the distance traveled, 2000 mi; $d \approx 133.33$; about 133.3 days.
31. about 0.22 mi/min   **35.** 26   **37.** $\frac{2}{3}$ **39.** 6
41. $6x + 40$; 100

Page 63   **Checking Key Concepts**

1. $4x + 3 = 11$; $x = 2$ **3.** $\frac{7}{5} = 1.4$ **5.** 9   **7.** $-\frac{78}{11} \approx -7.09$
9. -9 **11.** 10

Pages 64–65 Exercises and Applications

1. 31 **3.** −1 **5.** −6 **7.** 2 **9.** −4 **11.** −42 **13.** 72
15. −67.5 **17. a.** subtraction of 13; division by 6
19. 18 mi **23.** 2.5 **25.** −3 **27.** 30 **29.** −49 **31.** 130
33. −100 **35.** $x = 1$; 5 in. **37.** $x = 7.5$; 7.5 mm, 10.5 mm,
12.5 mm, and 12.5 mm **41.** $s + 5$; Rob's sister's height in
inches **43.** $25t − 7$ **45.** $−8m + 28$ **47.** $−2x$

Page 68 Checking Key Concepts

1. The steering wheel is like a function because the angle
through which the car turns depends on how far you turn the
steering wheel. For each amount you turn the wheel, the car
will turn in a predictable way. A TV channel selector is like a
function. For each number you select, one channel will be
chosen.

Pages 69–71 Exercises and Applications

1. No; for some values in the domain (13 and 14), there is
more than one value in the range. **3.** Yes; for each value in
the domain, there is exactly one value in the range.
5. C; p = the price of the book in dollars and C = her change
in dollars. **7.** B; n = the number of friends and p = the price
in dollars each friend pays. **9.** Ex. 5: the domain is 0.01,
0.02, 0.03, …, 10.00 and the range is 0, 0.01, 0.02, …, 9.99,
since she must pay at least \$.01 for the book and can spend no
more than \$10. Ex. 6: the domain is 1, 2, 3, … and the range
is 10, 20, 30, …, since the bike is rented by the day. Ex. 7: the
domain is 1, 2, 3, 4, … and the range is 10, 5, 3.33, 2.50, …,
since the number of friends is a whole number greater than 0.
Ex. 8: the domain is 10.01, 10.02, 10.03, … and the range is
0.01, 0.02, 0.03, since the original price of the shoes must be
at least \$10.01. **11.** 4 tickets **13. a.** $p = \dfrac{n}{5{,}021{,}000}$, where
n = the number of teens associated with a portion of the circle
graph and p = the related percent. **b.** Yes; for each value of
n, there is exactly one value of p. **15. a.** \$12.50 **b.** \$18.75
c. 8 h **17.** $P = 2s + 16$; the domain is all positive numbers
and the range is all numbers greater than 16. **25.** 52
27. about 6.64 **29.** 2.5 **31.** 15 **33.** −51 **35.** 21

Page 71 Assess Your Progress

1. $−\dfrac{1}{6}$ **2.** −8 **3.** −41 **4.** 8 **5.** −20 **6.** 14 **7.** −48
8. 78 **9.** 4.5 min **10. a.** $C = 3 + 2p$, where p is the number
of plants bought and C is the total cost in dollars. The domain
is the positive whole numbers and the range consists of these
amounts in dollars: 5, 7, 9, …. **b.** 21 plants

Page 74 Checking Key Concepts

1. about 400 ft below the surface **3.** C; I **5.** A; II **7.** G;
none **9.** D; III

Pages 74–77 Exercises and Applications

1–15.

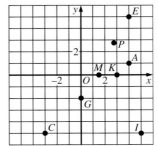

1. I **3.** III **5.** I **7.** none **9.** IV **11.** none **13.** none
15. I **17.** the number of cars washed; the amount of money
raised **19.** 40 cars **21.** March and September; about 12 h
per day **25. a.** \$100; The point (0, 100) is on the graph, so 0
weeks after the money is borrowed, \$100 is owed. **b.** \$40
c. 10 weeks; The point (10, 0) is on the graph, so 10 weeks
after the money is borrowed, the amount owed is \$0.
27. Estimates may vary; about 1820 mi. The band covers
about 13 grids horizontally. Each grid is about 140 mi on a
side. **31.** $A(−5, −5)$; $B(−2, −3)$, $C(3, −3)$; $D(5, 5)$; $E(1, 3)$;
$F(−2, 2)$; $G(−5, −1)$ **33.** B and C **35.** C and F **37.** about
5.66; about 5.88; 6.75 **39.** A: 3×2; B: 2×3; C: 3×3;
D: 1×4; E: 2×2; F: 1×3; G: 3×2; H: 1×4; I: 3×3;
A and G, C and I, D and H;

$$A + G = \begin{bmatrix} 5 & 3 \\ 13 & 15 \\ -11 & 19 \end{bmatrix}; C + I = \begin{bmatrix} 2 & 1 & 13 \\ 1 & 0 & -4 \\ 5 & 7 & 4 \end{bmatrix}; D + H = \begin{bmatrix} 16 & 5 & 3 & 7 \end{bmatrix}$$

Page 81 Checking Key Concepts

1. B; $1 = \frac{1}{2}(2)$ and $−1 = \frac{1}{2}(−2)$ **3.** A; $−2 = −(2)$ and $2 = −(−2)$
5. (1) an equation: $C = 8 + 0.32n$

(2) a table of values:

number of invitations sent	total cost (dollars)
0	8
1	8.32
2	8.64
…	…
15	12.80

(3) a graph:

Pages 81–83 Exercises and Applications

1. Yes.

3. No.

5. No.

7. Yes.

11. a.

Wheelchair Turning Radius

b. Estimates may vary; about 50 ft   **c.** The domain is the positive numbers less than or equal to 15 and the range is the positive numbers greater than 14.

13. Estimates and methods may vary. Examples are given. I estimate there were about 15 male wheelchair athletes and about 3 female wheelchair athletes in 1983. I estimated by using the values in the table, giving special attention to the years 1984 to 1993, and following the trends backward from 1984 to 1983.   **15.** D; –57   **17.** B; –28   **19.** A; B

21.

	A	B
1	–5	–24
2	–4	–23
3	–3	–22
4	–0.5	–19.5
5	0	–19
6	1	–18
7	2	–17
8	5	–14

23.

	A	B
1	1	0.5
2	2	1
3	3	1.5
4	4	2
5	5	2.5
6	6	3
7	7	3.5
8	8	4
...	...	...

29. $\frac{17}{12}$ **31.** $19s + 4t - 7$ **33.** $-6(2m - 3n) + 12m =$
$-12m + 18n + 12m = 18n + 0 = 18n$

1. a. $19 + 9d = 126$; $d \approx 12$; about 12 days

b.

Pages Read

1. 10 T-shirts   **3.** 2.5 mi, 0.5 mi, 0.5 mi   **5.** Estimates may vary.   **a.** about 30°F   **b.** about 70°F   **c.** about –5°F   **d.** about –20°F   **7.** Ex. 5: 32°F; 68°F; –4°F; –22°F; Ex. 6: 0°C; about 32.2°C; about –17.8°C; –40°C   **11.** 63°F
13. a. 45°F   **b.** 50°F   **c.** 60°F   **d.** 65°F
17, 19. Answers may vary. Examples are given.
17. $x \approx 4.8$

19. $x \approx 18.1$

21. B   **23.** No.   **25.** No.
27. (1) an equation: $y = 2x + 3$

(2) a table of values:

x	y
–1	1
0	3
1	5
...	...

(3) a graph:

29. (1) an equation: $y = 4 - x$

(2) a table of values:

x	y
–1	5
0	4
1	3
...	...

(3) a graph:

Page 89 Assess Your Progress

1–8.

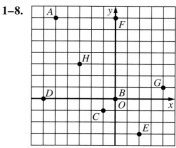

1. II **2.** none **3.** III **4.** none **5.** IV **6.** none **7.** I
8. II

9.

10.

11.

12. $P = 8n - 780$

Estimates may vary; the actual number of tickets is 160.

Pages 94–95 Chapter 2 Assessment

1. domain; range **2.** multiplication or division; inverse
operations **3.** 26 **4.** $-\frac{15}{16}$ **5.** –23 **6.** 32 **7.** –4 **8.** 5
9. 104 **10.** 54 **11.** –3 **12.** about 16,456.5 mi/h
13. a. $C = 50 + 12n$, where n is the number of times she plays
golf and C is the total cost in dollars **b.** The domain is 0, 1,
2, 3, … and the range is 50, 62, 74, 86, …. **c.** 16 times
14. a. Estimates may vary. highest temperature: about –30°C
after 15 h; lowest temperature: about –85°C after 5 h **b.** IV

15–18.

15. none **16.** III **17.** IV **18.** II

19.

20.

21.

22.

23.

24.

25. the equation $y = 4x - 3$ in Ex. 21 **27. a.** $C = 450 + 20n$,
where n is the number of people at the party and C is the total
cost in dollars

b.

c. 77 people

CHAPTER 3

Page 102 Checking Key Concepts

1. $\frac{\$54}{1 \text{ day}}$ **3.** 315 mi **5.** about 34.2 km/h; about 21.2 mi/h
7. Estimates may vary. about 2800 ft/min

1. $\dfrac{36 \text{ students}}{1 \text{ dormitory}}$ **3.** $\dfrac{60 \text{ s}}{1 \text{ min}}$ **5.** 0.9   **7.** 1760   **9.** 369.6

11. He multiplied by $\dfrac{1 \text{ h}}{60 \text{ min}}$ when he should have multiplied by $\dfrac{60 \text{ min}}{1 \text{ h}}$ to get 36 as the average number of shooting stars per hour.   **13.** Answers may vary. Examples are given. 4 wins out of the first 5 games, 1 touchdown per half   **15.** about 84 yards per game   **17.** Estimates may vary.

	Rate of change for LP sales (LPs/y)	Rate of change for CD sales (CDs/y)
a.	about −40 million	about 20 million
b.	about −40 million	about 30 million
c.	about −20 million	about 50 million
d.	about −30 million	about 50 million

23. Answers may vary. An example is given using centimeters per minute. Unionville: 3.1 cm/min; Curtea-de-Arges: 1.025 cm/min; Holt: about 0.73 cm/min; Cilaos: about 0.13 cm/min; Cherrapunji: about 0.02 cm/min   **25.** 300 min, or 5 h; 930 cm $\cdot \dfrac{1 \text{ min}}{3.1 \text{ cm}} = 300$ min   **27.** C   **29.** 0.375

31. 1.25   **33, 35.**

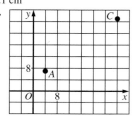

33. I   **35.** I

1. No; the graph is not a line through the origin.   **3.** about 0.8; If x = drop height in inches and y = bounce height in inches, then $y = 0.8x$.

1. No.   **3.** No.   **5.** Yes; the points lie on a line that goes through the origin.

7.

The graph is a line through the origin, so the data show direct variation.

9. $l = \dfrac{1}{6}f$ **13.** the graph on the right; This graph shows that Mina receives no commissions on sales at or below $1000.
15. square: $P = 4x$; rectangle: $P = 2x + 20$; The perimeter of the square varies directly with x because the equation $P = 4x$ is in the form of a direct variation equation. The perimeter of the rectangle does not vary directly with x because the equation for the perimeter cannot be written in the form $P = kx$ and because the graph of $P = 2x + 20$ does not pass through the origin.   **17.** mean, $60,000; median, $42,500; mode, $25,000

19. No.

21. Yes.

1. 21 km/h   **2.** about 4.17 km/h   **3.** 3.6 km/h   **4.** No.
5. Yes; 3; $y = 3x$, where x = the number of tickets sold and y = profit in dollars.

1. E and F **3.** A **5.** E

1. $\dfrac{1}{2}$ **3.** $-\dfrac{1}{3}$ **5.** −1

7. $\dfrac{1}{2}$

9. −2

11. $-\dfrac{3}{5}$ **13.** undefined   **15.** $-\dfrac{32}{15}$ **17.** $-\dfrac{15}{121}$

27. a. 4; −7; $\dfrac{3}{5}$ **b.** 4; −7; $\dfrac{3}{5}$

c. The constant of variation in each direct variation equation is equal to the slope of the graph of the equation.

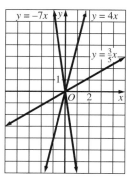

31. No; for each value in the domain, there are two values in the range.   **33.** 0.0005   **35.** 120,000

Page 122 Checking Key Concepts

1. 1; 4   3. 1; 0

7. $\frac{1}{3}, -\frac{4}{3}$

Pages 122–124 Exercises and Applications

1. $y = \frac{1}{3}x + 1$ 3. $y = -\frac{2}{3}x - 1$ 5. $y = -3x$

7.   9.

11.   13.

15.

21. −3; the rate at which the height of the candle changes, −3 in./h; 12; the original height of the candle, 12 in.

23. $h = -3t + 12$ 27. $\frac{8}{3}$ 29. 15 31. 2 33. −6

Page 124 Assess Your Progress

1. 2 2. undefined 3. −1 4. a. $y = -50x + 350$ b. The slope, −50, is the rate of change of Sue's distance from Tampa, −50 mi/h. The vertical intercept, 350, is Sue's distance from Tampa at $t = 0$, that is, the distance from her home to Tampa.

5. 3; −4   6. −1, 2

Page 127 Checking Key Concepts

1. $y = 2x - 3$ 3. $y = 4$ 5. $y = -x + 2$

Pages 127–129 Exercises and Applications

1. $y = x + 3$ 3. $y = -3x + 31$ 5. $y = -2x$ 7. $y = 5$
9. $x = 1$ 11. a. $y = -\frac{4}{5}x + 4$

b. $y = -\frac{4}{5}x + 6$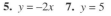

13. $y = 18$ 15. $y = 2x - 3$ 17. $y = -\frac{5}{3}x + \frac{5}{3}$
19. $y = -\frac{19}{7}x + \frac{65}{7}$ 23. Answers may vary, depending on rounding. $t = -0.004e + 109.2$; about 76° 25. Answers may vary. Examples are given. about 91°, about 60°; about 88°, about 56° 29. 22 students 31. about 14.7%
33. $15p + 5a$, where p is the cost of a postcard stamp and a is the cost of an airmail stamp

Page 132 Checking Key Concepts

1. Yes. Equations may vary. An example is given. $y = 5x$
3. about 102.6 min 5. about 106.2 min

Pages 133–135 Exercises and Applications

Equations based on line of fit may vary throughout. Examples are given. 1. $y = \frac{3}{16}x + \frac{3}{8}$ 5. Answers may vary. Examples are given. a. about 73.4 years b. about 78.2 years
c. about 62.6 years 9. D 11. $14b + 3$ 13. $9 - 9z$
15. $-7n + 24$ 17. at least 260 tickets

19.

at least 220 tickets; at least 110 tickets

Page 135 Assess Your Progress

1. $y = x - 5$ **2.** $y = 3$ **3.** $x = 6$ **4.** $y = -2x - 2$

5. $y = \frac{2}{3}x + 5$ **6.** $y = 5.5x - 30.5$

7. a. 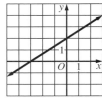 $y = \frac{2}{3}x + 2$

b. $y = \frac{2}{3}x - \frac{4}{3}$

8–10. Answers may vary. Examples are given.

8, 9.

$y = 560x + 4000$, where x is the number of years since 1980 and y is the number of cable systems

10. about 18,000 cable systems

Pages 140–141 Chapter 3 Assessment

1. Answers may vary. An example is given. The data are from Example 1 on page 107. The constant variation is 5.75.

2. Answers may vary. Examples are given.

3. $\frac{7 \text{ days}}{1 \text{ week}}$ **4.** $\frac{\$58}{1 \text{ day}}$ **5.** 134 **6.** 22.5 **7.** B

8. about 25,000 tons of garbage **9.** Yes; the constant of variation is about 0.159; $r = 0.159C$, where C is the circumference in centimeters and r is the radius in centimeters.

11. Let x = number of miles towed and y = cost in dollars; $y = 2x + 34$; the slope indicates the rate of change of the cost, \$2/mi. The vertical intercept indicates the basic charge of the trip excluding the cost per mile. The tow costs \$34 plus the cost per mile. **12.** –1 **13.** $-\frac{3}{4}$ **14.** $y = -x + 1$

15. $y = \frac{3}{4}x - \frac{1}{4}$ **16.**

17.   **18.**

19. $y = \frac{3}{2}x$ **20.** $y = x - 1$ **21.** $y = 7$ **22.** $y = 3x - 5$

23. Answers may vary. Examples are given.

a, b.

Average Weight and Height

c. Answers may vary. $h = 2.2w + 68$ **d.** about 23.6 kg

CHAPTERS 1–3

Pages 142–143 Cumulative Assessment

1. 3500 **3.** 20–24 **5.** –5 **7.** –10 **9.** $\frac{22}{5}$, or 4.4

11. mean 23.4; median: 21; mode: 21 **17.** –208 **19.** 11

21. 32

23, 25.

23. B: I **25.** D: III

27. about 2.5 s **31.** $\dfrac{8\ \text{h}}{1\ \text{day}}$ **33. a.** $\dfrac{3}{2}$ **b.** $y = \dfrac{3}{2}x + 6$ **c.** $y = 6$

CHAPTER 4

Page 150 Checking Key Concepts

1. $y = 60$; The cost is constant no matter how many visits a person makes to the pool. **3.** No; the cost for a pass is $60 and the cost for 10 visits without a pass is $40.

Pages 150–152 Exercises and Applications

1. Club A, since at 0 months Club A has a cost of $100 and Club B has a cost of $0; Club B, since the cost of the first month is $50 – 0$ or $50 and the cost at Club A is $140 – 100$ or $40 **3.** during the tenth month of membership

5.

No. of hours of overtime per year	Total earnings from Job A (dollars)	Total earnings from Job B (dollars)
50	30,000	24,600
100	30,000	25,200
150	30,000	25,800
200	30,000	26,400
250	30,000	27,000
300	30,000	27,600
350	30,000	28,200
400	30,000	28,800
450	30,000	29,400
500	**30,000**	**30,000**
550	30,000	30,600

Job Earnings

(Graph: Earnings (dollars) vs. Overtime worked (hours). Lines labeled "Earnings from Job A" and "Earnings from Job B".)

7. If you work no more than 500 h of overtime per year, then Job A would pay more than Job B. If you work more than 500 h of overtime per year, then Job B would pay more.
9. The cost of using a fluorescent bulb increases sharply after each 9000 hour period; about $20. **11.** Answers may vary.

13. If a customer plans to attend up to 15 shows, then it is less expensive to pay the nonmember rate. If a customer plans to attend more than 16 shows, buying a membership is the better choice. At 16 shows, the costs are equal. **17.** 10 **19.** 2 **21.** –5 **23.** No.

Page 155 Checking Key Concepts

1. $\dfrac{10}{9}$ **3.** $\dfrac{1}{8}$ **5.** $\dfrac{2}{3}n$ **7.** $-\dfrac{6}{5}c$ **9.** 20 **11.** 50

Pages 156–157 Exercises and Applications

1. $\dfrac{9}{2}$; 36 **3.** $\dfrac{6}{5}$; $\dfrac{72}{5} = 14\dfrac{2}{5}$ **5.** $\dfrac{7}{16}c$ **7.** $-\dfrac{5}{11}x$ **9.** –45
11. $\dfrac{21}{2} = 10\dfrac{1}{2}$ **13.** $\dfrac{18}{5} = 3\dfrac{3}{5}$ **15.** –80 **17.** E **19. a.** $l = \dfrac{3}{4}h$, where l = the frequency of the lower note and h = the frequency of the higher note **b.** $586\dfrac{2}{3}$ vibrations per second
c. 330 vibrations per second **21.** $2\dfrac{2}{3}$ yd **23.** $1\dfrac{1}{3}$ ft **25.** 12
27. 55 **29.** –36 **31.** 10 **33.** $-10\dfrac{4}{5}$ **35.** 350 mi **39.** 2
41. 5 **43.** $16\dfrac{2}{5}$ **45.** $5a^2 + 5a - 2$ **47.** $-8m + 12n$
49. $8p + 56$

Page 161 Checking Key Concepts

1. $2x$; 3; 3; 2 **3.** Subtract $2x$ from both sides; $x = \dfrac{9}{2}$.
5. Subtract $5n$ from both sides; $n = -10$. **7.** Rewrite $2(6x + 4)$ as $12x + 8$; no solution.

Pages 162–164 Exercises and Applications

3. –1 **5.** 3 **7.** 5 **9.** $\dfrac{1}{3}$ **11.** –1 **13.** 3 **15.** Let m = the number of miles they must drive; $30 + 0.6m = 55 + 0.35m$; $m = 100$; after 100 miles. **17.** about 50 m **19.** $8(t - 2) = 6t$; $t = 8$; The faster runner, Pam, caught up to the slower runner, Beth, 8 s after the race began. At that time, they were 48 m from the starting line. **21.** Assume the quantity is 3 because $\dfrac{1}{3}$ of 3 is an integer: $3 + \dfrac{1}{3} \cdot 3 = 4$, not 36. If you multiply both sides by 9, you get $3 \cdot 9 + \dfrac{1}{3} \cdot 3 \cdot 9 = 4 \cdot 9$, or $27 + 9 = 36$. Since $9 = \dfrac{1}{3} \cdot 27$, the unknown quantity is 27.
23. Ex. 20: $x + \dfrac{1}{4}x = 25$; $\dfrac{5}{4}x = 25$; $x = \dfrac{4}{5} \cdot 25 = 20$. Ex. 21: $x + \dfrac{1}{3}x = 36$; $\dfrac{4}{3}x = 36$; $x = \dfrac{3}{4} \cdot 36 = 27$. Ex. 22: $x + \dfrac{1}{12}x = 65$; $\dfrac{13}{12}x = 65$; $x = \dfrac{12}{13} \cdot 65 = 60$ **25.** –4 **27.** identity **29.** 11 **31.** $\dfrac{1}{4}$
37. 30 **39.** 60 **41.** The domain consists of the nonnegative real numbers. The range consists of the numbers 0, 0.20, 0.40, 0.60, ...

1. a.

No. of Checks	Minimum Account Cost ($)	Maximum Account Cost ($)
10	2.50	7.00
11	3.25	7.00
12	4.00	7.00
13	4.75	7.00
14	5.50	7.00
15	6.25	7.00
16	7.00	7.00

b. The monthly cost will be the same when 16 checks are written in a month.   **2.** 9   **3.** –7   **4.** –12   **5.** –65   **6.** 8   **7.** –1   **8.** 3   **9.** no solution   **10.** identity

Page 167 **Checking Key Concepts**

1. 30   **3.** 72   **5.** $10^2 = 100$ **7.** $\frac{6}{5}$ **9.** 16   **11.** 3

Pages 168–170 **Exercises and Applications**

1. 6   **3.** 90   **5.** 10   **7.** 21   **9.** –7   **11.** 20

13.

15. The equations in Ex. 14 are obtained by multiplying the right-hand side of the equations in Ex. 13 by the LCD, 6, of the fractions in the equations in Ex. 13. The intersections of both graphs have the same x-coordinate, 10.

17. a.

	Distance across (km)	Rate (km/h)	Time (hours)
Wind at his back	d	14	$\frac{d}{14}$
Into the wind	d	12	$\frac{d}{12}$

b. $\frac{d}{14} + \frac{d}{12} = \frac{1}{2}$; $84\left(\frac{d}{14} + \frac{d}{12}\right) = 84\left(\frac{1}{2}\right)$; $6d + 7d = 42$; $d \approx 3.2$; The distance across the lake is about 3.2 km.   **19.** –25   **21.** 0.2   **23.** –2000   **25.** 2   **27.** E   **29.** about 380 km

33.

	A	B	C
1	Number of pages	Cost (ordinary)	Cost (special)
2	1,000	1278.82	646.14
3	10,000	1322.20	863.43
4	100,000	1756.00	3036.29

37.

$x = 2.0$

39.

$x = -3.0$

Page 173 **Checking Key Concepts**

1. $x < 8$ **3.** $x > 2$ **5.** $x > 60$, where x = the number of dances attended in a year

Pages 173–175 **Exercises and Applications**

1. a. $20 < s < 65$
3. a.

b. $x < 0$

5. Estimates may vary. $-55 \le t \le 15$; $-85 \le t \le 0$
7. Estimates may vary. $78 < a < 102$ **11. a.** Company A charges $12 for the first 60 min and $.20 for each additional minute. Company B charges $10 for the first 60 min and $.25 for each additional minute.   **b.** $t > 100$, where t = the number of minutes of calling per month   **13.** $d > 18$ or $d \ge 19$, where d is the number of days   **15.** $y = 2x - 7$ **17.** $y = -12$

19.

21.

Page 178 **Checking Key Concepts**

1. Yes; $x < -2$.   **3.** No; $a \ge 1$.   **5.** Yes; $x \le -20$.
7. $x < -5$
9. $t < -8$
11. $x < 2.25$
13. if he rents skates 16 or more times

Pages 179–180 **Exercises and Applications**

1. $x > 5$
3. $k < 1$
5. $c < -6$
7. $x < 0$
9. $n \le -4$
11. $a > \frac{20}{3}$

13. Let g be the number of guests already invited; $g + 8 > 40$; $g > 32$; more than 32 people have already been invited
15. 9 or more days   **17.** never   **19.** B   **21.** A
23. $n < 1$

(number line: -4 -2 0 2 4 6)
25. $c \le -\dfrac{3}{4}$ (number line: -1 0 1)
27. $t \ge -4$ (number line: -8 -6 -4 -2 0 2)
29. $a < -6$ (number line: -8 -6 -4 -2 0 2)
31. $x \le \dfrac{1}{2}$ (number line: -2 -1 0 1 2 3)
33. all numbers (number line: -4 -2 0 2 4)
35. Let s be Veronica's annual sales. $19{,}500 + 0.05s > 43{,}000$; $s > 470{,}000$; Her annual sales last year were more than \$470,000.   **39.** 28 trips   **41.** 5280 ft/mi; 88   **43.** $-\dfrac{4}{9}$

Page 181   Assess Your Progress
1. $\dfrac{15}{2}$ **2.** 2   **3.** $-\dfrac{5}{4}$ **4.** 2   **5.** 1.25   **6.** 3.2   **7.** Let d be the distance of the ball from the node.   **a.** $-4 \le d \le 7$
b. $-2.8 \le d \le 4.8$
8. $n > -3$ (number line: -6 -4 -2 0 2 4)
9. $x \le -40$ (number line: -80 -60 -40 -20 0 20)
10. $a > -2$ (number line: -6 -4 -2 0 2 4)
11. $t \le \dfrac{1}{2}$ (number line: -2 -1 0 1 2 3)
12. $k \ge -2$ (number line: -6 -4 -2 0 2 4)
13. all numbers (number line: -4 -2 0 2 4)

Pages 186–187   Chapter 4 Assessment
1. reciprocal   **2.** identity
3.

Golf Costs

Estimates may vary; after about 55 rounds.

(Graph: Cost (dollars) vs. Number of rounds n; lines labeled "Buy clubs" and "Rent clubs"; axis marks 800, 1600; 0, 20, 40, 60)

4. Let $n =$ the number of rounds and $C =$ the cost in dollars; $35n = 25n + 525$.   **5.** $\dfrac{5}{3}$ **6.** $-\dfrac{8}{7}$ **7.** $\dfrac{2}{3}$ **8.** 25   **9.** 18
10. 42   **11.** 9   **12.** -5 **13.** $\dfrac{1}{3}$ **14.** A; D   **15.** Answers may vary. An example is given. First use the distributive property to simplify both sides: $3x + 6 = 5x - 15 + 13$; $3x + 6 = 5x - 2$. Next subtract $3x$ from both sides to get the variables

on one side of the equation: $3x + 6 - 3x = 5x - 2 - 3x$; $6 = 2x - 2$. Add 2 to both sides: $6 + 2 = 2x - 2 + 2$; $8 = 2x$. Finally, divide both sides by 2: $\dfrac{8}{2} = \dfrac{2x}{2}$; $x = 4$.   **16.** $-\dfrac{2}{3}$
17. identity   **18.** no solution   **19.** 2   **20.** Let $n =$ the number of performances; $35n = 140$; $n = 4$.   **21.** $10^3 = 1000$
22. 90   **23.** $-\dfrac{5}{9}$ **24.** -5 **25.** 2   **26.** 0.5   **27.** Let $s =$ the speed of the car in mi/h. Estimates may vary; $15 \le s < 20$ and $65 < s \le 75$.   **28.** Let $s =$ the speed of the car in mi/h. Estimates may vary; $25 < s < 55$.   **29.** Yes; No; Yes; No; if both sides of a true inequality are multiplied by the same positive number, the result is a true inequality. If both sides of a true inequality are multiplied by the same negative number, the result is not a true inequality. The order of the inequality sign must be reversed to make the resulting inequality true.
30. B   **31.** $c > 7$ (number line: 0 2 4 6 8 10)
32. $x \le -4$ (number line: -8 -6 -4 -2 0 2)
33. $f \ge 9$ (number line: 0 4 8 12 16 20)
34. $x \ge 3$ (number line: 0 2 4 6 8 10)
35. $x \le -15$ (number line: -20 -16 -12 -8 -4 0)
36. $x < -3$ (number line: -6 -4 -2 0 2 4)

CHAPTER 5

Page 194   Checking Key Concepts
1. $\dfrac{1}{3}$ **3.** $\dfrac{4}{1}$ **5.** 24   **7.** about 387.62 Cal

Pages 194–197   Exercises and Applications
1. $\dfrac{9}{4}$ **3.** $\dfrac{1}{375{,}000}$ **5.** 2   **7.** 12   **9.** 24   **11.** 4
13. $\dfrac{33}{7} \approx 4.7$ **17.** about $\dfrac{4}{1}$ **19.** The portrait is about 4 times as large as the girl. The ratio in the photo is about 4 to 1, so the ratio of the actual sizes is about 4 to 1.   **23.** $\dfrac{13}{18}$ **25.** 7   **27.** $-\dfrac{14}{33}$ **29.** 2   **31.** 5
33. a. Estimates may vary due to rounding; about 61 home runs.   **35.** Answers may vary. An example is given for a school with a population of 625 students. Mandarin: about 95; Hindi: about 38; English: about 37; Spanish: about 38; Bengali: about 21; Arabic: about 21   **37.** $y = 3x - 7$
39. $y = \dfrac{3}{4}x$ **41.** $x = 0$ **43.** -0.94 **45.** 1.13

Page 201   Checking Key Concepts
1. 11.25 in. by 25 in.   **3.** $\dfrac{5}{24}$ **5.** Answers may vary. An example is given. Let each of the given intervals on the horizontal axis represent 10 years instead of 5.

Pages 201–204 Exercises and Applications

1. Estimates may vary; about 120 ft. **3.** $\frac{1}{2}$; $\frac{3}{4}$ in. **5.** $\frac{2}{3}$; $1\frac{1}{3}$ in.

11. 5 to 1; The scale factor is $\frac{\text{distance on Montgomery map}}{\text{distance on Alabama map}}$.
Consider a distance of 4 mi. The corresponding distance on the Montgomery map is 1 in. and on the Alabama map is $\frac{4}{20}$ in. $= \frac{1}{5}$ in.; $\frac{1}{\frac{1}{5}} = 1 \cdot \frac{5}{1} = \frac{5}{1}$ **13.** Answers may vary.

15. Answers may vary. Moving the projector forward will make an image smaller. Moving the projector back will make the image larger. **17.** C

19. $x < 3$

21. $x \geq -1.2$

23. $x \leq -\frac{5}{7}$

Page 204 Assess Your Progress

1. $\frac{1}{24}$ **2.** $\frac{15}{1}$ **3.** 10 **4.** $33.\overline{3}$ **5.** 8 **6.** Estimates may vary due to rounding; about 9347 votes. **7. a.** $1\frac{7}{8}$ in.
b. 20 ft by $29\frac{1}{3}$ ft

Page 208 Checking Key Concepts

1. a. $\angle F$ **b.** $\overline{BC}$ **3.** $155°$ **5.** 3

Pages 208–211 Exercises and Applications

1. a. $\angle Q$ **b.** $\overline{MO}$ **c.** PQ; QR **3.** Answers may vary. Because the two figures are rectangles, there are several ways to write the similarity. For example, $MNOP \sim QTSR$ and $MNOP \sim RSTQ$. The examples are based on the similarity $MNOP \sim QTSR$; $\angle M$ and $\angle Q$, $\angle N$ and $\angle T$, $\angle O$ and $\angle S$, $\angle P$ and $\angle R$, $\overline{MN}$ and $\overline{QT}$, $\overline{NO}$ and $\overline{TS}$, $\overline{PO}$ and $\overline{RS}$, $\overline{MP}$ and $\overline{QR}$.
5. $\angle V$ and $\angle V$, $\angle VWX$ and $\angle Y$, and $\angle VXW$ and $\angle Z$, and $\overline{VW}$ and $\overline{VY}$, $\overline{WX}$ and $\overline{YZ}$, $\overline{VX}$ and $\overline{VZ}$ **7. a.** True; the corresponding angles have equal measure and the ratios of the lengths of corresponding sides are equal. **b.** True; the corresponding angles have equal measure and the ratios of the lengths of corresponding sides are equal. **c.** False; the corresponding angles do not have equal measure and the ratios of the lengths of corresponding sides are not equal. **9. a.** Yes; all four angles of a square are right angles, so corresponding angles of the two squares have equal measure. Let x be the length of a side of one square and y the length of a side of the other. The ratio of the length of any side of the first square to any side of the second is $\frac{x}{y}$. **b.** Yes; the measure of each angle of an equilateral triangle is $60°$, so corresponding angles of the two triangles have equal measure. Let x be the length of a side of one equilateral triangle and y the length of a side of the other. The ratio of the length of any side of the first triangle to any side of the second is $\frac{x}{y}$. **c.** Yes; Yes. Corresponding angles of the two pentagons have equal measure. Let x be the length of a side of one pentagon and y the length of a side of the

other. The ratio of the length of any side of the first pentagon to any side of the second is $\frac{x}{y}$. **d.** Regular polygons with the same number of sides are similar. **13.** $\triangle AFG$ and $\triangle ADE$ share $\angle A$ and $\overline{ED}$ and $\overline{GF}$ are both perpendicular to the ground, so $\angle D$ and $\angle F$ are both right angles. Two angles of $\triangle AFG$ have the same measures as two angles of $\triangle ADE$, so the triangles are similar. **17.** about 1640 ch'ih; Use the fact that $\triangle AFG \sim \triangle ABC$ to write the proportion $\frac{y}{y + 56} = \frac{7}{x}$. Substitute the value of y found in Ex. 14 and solve for x.
19. A **23.** 3840 **25.** No. **27.** No.

Page 214 Checking Key Concepts

1. 35.2; 66.56

Pages 215–217 Exercises and Applications

1. $\frac{4}{9}$; $\frac{16}{81}$ **3.** 132.3; 36 **5.** 225π; 15; 5 **7. a.** $\frac{25}{49}$
b. \$9.80; The area of the large pizza is $\frac{49}{25}$ times that of the small pizza, so the large pizza should cost $\frac{49}{25}$ times as much as the small pizza. **11.** Answers may vary. An example is given. In the upper pictograph, I would choose the 22¢ stamp. In the lower pictograph, I would choose the 29¢ stamp. Both appear to have area about twice that of the May 29, 1978 stamp. **13.** about 64 mm²; the 20¢ stamp (upper pictograph), 29¢ stamp (lower pictograph); No, Yes; No, Yes.
15. D **17.** 10 **19.** $-\frac{14}{5}$ **21.** 45

Page 217 Assess Your Progress

2. 7.5 cm **3.** 3.3 cm **4.** $\frac{5}{3}$ **5.** $\frac{25}{9}$ **6.** The ratio of the lengths of the sides should be about $\frac{3}{2}$.

Page 221 Checking Key Concepts

1. a. 0 **b.** 1 **c.** $\frac{1}{3}$ **3.** theoretical

Pages 221–223 Exercises and Applications

1, 3. Answers may vary. Examples are given.

1. 0 $\frac{1}{2}$ 1   **3.** 0 $\frac{1}{2}$ 1

5. 0 $\frac{1}{2}$ 1

9. about 9.9%; Divide the number of people with less than a ninth-grade education (9,060,000) by the total number (91,367,000). **13.** The domain is the positive even numbers greater than 2, and the range is the reciprocals of numbers in the domain. Let $x =$ the number of faces on the die and let $y =$ the probability of rolling a 1; $y = \frac{1}{x}$; for any number x of faces, there are x possible outcomes, one of which is the desired outcome. **15.** No; for any number of faces, the probability of rolling a 1 is the same as the probability of rolling a 2. **17.** D

21, 23, 25.

21. none   **23.** IV
25. I

27. $-\frac{5}{7}$ **29.** $\frac{6}{11}$ **31.** no solution

Page 226   Checking Key Concepts

1. $\frac{5}{9} \approx 56\%$ **3.** 75%

Pages 226–229   Exercises and Applications

1. 50%   **3.** 50%   **5.** 50%   **9.** $\frac{4}{9} \approx 44\%$

15. a.

Stage	Probability of hitting the target's unshaded area
0	1
1	$\frac{8}{9}$
2	$\frac{64}{81}$

b. $\frac{512}{729}$ **17. a.** $x \le 0.5$ and $y \le 0.5$ **b.** 25%   **c.** Answers may vary.   **19.** 104   **21.** 1636   **23.** $7\frac{3}{4}$

Page 229   Assess Your Progress

1. $\frac{2}{3}$ **2.** $\frac{1}{6}$ **1, 2.** Answers may vary. An example is given. A die is rolled 30 times. It lands with a number less than 5 showing 25 times. The experimental probability of rolling a die and having it land with a number less than 5 showing is $\frac{5}{6}$.   **3.** about 21.5%   **4.** 25%   **5.** 50%

Pages 234–235   Chapter 5 Assessment

1. proportion; means; extremes   **2.** similar; angle measures; sides   **3.** theoretical; experimental   **4.** 1; 0   **5.** $\frac{1}{11}$ **6.** $\frac{17}{2}$
7. a. $\frac{5}{28}$ **b.** $\frac{5}{16}$ **c.** $\frac{3}{28}$ **8.** 192.5   **9.** 1   **10.** 30   **11.** 18
12. 56   **13.** $4\frac{4}{9}$ **15.** 11 in. = 20 ft, or 1 in. ≈ 1 ft 10 in.; $\frac{11}{240}$ **18.** 6.8; 5   **19.** 90°; 127°; 53°; 90°   **20.** 24
21. 20.16   **22. a.** True.   **b.** False.   **23.** $\frac{1}{3}; \frac{11}{25}$
24. $\frac{\pi}{16} \approx 0.2$

CHAPTER 6

Page 242   Checking Key Concepts

1. 17   **3.** 2   **5.** No.   **7.** Yes.

Pages 242–244   Exercises and Applications

1. 5   **3.** ±10   **5.** ±20   **7.** 29   **9.** 24   **11.** $a = 80$
13. $b = 35$ **15.** No.   **17.** Yes.
27. $y = 1.22x$

Gasoline Prices

29. $-\frac{36}{13} \approx -2.77$ **31.** $-\frac{1}{11} \approx -0.09$

Page 247   Checking Key Concepts

1. $-5, -\sqrt{16}, 0.0, 3$ **3.** $-\sqrt{29}, \sqrt{11}$ **5.** $\frac{2}{5}$ **7.** $\frac{63}{50}$
9. about 44.90 mi/h

Pages 248–250   Exercises and Applications

1. $-6, -\sqrt{25}, 0, 8, \sqrt{961}$ **3.** $-\sqrt{29}, \sqrt{19}$ **5.** $\frac{7}{20}$ **7.** $\frac{7}{11}$
9. $\frac{5}{11}$ **11, 13.** Estimates may vary. Answers are given to the nearest hundredth.   **11.** about 211.31 mi   **13.** about 40.96 mi   **15.** Answers may vary. An example is given. 1.07   **17.** not possible   **19.** $0.2, 0.\overline{2}, 0.26, 0.\overline{26}, 0.265, 0.2\overline{65}, 0.2\overline{6}$
21. $a \approx 10.95$ **23.** $b \approx 17.66$ **25.** $b \approx 73.97$
35. 5 : 8; 25 : 64   **37.** $c = 13$ cm   **39.** $b = 12$ ft

Page 254   Checking Key Concepts

1. $2\sqrt{3}$ **3.** $\frac{\sqrt{13}}{6}$ **5.** $3\sqrt{5}$ **7.** $2 < \sqrt{5} < 3$
9. $9 < \sqrt{84} < 10$

Pages 254–257   Exercises and Applications

1. $2\sqrt{10}$ **3.** $4 + 2\sqrt{10}$ **5.** $6\sqrt{5}$ **7.** $3\sqrt{7}$ **9.** 75
11. $\frac{\sqrt{5}}{4}$ **13.** Nina; Answers may vary. An example is given. You could evaluate $11 + 6\sqrt{2}$ and $17\sqrt{2}$ using a calculator to show they are not equal, or you could point out that $11 + 6\sqrt{2}$ cannot be simplified because only one term contains $\sqrt{2}$.
15. $20 + 6\sqrt{2}; 30\sqrt{2}$ **17.** about 1.81 m²   **19. a.** Answers may vary.   **b.** Answers may vary.   **21. a.** The length of the hypotenuse of the second triangle is $\sqrt{3}$ units.   **b.** The length of the hypotenuse of the third triangle is $\sqrt{4} = 2$ units.
c. The lengths of the hypotenuses of the succeeding triangles are $\sqrt{5}$ units, $\sqrt{6}$ units, $\sqrt{7}$ units, and so on. The pattern is that the nth triangle has a hypotenuse that is $\sqrt{n + 1}$ units long.   **23.** C   **25.** 7   **27.** 4   **29.** 5   **31.** $\angle B = 101°$; $\angle C = 24°$; $\angle P = 55°$; $PQ = 4.5$ m; $AC = 6$ m

Page 257   Assess Your Progress

1. $b = 12$ **2.** $c = \sqrt{261} \approx 16.16$ **3.** $b = \sqrt{259} \approx 16.09$
4. $\frac{73}{20}$ **5.** $\frac{118}{33}$ **6.** $\frac{23}{9}$ **7.** $3\sqrt{5}$ **8.** $5\sqrt{5}$ **9.** $\frac{2\sqrt{7}}{11}$

Page 261 Checking Key Concepts

1. $x(2x) = 2x^2$ **3.** $(x + 2)x = x^2 + 2x$
5. $5x^2$ **7.** $x^2 + 2x + 1$

9. binomial **11.** monomial

Pages 261–263 Exercises and Applications

1. $(x + 1)2x = 2x^2 + 2x$ **3.** $(x + 2)(x + 4) = x^2 + 6x + 8$
5. $3x^2$ **7.** $3x + 12$

9. $2x^2 + 7x + 5$

11. $4x + 4$ **13.** $4x^2 + 12x$ **15.** $2x^2 + 7x - 4$
17. $15 + 3\sqrt{2}$ **19.** $3\sqrt{6} + 2\sqrt{2} + 3\sqrt{3} + 2$
21. **a.** $(a + b)(a + b) = a^2 + 2ab + b^2$ **b.** $4\left(\frac{1}{2}ab\right) + c^2 =$
$2ab + c^2$ **27.** $\frac{1}{2}$ **29.** Answers may vary. An example is given. $y = 0.72x + 0.75$

Page 267 Checking Key Concepts

1. $x^2 + 16x + 64$ **3.** $x^2 - 81$ **5.** $52 - 14\sqrt{3}$
7. $(50 + 4)^2 = 50^2 + 2(50)(4) + 4^2 = 2500 + 400 + 16 = 2916$

Pages 267–269 Exercises and Applications

1. $w^2 - 4w + 4$ **3.** $x^2 - 6x + 9$ **5.** $y^2 - 16$
7. $4r^2 + 24r + 36$ **9.** $100t^2 - 36$ **11.** Let $x =$ the width of the path. The greater the width, the greater the value of x and the greater the area of the path; the cost depends on the area. The area of the path is equal to the combined area of the path and garden minus the area of the garden alone. Then the area of the path =
$(2x + 20)^2 - 20^2 = 4x^2 + 80x$. **13.** $(40 - 1)^2 = 40^2 - 2(40)(1) + 1^2 = 1600 - 80 + 1 = 1521$ **15.** $(2 + 0.1)^2 = 2^2 + 2(2)(0.1) + (0.1)^2 = 4 + 0.4 + 0.01 = 4.41$
17. $\left(10 - \frac{1}{2}\right)\left(10 + \frac{1}{2}\right) = 10^2 - \left(\frac{1}{2}\right)^2 = 100 - \frac{1}{4} = 99\frac{3}{4}$
19. $x^2 - 2xy + y^2$ **21.** $x^2 - y^2$ **23.** $16x^2 - 49$
25. $28 + 16\sqrt{3}$ **27.** B **33.** $y = 4x$ **35.** $y = \frac{1}{2}x - 4$
37. $y = \frac{2}{3}x + 4$ **39.** $6m^2 - 15m$ **41.** $-25x^2 + 10x - 1$
43. $15s^2 - 20s$

Page 269 Assess Your Progress

1. $5x^2 + 15x$ **2.** $10 + 5\sqrt{3}$ **3.** $x^2 - 25$ **4.** $46 + 12\sqrt{10}$
5. $6x^2 + 9x$ **6.** $x^2 - 8x + 16$ **7.** $15x^2$ **8.** $2x^2 + 11x + 5$
9. $x^2 + 4\sqrt{2}x + 8$ **10.** $(n + 8)^2 = n^2 + 16n + 64$
11. $4x$; $x^2 - 49$

Pages 274–275 Chapter 6 Assessment

1. 90; hypotenuse; legs; $a^2 + b^2 = c^2$, that is, the square of the length of the hypotenuse is equal to the sum of the squares of the lengths of the legs. **2.** a number whose square root is an integer; examples: 100, 12^2, 0 **3.** any number that can be written as a ratio of two integers $\frac{a}{b}$, where $b \neq 0$; examples: $1.5, \frac{7}{6}, 2$; any number that cannot be written as a ratio of two integers, and whose decimal form never terminates or repeats; examples: $\sqrt{7}$, 1.010010001..., $\sqrt{3}$ **4.** $m = 120$ **5.** $p = 25$
6. $r = 7.5$ **7.** Yes. **8.** No. **9.** 13.75 in. **10.** $\frac{171}{250}$ **11.** $\frac{7}{9}$
12. $\frac{95}{99}$ **13.** $4\sqrt{3}$ **14.** $3 + 10\sqrt{2}$ **15.** $8\sqrt{2}$ **16.** $\frac{3}{5}$
17. $24\sqrt{5}$ **18.** $\frac{8\sqrt{3}}{11}$ **19.** $7 < \sqrt{54} < 8$
20. $10 < \sqrt{107} < 11$ **21.** $8 < \sqrt{72} < 9$ **22.** $16\sqrt{5}$; 75; $\sqrt{170}$ **23.** $14x + 35$ **24.** $3y^2 - 27y$ **25.** $6m^2 - 11m - 35$
26. $8\sqrt{3} + 3$ **27.** 13 **28.** $195 + 47\sqrt{5}$
29. **a.** $(2x - 6)(x - 5) = 2x^2 - 16x + 30$ **b.** 6
30. $n^2 - 16n + 64$ **31.** $4a^2 + 12a + 9$ **32.** $61 + 28\sqrt{3}$
33. $c^2 - 81$ **34.** $36x^2 - \frac{1}{25}$ **35.** 20 **36.** 24.75 **37.** 851
38. 784

CHAPTERS 4–6

Pages 276–277 Cumulative Assessment

1. -20 **3.** 1.5 **5.** 2.5
7. $x < 4$

9. $p \leq \frac{1}{3}$

11. $m > 100$ **15.** 15 **17.** 9 **19.** $\frac{3}{4}$ **21.** 22.5 **23.** 4:5
29. Yes; $15^2 + 36^2 = 39^2$. **31.** No; $8.2^2 + 9.1^2 \neq 11.9^2$.
33. $\frac{1}{33}$ **35.** $\frac{961}{999}$ **37.** $19\sqrt{2}$ **39.** $20\sqrt{3}$ **41.** $8\sqrt{7}$
43. $86 + 14\sqrt{3}$ **45.** $36x^2 - 5$

CHAPTER 7

Page 283 Checking Key Concepts

1. **a.** $8a + 5c$ **b.** $8a + 5c = 40$ **3.** No; the variable terms are not on one side. **5.** 10; 6 **7.** Answers may vary. Examples are given. One method is to graph the x- and y-intercepts and then draw the line through the points. Another method is to rewrite the equation in slope-intercept form and enter this equation into a graphing calculator.

Pages 284–286 Exercises and Applications

1. 15 points **3.** 5 points **5. a.** $2w + t = 13$

b.

The *t*-intercept 13 represents the number of ties it would take to have 13 points without any wins. The graph has no *w*-intercept since the domain and range of this function include only whole numbers. If the graph did intersect the *w*-axis, the *w*-intercept would represent the number of wins it would take to have 13 points without any ties.

7. $123 **9.** $15c + 9t$

11.

13.

15.

17.

19.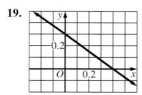

21. $y = -1.5x + 5$

23. $y = 0.8x + 2.4$

25. $y = -\dfrac{4}{3}x + \dfrac{8}{3}$

27. $y = -1.5x + 6$

29. the equations in Exs. 17, 18, 19, 27, and 28 because *a*, *b*, and *c* are not all integers and the equation in Ex. 25 because it does not have both variable terms on one side

31. $1.5x + y = 10$ **35.** $54c + 100r = 100$, where $c =$ the number of cups of oatmeal and $r =$ the number of cups of raisins; Solutions may vary. Examples are given. 1 cup of raisins alone, about 1.85 cups of oatmeal alone, and 1 cup of oatmeal with 0.46 cup of raisins

41. 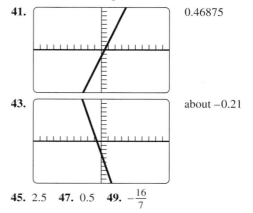   0.46875

43.   about -0.21

45. 2.5 **47.** 0.5 **49.** $-\dfrac{16}{7}$

Page 290 **Checking Key Concepts**

1. $\left(\dfrac{2}{7}, \dfrac{24}{7}\right)$; $\left(7, \dfrac{8}{3}\right)$; $(3, -2)$

Pages 290–293 **Exercises and Applications**

1. 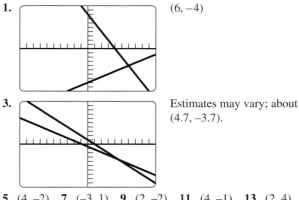 $(6, -4)$

3.   Estimates may vary; about $(4.7, -3.7)$.

5. $(4, -2)$ **7.** $(-3, 1)$ **9.** $(2, -2)$ **11.** $(4, -1)$ **13.** $(2, 4)$
17. $L = 2N$ and $L = N + 10$; The normal rate is 10 pints/min and the launch rate is 20 pints/min. **19.** Let $b =$ the cost in dollars of a beef meal and $c =$ the cost in dollars of a chicken meal. $98b + 66c = 1000$; $c = 0.8b$; about $6.63 for a beef meal and $5.31 for a chicken meal **23. a.** about 1030 ft
b. about 11.7 mi/h

Page 293 **Assess Your Progress**

1. −3, −4

2. −4, $\frac{3}{2}$

3. $y = 5x - 5$

4. $y = -\frac{2}{3}x + 3$

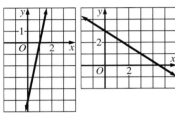

5. Let c = the number of correct answers and w = the number of wrong answers. **a.** $4c - w = 30$ **b.** $c + w = 20$

c.

10 correct answers and 10 wrong ones

6. (1, −1) **7.** (5, 2)

Page 296 **Checking Key Concepts**

1. addition; (3, 1) **3.** subtraction; (−14, −25)
5. either; (−3, −2)

Pages 296–298 **Exercises and Applications**

3. (2, 1) **5.** (4, −1) **7.** (−2, 3) **9.** (−1, −2) **11.** $495
13. $\left(-3, -\frac{2}{5}\right)$ **15.** (1, −2) **17.** (7, 5)

19.

The graph of the third equation contains the intersection point of the system.

21. 2 cups **27.** −2 **29.** 11
31. $x \le 3$

33. $x \ge \frac{1}{2}$

Page 303 **Checking Key Concepts**

1, 3. Answers may vary. Examples are given. **1.** addition or subtraction **3.** graphing or multiplication **5.** (5, −1)

Pages 303–305 **Exercises and Applications**

1. (3, −1) **3.** (1, 2) **5.** many solutions (all (x, y) such that $3x + 4y = 5$) **7.** $\left(6\frac{1}{3}, 2\right)$ **9.** (2, −1) **11.** $\left(-\frac{22}{29}, \frac{32}{29}\right)$

13. (6, 0) **15.** $\left(\frac{17}{8}, \frac{1}{2}\right)$ **21.** She can buy any number of Freesia bulbs from 0 to 15 in combination with any number of dahlia bulbs from 15 to 0, just so the number of bulbs totals 15.

25. No.

27.

29.

Page 305 **Assess Your Progress**

1. (−2, 3) **2.** (−4, 7) **3.** (−8, −1) **4.** all (s, t) such that $5 = 2t - s$ **5.** all (s, t) such that $4t - 6s = 2$ **6.** no solution
7. Emily: 8 visits; Adam: 20 visits

Page 308 **Checking Key Concepts**

1. C **3.** B **5.**

Pages 308–310 **Exercises and Applications**

1. D **3.** B
5.

7.

9.

11.

b. 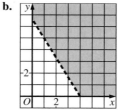 Answers may vary. Examples are given. (8, 0) and (4, 4)

13.

15.

33. B **37.** $2x^2 + 14x$ **39.** $a^2 + 2a - 35$ **41.** $-6 + 2\sqrt{5}$
43. (1, 1) **45.** (4, 11)

Page 313 **Checking Key Concepts**

1.

3.

17.

5. B: $y > 7 - x$ and $y > x - 3$; C: $y > 7 - x$ and $y < x - 3$;
D: $y < 7 - x$ and $y < x - 3$

19. Answers may vary. An example is given. (1) I solved for y to get $y \leq \frac{4}{3}x - 4$. Then I graphed $y = \frac{4}{3}x - 4$, using a solid line. Since y is less than or equal to $\frac{4}{3}x - 4$, I shaded the points below the line. (2) I graphed the solid line $4x - 3y = 12$ by using the intercepts. I chose (3, −2) as a test point that does not lie on the line and checked to see if the coordinates make the inequality $4x - 3y \geq 12$ true or false. Since the coordinates make the inequality true, I shaded the part of the graph that contains the point (3, −2). I prefer to use the method in Ex. 18 because it is hard to solve for y and sometimes I forget when to reverse the inequality sign. Using a test point seems easier.

Pages 313–315 **Exercises and Applications**

1.

3.

5.

7.

21.

23.

9.

25.

27.

11.

29. a. $300x + 180y > 1200$

While the system has many solutions, there are only three with whole-number coordinates: (0, 2), (4, 1), and (5, 1).

13.

4. 6; –8

15. a. $c \geq 0.6t$ **b.** $c \leq 0.7t$ **21.** B **25.** $\frac{6}{7}$ **27.** 20

29. $\frac{14}{3}$ **31.** $x^2 + 4x + 4$ **33.** $x^2 - 25$

5. $8\frac{1}{2}$; $5\frac{2}{3}$

Page 315 Assess Your Progress

1.

2.

6. $y = \frac{5}{2}x - \frac{11}{2}$

3.

4.

7. $y = \frac{2}{3}x - 4$

5.

6.

7. Let t = the number of T-shirts and s = the number of sweatshirts; $t \geq 10$; $s \geq 10$; $5t + 10s \geq 200$.

8. Let x = the number of pounds of Sunrise Tea and y = the number of pounds of Great Taste Tea.
a. $6x + 4y = 20$ **b.** $x + y = 4$ **c.** $(2, 2)$; Anita buys two pounds of each kind of tea. **9.** $(3, 4)$ **10.** $\left(1\frac{2}{3}, \frac{2}{3}\right)$

11. $(3, 1)$ **12.** $(1, -2)$ **13.** $(1, 2)$ **14.** $\left(0, 3\frac{1}{3}\right)$

15. all (x, y) such that $2x - 3y = 11$ **16.** $(-14, 11)$

17. 250 children's tickets

Pages 320–321 Chapter 7 Assessment

1. Any equation $ax + bx = c$, where a, b, and c are integers and a and b are not both zero is in standard form. For example, $2x + 5y = 7$ is in standard form. Any equation $y = mx + b$ is in slope-intercept form. For example, $y = \frac{3}{4}x - \frac{1}{8}$ is in slope-intercept form.

2. Answers may vary. Examples are given. $y \leq 2x - 1$; $y < 7$; $y \geq 3x + \frac{1}{5}$; $x > 5$ **3.** A horizontal intercept is on the x-axis and has coordinates $(x, 0)$ for some number x. A vertical intercept is on the y-axis and has coordinates $(0, y)$ for some number y.

19.

20.

21.

22.

23.

24. no solution

25.

26.

27.

28. a. $x + y \geq 10$ **b.** $2x + 5y \leq 25$

c.

d. $1\frac{2}{3}$ lb; The greatest amount is the maximum y-value, which occurs where the graphs of $x + y = 10$ and $2x + 5y = 25$ intersect.

CHAPTER 8

Page 327   Checking Key Concepts

1. nonlinear   **3.** linear

5. nonlinear

7. nonlinear; The graph is not a line.

Pages 328–330   Exercises and Applications

1. linear

3. nonlinear

5. linear

7. linear

9. nonlinear

11. linear; The graph of the equation is a line; also, the equation is of the form $y = mx + b$, where $m = 2\pi$ and $b = 0$.

13. nonlinear; The graph of the equation is not a line; also, the equation cannot be written in the form $y = mx + b$.

15.

Ticket price (dollars)	Number of tickets sold	Income (dollars)
3.00	500	1500
3.50	450	1575
4.00	400	1600
4.50	350	1575
5.00	300	1500

17. nonlinear; The points on the scatter plot do not lie on a line.

19.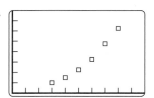

21. linear; Answers may vary. An example is given. In the scatter plot, numbers of servings is on the horizontal axis.

23. a.

b.

c. Answers may vary. An example is given. As the coefficient increases, the sides of the "V" become steeper. As the coefficient decreases toward 0, the sides of the "V" flatten out.

25. Graph $y = -2|x|$, $y = -3|x|$, $y = -0.5|x|$, and $y = -0.25|x|$ on your graphing calculator along with the graph of $y = |x|$. The "V" opens in the downward direction for negative values of a, and as $|a|$ increases, the sides of the "V" become steeper.
29. $p^2 + 4p - 21$ **31.** $6r^2 + 21r + 15$
33. $2 + 2\sqrt{3} + \sqrt{5} + \sqrt{15}$

35. $y = -\frac{3}{5}x + \frac{7}{5}$

37. $y = -\frac{5}{7}x + \frac{3}{7}$

39. $y = \frac{3}{7}x - \frac{5}{7}$

Page 333　Checking Key Concepts

1. narrower; up; The function reaches a minimum at the vertex. **3.** wider; up; The function reaches a minimum at the vertex. **5.** wider; up; The function reaches a minimum at the vertex.

7.

9.
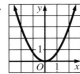

Pages 334–336　Exercises and Applications

1. C **3.** B **5.** Ex. 1: minimum; Ex. 2: maximum; Ex. 3: minimum; Ex. 4: minimum **7. a.** Since $6 > 1$, the graph will be narrower. Since $6 > 0$, the graph will also open up. **b.**

9. a. Since $|-10| > 1$, the graph will be narrower. Since $-10 < 0$, the graph will open down. **b.**

11. a. Since $\left|-\frac{1}{3}\right| < 1$, the graph will be wider. Since $-\frac{1}{3} < 0$, the graph will open down. **b.**

13. a. Since $0.5 < 1$, the graph will be wider. Since $0.5 > 0$, the graph will also open up. **b.**

15. a. Since $|-2.5| > 1$, the graph will be narrower. Since $-2.5 < 0$, the graph will open down. **b.**

19. No; if the graph were folded along the y-axis, the two halves would not match exactly. **21.** C; The coefficient of x^2 is less than the coefficients in the equations in both Ex. 22 and Ex. 23, so the parabola is wider than both those parabolas. **23.** B; The coefficient of x^2 is greater than that in the equation in Ex. 21 and less than that in Ex. 22, so it is narrower than the first parabola and wider than the second. **25.** The area of each circle should be a rough approximation of πr^2. **27.** The graphs should be similar. **31.** $y^2 + 6y + 9$ **33.** $4p^2 + 4p + 1$ **35.** $x^2 - 36$ **37.** 500 **39.** 36

Page 340　Checking Key Concepts

1. 0 **3.** ± 4 **5.** ± 4 **7.** $\pm\sqrt{41}$, or about 6.4 and about -6.4 **9.** $\pm 2\sqrt{3}$, or about 3.46 and about -3.46

Pages 340–343　Exercises and Applications

1. ± 9 **3.** ± 3 **5.** ± 2 **7.** ± 12 **9.** ± 0.5 **11, 13.** Estimates may vary. Examples are given. **11.** about -1.75, about 1.75 **13.** about -4.5, about 4.5 **15.** about 30 knots
17. a. Vertices: $(0, 0)$, $(0, 1)$, $(0, 2)$, $(0, 3)$

b. Vertices: $(0, 0)$, $(0, -1)$, $(0, -2)$, $(0, -3)$

c. The graphs of $y = x^2 + 1$, $y = x^2 + 2$, and $y = x^2 + 3$ are the graph of $y = x^2$ shifted up 1, 2, and 3 units, respectively. The graphs of $y = x^2 - 1$, $y = x^2 - 2$, and $y = x^2 - 3$ are the graph of $y = x^2$ shifted down 1, 2, and 3 units, respectively.

19. Answers may vary. Examples are given. Graph equations using different values for c and compare the graphs to the original graph. Use both positive and negative values. The graph of $y = |x| + c$ is the graph of $y = |x|$ shifted up c units if c is positive, and shifted down $|c|$ units if c is negative.
21. ± 2 **23.** $\pm\sqrt{3}$, or about 1.73 and about -1.73 **25.** ± 2
27. $\pm\frac{4}{3}$ **29.** $\pm\sqrt{39}$, or about 6.24 and about -6.24

33.

Distance Fallen During Free Fall

The scatter plot falls below the graph of the equation. That means the skydiver does not fall as far during each time interval as the equation would indicate.

37. $13\frac{1}{3}$ in. **39.** 128; 36

Page 343 Assess Your Progress

1. linear

2. nonlinear

3. nonlinear

4. Since $7(0) = 0$, $7 > 0$, and $7 > 1$, the graph will have the same vertex, will open in the same direction but will be narrower and steeper.

5. Since $-7(0) = 0$, $-7 < 0$, and $|-7| > 1$, the graph will have the same vertex, will open down, and will be narrower and steeper.

6. Since $0.7(0) = 0$, $0.7 < 1$, and $0.7 > 0$, the graph will have the same vertex, will open up but will be wider and flatter.

7. ± 12 **8.** ± 3 **9.** ± 2 **10.** ± 2 **11.** $\pm 2\sqrt{2}$, or about 2.83 and about -2.83 **12.** $\pm\sqrt{6}$, or about 2.45 and about -2.45

Page 346 Checking Key Concepts

1. $x^2 + 3x - 8 = 0$ **3.** $5x^2 + x - 14 = 0$ **5.** no solution
7. $1, -\frac{2}{3}$ **9.** 2

Pages 347–349 Exercises and Applications

1. 1, 3 **3.** about $-4\frac{1}{4}$, about $\frac{1}{4}$ **5, 7.** Estimates may vary.
Examples are given. **5.** about 34 mi/h to about 44 mi/h
7. $\sqrt{\frac{15}{0.013}}$ or about 33.97 mi/h to $\sqrt{\frac{25}{0.013}}$ or about 43.85 mi/h
9. $-4, 2$ **11.** ± 3 **13.** 4 **15.** $-\frac{3}{2}$ **17, 19.** Estimates may vary. **17.** about -3.68, about 0.68 **19.** about -1.27, about 6.27 **21.** A **23.** about 122 km/h; Germany and some places in France **33.** $\frac{2}{3}$ **35.** $\frac{1}{999}$ **37.** $\frac{7}{3}$ **39.** $4\sqrt{3}$
41. $5\sqrt{2}$

Page 352 Checking Key Concepts

1. There is only one x-intercept so the equation has only one solution. **3.** $1, -3, -10$ **5.** $35, 3, 0$ **7.** $4x^2 - 2x - 1 = 0$; $\frac{1 \pm \sqrt{5}}{4}$, or about 0.81 and about -0.31 **9.** $4.9t^2 - 7t = 0$; $0, \frac{10}{7}$

Pages 353–354 Exercises and Applications

1. $-3, 6$ **3.** $\frac{1}{3}, 3$ **5.** $2, -\frac{1}{5}$ **9.** $0, 4$ **11.** 1, 2
13. $\frac{1 \pm \sqrt{19}}{6}$, or about 0.89 and about -0.56 **15.** $1, \frac{1}{4}$
17. about 25 ft **21.** 180 ft **23.** The ball achieves a height of 8 ft twice, once about 10 ft before reaching the goal, and again about 118 ft on the far side of the goal. **25.** irrational
27. irrational **29.** $(1, -1)$ **31.** $\left(\frac{1}{3}, \frac{1}{4}\right)$ **33.** $(11, 23)$

Page 357 Checking Key Concepts

1. 1; The equation has two solutions. **3.** 0; The equation has one solution. **5.** 0; The equation has one solution.
7. $a = -0.0069$, $b = 0.0549$, $c = 999.8266$
9. 999.8266 mg/cm³

Pages 357–359 Exercises and Applications

1. two solutions **3.** two solutions **5.** two solutions
7. no solutions **9.** two solutions **11.** two solutions; about -8, about 208 (Note that while both are solutions of the equation and can be used to sketch its graph, only the positive solution makes sense as a horizontal distance.)

13.

15. two solutions; $-6, -1$ **17.** one solution; 1
19. two solutions; $1 \pm \sqrt{3}$, or about -0.73 and about 2.73
21. one solution; 4   **23.** two solutions; $\dfrac{3 \pm 3\sqrt{2}}{2}$, or about
-0.62 and about 3.62 **27.** B   **31.** $(-1, -4)$ **33.** ± 2
35. ± 1 **37.** $\pm \sqrt{5}$, or about -2.24 and about 2.24

Page 359   Assess Your Progress

1. about -0.7, about 5.7 **2.** $1, 2$ **3.** no solutions   **4.** $-3, 8$
5. $-3, \dfrac{5}{2}$ **6.** $\dfrac{1 \pm \sqrt{17}}{4}$, or about 1.28 and about -0.78
7. two solutions   **8.** no solutions   **9.** one solution

Pages 364–365   Chapter 8 Assessment

1. Answers may vary. Examples are given; $y = 3x$; $y = 3x^2$.
2. Sketches may vary. An example is given; $y = x^2$.

Symmetry is a correspondence in size, shape, and position of parts on opposite sides of a dividing line, plane, or point.

3. All quadratic equations must contain an x^2-term whose coefficient is not equal to zero. All graphs of quadratic equations are parabolas with a vertex and an axis of symmetry.
4. The discriminant is the value of $b^2 - 4ac$ in the quadratic formula. The discriminant gives you information about the number of solutions when you solve a quadratic equation.
5. nonlinear   **6.** linear   **7.** nonlinear   **8.** nonlinear
9. linear   **10.** The graph opens in the same direction, but it is narrower and steeper.   **11.** The graph opens in the opposite direction and is narrower and steeper.   **12.** The graph opens in the opposite direction and is wider and flatter.
13. C   **14.** A   **15.** B   **16.** $-4, 4$ **17.** $-5, 5$ **18.** $-3, 3$
19. $-4, 4$ **20.** $-6, 6$ **21.** $\pm\sqrt{26}$, or about -5.1 and
about 5.1 **22.** about -1.15, about 1.15 **23.** about -1.6,
about 1.6 **24.** $-0.5, 3$ **25.** about -2.76, about -0.24
26. no solutions   **27.** $-11, -1$ **28.** $\dfrac{1}{4}$ **29.** $\dfrac{3 \pm \sqrt{29}}{10}$,
or about 0.84 and about -0.24 **30.** two solutions
31. no solutions   **32.** one solution   **33. a.** $4x^2 = 4x + 24$
b. two solutions

CHAPTER 9

Page 373   Checking Key Concepts

1. 5^4 **3.** $2^1 \cdot y^3$ **5.** $\dfrac{1}{343}$ **7.** 54   **9.** 512

Pages 373–375   Exercises and Applications

1. $(4.1)^5$ **3.** $\left(-\dfrac{3}{8}\right)^3$ **5.** $pq(4r)^4$ **7.** 64   **9.** 1331   **11.** $\dfrac{1}{32}$
13. 7.84   **15.** 6^7 **17.** y^{10} **19. a.** 390,625
b. 1829.19   **c.** $-40,353,607$ **25.** 106   **27.** -108
29. 250   **31.** 641   **33.** 25   **35.** -270

37.

Radius	Surface area	Volume
1	4π	$\dfrac{4}{3}\pi$
2	16π	$\dfrac{32}{3}\pi$
4	64π	$\dfrac{256}{3}\pi$
8	256π	$\dfrac{2048}{3}\pi$
n	$4\pi n^2$	$\dfrac{4}{3}\pi n^3$
$2n$	$16\pi n^2$	$\dfrac{32}{3}\pi n^3$

45. Answers may vary. An example is given. The total price of 4.5 lb of a commodity that costs p dollars per pound; $9
47. $(r + 5)^2 = r^2 + 10r + 25$; $9 = 9$
49. $(7 - 3r)^2 = 49 - 42r + 9r^2$; $169 = 169$

Page 378   Checking Key Concepts

1. Answer is given to the nearest hundredth. 773.18
3. 14,592   **5.** the annual growth rate of the number of cellular telephone subscribers after 1983

Pages 379–382   Exercises and Applications

1, 3, 5, 7, 9. When necessary, answers are rounded to the nearest hundredth.   **1.** 2430   **3.** 32,659.2   **5.** 29.48
7. 722.11   **9.** 7,100,000   **11, 13, 15.** Answers are obtained from the equation $y = a(1 + r)^x$, where $y =$ the population in thousands, $x =$ the number of years after 1980, a is the initial population, and r is the growth rate.   **11. a.** 7,549,000
b. 0.002   **c.** 7,686,000   **13. a.** 4,780,000   **b.** 0.004
c. 4,955,000   **15. a.** 56,451,000   **b.** 0.002   **c.** 57,475,000
17. a. $y = 200(1.035)^x$, where $y =$ the amount in dollars and $x =$ the number of years from now.   **b.** $237.54

23. a–c.

The graphs all show exponential growth with the same growth factor but different initial values.

29. a. $y = 1653(1.04)^x$, where y is the number of vehicle-miles in billions x years after 1983.   **b.** about 2092 billion vehicle-miles to the nearest billion; $x = 6$ since 1989 is 6 years after 1983.   **33.** $\left(0, -\dfrac{11}{7}\right)$ **35.** $(1, 3)$ **37.** $\left(1, -\dfrac{1}{2}\right)$
39. $\dfrac{1}{2} \pm \dfrac{\sqrt{5}}{2}$ **41.** $-1; 3\dfrac{1}{2}$ **43.** $\dfrac{1}{4} \pm \dfrac{\sqrt{5}}{4}$

Page 385 Checking Key Concepts

Where necessary, answers are rounded to the nearest hundredth. **1.** 82.71 **3.** $\frac{4}{3}$ **5.** 5%; The decay factor, $1 - r$, is 0.95, so $r = 0.05$, or 5%.

Pages 386–389 Exercises and Applications

1. 0.03 **3.** 5970.74 **5.** 1351.77 **7.** 0.75
9. 262.14 **11.** $y = 100(0.83)^x$, where y = the number of grams remaining and x = the number of hours after 3:00 P.M.
15. exponential growth; The value of y increases exponentially as x increases. **17. a.** decays **b.** 14 **c.** 0.23
19. a. grows **b.** 0.25 **c.** 1.88 **21. a.** grows **b.** 0.99
c. 1.01 **23.** 133.12 mg **27. a.** about 0.46 h **b.** about 1.83 h **c.** about 7.37 h **29.** $n = 2^x$ **33.** $\frac{3}{5}$ **35.** $5\frac{3}{4}$
37. $-\frac{3}{2}$; -6 **39.** $-\frac{7}{3}$; 49 **41.** 8; 64

Page 389 Assess Your Progress

1. 64 **2.** 64 **3.** $\frac{1}{625}$ **4.** 80 **5.** 343 **6.** 42 **7.** -35
8. -59 **9.** 0 **10. a.** $y = 250(1.05)^x$ **b.** $289.41

11. a. $y = 591(0.98)^x$, where y = fuel consumption in gallons and x = the number of years since 1980.

b.

Page 393 Checking Key Concepts

1. $\frac{1}{121}$ **3.** $\frac{1}{49}$ **5.** $\frac{7}{m^3}$ **7.** $\frac{1}{a^4 c^6}$ **9.** 1 **11.** $-\frac{1}{8}$

Pages 393–395 Exercises and Applications

1. 1 **3.** $\frac{1}{32}$ **5.** 1 **7.** $\frac{81}{16}$ **9.** $\frac{1}{a^2}$ **11.** $\frac{4}{b^6}$ **13.** $\frac{b^3}{a^3}$ **15.** $\frac{2}{w^8}$
17. $\frac{1}{y^2 z^2}$ **19.** $\frac{1}{4}$ **21.** $\frac{1}{16}$ **23.** $\frac{53}{27}$ **25.** $\frac{1}{9}$ **31. a.** $502.53
b. $349.80 **c.** $273.74 **d.** $228.34 **35.** A **39.** $m \geq -3$
41. $q > -\frac{14}{3}$ **43.** $y \geq \frac{1}{6}$

Page 398 Checking Key Concepts

1. 600,000 **3.** 5,030,000 **5.** 0.000772 **7.** 8×10^6
9. 3.40056×10^{10} **11.** 7.4×10^{-4}

Pages 398–400 Exercises and Applications

1. 2000 **3.** 4,900,000 **5.** 676,100,000 **7.** 7.8×10^2
9. 2.1×10^4 **11.** 9.7×10^{12} **13.** 3.2×10^6
15. 8.023×10^{-2} **19.** 7 to 12 microns; 0.000007 = 7×10^{-6} and 0.000012 = $1.2 \times 10^{-5} = 12 \times 10^{-6}$
23. 9.54×10^{-7} **25.** 3.6288×10^{11} ft^3 **29.** B **33.** 1296
35. -125 **37.** 16

Page 400 Assess Your Progress

1. 1 **2.** $\frac{1}{1024}$ **3.** $\frac{27}{64}$ **4.** $\frac{1}{32}$ **5.** $\frac{9}{16}$ **6.** 4 **7. a.** about 13.81 million **b.** 14.6 million **c.** about 13.06 million
8. 7,003,000,000 **9.** 0.0000421 **10.** 0.000008
11. 5.4×10^{10} **12.** 3.28×10^0 **13.** 7.09×10^{-6}

Page 403 Checking Key Concepts

1. 4^8 **3.** 2^0 **5.** t^4 **7.** $\frac{2}{3}$ **9.** $15p^6 q^6$

Pages 404–405 Exercises and Applications

1. 2^7 **3.** 3^6 **5.** 9^{13} **7.** a^5 **9.** $\frac{1}{c^7}$ **11.** $40d^{10}$ **13.** $p^2 q^2$
15. $\frac{b}{3a^6}$ **17.** 6750 people

19.

Country	Population	Area (square miles)
China	1.192×10^9	3.696×10^6
France	5.8×10^7	2.10×10^5
Iran	6.0×10^7	6.32×10^5
United States	2.61×10^8	3.679×10^6

21. a. 1 light-year = $\frac{186{,}282 \text{ mi}}{\text{s}} \cdot \frac{60 \text{ s}}{\text{min}} \cdot \frac{60 \text{ min}}{\text{h}} \cdot \frac{24 \text{ h}}{\text{d}} \cdot \frac{365.25 \text{ d}}{\text{y}}$
1 y $\approx 5.88 \times 10^{12}$ mi **b.** 2.352×10^{16} mi **23.** 1 **25.** $\frac{343}{64}$
27. $-\frac{2048}{177{,}147}$

Page 408 Checking Key Concepts

1. 1296 **3.** 64 **5.** $\frac{49}{64}$ **7.** $a^9 b^9$ **9.** z^{16} **11.** $25c^2$

Pages 409–411 Exercises and Applications

1. 9×10^8 **3.** 6.25×10^{30} **5.** $\frac{64}{343}$ **7.** $a^5 b^5$
9. $\frac{16a^6 c^{12} d^8}{b^6}$ **11.** $81a^4 b^4$ **13.** $\frac{q^{35}}{p^{14}}$ **15.** $\frac{b^{30} c^5}{d^{15}}$
21. a. 28.544 g **b.** Let x = length of short snake. Then $2x$ = length of longer snake. Weight of longer snake = $446(2x)^3 = 446 \cdot 8x^3 = 8(446x^3) = 8$(weight of shorter snake).
23. B **25.** $-14u - 9v$ **27.** $6 - m$ **29.** $3q + 5$ **31.** $f = 36$
33. $p = 42\frac{6}{7}$ **35.** $x = 2$

Page 411 Assess Your Progress

1. a^{16} **2.** b^4 **3.** $\frac{a^7}{b^4}$ **4.** $a^4 b^4$ **5.** c^{56} **6.** $\frac{bc^{10}}{d^{10}}$ **7.** $\frac{p^6 q^{18}}{r^{12}}$
8. $\frac{x^8}{y^8}$ **9.** about 1.53×10^{12} mi **10.** 4; Let n_A = the number of seconds that Rock A has been falling; then $2n_A$ = the time in seconds that Rock B has been falling. Then the distance traveled by Rock $B = 16(2n_A)^2 = 16(2^2)(n_A^2) = 16(4)n_A^2 = 4(16n_A^2) = 4$ times the distance traveled by Rock A.

Pages 416–417 Chapter 9 Assessment

1. base; exponent; power **2.** exponential growth; growth rate; growth factor **3.** exponential decay; decay factor
5. 729 **6.** 81 **7.** $\frac{1}{1024}$ **8.** 0.125 **9.** 2^3; 8 **10.** 3^2; 9

11. -79 **12.** -200 **13.** 8 **14. a.** $y = 2000(1.075)^x$
b. $\$36{,}088.48$ **15. a.** $y = 13{,}000(0.88)^x$ **b.** $\$3620.51$

c. $(0, 13{,}000)$; the value of the car when it was new

16. 1 **17.** $\dfrac{1}{32}$ **18.** $\dfrac{1}{729}$ **19.** $\dfrac{125}{8}$ **20.** $-\dfrac{1}{125}$ **21.** $\dfrac{25}{8}$

22. $-\dfrac{5}{6}$ **23.** $\dfrac{103}{4}$ **24. a.** about 1.15 g **b.** 2.1 g

c. about 0.057 g **25.** $45{,}700{,}000$ **26.** 0.0000038
27. $70{,}230$ **28.** 6.78×10^{-9} **29.** 2.3×10^{10}
30. 5.0004×10^0 **32.** $\dfrac{m^2}{d^3}$ **33.** $16b^8$ **34.** $-6c^7d^{15}$

35. $\dfrac{yz}{x^3}$ **36.** $-\dfrac{8a^3}{27b^3}$ **37.** $\dfrac{s^{12}}{t^4}$ **38.** about $\$10.44$

CHAPTERS 7–9

Pages 418–419 Cumulative Assessment

1. horizontal intercept: -2; vertical intercept: 10; $y = 5x + 10$
3. many solutions **5.** Answers may vary. Methods may be graphical or algebraic. **7.** $2.5i + 2m = 20$

9.   **11.**

13. nonlinear

15. ± 20 **17.** C **19.** B **21.** about -0.47 and about 2.14;
$\dfrac{-5 \pm \sqrt{61}}{-6}$ **25.** -8 **27.** $a^{14}b^6$ **29.** $128m^{11}$ **31.** $\dfrac{32r^5}{243s^5}$

33. $y = 14{,}952{,}000(1.006)^x$ **35.** $15{,}874{,}000$

CHAPTER 10

Page 426 Checking Key Concepts

1. Yes. **3.** No; the exponent of n^{-3} is not a whole number.
5. Yes. **7.** $5x^2 - 2x + 1$; quadratic **9.** $-x^4 + 1$;
none of these **11.** $-\dfrac{3}{4}h + 14$; linear **13.** $6x^2 - 9$
15. $8x + 11xy + 2y$

Pages 426–428 Exercises and Applications

1. E **3.** A **5.** B **7.** $6x^2 - x - 4$; 2 **9.** $2r^3 - 3r^2 - 2r + 3$;
3 **11.** $17n^2 + 2n - 3$ **13.** $y^2 - 4y - 7$ **15.** $-a - 4b + 7$
17. $2b^3 + b^2 - 6b + 1$ **19.** $x^2 + 8x - 8$ **21.** 30 cm, 40 cm,
and 50 cm **25.** $c^2 + 3c + 4$ **27.** $4z^3 - 8z^2 - 8z + 8$

29. $-8m^2 - 5m + 4$ **31.** $-2x^2 + x + 7$ **33.** $-3y^3 - y^2 + 9$
35. $-a^2 + 12ab - 18b^2$ **39.** $x^2 - 1$ **41, 43.** Estimates may
vary. Examples are given. **41.** about 1.4, about -1.4
43. about 2.1, about -2.1

Page 432 Checking Key Concepts

1. x, x; $x, 8$; $4, x$; $4, 8$ **3.** $2n, n$; $2n, -8$; $4, n$; $4, -8$ **5.** $4ab$
7. $m^2 + 9m + 20$ **9.** $k^3 + 2k^2 - 24k$ **11.** $3y^3 - 11y^2 + 4y + 4$

Pages 433–435 Exercises and Applications

1. $c^2 + 12c + 35$ **3.** $s^2 - 6s + 8$ **5.** $12a^2 + 38a + 6$
7. $n^3 + 10n$ **9.** $45b^3 + 30b$ **11.** $-z^3 + 5z^2 - 4z$
15. a. The width is $(5x + 6y + 20)$ in. The length is
$(6x + 7y + 20)$ in. **b.** The perimeter is $(22x + 26y + 80)$ in.
The area is $(30x^2 + 220x + 71xy + 260y + 42y^2 + 400)$ in^2.
17. Substituting $x = 12$ and $y = 3$ in the expressions for
perimeter and area gives $P = 422$ and $A = 11{,}074$. Since
$P = 2(98) + 2(113) = 422$ and $A = (98)(113) = 11{,}074$, the
polynomials check. **19.** $18x^2 + 15x + 2$ **21.** $3y^2 + 10y + 3$
23. $6n^3 - 10n^2 - 15n - 27$ **25.** $5x^3 - 16x^2 - 47x + 10$
27. $42j^3 - 13j^2 + 25j - 4$ **29.** $6x^3 + 23x^2 - 30x + 8$
33. length $= (4 - 2x)$ in.; width $= (3 - 2x)$ in.; height $= x$ in.
35. $y = 4x^3 - 14x^2 + 12x$; The maximum volume, y, is created
when x is about 0.57 in.

39. $(2, 3)$ **41.** The graph will open down and be narrower
than the graph of $y = x^2$. The prediction results from the fact
that $-4 < 0$ and $|-4| > 1$. **43.** The graph will open down and
be narrower than the graph of $y = x^2$. The prediction results
from the fact that $-1.2 < 0$ and $|-1.2| > 1$.

Page 435 Assess Your Progress

1. No. The term $\dfrac{-1}{x}$, or $-1x^{-1}$, is not a monomial with a whole-
number exponent. **2.** $5x^2 - 5x - 3$; 2 **3.** $-3y^2 + 7y - 3$; 2
4. $2a$; 1 **5.** $z^3 - 2z^2 + 7z + 3$; 3 **6.** $8y^2 - 34y + 35$
7. $7n^3 + 21n^2 - 56n$ **8.** $5j^3 - 17j^2 + 18j + 18$ **9.** Methods
may vary. Examples are given. Subtract the area of the center
square from the total area: $(x + 24)^2 - x^2 = 48x + 576$. Divide
the border into eight parts—four that are 12 by x and four that
are 12 by 12. The add the areas: $4(12x) + 4(12^2) = 48x + 576$.

Page 439 Checking Key Concepts

1. x **3.** $2y$ **5.** $7x(2x + 1)$ **7.** $15n(2n - 3)$
9. $3t(t^2 - 2t + 3)$ **11.** $\dfrac{2}{5}, -\dfrac{5}{2}$ **13.** $0, -\dfrac{3}{4}$

Pages 439–441 Exercises and Applications

1. $x(x - 4)$ **3.** $4(y^2 + 1)$ **5.** $2x(x + 3)$ **7.** $5n(3n + 2)$
9. $a(a^2 - 4a + 1)$ **11.** $2x(27x^2 + 9x + 1)$ **15.** $0, 2$
17. $0, 4$ **19.** $0, \dfrac{1}{3}$ **21.** $-4, 7$ **23.** $-\dfrac{3}{4}, -\dfrac{1}{2}$

25. As the price increases from $0 to $1500, net sales increase from $0 to $225 million. As the price continues to increase to $3000, net sales decrease back to $0. **29.** $0, -\frac{7}{3}$
31. $0, -1$ **33.** $0, 5$ **35.** $0, 2$ **39. a.** 2 **b.** -4 **c.** -5
d. 3 **41.** The graph is above the x-axis except that its vertex is $(-k, 0)$. The new graph is shifted $|k|$ units to the left if k is positive and $|k|$ units to the right if k is negative. **43.** The graph of $y = |x - 2|$ is obtained by shifting the graph of $y = |x|$ two units to the right. The graph of $y = |x + 2|$ is obtained by shifting the graph of $y = |x|$ two units to the left. The graph of $y = |x + k|$ is the graph of $y = |x|$ shifted horizontally so that its vertex is $(-k, 0)$. **45.** 1 **47.** 1 **49.** $\frac{121}{25}$, or $4\frac{21}{25}$
51. $3x^2 + 7x + 2$ **53.** $2z^2 - 17z + 21$ **55.** $4a^2 - 1$

Page 445 Checking Key Concepts

1. 10, 1; $-10, -1$; 2, 5; $-2, -5$; $2 + 5 = 7$ **3.** 12, 1; $-12, -1$; 6, 2; $-6, -2$; 4, 3; $-4, -3$; $4 + 3 = 7$ **5.** $(y - 2)^2$
7. $(a + 9)(a + 2)$ **9.** $(3x - 2)(x - 4)$ **11.** $(5p + 3)^2$
13. The factors of 4 are 1 and 4, -1 and -4, 2 and 2, or -2 and -2. None of these pairs add to -3.

Pages 445–446 Exercises and Applications

1. $(x + 3)(x + 1)$ **3.** $(k - 4)(k - 2)$ **5.** $(x + 2)(x + 10)$
7. $(y - 9)(y - 4)$ **9.** $(s - 5)(s - 4)$ **11.** $(t - 8)(t - 3)$
15. Area $= 2x^2 + 9x + 10 = (x + 2)(2x + 5)$
17. $(5k + 1)(k + 1)$ **19.** $(3n + 2)(n + 2)$ **21.** $(4a - 9)(a - 1)$
23. $(5x - 8)(x - 1)$ **25.** $(2g - 1)(5g - 3)$
27. $(3p - 4)(3p - 2)$ **31.** $1000 is the amount invested. The growth factor is $(1 + r)$. At the end of year one, the value is $1000(1 + r)$. At the end of year two, the value is $1000(1 + r) \cdot (1 + r)$ or $1000(1 + r)^2$. **33. a.** Answers may vary. The percent of the population 65 and over will increase substantially between 1995 and 2040. **b.** Estimates may vary. An example is given: In 50 years, the investment will be worth $5000(1 + 0.07)^{50}$, or about \$147,285. **41.** ± 7 **43.** ± 15
45. $\pm \frac{3}{5}$ **47.** $-3, -4$ **49.** $\frac{5 \pm \sqrt{33}}{4}$, or about 2.69 and about -0.19 **51.** $3, -1$

Page 450 Checking Key Concepts

1. No. The factors of $2x^2$ are $2x$ and x; the factors of 1 are 1 and 1. To get the right quadratic and constant terms, the product would have to be $(2x + 1)(x + 1)$. But this product gives the wrong linear term. **3.** B; $(3y - 1)(y - 2)$ **5.** A; $(2r + 3)^2$
7. C; $(p + 10)(p - 10)$ **9.** Methods may vary. An example is given. $2 \pm \sqrt{3}$, or about 3.73 and about 0.27; The polynomial is not factorable, so use the quadratic formula.

Pages 451–453 Exercises and Applications

1. +; − **3.** −; − **5.** $(t - 8)(t + 1)$ **7.** $(z - 9)(z + 4)$
9. $(y + 6)(y - 6)$ **11.** $(2d + 11)(d - 1)$ **13.** $(2x + 3)(x - 2)$
15. $(2y - 1)(3y - 4)$ **17.** C **19.** 1, -1, 11, -11, 19, -19, 41, -41 **23.** 2, -1 **25.** 4, -1.5 **27.** $\frac{2}{3}, -3$ **29.** $\frac{2}{3}, -\frac{2}{3}$

31. a. 0.75 s after release **b.** No. Explanations may vary. An example is given. The graph of the equation shows that 10 ft is the maximum height. **33.** $\frac{3 \pm \sqrt{7}}{2}$, or about 2.82 and about 0.18 **35.** $\frac{1}{2}, -3$ **37.** $3, -\frac{2}{3}$ **39.** about 4.43 s; 17.82 ft; Yes, it could clear the centerfield wall or the right-field wall.

47, 49, 51.

47. I **49.** IV **51.** III

53. Yes; the points lie on a straight line that passes through the origin.

Page 453 Assess Your Progress

1. $\frac{3}{2}, -\frac{5}{4}$ **2.** 0, 3 **3.** 0, -4 **4.** 0, -20; The car must be traveling at 0 mi/h in order to stop in 0 ft. The negative solution does not make sense in this real-world problem. Speed cannot be negative. **5.** $(x + 2)(x + 5)$ **6.** $(k + 7)(k - 4)$
7. $(k - 14)(k + 2)$ **8.** $(2a - 1)(a - 5)$ **9.** $(5z + 1)^2$
10. $(2m - 1)(2m - 49)$ **11.** $(3c - 2)(3c - 5)$
12. $(9t + 8)(9t - 8)$ **13.** $(9d - 10)(d + 1)$ **14.** 4, -7
15. $-2, -\frac{7}{4}$ **16.** $\frac{1}{3}, -\frac{1}{8}$ **17.** 10 mi/h

Pages 458–459 Chapter 10 Assessment

1. Answers may vary. An example is given. $2x^3 - 4x + 6$
2. Answers may vary. Examples are given. linear: $4x - 6$; quadratic: $x^2 - 2$; cubic: $4x^3 + 8x^2 + x - 5$ **3.** To factor a polynomial means to write it as a product of two or more polynomials. **4.** Yes. **5.** Yes. **6.** Yes. **7.** Yes.
8. No; the power of x is not a whole number. **9.** No; the power of x in the term $-\frac{2}{x^2} = -2x^{-2}$ is not a whole number.
10. $8x + 10$; 1 **11.** $7a^2 - a + 2$; 2 **12.** $n^3 + 6n^2 - 4n + 4$; 3
13. $8x^2 + 2x + 11$ **14.** $4r^2 - 14r + 2$
15. $6x^3 - 14x^2 - 10x + 14$ **16.** $3b - 2c - 5$
17. $x^2 - 7x + 12$ **18.** $b^2 - 4b - 45$ **19.** $5r^2 - 21r + 18$
20. $6c^2 + 22c + 16$ **21.** $6a^3 + 27a$ **22.** $-d^3 + 3d^2 + 6d$
23. $2h^3 - h^2 - 16h + 42$ **24.** $4x^3 - 29x^2 - 22x - 16$
25. $5x^2 - 4x$ **26.** $y(13y - 4)$ **27.** $2a(3a - 5)$
28. $2x(3x^2 + 2x - 4)$ **29.** 0, 7 **30.** 0, $-\frac{1}{4}$ **31.** 1, $-\frac{8}{5}$
32. 0, 5 **33.** 0, -1 **34.** 0, 2 **35.** 2 in.
36. $(x + 10)(x + 1)$ **37.** $(y - 8)(y - 3)$ **38.** not factorable
39. $(q + 7)(q - 1)$ **40.** $(y - 8)(y - 2)$ **41.** $(a + 4)(a - 3)$
42. $(m + 3)(m - 3)$ **43.** $(x - 1)(2x - 1)$
44. $(2n + 1)(2n + 5)$ **45.** $-1, \frac{11}{3}$ **46.** $-\frac{8}{7}, 1$ **47.** $-\frac{5}{4}, 2$

CHAPTER 11

Page 466   Checking Key Concepts

1. inverse variation; $y = \dfrac{60}{x}$ **3.** neither

5. $y = -\dfrac{150}{x}$ or $xy = -150$

Pages 466–468   Exercises and Applications

1. No   **3.** Yes. 12

5.

 Yes; the product of x and y is constant.

7.

 Yes; the equation has the form $y = \dfrac{k}{x}$.

9. B   **11.** 750 ft; 3000 ft   **15.** $g = \dfrac{840}{n}$ **17.** 30

19. Answers may vary. An example is given. When k is negative, the two parts of the graph lie in Quadrants II and IV, rather than in Quadrants I and III. I found this answer by graphing a few equations such as $xy = -1$ and $xy = -4$.

21. 84; 85; 87   **23.** 13.5; 13; 13   **25.** $(6, -2)$

Page 472   Checking Key Concepts

1. Daniel, 80; Peter, 77.5; Sarah, 80   **3.** In Question 1, the students' grades are being averaged, and the weights are 1 (for the tests) and 2 (for the term paper). In Question 2, the meal costs are being averaged, and the weights are the numbers of people who ordered each meal choice.

Pages 472–475   Exercises and Applications

1. a. $73.48   **b.** 62 gal   **c.** about $1.19 per gal   **3.** 92 or better   **5.** Baiul, Kerrigan, Chen, Bonaly, Sato and Szewczenko, Witt, Harding.   **7.** Baiul and Kerrigan tied, Bonaly and Chen tied, Szewczenko, Sato, Witt, Harding

9. 6   **11.** 1

15. a.

b.

c.

d.

 Answers may vary. An example is given. The graph of $y = \dfrac{kx}{x+1}$ is the graph of $y = \dfrac{x}{x+1}$ stretched vertically by a factor of $|k|$. If $k < 0$, the graph is also reflected in the x-axis.

21. 25   **23.** Yes.   **25.** No; the first two terms cannot be written as monomials with exponents that are whole numbers.

Page 475   Assess Your Progress

1. No.   **2.** Yes; $y = \dfrac{120}{x}$.   **3.** 16 cm   **4.** 86.8

5. 25 executive suites

Page 479   Checking Key Concepts

1. Yes; both sides are rational expressions.   **3.** Yes; both sides are rational expressions.   **5.** 24   **7.** $10x(x + 1)$

9. no solution

Pages 479–481   Exercises and Applications

1. Yes.   **3.** Yes.   **5.** 8.5   **7.** 4.8   **9.** 4.8   **13. a.** Yes, both sides of the equation are ratios.   **b.** 17   **15.** -3

17. -0.75 **19.** 24 shots   **21.** no solution   **23.** $\dfrac{73}{9}$

25. 100 m

27. about 1 mi/h; about 2 mi/h

33.

35. $-3; \dfrac{1}{2}$ **37.** $-4; \dfrac{1}{3}$

Page 485 **Checking Key Concepts**

1. $z = \frac{2}{5}y$ **3.** $m = 3 - \frac{1}{2}d$ **5.** $b = \frac{1}{2c}$

7.

9. $t = \frac{d}{r}$

Pages 485–487 **Exercises and Applications**

1. a. $h = \frac{3V}{B}$ **b.** 10 cm **3.** $b = P - a - c$ **5.** $C = P + S$

7. $y = 24 - 2x$ **9.** $y = \frac{c - ax}{b}$ **11.** $x = \frac{1}{y}$

13.

15.

17.

27. 3 **29.** $64x^9$ **31.** $\frac{m^{10}}{n^4 p^6}$

Page 487 **Assess Your Progress**

1. 7 **2.** 2 **3.** 5 **4.** $y = \frac{1}{2}x - 4$ **5.** $x = \frac{y^2 + 7}{2}$

6. $b = \frac{a - 2cd}{2c - 1}$ **7.** $F = \frac{9}{5}C + 32$; $F = 68°$

Page 490 **Checking Key Concepts**

1. $\frac{1}{2x + 4}$ **3.** $\frac{3x^2}{2}$ **5.** simplest form **7.** $4x$ **9.** $4x$ **11.** $\frac{4}{3}$

Pages 491–493 **Exercises and Applications**

1. $\frac{b}{2}$ **3.** simplest form **5.** simplest form **7.** 4 **11. a.** $\frac{1}{n}$

b. The probability that you will get a green light is 1 minus the probability that you will get a red light; $\left(1 - \frac{1}{n}\right)\left(1 - \frac{1}{n}\right) = \frac{n - 1}{n} \cdot \frac{n - 1}{n} = \left(\frac{n - 1}{n}\right)^2$. **13.** $\frac{6y}{5}$ **15.** $3y$

17. $\frac{2a + 2}{a^2 + 2a}$ **19.** $\frac{10k}{3}$ **21.** $\frac{n}{n - 2}$ **23.** $2n^2(n + 1)$ **27.** $\frac{4}{25}$

35. $20x^2 - 6x - 8$ **37.** a^9 **39.** c^{-4}, or $\frac{1}{c^4}$

Page 497 **Checking Key Concepts**

1. $\frac{4}{y}$ **3.** -2 **5.** $\frac{12m + 5}{15m^2}$ **7.** rational **9.** linear

Pages 497–499 **Exercises and Applications**

1. $\frac{t - 2}{t - 1}$ **3.** 2 **5.** $\frac{25x - 6y}{20x^2y^2}$ **7.** quadratic **9.** exponential

11. rational **13.** $\frac{5}{6}$ **19.** $\frac{8x + 42}{(x + 8)(x - 3)}$ **21.** $\frac{3x + 5}{x + 1}$

23. $\frac{b^2 - 4c}{4a}$ **25. a.** $\frac{250}{s} - \frac{250}{s + 5} = \frac{1250}{s(s + 5)}$ **b.** about 27 min

31. $-\frac{1}{4}$ **33.** no solution **35.** $y = \frac{x}{2x + 1}$

Page 499 **Assess Your Progress**

1. $\frac{2}{5}$ **2.** $\frac{2}{5x}$ **3.** $\frac{3y}{y + 1}$ **4.** $\frac{y + 5}{y + 1}$ **5.** $\frac{6z - 5}{2(z + 1)}$ **6.** $\frac{3}{m}$

Pages 504–505 **Chapter 11 Assessment**

1–3. Answers may vary. Examples are given.

1. A weighted average is one in which some numbers are given more weight than others. Examples: calculating roommates' shares of utility bills when one roommate is home more than others; calculating grades when certain tests or test items count for more than others. **2.** $y = \frac{12}{x}$ or $xy = 12$

3. $\frac{2x + 4}{6x^2 + 24x + 24}$

4.

Cost per calculator	Number of calculators
$100	60
$60	100
$50	120
$30	200

5. *Ex.*: The product of the cost per calculator and the number of calculators is a constant.

6. Let y = the number of calculators the department can buy and x = the cost per calculator; $y = \frac{6000}{x}$.

7.

Only the portion of the graph in Quadrant I makes sense, since both the number of calculators and the cost per calculator must be positive.

8. $24 **9.** 94 or better **10.** Estimates may vary. An example is given. about 26,800 ft^2 **11.** $-\frac{5}{11}$ **12.** 7 **13.** no solution **14.** 2 **15.** $x = -\frac{4}{3}y + 4$ **16.** $n = \frac{m}{2m - 1}$

17. $y = \dfrac{2x}{2-x}$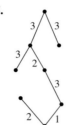

18. $n = \dfrac{S}{180} + 2$ or $n = \dfrac{S+360}{180}$; 9   **19.** $\dfrac{5b^3}{2a}$ **20.** $\dfrac{4c}{5}$ **21.** $\dfrac{1}{27t}$

22. $\dfrac{(p-2)^2}{8}$ **23.** 2   **24.** $\dfrac{k+5}{k+2}$ **25.** $\dfrac{3y+2}{6y^2}$ **26.** $\dfrac{-2z-5}{z(z+1)}$

27. $\dfrac{49}{n(n-7)}$ **28.** Answers may vary. Examples are given.

linear: $y = \dfrac{1}{2}x - 5$; quadratic: $y = x^2 - 2x + 1$;

polynomial: $y = x^3 + \dfrac{3}{4}x^2 - 11$; rational: $y = \dfrac{1}{x+1} + x(x-2)$;

exponential: $y = 2.5(1.08)^x$

CHAPTER 12

Page 511   Checking Key Concepts

1, 3. Answers may vary. Examples are given.   **1.** You might be trying to cook the eggs in their shells. You might end up eating uncooked eggs.   **3.** The following is an example of an algorithm for cooking scrambled eggs on a conventional stovetop. Step 1: Take the eggs out of the refrigerator and put them in a bowl. Step 2: Break the eggs into a bowl and stir until the yolks are broken and the eggs are well mixed. Step 3: Put a frying pan on a stove burner over medium heat. If the pan does not have a non-stick surface, oil the pan before putting it on the stove. Step 4: When the pan is heated, pour the eggs into the pan. Step 5: Stir the eggs as they cook. When the eggs are cooked to the degree you prefer, remove the pan from the heat.

Pages 512–515   Exercises and Applications

1, 3. Answers may vary. Examples are given.   **1.** Step 1: Take out two slices of bread. Step 2: Spread peanut butter on one slice. Step 3: Spread jelly on the other slice. Step 4: Put the slices together with unspread sides out.   **3.** Step 1: If the range is not set on standard, enter ZOOM 6. Step 2: Enter $Y=1/(X+2)$. Step 3: Enter GRAPH. Step 4: If necessary, adjust the range.   **7.** Step 1: Subtract 52 from 326. Step 2: Subtract 52 from the remainder. Step 3: Repeat Step 2 until the remainder is less than 52. Step 4: Count how many times 52 was subtracted. That number is the quotient and the number left is the remainder.   **9.** Step 1: Multiply 4 by 4. Step 2: Multiply the product by 4. Step 3: Repeat Step 2 five more times.

17.

6	5	4		2	4	6		3	6	2		1	2	3
3	2	1	→	1	3	5	→	5	1	4	→	4	5	6

21. A: 03–b–01; B: 03–b–03; C: 03–a–03; D: 03–c–02

25.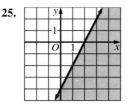

27. $-\dfrac{3}{5y}$ **29.** $\dfrac{9s}{4}$ **31.** $\dfrac{y(4x+y)}{2x(x-y)}$

Page 519   Checking Key Concepts

1. a tree   **3.** The shortest path is 9 units long.

5. 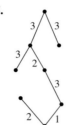 17 units; Yes; the seven edges are the seven shortest edges of the graph.

Pages 519–522   Exercises and Applications

1. both; The edges allow you to travel from one vertex to the other without repeating any vertex or edge, so the graph is a path. The vertices are connected with the smallest possible number of edges, so the graph is a tree.   **3.** The graph is a tree because the vertices are connected with the smallest possible number of edges. The graph is not a path because you cannot travel the graph without repeating any edges or vertices.   **5. a.** Answers may vary. An example is given.

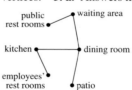

b. Other graphs are possible, but all would have 6 vertices and 5 edges and would be essentially the same except for the position of the vertices.

7, 9. Drawings may vary. Examples are given.

7.

9.

11. The edges along the bottom of the graph form a path 17 units long.

13. 31 units

15. B

19. Answers may vary. An example is given.

23. 1.4   **25.** $\frac{3}{11}$ **27.** $-\frac{1}{2}, 3$ **29.** $\frac{5}{2}, -\frac{5}{2}$ **31.** $\frac{3}{2}, -\frac{1}{2}$

Page 526   Checking Key Concepts

1.

Candidate	Number of votes
Jack	51
Denzel	52
Stacey	52
Mary	44
Renee	33

Denzel and Stacey (tie); Jack; Mary; Renee
3. One person-one vote elections: Denzel and Stacey (tie); Mary; Renee; Jack and new person (tie); Approval election: Denzel and Stacey (tie); Mary; Jack and new person (tie); Renee

Pages 526–529   Exercises and Applications

1. You should divide the piece into two smaller pieces, either of which would satisfy you. Your sister should choose one of the pieces. You get the remaining piece.   **15.** $\frac{3}{4}$
17. 3.4×10^{-3} **19.** 1.0078×10^{-3} **21.** 3×10^{-4}

Page 529   Assess Your Progress

1, 3. Answers may vary. Examples are given.   **1.** Step 1: If necessary, write both trinomials with the square term first, the linear term next, and the constant term last. Step 2: Combine like terms.   **3.** The first person should pick up donations weighing 5 lb, 10 lb, 25 lb, 30 lb, and 60 lb. The second person should pick up donations weighing 10 lb, 10 lb, 20 lb, 40 lb, and 50 lb. I think this is fair because each person has the same number of stops and the same weight to pick up, with the times distributed fairly evenly.

Page 532   Checking Key Concepts

1. $1 \cdot 2 \cdot 4 = 8$; The number of branches at each stage of the tree diagram is equal to the product of the number of items at that stage.   **3.** 5! or 120 paths   **5.** 6! = 720 passwords, if you include Claire

Pages 533–535   Exercises and Applications

1. 12 groups

7. 6720 councils   **11.** 362,880 ways; Answers may vary. Examples might include the abilities of the players or the use of designated hitters.   **13.** 2730 orders   **17.** (–2, –3)
19. (3, –2)   **21.** $\frac{5}{2} - \frac{\sqrt{29}}{2}; \frac{5}{2} + \frac{\sqrt{29}}{2}$ **23.** $-\frac{3}{10} - \frac{\sqrt{109}}{10};$ $-\frac{3}{10} + \frac{\sqrt{109}}{10}$ **25.** $\frac{7}{4} - \frac{\sqrt{113}}{4}; \frac{7}{4} + \frac{\sqrt{113}}{4}$

Page 538   Checking Key Concepts

1. 28 handshakes   **3.** 4950 handshakes   **5.** 21   **7.** 55
9. 15 combinations   **11.** 3276 ways

Pages 539–541   Exercises and Applications

1. 66 delegations   **3.** 36 delegations; The number of such delegations is the number of two-person delegations that can be chosen from a group of 9, or $_9C_2$.   **5.** 21 pairs; If the colors alternate, there are 42 possible flags.   **7.** 78   **9.** 680
11. a. 10 choices   **b.** 10 combinations   **13.** The entry in column B is the sum of the entries in columns A and B of the preceding row. Since the new person does not shake hands with himself or herself, the total number of added handshakes each time is the number of people in the group before the new person was added. Then the total number of handshakes is the previous total plus the number of people before the new person was added.

15.

number of points	number of triangles
3	1
4	4
5	10
6	20

The number of triangles that can be drawn by connecting n points is $_nC_3$.

17.

number of toppings	number of pizzas
1	9
2	36
3	84
4	126
5	126
6	84
7	36
8	9
9	1

$_9C_1 = {}_9C_8$; $_9C_2 = {}_9C_7$; In fact, for any whole number r between 1 and 9, $_9C_r = {}_9C_{9-r}$. For example, the number of ways of choosing three things from among nine is the same as the number of ways of choosing six things from among nine. When you choose r items from among n items, you automatically chose the other $n - r$ items as well. **19.** The 9 teams will play $_9C_2$, or 36 matches in all. It will take 4 days to schedule them all. There could be 9 matches each day, or 10 matches on each of the first 3 days and 6 matches on the last day. **23.** decays; 0.52 **25.** grows; 2 **27.** grows if $b > 1$ and decays if $0 < b < 1$; b **29.** No.

Page 544 Checking Key Concepts

1. HHHH, HHHT, HHTH, HHTT, HTHH, HTHT, HTTH, HTTT, THHH, THHT, THTH, THTT, TTHH, TTHT, TTTH, TTTT **3.** $\frac{1}{16}$ **5.** $\frac{1}{55}$

Pages 544–547 Exercises and Applications

1. $\frac{3}{8}$ **3.** 64 possible results **7.** $\frac{1}{8000}$ **9.** Answers may vary. **15.** 2^{92}; On each toss, there are two possible outcomes, so the total number of outcomes is $2 \cdot 2 \cdot 2 \cdot \cdots \cdot 2$ (92 times) $= 2^{92} \approx 4.95 \times 10^{27}$. **17.** 1; Doubling any number produces an even number.
19. $5x - 10$ **21.** $4 - 2n - 8n^2$ **23.** $-\frac{1}{6ab}$ **25.** $\frac{3t + 37}{20}$
27. $\frac{3b - 2a - 2a^2}{2ab}$

Page 547 Assess Your Progress

1. 60 ways **2.** The number of pairs of initials (in the same order) is the number of two-letter permutations of the 26 letters of the alphabet, which is $26 \cdot 26$, or 676. There are not enough pairs to allow all 713 employees to have unique pairs of initials. **3. a.** 11,175 pairs **b.** 828 pairs **4.** $\frac{11}{36} \approx 0.31$

Pages 552–553 Chapter 12 Assessment

1. A permutation is an arrangement of a group of items in a definite order. A combination is a selection of items chosen from a group in which order does not matter.

2. Answers may vary. An example is given. In the diagram, a greedy algorithm would choose the path from A to B to C as the best path from A to C.

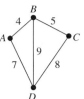

3. Answers may vary. Examples are given. A tree could be used to join a series of computers together with the least amount of cable. A path could be used to determine a good way to visit a collection of sites, such as pickups or drops on a delivery route, or vacation destinations. **4.** Answers may vary. Examples are given. (1) You might not add detergent or fabric rinse. (2) You might not dry the laundry. **5.** Answers may vary. Examples, based on the answers to Ex. 4 above, are given. (1) Add "Add detergent and fabric conditioner to the washing machine" to Step 2. (2) Add "Put the clean laundry into the dryer or hang it on a line until it is dry" to Step 4.
6. Step 1: Find the least common denominator (LCD) of the fractions. Step 2: Divide the LCD by the denominator of the first fraction. Multiply the numerator and denominator of the first fraction by the quotient. Step 3: Repeat Step 2 for the second fraction. Step 4: Add the numerators of the fractions obtained in Steps 2 and 3. Step 5: Write the fraction whose numerator is the result from Step 4 and whose denominator is the LCD. This is the sum of the original fractions. **7.** A to F to E to D **8.** D to C to E to F to A to B **9.** D to C to E to F to A and E to B **11.** Answers may vary. An example is given. Two sisters want to share a small pizza. One should divide the pizza into two pieces, either of which would satisfy her. The second sister should choose one of the pieces. The first sister would get the remaining piece. **12.** 120 groups
13. 120 ways **14.** 91 games **15.** $\frac{3}{8}$ **16.** $\frac{1}{136}$

CHAPTERS 10–12

Pages 554–555 Cumulative Assessment

1. $-2x^2 + 3x - 2$; 2 **3.** $4y^3 - 2y^2 - 2y + 12$; 3
5. $80x^3 - 32x^2 + 56x$; 3 **7.** $(x + 7)(x - 3)$ **9.** $(a - 6)(a - 3)$
11. $(3z - 2)(z + 9)$ **13.** $-4, 3$ **15.** $\pm\frac{5}{9}$ **17.** $-\frac{1}{3}, 3$ **19.** 12
23. direct variation; $y = \frac{1}{4}x$ **25.** -1 **27.** 85 **29.** $\frac{21 - 5b}{9ab}$
31. 2 **35.** $\overline{EA}$ or $\overline{EF}$, and $\overline{AB}$, $\overline{BC}$, $\overline{CD}$, and $\overline{DF}$
37. 362,880 **39.** 84 **41.** 240 **43.** Answers may vary. An example is given. Each handshake involves two people so the number of possible handshakes is related to the total number of people involved, taken two at a time.

EXTRA PRACTICE

Pages 557–559 Chapter 1

1. 31 days **3. a.** 15 days **b.** 12 days **5.** Answers may vary. Example: Let t = the daily high temperature; $t \geq 90$ and $t \leq 99$. **7.** 11 **9.** 3 **11.** $2\frac{2}{3}$ **13.** 17.04 **15.** 0.3
17. $3\frac{25}{68}$ **19.** 27 **21.** 34 **23.** 6 **25.** $\frac{1}{10}$ **27.** 1 **29.** 48
31. $13\frac{3}{7}$; 13; 13 **33.** $19\frac{13}{24}$; 20; 22 **35.** $9\frac{4}{17}$; 9; 8 **37.** −5
39. −21 **41.** −5 **43.** 3.2 **45.** 1 **47.** 5 **49.** −4 **51.** $\frac{1}{2}$
53. 8.4 **55.** $\frac{1}{3}$ **57.** −7 **59.** 9 **61.** 3 **63.** 3 **65.** 16
67. 28 **69.** −9 **71.** 6 **73.** −48 **75.** −5 **77.** −23
79. $-\frac{26}{7}$ **81.** −5 **83.** −12 **85.** 8 **87.** −24 **89.** −23
91. Let x = the number of minutes a call lasts and y = the cost in dollars; $y = 0.35 + 0.10(x - 1)$. **93.** Let x = the number of years from now and y = the deer population; $y = 2200 + 130x$.
95. $32 - 16x$ **97.** $4a + 3b$ **99.** $19n - 8n^2$ **101.** $-6f + 2$
103. $-13y - 8$ **105.** $11a^2$ **107.** E: 3×4; F: 3×4; G: 4×3; H: 4×3 **109.** E and G, E and H, F and G, F and H; They do not have the same dimensions.

Pages 559–561 Chapter 2

1. 1.9 **3.** −1.3 **5.** −16 **7.** −10 **9.** 4 **11.** 0.36
13. −0.07 **15.** $4\frac{1}{2}$ **17.** 6 **19.** 30 **21.** 5 **23.** −60
25. 3 **27.** −108 **29.** 43.8 **31.** −7.6 **33.** Let x = the cost in dollars of a juice box; $9x + 2.29 = 5.17$; $x = 0.32$; 32¢.
35. a. Let C = the total cost in dollars and n = the number of days the car is rented; $C = 27n$. **b.** The domain consists of the whole numbers and the range consists of the non-negative multiples of 27. **c.** $216 **37. a.** Let C = the total cost in dollars and n = the number of pens purchased; $C = 3n + 5.75$. **b.** The domain consists of the whole numbers and the range consists of the numbers 5.75, 8.75, 11.75, … . **c.** $23.75 **39.** Point F lies in Quadrant I, point I lies in Quadrant II, points C, E, and J lie in Quadrant III, and points A and H lie in Quadrant IV. Points B, D, and G do not lie in any quadrant; they lie on axes.

41, 43, 45.

47, 49, 51, 53. Solutions may vary. Examples are given.
47. (−1, 4)   **49.** (−1, −3)

51. (1, 3)   **53.** (−1, 2)

55. Let n = the number of payments he makes and A = the amount in dollars of the loan; then the equation $A = 125n + 1625$ describes the situation. To solve graphically, graph the equation and find 7500 on the vertical axis. Then move across to the graph of the equation. Move down to the horizontal axis to find the number of payments, 47.

Loan Repayment

To solve using an equation, substitute 7500 for A in the given equation and solve for n: $7500 = 125n + 1625$; $n = 47$; 47 payments.

Pages 561–563 Chapter 3

1. 3000 cycles/min **3.** $27.54/tank **5.** 60 min/h
7. 6.5 mi/min **9.** about 24.44 **11.** 200,000 **13.** $3480
15. Yes; 7.5; $y = 7.5x$. **17.** No. **19.** No.
21. −2   **23.** 3

25. 0   **27.** −1

29. 2   **31.** 5

33. 1 **35.** 5 **37.** $\frac{6}{5}$ **39.** $\frac{1}{4}$ **41.** $\frac{5}{4}$ **43.** $-\frac{15}{19}$ **45.** $-\frac{10}{7}$

47. $\frac{18}{19}$ **49.** $-2; 8; y = -2x + 8$

51. $1; -2$

53. $-3; 1$

55. $8; 0$

57. $\frac{1}{2}; 6$

59. $\frac{1}{3}; -2$

61. $\frac{3}{2}; 0$

63. $-5; 1$

65. $-7; -8$

67. $y = 2x + 1$ **69.** $y = -3x + 7$ **71.** $y = 2$ **73.** $x = 5$
75. $y = \frac{1}{4}x - \frac{11}{4}$ **77.** $y = -11x + \frac{11}{2}$ **79.** $y = -x + 6$
81. $y = \frac{2}{9}x - \frac{35}{9}$ **83.** $y = \frac{8}{3}x - \frac{19}{3}$ **85.** $y = -\frac{4}{3}x + 8$
87, 89. Answers may vary. Examples are given.
87. $y = x + 8$ **89.** $y = 2.7x + 26.4$

Pages 563–565 Chapter 4

1. a. \$3 **b.** $3x$ **3.** 23 daily rides; \$21 **5.** $-\frac{3}{2}a; -\frac{2}{3}$
7. $-\frac{5}{8}b; -\frac{8}{5}$ **9.** 21 **11.** 15 **13.** 6 **15.** -30 **17.** 6
19. 18 **21.** 24 **23.** 88 **25.** -56 **27.** 12 **29.** 7 **31.** 4
33. $-\frac{1}{2}$ **35.** -2 **37.** 5 **39.** no solution **41.** identity
43. $-\frac{29}{12}$ **45.** no solution **47.** no solution **49.** identity
51. $3\frac{4}{5}$ **53.** no solution **55.** no solution **57.** 3
59. 20; 120 **61.** $5; -\frac{5}{9}$ **63.** 15; 10 **65.** 8; -4
67. 10; 18 **69.** 100; 3 **71.** 10; 3 **73.** 1000; 359

75. 10; 5 **77.** 100; $\frac{1}{60}$

79.

$y < -1$

81.

$y > 2$

83. $v < 7$
85. $x > 4$
87. $x < 10$
89. $t > 4$
91. $n > -6$
93. $a < 10\frac{4}{5}$
95. $r \geq 0$
97. $n > -12$
99. $x < -\frac{6}{7}$
101. Let n = the number of books in a carton; $0.6n + 2 < 44$; $n < 70$; at most, 69 books can go in the carton.

Pages 566–568 Chapter 5

1. $\frac{1}{2}$ **3.** $\frac{1}{1000}$ **5.** $\frac{66}{25}$ **7.** $\frac{4}{1}$ **9.** $\frac{9}{5}$ **11.** $2\frac{1}{2}$ **13.** 36
15. 8.4 **17.** 1 **19.** $\frac{23}{2}$ **21.** $-\frac{11}{6}$ **23.** $\frac{4}{3}$
25. a. 6 in. by 8 in. **b.** 3.2 in. by 4 in. **c.** 4.8 in. by 5.2 in.
d. 5.6 in. by 6.4 in. **27. a.** $\frac{3}{2}$ **b.** $\frac{7}{3}$ **c.** either $\frac{17}{22}$ or $\frac{11}{14}$
d. $\frac{1}{2}$ **29.** 105° **31.** 3 to 2 **33.** $4\frac{2}{3}$ **35.** Yes; $\angle C$ and
$\angle C, \angle CBD$ and $\angle A, \angle CDB$ and $\angle E; \overline{CB}$ and $\overline{CA}; \overline{CD}$ and
$\overline{CE}; \overline{BD}$ and $\overline{AE}$. **37.** 120° **39.** $P = 36; A = 74\frac{2}{3}$
41. $P = 70; A = 800$ **43.** $\frac{1}{2} = 50\%$ **45.** $\frac{1}{2} = 50\%$
47. $\frac{1}{3} < 33.33\%$ **49.** $\frac{1}{8} = 12.5\%$ **51.** $\frac{1}{3} < 33.33\%$

Pages 568–570 Chapter 6

1. 8 **3.** 7 **5.** No. **7.** Yes. **9.** Yes. **11.** Yes.
13. $c = 15$ **15.** $a = 15$ **17.** $c = 34$ **19.** $\frac{1}{9}$ **21.** $\frac{8}{11}$ **23.** $\frac{2}{3}$
25. $\frac{16}{9}$ **27.** $\frac{2}{333}$ **29.** rational **31.** rational **33.** rational
35. $9 < \sqrt{95} < 10$ **37.** $11 < \sqrt{135} < 12$ **39.** $3\sqrt{5}$

41. $4\sqrt{5}$ **43.** $3\sqrt{10}$ **45.** $3\sqrt{14}$ **47.** $15\sqrt{3}$ **49.** $\dfrac{\sqrt{3}}{2}$

51. $\dfrac{1}{6}$ **53.** $8\sqrt{3}$ **55.** $15\sqrt{5}$ **57.** 5 **59.** $12\sqrt{2}$

61. $18\sqrt{10}$ **63.** $(x+3)(x+2) = x^2 + 5x + 6$ **65.** $6x - 12$

67. $5x^2$ **69.** $8x^2 + 28x$ **71.** $10x + 20x^2$

73. $5x^2 + 34x - 48$ **75.** $10x^2 + 34x - 24$ **77.** $24 + 4\sqrt{5}$

79. $15 - 3\sqrt{14} - 5\sqrt{2} + 2\sqrt{7}$ **81.** $x^2 + 8x + 16$

83. $a^2 - 16a + 64$ **85.** $4w^2 - 1$ **87.** $49k^2 + 5.6k + 0.16$

89. $36b^2 - 24bc + 4c^2$ **91.** $11 - 4\sqrt{7}$ **93.** 961

95. 27.04 **97.** $14 + 6\sqrt{5}$; $12 + 4\sqrt{5}$

Pages 570–572   Chapter 7

1, 3, 5, 7, 9, 11. Tables may vary.

1.

3.

5.

7.

9.

11.
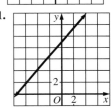

13. C   **15.** B   **17.** $\left(\dfrac{1}{3}, -\dfrac{1}{3}\right)$ **19.** $\left(\dfrac{4}{5}, \dfrac{7}{5}\right)$ **21.** $(1, -1)$

23. $(2, 1)$ **25.** $(0, 0)$ **27.** $\left(\dfrac{17}{5}, \dfrac{4}{5}\right)$ **29.** $\left(-\dfrac{8}{5}, -\dfrac{7}{5}\right)$

31. $\left(-4, -\dfrac{5}{2}\right)$ **33.** $(2, -4)$ **35.** $(1, 5)$ **37.** $(3, 4)$

39. $\left(-\dfrac{1}{3}, 7\right)$ **41.** $(3, 1)$ **43.** $(9, 6)$ **45.** $(3, 1)$ **47.** $(-2, 0)$

49. $(0, 2)$ **51.** $(3, 15)$ **53.** no solution

55. many solutions   **57.** A   **59.** B

61.

63.

65.

67.

69.

71.

73.

75.
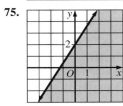

77. D   **79.** B

81.

83.

85.

87.

Pages 572–574   Chapter 8

1. linear

3. nonlinear

5. nonlinear

7. linear

9. nonlinear   11. linear

13. nonlinear 15. linear 17. a. wider b. up
19. a. wider b. down 21. a. wider b. up
23. a. narrower b. down 25. B 27. ± 12 29. ± 3
31. ± 6 33. ± 7 35. ± 9 37. ± 0.1
39. $\pm\sqrt{2} \approx \pm 1.41$ 41. $\pm 2\sqrt{2} \approx \pm 2.83$ 43. -3
45. about 1.70; about -4.70 47. 1; 3 49. -1; 2.5
51. about ± 0.49 53. no solution 55. no solution
57. 4; 5 59. -3; 5 61. 3; 2 63. -5; $\frac{1}{2}$ 65. 8; -2
67. 1 69. -1; 3 71. -6; 7 73. two solutions
75. no solution 77. no solution 79. two solutions
81. one solution; -2 83. two solutions; -3; 8
85. two solutions; -4; 5 87. one solution; -5
89. no solution 91. one solution; -8

Pages 574–576   **Chapter 9**

1. $(ab)^6$ 3. $\left(-\frac{1}{2}\right)^5$ 5. $3\left(\frac{2}{5}\right)^2$ 7. 343 9. 25 11. $\frac{1}{27}$
13. -3.375 15. 16 17. -80 19. 8 21. 864 23. a. 26
b. 1.25; 0.25 c. 3 25. a. 300 b. 1.35; 0.35 c. 10
27. a. 1500 b. 1.8; 0.8 c. 4 29. 40.20
31. 155,271,168 33. 1387.84 35. about 622 37. 0.02
39. 1.69 41. 12.8 43. 40.63 45. 0.60 47. a. grows
b. 455 c. 1.15 49. a. grows b. 335 c. 1.01
51. a. grows b. 100 c. 2.3 53. $\frac{1}{x^3}$ 55. $\frac{1}{b^2c^5}$ 57. $\frac{z^7}{y^4}$
59. $\frac{w^2z^3}{x}$ 61. $\frac{1}{4b^5}$ 63. $\frac{1}{m^5n^6}$ 65. $\frac{1}{9c^2}$ 67. $\frac{2z^8}{x^5y^4}$ 69. $\frac{1}{32}$
71. $\frac{1}{4}$ 73. $\frac{16}{9}$ 75. $\frac{19}{48}$ 77. 8 79. $-\frac{29}{96}$ 81. 300,000
83. 5.341 85. 0.00004503 87. 30,010,000,000
89. 0.0000099 91. 600,320,000,000 93. 9.675×10^3
95. 5.8×10^{-3} 97. 2.7×10^6 99. 8.7492×10^{10}
101. 9×10^{-6} 103. 4.000562×10^{-4} 105. 4.506×10^{-3}
107. 6.8×10^6 109. 10^9 111. 4^3 113. $\frac{1}{9^7}$ 115. 3^{11}
117. 2^4 119. 2^6 121. a^9 123. $\frac{a^2}{b^3}$ 125. $\frac{1}{c^7}$ 127. $12b^3$
129. $14c^5$ 131. $\frac{5c^8}{b}$ 133. 3×10^3 135. $\frac{2y^{10}}{x^6}$
137. 1.6×10^{13} 139. 2.25×10^{-8} 141. 8.1×10^{-11}
143. b^{15} 145. $27x^3$ 147. $\frac{a^6}{b^{12}}$ 149. $4y^{12}$ 151. $\frac{b^{12}}{c^{16}}$
153. $49a^4b^6c^{10}$

Pages 576–578   **Chapter 10**

1. $-3x^2 + 2x + 7$; quadratic 3. $-4c^2 + 5c$; quadratic
5. $\frac{1}{2}b^3 + 2b - 4$; cubic 7. $\frac{1}{2}y + 22$; linear 9. $5t^2 + 9t - 8$;
quadratic 11. $2a^2 - 8a - 2$ 13. $-b^2 - 4b - 10$
15. $5x + 4y - 11$ 17. $3a^2 - a + 3b - 6$

19. $6r^3 - 2r^2 + 5r + 2$ 21. $18n^2 + 48n$ 23. $x^2 + 8x + 15$
25. $4b^2 - 31b + 21$ 27. $16x^2 - 25$ 29. $0.8a^3 - 0.6a^2 - a$
31. $-6d^2 + d + 70$ 33. $3s^3 + 7s^2 - 32s - 48$
35. $-18y^3 - 21y^2 + 9y + 12$ 37. $x^2 + 8xy - x + 15y^2 - 5y$
39. $2t^3 + 5t^2 - 8t - 6$ 41. $x^3 + 3x^2 + 3x + 1$
43. $x^3 - 5x^2 + 6x$ 45. $7(n + 2)$ 47. $x(x^2 - 5)$
49. $6t(4t - 1)$ 51. $3v(v - 2)$ 53. $3(5a^2 + 3a - 3)$
55. $4z(2z^2 - 1)$ 57. $6x(3x^2 + x + 2)$ 59. $x(18x^2 - 14x + 9)$
61. 0; -9 63. 0; $-\frac{1}{2}$ 65. 0; 6 67. 8; -3 69. -1; $-\frac{4}{3}$
71. $-\frac{3}{4}$; $\frac{5}{7}$ 73. 0; 3 75. 0; $\frac{1}{2}$ 77. 0; 16
79. $2x^2 + 4x$; $2x(x + 2)$ 81. $(x + 2)^2$ 83. $(b - 7)(b - 1)$
85. $(c - 5)(c - 2)$ 87. $(b - 7)^2$ 89. not factorable
91. $(5n - 2)(n - 1)$ 93. $(2c + 1)(2c + 3)$
95. $(3n - 1)(2n - 1)$ 97. $(9t - 1)(t - 2)$ 99. $(x + 4)(x - 1)$
101. $(y + 3)(y - 2)$ 103. $(t + 7)(t - 4)$ 105. $(3x - 2)(x + 1)$
107. $(7d - 4)(d - 3)$ 109. $3(x + 1)(x - 5)$
111. $(5c + 2)(3c - 1)$ 113. $(3r + 1)(r - 3)$ 115. 5; 6
117. -3; -7 119. -6; -7 121. $\frac{4}{3}$; -1 123. $-\frac{1}{2}$; 3
125. $\frac{3}{2}$; $-\frac{7}{6}$ 127. $\frac{2}{3}$; -3 129. $\frac{3}{4}$; $-\frac{2}{3}$ 131. $\frac{7}{5}$; $-\frac{9}{2}$
133. $-\frac{7}{3}$; $\frac{2}{3}$

Pages 579–581   **Chapter 11**

1. Yes; 1.5. 3. Yes; 504. 5. 45.2 7. a. about $7.29
b. about $3.67 9. $12x^2(x + 1)$ 11. $30t(t + 1)$
13. $12y(y - 2)(y + 2)$ 15. no solution 17. 4 19. -1
21. -4 23. no solution 25. -9 27. -2 29. $t - v - x$
31. $\frac{cd}{f}$ 33. $\frac{y - b}{c}$ 35. $\frac{2 - t}{t}$ 37. $a^2 - 2$ 39. $-\frac{11}{6}$
41. $\frac{7y - 6}{3}$ 43. $\frac{-2ax + 5}{4a}$ 45. $\frac{ad}{ad - c}$ 47. $\frac{x^2}{5}$
49. simplest form 51. $\frac{3t^2}{4y}$ 53. $\frac{1}{4n}$ 55. $2b + 1$ 57. $\frac{9y}{x}$
59. 6 61. $\frac{1}{2b}$ 63. $\frac{1}{2y + 1}$ 65. $\frac{2}{y}$ 67. $\frac{1 - y}{xy}$ 69. $\frac{y + 6}{y(y + 2)}$
71. $\frac{(x + 2)(x - 1)}{x(x - 2)}$ 73. $\frac{2x + 1}{x + 1}$ 75. $\frac{5(x - 2)}{x(x - 5)}$ 77. $\frac{x - 2}{x + 1}$
79. $\frac{x - y}{x^2y^2}$

Pages 581–583   **Chapter 12**

1, 3, 5, 7. Algorithms may vary. Examples are given.
1. Step 1: Enter 52. Step 2: Add 52. Step 3: Repeat Step 2
thirteen times.   3. Step 1: Choose a number and square it.
If the result is less than 30, go to Step 2. If the result is greater
than 30, go to Step 3. Step 2: Choose a larger number and
square it. If the result is less than 30, repeat Step 2. If the
result is greater than 30, go to Step 3. Step 3: Choose a smaller number and square it. If the result is less than 30, go to
Step 2. If the result is greater than 30, repeat Step 3.
Step 4: Continue until the number you find is satisfactorily
close to $\sqrt{30}$.   5. Step 1: Move the decimal point one place
to the left to find 10% of 132, 13.2. Step 2: Multiply the result
by 3 to find 30% of 132, 39.6.   7. Step 1: Choose three values for x. Step 2: Substitute each of the chosen values into the

equation and find the corresponding values of y. Step 3: Draw coordinate axes. Graph the points corresponding to the ordered pairs you found in Steps 1 and 2. Step 4: Draw a line through the three points.   **9.** Answers may vary. One possible path is $4 + 3 + 3 + 3 + 4 + 2 + 3 + 4 = 26$. It is longer. **11.** $2 + 4 + 3 + 2 + 3 + 3 + 2 + 4 = 23$ **13.** Yearbook; Community Service and Computer Club (tie); Newspaper. The Yearbook had the most 1's, Community Service and Computer Club had one each 1, 2, 3, and 4, and the Newspaper had zero 1's.   **15.** One of you should break or cut the ear of corn into two pieces, either of which you would be willing to take. The other should choose one of the pieces. **17.** One of you should divide the task into two tasks, either of which you would be willing to do. (The task may be divided by area or by time, for example.) The other should choose one of the tasks. **19.** 32 sandwiches   **21.** 28 phones   **23.** 720 ways **25.** 40,320 orders   **27. a.** 15 pairs   **b.** 1, 2; 1, 3; 1, 4; 1, 5; 1, 6; 2, 3; 2, 4; 2, 5; 2, 6; 3, 4; 3, 5; 3, 6; 4, 5; 4, 6; 5, 6 **c.** 9 pairs   **29.** 1540 ways   **31. a.** 19,600 combinations **b.** 117,600 ways   **33.** $\frac{1}{1000}$

TOOLBOX

Page 584   Practice

1. GCF: 5; LCM: 90   **2.** GCF: 7; LCM: 196   **3.** GCF: 5; LCM: 210   **4.** GCF: 4; LCM: 120   **5.** GCF: 18; LCM: 216 **6.** GCF: 1; LCM: 450   **7.** GCF: 16; LCM: 64   **8.** GCF: 14; LCM: 210   **9.** You can buy 3 packages of pencils and 5 packages of pencil-top erasers.

Page 585   Practice

1. 25,000   **2.** 0.184   **3.** 194,000   **4.** 0.3   **5.** 0.00368 **6.** 0.5   **7.** 0.0000781   **8.** 0.09346   **9.** 0.293   **10.** 0.1083 **11.** 46.8   **12.** 0.413   **13.** 0.0006   **14.** 0.009   **15.** 0.1995 **16.** 0.00021973

Page 586   Practice

1–8. Forms of ratios may vary.   **1.** 3 to 1   **2.** 2 : 9   **3.** $\frac{21}{32}$ **4.** 17 to 3   **5.** 1 to 9   **6.** 3 : 2600   **7.** 3 : 16   **8.** 3 to 4 **9.** 50 miles per hour   **10.** $2.50 per hat   **11.** 84 vibrations per second   **12.** 9 hours per day   **13.** $43.50 per month **14.** $.45 per pen

Page 587   Practice

1. 0.04; $\frac{1}{25}$ **2.** 0.87; $\frac{87}{100}$ **3.** 1.75; $\frac{7}{4}$ **4.** 0.824; $\frac{103}{125}$ **5.** 0.005; $\frac{1}{200}$ **6.** 75%; $\frac{3}{4}$ **7.** $66\frac{2}{3}$%; $\frac{2}{3}$ **8.** 2.5%; $\frac{1}{40}$ **9.** 160%; $\frac{8}{5}$ **10.** 51.3%; $\frac{513}{1000}$ **11.** 0.35; 35% **12.** 0.8; 80%   **13.** 0.3375; 33.75%   **14.** 0.56; 56% **15.** $0.5\overline{3}$; $53\frac{1}{3}$%

Page 589   Practice

1. 5   **2.** –9   **3.** –5   **4.** 5   **5.** –8   **6.** –6   **7.** –13 **8.** –11   **9.** 11   **10.** –5   **11.** 10   **12.** 9   **13.** 2   **14.** 6 **15.** –12   **16.** –18   **17.** –2   **18.** –18   **19.** 0   **20.** 20 **21.** –1   **22.** –3   **23.** –3   **24.** –16   **25.** 1   **26.** 0   **27.** –1 **28.** –10   **29.** –21   **30.** –2

Page 590   Practice

1. all numbers greater than 6   **2.** all numbers less than or equal to 9   **3.** all numbers less than –2   **4.** all numbers greater than or equal to 0   **5–8.** Let n = any number. **5.** $n \geq 7$ **6.** $n < 20$ **7.** $n > 3$ **8.** $n \geq 24$ **9.** Let p = a price; $p < 30$.   **10.** Let h = a hiking distance; $h \leq 12$.   **11.** The integers 8, 9, 10, … make $n > 7$ a true statement.   **12.** The integers –1, 0, 1, 2, 3, … make $r \geq -1$ a true statement.   **13.** The integers 4, 3, 2, 1, 0, –1, –2, … make $v < 5$ a true statement.   **14.** The integers 3, 2, 1, 0, –1, –2, … make $t \leq 3$ a true statement.

Page 591   Practice

9. $x > 0$ **10.** $x \leq -4$

Page 592   Practice

1. 7 ft 3 in.   **2.** 5000 lb   **3.** 32 fl oz   **4.** 55,000 cm **5.** 3.4 kg   **6.** $21\frac{3}{4}$ ft   **7.** 11 pt   **8.** 250,000 mL

Page 593   Practice

1. 51 m   **2.** about 6.59 ft   **3.** 18 cm   **4.** about 50.24 in. **5.** 16.8 cm   **6.** about 20.72 mm   **7.** $16\frac{1}{2}$ in. **8.** about 43.96 yd   **9.** about 69 in.

Page 594   Practice

1. about 78.5 ft^2 **2.** 17.5 m^2 **3.** 133 in.^{2} **4.** 37.5 cm^2 **5.** 162 in.^{3} **6.** 200 m^3 **7.** 15,625 in.^{3} **8.** 249.375 cm^3 **9.** 56 yd^3 **10.** The rectangular pizza has the greater area.

Page 595 Practice

1. about 30°; acute **2.** about 90°; right **3.** about 140°; obtuse **4.** about 65°; acute

5.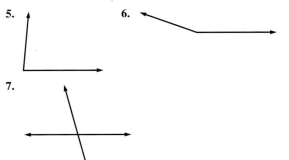

6.

7.

Pages 596–599 Practice

1. 34 people **2.** black **3.** 2 people **4.** Burlington: about 5 in.; Portland: about 12 in. **5.** July/August
6. about 5 in. **7.** about 21,500 bicycles **8.** Tandem Tiger
9. 9.5 symbols **10.** Strings: about 77; Woodwinds: about 27; Brass: about 25; Percussion: about 11
11. about 118 players **12.** 9 racers **13.** 32 min; 71 min
14. 34 students **15.** 8 students **16.** 90–99 laps; 2 students

17.

Page 601 Practice

1. $128 + 367 = 495$ or $129 + 357 = 486$ **2.** 15 dimes;
10 dimes and 2 quarters; 5 dimes and 4 quarters; 6 quarters
3. 48; 60; 72; 84 **4.** 24 ft^2 **5.** about 5.2 gal **6.** 400 cans
7. about 25 hours **8.** 60°

TECHNOLOGY HANDBOOK

Pages 603–610 Calculator Practice

1. −8 **2.** 30 **3.** about 6.28 **4.** 3.5 **5.** 32 **6.** 0.01872
7. 40,320 **8.** 21

9. a.

b.

c.

d.

e. Zoom windows may vary. An example is given.

10. Answers may vary. An example is given; $-75 \le x \le 75$,
$-20 \le y \le 80$ **11. a.** about 4.9 **b.** between 4.85 and 4.86
12. about 56.55 **13.** x is about 0.15 and y is about 4.22.

14. about −1.70 and about 2.20 **15.** $\begin{bmatrix} -2 & -1 & 5 \\ -1 & 4 & 5 \end{bmatrix}$

16. Equations may vary. An example is given;
$y = 3.11x + 0.1$; about 9.7 in.

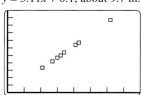

17. Possible table:

X	Y$_1$
0	7
1	2
2	−1
3	−2
4	−1
5	2
6	7

X=3

The y-values stop decreasing at $x = 3$.

18. Monday: 60 min; Tuesday: 90 min; Wednesday: 120 min;
Thursday: 60 min; Friday: 0 min; Saturday: 285 min; Sunday:
0 min **19.** You will owe $45 at 15 weeks. The loan will be
paid back in 22 weeks.

APPENDIX 1

Page 617   **Exercises and Applications**

1. a. –22, –16, –4, 2   **b.** …, –18, –14, –10, –6, …   **3.** The domain is real numbers greater than or equal to 0 that can be expressed in hundredths: 0.00, 0.01, 0.02, …. The range is the same as the domain.   **5.** The domain is the integers –2, –1, 0, 1, 2, 3, 4. The range is the integers 0, 1, 4, 9.   **7.** Answers may vary. An example is given. The domain could be real numbers from 0.02 to 1. This gives a range of real numbers from 0.0036 to 446.

15. a–c.

d. All three are the same width. (a) and (c) have the same y-intercept, but open in different directions. (b) and (c) both open down, but have different y-intercepts.

17. Both have the same shape, but the vertex of the graph of $h = 10 - 16t^2$ is higher and the graph opens down rather than up.

APPENDIX 2

Page 620   **Exercises and Applications**

1.

The graph of $y = 4x + 3$ is steeper than the graph of $y = x$ and crosses the y-axis above the graph of $y = x$.

3.

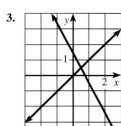

The graph of $y = -2x + 1.5$ is steeper than the graph of $y = x$, crosses the y-axis above the graph of $y = x$, and slants down instead of up.

5. a. greater than   **b.** greater than   **7. a.** greater than   **b.** less than   **9. a.** equal to   **b.** less than   **11. a.** $g = 12.7 - \dfrac{d}{28}$; $g = 12.7 - \dfrac{d}{37}$ **b.** $g = 20 - \dfrac{d}{17}$; $g = 20 - \dfrac{d}{25}$ **c.** The gas tank capacity affects b, the y-intercept. The greater the capacity of the tank, the higher the graph crosses the y-axis. Fuel efficiency affects m, the slope. The greater the fuel efficiency, the smaller $|m|$ is and the less steeply the line slants down.

13.

The graph of $y = 0.2x^2 - 3$ is wider than the graph of $y = x^2$ and crosses the y-axis below the graph of $y = x^2$.

APPENDIX 3

Page 623   **Exercises and Applications**

1. (1, 1)   **3.** (5, 2)   **5. b.** The values of y are always the same. The system of equations has infinitely many solutions.   **7.** (4, 3)